Stochastic Mechanics
Random Media
Signal Processing and Image Synthesis
Mathematical Economics and Finance
Stochastic Optimization
Stochastic Control

Applications of Mathematics

Stochastic Modelling and Applied Probability

6

Edited by I. Karatzas
M. Yor

Springer
Berlin
Heidelberg
New York
Barcelona
Hong Kong
London
Milan
Paris
Singapore
Tokyo

Applications of Mathematics

1 Fleming/Rishel, **Deterministic and Stochastic Optimal Control** (1975)
2 Marchuk, **Methods of Numerical Mathematics** (1975, 2nd. ed. 1982)
3 Balakrishnan, **Applied Functional Analysis** (1976, 2nd. ed. 1981)
4 Borovkov, **Stochastic Processes in Queueing Theory** (1976)
5 Liptser/Shiryaev, **Statistics of Random Processes I: General Theory** (1977, 2nd. ed. 2001)
6 Liptser/Shiryaev, **Statistics of Random Processes II: Applications** (1978, 2nd. ed. 2001)
7 Vorob'ev, **Game Theory: Lectures for Economists and Systems Scientists** (1977)
8 Shiryaev, **Optimal Stopping Rules** (1978)
9 Ibragimov/Rozanov, **Gaussian Random Processes** (1978)
10 Wonham, **Linear Multivariable Control: A Geometric Approach** (1974, 3rd. ed. 1985)
11 Hida, **Brownian Motion** (1980)
12 Hestenes, **Conjugate Direction Methods in Optimization** (1980)
13 Kallianpur, **Stochastic Filtering Theory** (1980)
14 Krylov, **Controlled Diffusion Processes** (1980)
15 Prabhu, **Stochastic Storage Processes: Queues, Insurance Risk, and Dams** (1980)
16 Ibragimov/Has'minskii, **Statistical Estimation: Asymptotic Theory** (1981)
17 Cesari, **Optimization: Theory and Applications** (1982)
18 Elliott, **Stochastic Calculus and Applications** (1982)
19 Marchuk/Shaidourov, **Difference Methods and Their Extrapolations** (1983)
20 Hijab, **Stabilization of Control Systems** (1986)
21 Protter, **Stochastic Integration and Differential Equations** (1990)
22 Benveniste/Métivier/Priouret, **Adaptive Algorithms and Stochastic Approximations** (1990)
23 Kloeden/Platen, **Numerical Solution of Stochastic Differential Equations** (1992, corr. 3rd. printing 1999)
24 Kushner/Dupuis, **Numerical Methods for Stochastic Control Problems in Continuous Time** (1992)
25 Fleming/Soner, **Controlled Markov Processes and Viscosity Solutions** (1993)
26 Baccelli/Brémaud, **Elements of Queueing Theory** (1994)
27 Winkler, **Image Analysis, Random Fields and Dynamic Monte Carlo Methods** (1995)
28 Kalpazidou, **Cycle Representations of Markov Processes** (1995)
29 Elliott/Aggoun/Moore, **Hidden Markov Models: Estimation and Control** (1995)
30 Hernández-Lerma/Lasserre, **Discrete-Time Markov Control Processes** (1995)
31 Devroye/Györfi/Lugosi, **A Probabilistic Theory of Pattern Recognition** (1996)
32 Maitra/Sudderth, **Discrete Gambling and Stochastic Games** (1996)
33 Embrechts/Klüppelberg/Mikosch, **Modelling Extremal Events for Insurance and Finance** (1997, corr. 2nd printing 1999)
34 Duflo, **Random Iterative Models** (1997)
35 Kushner/Yin, **Stochastic Approximation Algorithms and Applications** (1997)
36 Musiela/Rutkowski, **Martingale Methods in Financial Modelling** (1997)
37 Yin, **Continuous-Time Markov Chains and Applications** (1998)
38 Dembo/Zeitouni, **Large Deviations Techniques and Applications** (1998)
39 Karatzas, **Methods of Mathematical Finance** (1998)
40 Fayolle/Iasnogorodski/Malyshev, **Random Walks in the Quarter-Plane** (1999)
41 Aves/Jensen, **Stochastic Models in Reliability** (1999)
42 Hernandez-Lerma/Lasserre, **Further Topics on Discrete-Time Markov Control Processes** (1999)
43 Yong/Zhon, **Stochastic Controls. Hamiltonian Systems and HJB Equations** (1999)
44 Serfozo, **Introduction to Stochastic Networks** (1999)
45 Steele, **Invitation to Stochastic Calculus** (2000)

Robert S. Liptser Albert N. Shiryaev

Statistics of Random Processes

II. Applications

Translated by A. B. Aries
Translation Editor: Stephen S. Wilson

Second, Revised and Expanded Edition

Springer

Authors

Robert S. Liptser

Tel Aviv University
Department of
Electrical Engineering Systems
Ramat Aviv, P.O. Box 39040
69978 Tel Aviv, Israel
e-mail: liptser@eng.tau.ac.il

Albert N. Shiryaev

Steklov Mathematical Institute,
Russian Academy of Sciences
Gubkina 8, 117966 Moscow, Russia
e-mail: shiryaev@genesis.mi.ras.ru

Translation Editor

Stephen S. Wilson

31 Harp Hill
Cheltenham, Gloucestershire, GL52 6PY, United Kingdom
e-mail: techtrans@cwcom.net

Managing Editors

I. Karatzas
Departments of Mathematics
and Statistics
Columbia University
New York, NY 10027, USA

M. Yor
CNRS, Laboratoire de Probabilités
Université Pierre et Marie Curie
4 Place Jussieu, Tour 56
F-75230 Paris Cedex 05, France

Mathematics Subject Classification (2000):
60Gxx, 60Hxx, 60Jxx, 62Lxx, 62Mxx, 62Nxx, 93Exx, 94A05

Title of the Russian Original Edition: *Statistika sluchaĭnykh protsessov.*
Nauka, Moscow, 1974

Cover pattern by courtesy of Rick Durrett (Cornell University, Ithaca)

Library of Congress Cataloging-in-Publication Data
Liptser, R. Sh. (Robert Shevilevich) [Statistika sluchainykh protsessov. English] Statistics of random processes / Robert Liptser, Albert N. Shiryaev; translated by A. B. Aries; translation editor, Stephen S. Wilson. - 2nd, rev. and expanded ed. p. cm. - (Applications of mathematics, ISSN 0172-4568; 5-6) Includes bibliographical references and indexes. Contents: 1. General theory - 2. Applications.
ISBN 3-540-63929-2 (v. 1: acid-free paper) - ISBN 3-540-63928-4 (v. 2: acid-free paper)
1. Stochastic processes. 2. Mathematical statistics. I. Shiriaev, Al'bert Nikolaevich. II. Title
III. Series QA274.L5713 2000 519.2'3-dc21

ISSN 0172-4568
ISBN 3-540-63928-4 Springer-Verlag Berlin Heidelberg New York
ISBN 3-540-90236-8 1st Edition Springer-Verlag Berlin Heidelberg New York

Springer-Verlag Berlin Heidelberg New York
a member of BertelsmannSpringer Science+Business Media GmbH

Printed in Germany

Printed on acid-free paper SPIN: 10640137 41/3142CK - 5 4 3 2 1 0

Preface to the Second Edition

At the end of 1960s and the beginning of 1970s, when the Russian version of this book was written, the 'general theory of random processes' did not operate widely with such notions as *semimartingale, stochastic integral with respect to semimartingale, the Itô formula for semimartingales, etc.* At that time in stochastic calculus (theory of martingales), the main object was the *square integrable martingale.* In a short time, this theory was applied to such areas as nonlinear filtering, optimal stochastic control, statistics for diffusion-type processes.

In the first edition of these volumes, the stochastic calculus, based on square integrable martingale theory, was presented in detail with the proof of the Doob–Meyer decomposition for submartingales and the description of a structure for stochastic integrals. In the first volume ('General Theory') these results were used for a presentation of further important facts such as the Girsanov theorem and its generalizations, theorems on the innovation processes, structure of the densities (Radon–Nikodym derivatives) for absolutely continuous measures being distributions of diffusion and Itô-type processes, and existence theorems for weak and strong solutions of stochastic differential equations.

All the results and facts mentioned above have played a key role in the derivation of 'general equations' for nonlinear filtering, prediction, and smoothing of random processes.

The second volume ('Applications') begins with the consideration of the so-called conditionally Gaussian model which is a natural 'nonlinear' extension of the Kalman–Bucy scheme. The conditionally Gaussian distribution of an unobservable signal, given observation, has permitted nonlinear filtering equations to be obtained, similar to the linear ones defined by the Kalman–Bucy filter. Parallel to the explicit filtering implementation this result has being applied in many cases: to establish the 'separation principle' in the LQG (linear model, quadratic cost functional, Gaussian noise) stochastic control problem, in some coding problems, and to estimate unknown parameters of random processes.

The square integrable martingales, involved in the above-mentioned models, were assumed to be continuous. The first English edition contained two additional chapters (18 and 19) dealing with point (counting) processes which are the simplest discontinuous ones. The martingale techniques, based on the Doob–Meyer decomposition, permitted, in this case as well, the investigation of the structure of discontinuous local martingales, to find the corresponding version of Girsanov's theorem, and to derive nonlinear stochastic filtering equations for discontinuous observations.

Over the long period of time since the publication of the Russian (1974) and English (1977, 1978) versions, the monograph 'Statistics of Random Processes' has remained a frequently cited text in the connection with the stochastic calculus for square integrable martingales and point processes, nonlinear filtering, and statistics of random processes. For this reason, the authors decided not to change the main material of the first volume. In the second volume ('Applications'), two subsections 14.6 and 16.5 and a new Chapter 20 have being added. In Subsections 14.6 and 16.5, we analyze the Kalman–Bucy filter under wrong initial conditions for cases of discrete and continuous time, respectively. In Chapter 20, we study an asymptotic optimality for linear and nonlinear filters, corresponding to filtering models presented in Chapters 8–11, when in reality filtering schemes are different from the above-mentioned but can be approximated by them in some sense.

Below we give a list of books, published after the first English edition and related to its content:

- Anulova, A., Veretennikov, A., Krylov, N., Liptser, R. and Shiryaev, A. (1998) Stochastic Calculus [4]
- Elliott, R. (1982) Stochastic Calculus and Applications [59]
- Elliott, R.J., Aggoun, L. and Moore, J.B. (1995) Hidden Markov Models [60]
- Dellacherie, C. and Meyer, P.A. (1980) Probabilités et Potentiel. Théorie des Martingales [51]
- Jacod, J. (1979) Calcul Stochastique et Problèmes des Martingales [104]
- Jacod, J. and Shiryaev, A.N. (1987) Limit Theorems for Stochastic Processes [106]
- Kallianpur, G. (1980) Stochastic Filtering Theory [135]
- Karatzas, I. and Shreve, S.E. (1991) Brownian Motion and Stochastic Calculus [142]
- Krylov, N.V. (1980) Controlled Diffusion Processes [164]
- Liptser, R.S. and Shiryaev, A.N. (1986, 1989) Theory of Martingales [214]
- Meyer, P.A. (1989) A short presentation of stochastic calculus [230]
- Métivier, M. and Pellaumail, J. (1980) Stochastic Integration [228]
- Øksendal, B. (1985, 1998) Stochastic Differential Equations [250]
- Protter, P. (1990) Stochastic Integration and Differential Equations. A New Approach [257]
- Revuz, D. and Yor, M. (1994) Continuous Martingales and Brownian Motion [261]
- Rogers, C. and Williams, D. (1987) Diffusions, Markov Processes and Martingales: Itô Calculus [262]
- Shiryaev, A.N. (1978) Optimal Stopping Rules [286]
- Williams, D. (ed) (1981) Proc. Durham Symposium on Stochastic Integrals [308]
- Shiryaev, A.N. (1999) Essentials of Stochastic Finance [288].

The topics gathered in these books are named 'general theory of random processes', 'theory of martingales', 'stochastic calculus', applications of the stochastic calculus, etc. It is important to emphasize that substantial progress in developing this theory was implied by the understanding of the fact that it is necessary to add to the Kolmogorov probability space $(\Omega, \mathcal{F}, P)$ the increasing family (filtration) of σ-algebras $(\mathcal{F}_t)_{t \geq 0}$, where $\mathcal{F}_t$ can be interpreted as the set of events observed up to time t. A new filtered probability space $(\Omega, \mathcal{F}, (\mathcal{F}_t)_{t \geq 0}, P)$ is named the *stochastic basis*. The introduction of the stochastic basis has provided such notions as: 'to be adapted (optional, predictable) to filtration', semimartingale, and others. It is very natural that the old terminology also has changed for many cases. For example, the notion of the *natural* process, introduced by P.A. Meyer for the description of the Doob–Meyer decomposition, was changed to *predictable* process. The importance of the notion of '*local martingale*', introduced by K. Itô and S. Watanabe, was also realized.

In this publication, we have modernized the terminology as much as possible. The corresponding comments and indications of useful references and known results are given at the end of every chapter headed by 'Notes and References. 2'.

The authors are grateful to Dr. Stephen Wilson for the preparation of the Second Edition for publication. Our thanks are due to the member of the staff of the Mathematics Editorial of Springer-Verlag for their help during the preparation of this edition.

Table of Contents

Table of Contents of Volume 5

11. Conditionally Gaussian Processes

11.1 Assumptions and Formulation of the Theorem of Conditional Gaussian Behavior

11.1.1. Let $(\theta, \xi) = (\theta_t, \xi_t)$, $0 \le t \le T$, be a random process with unobservable first component and observable second component. In employing the equations of optimal nonlinear filtering given by (8.10) one encounters an essential difficulty: in order to find $\pi_t(\theta)$, it is necessary to know the conditional moments of the higher orders

$$\pi_t(\theta^2) = M(\theta_t^2 | \mathcal{F}_t^\xi), \quad \pi_t(\theta^3) = M(\theta_t^3 | \mathcal{F}_t^\xi).$$

This 'nonclosedness' of the equations given by (8.10) forces us to search for additional relations between the moments of higher orders so as to obtain a closed system.

In the case considered in the previous chapter the random process (θ, ξ) was Gaussian, which yielded the additional relation

$$\pi_t(\theta^3) = 3\pi_t(\theta)\pi_t(\theta^2) - 2[\pi_t(\theta)]^3, \tag{11.1}$$

enabling us to obtain from (8.10) the closed system of equations given by (10.10)–(10.11) for the a posteriori mean $\pi_t(\theta) = M(\theta_t | \mathcal{F}_t^\xi)$ and the a posteriori variance $\gamma_t(\theta) = \pi_t(\theta^2) - [\pi_t(\theta)]^2$.

The present chapter will deal with one class of random processes $(\theta, \xi) = (\theta_t, \xi_t)$, $0 \le t \le T$, which are not Gaussian but have the important property that (P-a.s.) the conditional distribution $F_{\xi_0^t}(x) = P\{\theta_t \le x | \mathcal{F}_t^\xi\}$ is Gaussian, yielding, in particular, (11.1).

For such processes (we call them *conditionally Gaussian* processes) the solution of problems of filtering, interpolation and extrapolation can be obtained as in the case of the Gaussian process (θ, ξ), considered in Chapter 10. A detailed investigation of these problems is given in the next chapter.

11.1.2. Let us now describe the processes involved and indicate the basic assumptions.

Let us consider as given some (complete) probability space $(\Omega, \mathcal{F}, P)$ with a nondecreasing right-continuous family of sub-σ-algebras $(\mathcal{F}_t)$, $0 \le t \le T$, and let $W_1 = (W_1(t), \mathcal{F}_t)$ and $W_2 = (W_2(t), \mathcal{F}_t)$ be mutually independent

Wiener processes. The random variables θ_0 and ξ_0 are assumed to be independent of the Wiener processes W_1 and W_2.

Let $(\theta, \xi) = (\theta_t, \xi_t)$, $0 \le t \le T$, be a (continuous) process of the diffusion type with

$$d\theta_t = [a_0(t,\xi) + a_1(t,\xi)\theta_t]dt + b_1(t,\xi)dW_1(t) + b_2(t,\xi)dW_2(t), \tag{11.2}$$

$$d\xi_t = [A_0(t,\xi) + A_1(t,\xi)\theta_t]dt + B(t,\xi)dW_2(t). \tag{11.3}$$

Each of the (measurable) functionals $a_i(t,x)$, $A_i(t,x)$, $b_j(t,x)$, $B(t,x)$, $i = 0, 1$, $j = 1, 2$, is assumed to be nonanticipative (i.e., $\mathcal{B}_t$-measurable where $\mathcal{B}_t$ is the σ-algebra in the space C_T of continuous functions $x = \{x_s, s \le T\}$ generated by the functions x_s, $s \le t$).

It is assumed that for each $x \in C_T$,

$$\int_0^T \left(\sum_{i=0,1} \{|a_i(t,x)| + |A_i(t,x)|\} + \sum_{j=1,2} b_j^2(t,x) + B^2(t,x) \right) dt < \infty. \tag{11.4}$$

Along with (11.4) assuring the existence of the integrals in (11.2) and (11.3), the following conditions will also be assumed:

(1) for each $x \in C_T$,

$$\int_0^T [A_0^2(t,x) + A_1^2(t,x)]dt < \infty; \tag{11.5}$$

$$\inf_{x \in C} B^2(t,x) \ge C > 0, \quad 0 \le t \le T; \tag{11.6}$$

(2) for any $x, y \in C_T$,

$$|B(t,x) - B(t,y)|^2 \le L_1 \int_0^t |x_s - y_s|^2 dK(s) + L_2 |x_t - y_t|^2, \tag{11.7}$$

$$B^2(t,x) \le L_1 \int_0^t (1 + x_s^2) dK(s) + L_2(1 + x_t^2), \tag{11.8}$$

where $K(s)$ is a nondecreasing right-continuous function, $0 \le K(s) \le 1$;

(3)

$$\int_0^T M|A_1(t,\xi)\theta_t|dt < \infty, \tag{11.9}$$

$$M|\theta_t| < \infty, \quad 0 \le t \le T, \tag{11.10}$$

$$P\left\{ \int_0^T A_1^2(t,\xi) m_t^2 dt < \infty \right\} = 1, \tag{11.11}$$

where $m_t = M(\theta_t | \mathcal{F}_t^\xi)$.

The following result is the main one in this chapter.

Theorem 11.1. *Let (11.4)–(11.11) be fulfilled and let (with probability 1) the conditional distribution* $F_{\xi_0}(a) = P(\theta_0 \le a|\xi_0)$ *be Gaussian,* $N(m_0, \gamma_0)$, *with* $0 \le \gamma_0 < \infty$. *Then the random process* $(\theta, \xi) = (\theta_t, \xi_t)$, $0 \le t \le T$, *satisfying Equations (11.2) and (11.3) is conditionally Gaussian, i.e., for any* t, *and* $0 \le t_0 \le t_1 < \cdots < t_n \le t$, *the conditional distributions*

$$F_{\xi_0^t}(x_0, \ldots, x_n) = P(\theta_{t_0} \le x_0, \ldots, \theta_{t_n} \le x_n|\mathcal{F}_t^\xi)$$

are (P-a.s) Gaussian.

The proof of this theorem (see Section 11.3) is based on a number of auxiliary lemmas which will be given in the following section.

11.2 Auxiliary Lemmas

11.2.1. Let $\eta = (\eta_t, \mathcal{F}_t)$, $0 \le t \le T$, denote any of the processes $\xi = (\xi_t, \mathcal{F}_t)$ or $\tilde{\xi} = (\tilde{\xi}_t, \mathcal{F}_t)$, where ξ is an observable component of a process (θ, ξ) with the differential

$$d\xi_t = [A_0(t, \xi) + A_1(t, \xi)\theta_t]dt + B(t, \xi)dW_2(t), \tag{11.12}$$

and $\tilde{\xi}$ is a solution of the equations

$$d\tilde{\xi}_t = B(t, \tilde{\xi}_t)dW_2(t), \quad \tilde{\xi}_0 = \xi_0. \tag{11.13}$$

By virtue of (11.4)–(11.11)[1] and Theorem 4.6, this equation has a unique continuous solution.

Write

$$\tilde{a}_0(t, x) = a_0(t, x) - \frac{b_2(t, x)}{B(t, x)}A_0(t, x), \tag{11.14}$$

$$\tilde{a}_1(t, x) = a_1(t, x) - \frac{b_2(t, x)}{B(t, x)}A_1(t, x), \tag{11.15}$$

and consider the equation (with respect to $\tilde{\theta}_t$, $0 \le t \le T$)

$$\tilde{\theta}_t = \theta_0 + \int_0^t [\tilde{a}_0(s, \eta) + \tilde{a}_1(s, \eta)\tilde{\theta}_s]ds + \int_0^t b_1(s, \eta)dW_1(s) + \int_0^t \frac{b_2(s, \eta)}{B(s, \eta)}d\eta_s. \tag{11.16}$$

Lemma 11.1. *For each* t, $0 \le t \le T$, *Equation (11.16) has a (unique) continuous,* $\mathcal{F}_t^{\theta_0, W_1, \eta}$*-measurable solution* $\tilde{\theta}_t$, *given by the formula*

[1] Throughout this section (11.4)–(11.11) are assumed to be satisfied.

$$\tilde{\theta}_t = \Phi_t(\eta)\left[\theta_0 + \int_0^t \Phi_s^{-1}(\eta)\tilde{a}_0(s,\eta)ds + \int_0^t \Phi_s^{-1}(\eta)b_1(s,\eta)dW_1(s) + \int_0^t \Phi_s^{-1}(\eta)\frac{b_2(s,\eta)}{B(s,\eta)}d\eta_s\right] \tag{11.17}$$

where

$$\Phi_t(\eta) = \exp\left\{\int_0^t \tilde{a}_1(s,\eta)ds\right\}. \tag{11.18}$$

PROOF. It is not difficult to show that by virtue of (11.4)–(11.6) all the integrals in (11.17) and (11.18) are defined.

Applying now the Itô formula we convince ourselves that the process $\tilde{\theta}_t$, $0 \le t \le T$, given by the right-hand side of (11.17) satisfies Equation (11.16). Thus, to complete the proof of the lemma it is only necessary to establish the uniqueness of the solution.

Let $\Delta_t = \tilde{\theta}_t - \tilde{\theta}_t'$ be the difference of two continuous solutions of Equation (11.16). Then

$$\Delta_t = \int_0^t \tilde{a}_1(s,\eta)\Delta_s ds,$$

and, therefore,

$$|\Delta_t| \le \int_0^t |\tilde{a}_1(s,\eta)||\Delta_s|ds.$$

From this, by Lemma 4.13, we obtain: $|\Delta_t| = 0$ (P-a.s.) for any t, $0 \le t \le T$. Therefore,

$$P\left\{\sup_{0\le t\le T}|\Delta_t| > 0\right\} = 0. \qquad \square$$

Let $\eta = \xi$. In this case, $\tilde{\theta}_t$ is a $\mathcal{F}_t^{\theta_0,W_1,\xi}$-measurable random variable. According to Lemma 4.9, there exists a functional $Q_t(a,x,y)$ defined on $([0,T]\times\mathbb{R}^1\times C_T\times C_T)$ which, for each t and a, is $\mathcal{B}_{t+}\times\mathcal{B}_{t+}$-measurable such that for almost all t, $0 \le t \le T$,

$$\tilde{\theta}_t = Q_t(\theta_0, W_1, \xi) \quad (P\text{-a.s.}).$$

Following the notation in Equations (11.14) and (11.15), Equation (11.2) can be written as follows:

$$d\theta_t = [\tilde{a}_0(t,\xi) + \tilde{a}_1(t,\xi)\theta_t]dt + b_1(t,\xi)dW_1(t) + \frac{b_2(t,\xi)}{B(t,\xi)}d\xi_t.$$

Comparing this equation with (11.16) we note that, by virtue of Lemma 11.1, for almost all t, $0 \le t \le T$,

$$\theta_t = Q_t(\theta_0, W_1, \xi) \quad (P\text{-a.s}). \tag{11.19}$$

From this and from (11.3) it follows that the process $\xi = (\xi_t, \mathcal{F}_t)$, $0 \le t \le T$, yields the stochastic differential

$$d\xi_t = [A_0(t,\xi) + A_1(t,\xi)Q_t(\theta_0, W_1, \xi)]dt + B(t,\xi)dW_2(t). \tag{11.20}$$

11.2.2. Consider now the two random processes $(\alpha, \beta, \xi) = [(\alpha_t, \beta_t, \xi_t), \mathcal{F}_t]$ and $(\alpha, \beta, \tilde{\xi}) = [(\alpha_t, \beta_t, \tilde{\xi}_t), \mathcal{F}_t]$, $0 \le t \le T$, given by the equations

$$\begin{aligned} d\alpha_t &= 0, \quad \alpha_0 = \theta_0, \\ d\beta_t &= dW_1(t), \quad \beta_0 = 0, \\ d\xi_t &= [A_0(t,\xi) + A_1(t,\xi)Q_t(\alpha,\beta,\xi)]dt + B(t,\xi)dW_2(t) \end{aligned} \tag{11.21}$$

and

$$\begin{aligned} d\alpha_t &= 0, \quad \alpha_0 = \theta_0, \\ d\beta_t &= dW_1(t), \quad \beta_0 = 0, \\ d\tilde{\xi}_t &= B(t,\tilde{\xi})dW_2(t), \quad \tilde{\xi}_0 = \xi_0, \end{aligned} \tag{11.22}$$

respectively.

Let $\mu_{\alpha,\beta,\xi}$ $(= \mu_{\theta_0,W_1,\xi})$ and $\mu_{\alpha,\beta,\tilde{\xi}}$ $(= \mu_{\theta_0,W_1,\tilde{\xi}})$ be measures corresponding to the processes (α,β,ξ) and $(\alpha,\beta,\tilde{\xi})$.

Lemma 11.2. *The measures $\mu_{\alpha,\beta,\xi}$ and $\mu_{\alpha,\beta,\tilde{\xi}}$ are equivalent*

$$\mu_{\alpha,\beta,\xi} \sim \mu_{\alpha,\beta,\tilde{\xi}}. \tag{11.23}$$

Further,

$$\varphi_t(\alpha,\beta,\tilde{\xi}) = \frac{d\mu_{\alpha,\beta,\xi}}{d\mu_{\alpha,\beta,\tilde{\xi}}}(t,\alpha,\beta,\tilde{\xi}), \quad \psi_t(\alpha,\beta,\xi) = \frac{d\mu_{\alpha,\beta,\xi}}{d\mu_{\alpha,\beta,\tilde{\xi}}}(t,\alpha,\beta,\xi)$$

are given by the formulae

$$\begin{aligned} \varphi_t(\alpha,\beta,\tilde{\xi}) = \exp\Bigg\{ &\int_0^t \frac{A_0(s,\tilde{\xi}) + A_1(s,\tilde{\xi})Q_s(\alpha,\beta,\tilde{\xi})}{B^2(s,\tilde{\xi})} d\tilde{\xi}_s \\ &-\frac{1}{2}\int_0^t \frac{[A_0(s,\tilde{\xi}) + A_1(s,\tilde{\xi})Q_s(\alpha,\beta,\tilde{\xi})]^2}{B^2(s,\tilde{\xi})} ds \Bigg\}, \end{aligned} \tag{11.24}$$

$$\begin{aligned} \psi_t(\alpha,\beta,\zeta) = \exp\Bigg\{ &-\int_0^t \frac{A_0(s,\xi) + A_1(s,\xi)Q_s(\alpha,\beta,\xi)}{B^2(s,\xi)} d\xi_s \\ &+\frac{1}{2}\int_0^t \frac{A_0(s,\xi) + A_1(s,\xi)Q_s(\alpha,\beta,\xi)]^2}{B^2(s,\xi)} ds \Bigg\}. \end{aligned} \tag{11.25}$$

PROOF. Note first that (see Section 13.1)

$$\begin{pmatrix} 0 & 0 & 0 \\ 0 & 1 & 0 \\ 0 & 0 & B^2(t,x) \end{pmatrix}^+ = \begin{pmatrix} 0 & 0 & 0 \\ 0 & 1 & 0 \\ 0 & 0 & B^{-2}(t,x) \end{pmatrix}. \tag{11.26}$$

Since $\theta_t = Q_t(\theta_0, W, \xi) = Q_t(\alpha, \beta, \xi)$ for almost all t (P-a.s), $0 \le t \le T$, and since (11.2) and (11.3) are satisfied,

$$P\left\{\int_0^t \frac{[A_0(t,\xi) + A_1(t,\xi)Q_t(\alpha,\beta,\xi)]^2}{B^2(t,\xi)}dt < \infty\right\} = 1.$$

Then, by the multidimensional analog of Theorem 7.20, $\mu_{\alpha,\beta,\xi} \ll \mu_{\alpha,\beta,\tilde{\xi}}$.

According to Lemma 4.9, there exists a measurable function $\tilde{Q}_t(a,x,y)$ defined on $([0,T] \times \mathbb{R}^1 \times C_T \times C_T)$ which, for each t and a, is $\mathcal{B}_{t+} \times \mathcal{B}_{t+}$-measurable such that for almost all t, $0 \le t \le T$, (P-a.s.)

$$\tilde{\theta}_t^{\tilde{\xi}} = \tilde{Q}_t(\theta_0, W_1, \tilde{\xi}),$$

where $\tilde{\theta}_t^{\tilde{\xi}}$, $0 \le t \le T$, is a solution of Equation (11.16) with $\eta = \tilde{\xi}$.

By Lemma 4.10, for almost all t, $0 \le t \le T$, (P-a.s.)

$$Q_t(\alpha,\beta,\xi) = \tilde{Q}_t(\alpha,\beta,\xi).$$

Therefore, the process $\xi = (\xi_t, \mathcal{F}_t)$, $0 \le t \le T$, also has the differential (compare with (11.20))

$$d\xi_t = [A_0(t,\xi) + A_1(t,\xi)\tilde{Q}_t(\theta_0, W_1, \xi)]dt + B(t,\xi)dW_2(t).$$

Hence

$$P\left\{\int_0^T \frac{[A_0(t,\tilde{\xi}) + A_1(t,\tilde{\xi})\tilde{Q}_t(\alpha,\beta,\tilde{\xi})]^2}{B^2(t,\tilde{\xi})}dt < \infty\right\} = 1.$$

From this, by the multidimensional analog of Theorem 7.19 and Lemma 4.10 the proof follows. □

11.2.3. Let (θ, ξ) be a random process obeying Equations (11.2), (11.3). Denote by $(m_t(x), \mathcal{B}_{t+})$ a functional[2] such that for almost all t, $0 \le t \le T$,

$$m_t(\xi) = M(\theta_t | \mathcal{F}_t^{\xi}) \quad (P\text{-a.s.}),$$

and let

$$\overline{W}_t = \int_0^t \frac{d\xi_s}{B(s,\xi)} - \int_0^t \frac{A_0(s,\xi) + A_1(s,\xi)m_s(\xi)}{B(s,\xi)}ds. \tag{11.27}$$

[2] See Lemma 4.9.

Lemma 11.3. *The random process $\overline{W} = (\overline{W}_t, \mathcal{F}_t^\xi)$, $0 \le t \le T$, is a Wiener process.*

PROOF. From (11.27) and (11.3) we obtain

$$\overline{W}_t = W_2(t) + \int_0^t \frac{A_1(s,\xi)}{B(s,\xi)}[\theta_s - m_s(\xi)]ds. \tag{11.28}$$

From this, with the help of the Itô formula, we find that

$$\begin{aligned} e^{iz\overline{W}_t} &= e^{iz\overline{W}_s} + iz\int_s^t \frac{A_1(u,\xi)}{B(u,\xi)} e^{iz\overline{W}_u}[\theta_u - m_u(\xi)]du \\ &+ iz\int_s^t e^{iz\overline{W}_u} dW_2(u) - \frac{z^2}{2}\int_s^t e^{iz\overline{W}_u} du. \end{aligned} \tag{11.29}$$

As in the proof of Theorem 7.17, from (11.29) we deduce that (P-a.s.)

$$M(e^{iz(\overline{W}_t - \overline{W}_s)} | \mathcal{F}_s^\xi) = e^{-(z^2/2)(t-s)}. \qquad \square$$

11.2.4.
Lemma 11.4. *Let μ_ξ and $\mu_{\tilde{\xi}}$ be measures corresponding to the processes ξ and $\tilde{\xi}$ defined by (11.21) and (11.22). Then $\mu_\xi \sim \mu_{\tilde{\xi}}$ and the densities*

$$\varphi_t(\tilde{\xi}) = \frac{d\mu_\xi}{d\mu_{\tilde{\xi}}}(t,\tilde{\xi}), \quad \psi_t(\xi) = \frac{d\mu_{\tilde{\xi}}}{d\mu_\xi}(t,\xi)$$

are given by the formulae

$$\begin{aligned} \varphi_t(\tilde{\xi}) &= \exp\left\{\int_0^t \frac{A_0(s,\tilde{\xi}) + A_1(s,\tilde{\xi})m_s(\tilde{\xi})}{B^2(s,\tilde{\xi})} d\tilde{\xi}_s \right. \\ &\left. -\frac{1}{2}\int_0^t \frac{[A_0(s,\tilde{\xi}) + A_2(s,\tilde{\xi})m_s(\tilde{\xi})]^2}{B^2(s,\tilde{\xi})} ds \right\}, \end{aligned} \tag{11.30}$$

$$\begin{aligned} \psi_t(\tilde{\xi}) &= \exp\left\{-\int_0^t \frac{A_0(s,\xi) + A_1(s,\xi)m_s(\xi)}{B^2(s,\xi)} d\xi_s \right. \\ &\left. +\frac{1}{2}\int_0^t \frac{A_0(s,\xi) + A_2(s,\xi)m_s(\xi)]^2}{B^2(s,\xi)} ds \right\}. \end{aligned} \tag{11.31}$$

PROOF. From (11.27) we find

$$d\xi_t = [A_0(t,\xi) + A_1(t,\xi)m_t(\xi)]dt + B(t,\xi)d\overline{W}_t. \tag{11.32}$$

Let $\overline{\xi} = (\overline{\xi}, \mathcal{F}_t)$, $0 \le t \le T$, be a random process with the differential

$$d\overline{\xi}_t = B(t,\overline{\xi})d\overline{W}_t, \quad \overline{\xi}_0 = \xi_0. \tag{11.33}$$

By virtue of (11.7), (11.8), and Theorem 4.6, Equation (11.33) has a unique strong solution. Hence, the measures $\mu_{\overline{\xi}}$ and $\mu_{\tilde{\xi}}$ coincide (compare (11.33) with (11.22)).

The absolute continuity of the measure μ_ξ with respect to the measure $\mu_{\overline{\xi}}$ (and therefore with respect to $\mu_{\tilde{\xi}}$) follows from Theorem 7.20. It will also be shown that $\mu_{\overline{\xi}} \ll \mu_\xi$.

By Lemma 11.2,

$$\begin{aligned}\mu_{\overline{\xi}}(\Gamma) &= M[\chi_{(\xi\in\Gamma)}\psi_t(\alpha,\beta,\xi)] \\ &= M[\chi_{(\xi\in\Gamma)}M(\psi_t(\alpha,\beta,\xi)|\mathcal{F}_t^\xi)] = \int_\Gamma M[\psi_t(\alpha,\beta,\xi)|\mathcal{F}_t^\xi]_{\xi=x}d\mu_\xi(x).\end{aligned}$$

Hence $\mu_{\overline{\xi}} \ll \mu_\xi$. (11.30) and (11.31) follow from Theorem 7.20 and Lemma 6.8. □

11.2.5. Let

$$\overline{\rho}_t(\alpha,\beta,\overline{\xi}) = \frac{\varphi_t(\alpha,\beta,\overline{\xi})}{\varphi_t(\overline{\xi})},$$

and for each t, $0 \le t \le T$, let $\rho_t(a,b,x)$ denote a (measurable) functional

$$\rho_t(\alpha,\beta,\xi) = \overline{\rho}_t(\alpha,\beta,\overline{\xi})_{\overline{\xi}=\xi} \quad (P\text{-a.s.}).$$

Then, because the measures $\mu_{\overline{\xi}}$ and $\mu_{\tilde{\xi}}$ are the same, we deduce from Lemmas 4.10, 11.2 and 11.4 that

$$\begin{aligned}\rho_t(\alpha,\beta,\xi) &= \exp\left\{\int_0^t \frac{A_1(s,\xi)}{B,s,\xi)}[Q_s(\alpha,\beta,\xi) - m_s(\xi)]d\overline{W}_s\right. \\ &\quad \left. -\frac{1}{2}\int_0^t \frac{A_1^2(s,\xi)}{B^2(s,\xi)}[Q_s(\alpha,\beta,\xi) - m_s(\xi)]^2 ds\right\}. \qquad (11.34)\end{aligned}$$

Lemma 11.5. *Let $f_t(\theta_0, W_1, \xi)$ be a $\mathcal{F}_t^{\theta_0,W_1,\xi}$-measurable functional with $M|f_t(\theta_0,W_1,\xi)| < \infty$. Then we have the following (Bayes) formula:*

$$M[f_t(\theta_0,W_1,\xi)|\mathcal{F}_t^\xi] = \int_{-\infty}^{\infty} f_t(a,c,\xi)\rho_t(a,c,\xi)d\mu_W(c)dF_{\xi_0}(a), \qquad (11.35)$$

where $\mu_W(\cdot)$ is a Wiener measure on the measurable space $(C_T, \mathcal{B}_T)$ of the continuous functions $C_T = \{c_s, 0 \le s \le T\}$, and $F_{\xi_0}(a) = P\{\theta_0 \le a|\xi_0\}$.

Formula (11.35) is the *Bayes formula* (7.178), which was proved in Theorem 7.23[3].

[3] See also Notes 1 and 2 to this theorem.

Corollary. *Let*

$$f_t(\theta_0, W_1, \xi) = \exp\left\{ i \left[z_0\theta_0 + \sum_{j=1}^{k} z_j W_1(t_j) \right] \right\},$$

where $0 \le t_1 \le \cdots \le t_k \le t$. *Then the conditional characteristic function*

$$M\left(\exp\left\{ i \left[z_0\theta_0 + \sum_{j=1}^{k} z_j W_1(t_j) \right] \right\} \Bigg| \mathcal{F}_t^\xi \right)$$

$$= \int_{-\infty}^{\infty} \int_{C_T} \exp\left\{ i \left[z_0 a + \sum_{j=1}^{k} z_j c_{t_j} \right] \right\} \rho_t(a, c, \xi) d\mu_W(c) dF_{\xi_0}(a), \quad (11.36)$$

where c_{t_j} *are the values of the continuous functions* $c = (c_s)$, $0 \le s \le T$, *at points* t_j.

11.3 Proof of the Theorem of Conditional Gaussian Behavior

11.3.1. As a preliminary, let us prove the following:

Theorem 11.2. *Let (11.5)–(11.11) be satisfied, and with probability 1, let the conditional distribution*

$$F_{\xi_0}(a) = P(\theta_0 \le a | \xi_0)$$

be Gaussian, with parameters

$$m_0 = M(\theta_0 | \xi_0), \quad \gamma_0 = M[(\theta_0 - m_0)^2 | \xi_0], \quad 0 \le \gamma_0 < \infty.$$

Then the conditional distribution

$$G_{\xi_0^t}(a, c_1, \ldots, c_n) = P\{\theta_0 \le a, W_1(t_1) \le c_1, \ldots, W_1(t_n) \le c_n | \mathcal{F}_t^\xi\}$$

is Gaussian for any t, $0 \le t_1 \le \cdots \le t_n \le t$, *and* $n = 1, 2, \ldots$.

11.3.2.
PROOF OF THEOREM 11.2. The proof of this theorem is based on (11.36) for a conditional characteristic function.

From (11.36) it is seen that to prove the theorem it would suffice to show that for almost all ω the measure $\rho_t(a, c, \xi) d\mu_W(c) dF_{\xi_0}(a)$ is Gaussian. However, the verification of this fact is difficult.

We start by writing $\ln \rho_t(\alpha, \beta, \xi)$ in a more convenient form.

Using the notation given by (11.14), (11.15), and (11.18) we set

$$
\begin{aligned}
g_1(t,\xi) &= \frac{A_1 t,\xi)}{B(t,\xi)}\left\{\Phi_t(\xi)\left[\int_0^t \Phi_s^{-1}(\xi)\tilde{a}_0(s,\xi)ds\right.\right. \\
&\quad \left.\left. + \int_0^t \Phi_s^{-1}(\xi)\frac{b_2(s,\xi)}{B(s,\xi)}d\xi_s\right] - m_t(\xi)\right\}, \qquad (11.37)\\
g_2(t,\xi) &= \frac{A_1(t,\xi)}{B(t,\xi)}\Phi_t(\xi), \qquad (11.38)\\
g_3(t,\xi) &= \Phi_t^{-1}(\xi)b_1(t,\xi). \qquad (11.39)
\end{aligned}
$$

Then from (11.17), we find that

$$
\begin{aligned}
\frac{A_1(t,\xi)}{B(t,\xi)}[Q_t(\theta_0,W_1,\xi) - m_t(\xi)] &= g_1(t,\xi) + \theta_0 g_2(t,\xi) \\
&\quad + g_2(t,\xi)\int_0^t g_3(s,\xi)dW_1(s).
\end{aligned}
$$

By virtue of (11.34), this enables us to write

$$
\begin{aligned}
&\ln \rho_t(\theta_0, W_1, \xi) \\
&= \int_0^t \left\{g_1(s,\xi) + g_2(s,\xi)\left[\theta_0 + \int_0^s g_3(u,\xi)dW_1(u)\right]\right\} d\overline{W}_s \\
&\quad -\frac{1}{2}\int_0^t \left\{g_1(s,\xi)+g_2(s,\xi)\left[\theta_0+\int_0^s g_3(u,\xi)dW_1(u)\right]\right\}^2 ds. \qquad (11.40)
\end{aligned}
$$

For each t, $0 \le t \le T$, let $\Delta_i(t,x)$, $i = 1,2,3$ be $\mathcal{B}_t$-measurable functionals such that

$$
\begin{aligned}
\Delta_1(t,\xi) &= \int_0^t g_1(s,\xi)d\overline{W}_s - \frac{1}{2}\int_0^t [g_1(s,\xi)]^2 ds, \\
\Delta_2(t,\xi) &= \int_0^t g_2(s,\xi)d\overline{W}_s - \int_0^t g_1(s,\xi)g_2(s,\xi)ds, \\
\Delta_3(t,\xi) &= \left(\int_0^t [g_2(s,\xi)]^2 ds\right)^{1/2},
\end{aligned}
$$

and let $\Delta_j(t,x,y)$, $j = 4,5$, be $\mathcal{B}_t \times \mathcal{B}_t$-measurable functionals such that

$$
\begin{aligned}
\Delta_4(t,W_1,\xi) &= \int_0^t g_2(s,\xi)\int_0^s g_3(u,\xi)dW_1(u)d\overline{W}_s \\
&\quad - \int_0^t g_1(s,\xi)\int_0^s g_3(u,\xi)dW_1(u)ds, \\
\Delta_5(t,W_1,\xi) &= -\int_0^t g_2(s,\xi)\int_0^s g_3(u,\xi)dW_1(u)ds.
\end{aligned}
$$

With the help of the Itô formula and the relation

$$\overline{W}_t = W_2(t) + \int_0^t \frac{A_1(s,\xi)}{B(s,\xi)}(\theta_s - m_s)ds,$$

it is easy to obtain the following relations:

$$\begin{aligned}&\int_0^t g_2(s,\xi)\int_0^s g_3(u,\xi)dW_1(u)d\overline{W}_s\\ &= \int_0^t g_2(s,\xi)d\overline{W}_s \int_0^t g_3(s,\xi)dW_1(s)\\ &\quad - \int_0^t g_3(t,\xi)\int_0^s g_2(u,\xi)d\overline{W}_u dW_1(s),\end{aligned} \tag{11.41}$$

and $(i = 1, 2)$

$$\begin{aligned}&\int_0^t g_i(s,\xi)\int_0^s g_3(u,\xi)dW_1(u)ds\\ &= \int_0^t g_i(s,\xi)ds \int_0^t g_3(s,\xi)dW_1(s)\\ &\quad - \int_0^t g_2(s,\xi)\int_0^s g_i(u,\xi)du dW_1(s).\end{aligned} \tag{11.42}$$

Using (11.40), the definition of $\Delta_1(t,x)$, $\Delta_2(t,x)$, $\Delta_3(t,x)$, $\Delta_4(t,x,y)$, $\Delta_5(t,x,y)$ and Lemmas 4.10 and 11.2, we find that for $a \in \mathbb{R}^1$,

$$\begin{aligned}\ln \rho_t(a, W_1, \tilde{\xi}) &= \Delta_1(t,\tilde{\xi}) + a[\Delta_2(t,\tilde{\xi}) + \Delta_5(t, W_1, \tilde{\xi})]\\ &\quad + \Delta_4(t, W_1, \tilde{\xi}) - \frac{a^2}{2}\Delta_3^2(t,\tilde{\xi})\\ &\quad - \frac{1}{2}\int_0^t g_2^2(s,\tilde{\xi})\left(\int_0^s g_3(u,\tilde{\xi})dW_1(u)\right)^2 ds.\end{aligned} \tag{11.43}$$

Using the definitions of $\Delta_4(t,x,y)$ and $\Delta_5(t,x,y)$ as well as (11.41), (11.42), Lemmas 4.10 and 11.2, and the independence of the processes W_1 and $\tilde{\xi}$, we conclude that the conditional distribution of these variables (for a fixed $\tilde{\xi}$) is (P-a.s.) Gaussian.

To prove that the measure $\rho_t(a,c,\xi)d\mu_W(c)dF_{\xi_0}(a)$ is Gaussian it is enough to show (because of the equivalence $\mu_\xi \sim \mu_{\tilde{\xi}}$) that the measure

$$\rho_t(a,c,\tilde{\xi})d\mu_W(c)dF_{\xi_0}(a)$$

is Gaussian.

To this end we show that the characteristic function (see (11.36))

$$\begin{aligned}\varphi_t(z_0,\ldots,z_k) &= \int_{-\infty}^{\infty}\int_{C_T} \exp\left\{i\left[z_0 a + \sum_{j=1}^k z_j c_{t_j}\right]\right\}\\ &\quad \times \rho_t(a,c,\tilde{\xi})d\mu_W(c)dF_{\xi_0}(a),\end{aligned} \tag{11.44}$$

$0 \le t_t \le \cdots \le t_k \le t$, is the characteristic function of a Gaussian distribution. Let

$$I(a,t,\tilde{\xi},z_0,\dots,z_k) = \int_{C_T} \exp\left\{ i \left[z_0 a + \sum_{j=1}^{k} z_j c_{t_j} \right] \right\} \rho_t(a,c,\tilde{\xi}) d\mu_W(c).$$

Then the desired characteristic function is given by the formula

$$\varphi_t(z_0,\dots,z_k) = \int_{-\infty}^{\infty} I(a,t,\tilde{\xi},z_0,\dots,z_k) dF_{\xi_0}(a). \tag{11.45}$$

If we can show that $I(a,t,\tilde{\xi},z_0,\dots,z_k)$ has the form

$$I(a,t,\tilde{\xi},z_0,\dots,z_k) = \exp\{\Phi(t,a,\tilde{\xi},z_0,\dots,z_k)\} \tag{11.46}$$

where $\Phi(t,a,\tilde{\xi},z_0,\dots,z_k)$ is quadratic in the variables $a,z_0,\dots,z_k$, and is nonnegative definite in $z_1,\dots,z_k$, then the conclusion of the theorem will follow from (11.45), Gaussians of $F_{\xi_0}(a)$, and the fact that $\varphi_t(z_0,\dots,z_k)$ is a characteristic function.

(11.46), with $\Phi(t,a,\tilde{\xi},\gamma_0,\dots,z_k)$ having the above properties, follows from Lemma 11.6 below, which is of interest on its own merits.

Lemma 11.6[4]. *Consider a random vector $\beta = (\beta_1,\dots,\beta_n)$ and a Wiener process $W = (W_t) = (W_1(t),\dots,W_m(t))$, $t \le T$, with independent components and suppose that the system (β, W) is Gaussian.*

Let $b = (b_1,\dots,b_n)$ be a row vector and let $B(t)$ and $Q(t)$ be $(m \times m)$ matrices such that $Q(t)$ is symmetric and nonnegative definite and

$$\mathrm{Tr} \int_0^T [B(t)B^*(t) + Q(t)]dt < \infty.$$

Then

$$\begin{aligned} \mathcal{T}(b) &\equiv M \exp\left(b\beta - \int_0^T \left[\int_0^t B(s)dW_s\right]^* Q(t) \left[\int_0^t B(s)dW_s\right] dt \right) \\ &= \exp\left(bM\beta + \frac{1}{2} bRb^* + \frac{1}{2}\mathrm{Tr} \int_0^T B(t)B^*(t)\Gamma(t)dt \right), \end{aligned} \tag{11.47}$$

where R is a nonnegative definite matrix, and $\Gamma(t)$ is a nonpositive definite matrix defined by the Ricatti equation

$$\dot{\Gamma}(t) = 2Q(t) - \Gamma(t)B(t)B^*(t)\Gamma(t), \quad \Gamma(T) = 0. \tag{11.48}$$

PROOF. The existence of a unique continuous solution for (11.48) is established as in the proof of Theorem 7.21 for Equation (7.142).

---

[4] This lemma generalizes Theorem 7.21.

Define the random processes $\eta = (\eta_t)$ and $\xi = (\xi_t)$, $t \le T$, by $(\eta_0 = \xi_0 = 0)$ and

$$\begin{aligned} d\eta_t &= B(t)dW_t, \\ d\xi_t &= B(t)B^*(t)\Gamma(t)\xi_t dt + B(t)dW_t. \end{aligned} \tag{11.49}$$

According to the multidimensional analog of Theorem 7.19, the measures μ_ξ and μ_η which correspond to the processes ξ and η are equivalent. At the same time, using certain properties of pseudo-inverses, namely $A = AA^+A$ and $A^+ = A^*(AA^*)^+$ (see (1°) and (6°), Subsection 13.13), we obtain from (7.138)

$$\frac{d\mu_\xi}{d\mu_\eta}(\eta(\omega)) = \exp\left(\int_0^T \eta_s^*\Gamma(s)B(s)dW_s - \frac{1}{2}\int_0^T \eta_s^*\Gamma(s)B(s)B^*(s)\Gamma(s)\eta_s ds\right).$$

We now show that

$$\begin{aligned} \frac{d\mu_\xi}{d\mu_\eta}(\eta(\omega)) &= \exp\Bigg(-\int_0^T \left[\int_0^t B(s)dW_s\right]^* Q(t)\left[\int_0^t B(s)dW_s\right]dt \\ &\quad -\frac{1}{2}\operatorname{Tr}\int_0^T B(t)B^*(t)\Gamma(t)dt\Bigg). \end{aligned} \tag{11.50}$$

Indeed, using the identities $\eta_T^*\Gamma(T)\eta_T = 0$, $\eta_0^*\Gamma(0)\eta_0 = 0$, (11.48), and the Itô formula (see Example 2, Subsection 4.3.3), we find

$$\begin{aligned} 0 &= \eta_T^*\Gamma(T)\eta_T - \eta_0^*\Gamma(0)\eta_0 \\ &= 2\int_0^T \eta_t^*\Gamma(t)B(t)dW_t + 2\int_0^T \eta_t^*[Q(t) - \frac{1}{2}\Gamma(t)B(t)B^*(t)\Gamma(t)]\eta_t dt \\ &\quad -\operatorname{Tr}\int_0^T B(t)B^*(t)\Gamma(t)dt. \end{aligned}$$

This and (11.49) prove (11.50). On the other hand,

$$\begin{aligned} M[M(\exp\{b\beta\}|\mathcal{F}_T^\eta)_{\eta=\xi}] &= M\left[M(\exp\{b\beta\}|\mathcal{F}_T^\eta)\frac{d\mu_\xi}{d\mu_\eta}(\eta)\right] \\ &= M\exp\{b\beta\}\frac{d\mu_\xi}{d\mu_\eta}(\eta). \end{aligned}$$

Therefore, according to (11.50),

$$\mathcal{T}(b) = M[M(\exp\{b\beta\}|\mathcal{F}_T^\eta)_{\eta=\xi}]\exp\left(\frac{1}{2}\operatorname{Tr}\int_0^T B(t)B^*(t)\Gamma(t)dt\right). \tag{11.51}$$

Since (β, W) forms a Gaussian system, so does (β, η), and, hence, by the multidimensional analogs of Theorems 5.16 and 5.21, the martingale $(M(\beta|\mathcal{F}_t^\eta), \mathcal{F}_t^\eta)$ admits the representation

$$M(\beta|\mathcal{F}_t^\eta) = M\beta + \int_0^t g(s)d\eta_s, \quad t \le T,$$

where the $(n \times m)$ matrix $g(t)$ is such that

$$\mathrm{Tr} \int_0^T g(t)B^*(t)B(t)g(t)dt < \infty.$$

Let $D_T = M[(\beta - M(\beta|\mathcal{F}_T^\eta))(\beta - M(\beta|\mathcal{F}_T^\eta))^*]$.

Using the theorem on normal correlation (Theorem 13.1), it is possible to show that D_T coincides (P-a.s.) with the conditional covariance $M[(\beta - M(\beta|\mathcal{F}_T^\eta))(\beta - M(\beta|\mathcal{F}_T^\eta))^*\mathcal{F}_T^\eta]$. Therefore, since $P(\beta \le x|\mathcal{F}_T^\eta)$ is Gaussian and μ_ξ and μ_η are equivalent, by Lemma 4.10 we find that

$$M(\exp\{b\beta\}|\mathcal{F}_T^\eta)_{\eta=\xi} = \exp\left(b\left[M\beta + \int_0^T g(t)d\xi_t\right] + \frac{1}{2}b^* D_T b\right).$$

The random vector $\int_0^T g(t)d\xi_t$ is Gaussian and has zero mean. Denote by G_T its covariance matrix and set $R = \frac{1}{2}(D_T + G_T)$. Then

$$M[M(\exp(\{b\beta\}|\mathcal{F}_T^\eta)_{\eta=\xi}] = \exp(bM\beta + b^* Rb). \tag{11.52}$$

The above and (11.51) complete the proof of Lemma 11.6, and, therefore, also of Theorem 11.2. □

11.3.3.

PROOF OF THEOREM 11.1. Let $0 \le t_0 < t_1 < \cdots < t_n \le t \le T$ be some decomposition of the interval $[0, T]$. Then, considering (11.19), we have

$$M\left(\exp\left\{i\sum_{j=0}^n z_j\theta_{t_j}\right\}\middle|\mathcal{F}_t^\xi\right) = M\left(\exp\left\{i\sum_{j=0}^n z_j Q_{t_j}(\theta_0, W_1, \xi)\right\}\middle|\mathcal{F}_t^\xi\right),$$

where, according to Lemma 11.1,

$$\begin{aligned} Q_{t_j}(\theta_0, W_1, \xi) = \Phi_{t_j}(\xi)\Bigg\{\theta_0 &+ \int_0^{t_j} \Phi_s^{-1}(\xi)\tilde{a}(s,\xi)ds \\ &+ \int_0^{t_j} \Phi_s^{-1}(\xi)b_1(s,\xi)Dw_1(s) + \int_0^t \Phi_s^{-1}(\xi)\frac{b_2(s,\xi)}{B(s,\xi)}d\xi_s\Bigg\}. \end{aligned}$$

Hence

$$M\left[\exp\left\{i\sum_{j=0}^{n} z_j Q_{t_j}(\theta_0, W_1, \xi)\right\}\Big| \mathcal{F}_t^\xi\right]$$

$$= \exp\left\{i\sum_{j=0}^{n-1} z_j\left[\Phi_{t_j}(\xi)\left(\int_0^{t_j}\Phi_s^{-1}(\xi)\tilde{a}_0(s,\xi)ds + \int_0^{t_j}\Phi_s^{-1}(\xi)\frac{b_2(s,\xi)}{B(s,\xi)}d\xi_s\right]\right)\right\}$$

$$\times M\left(\exp\left\{i\sum_{j=0}^{n-1} z_j\Phi_{t_j}\left[\theta_0 + \int_0^{t_j}\Phi_s^{-1}(\xi)b_1(s,\xi)dW_1(s)\right]\right\}\Big| \mathcal{F}_t^\xi\right).$$

Applying Lemma 11.6. we find that

$$M\left[\exp\left\{i\sum_{j=0}^{n} z_j\theta_{t_j}\right\}\Big| \mathcal{F}_t^\xi\right]$$

$$= \exp\left\{i\sum_{j=0}^{n} z_j\tilde{\delta}_j(t,\xi) - \frac{1}{2}\sum_{k,j=0}^{n} z_k z_j\tilde{\gamma}_{kj}(t,\xi)\right\}, \qquad (11.53)$$

where $\|\tilde{\gamma}_{kj}(t,\xi)\|$ is some nonnegative definite symmetric matrix.

Because of the arbitrariness of $z_0, z_1, \dots, z_n$, it follows from (11.53) that the conditional distribution

$$P(\theta_{t_0} \le a_0, \dots, \theta_{t_n} \le a_n | \mathcal{F}_t^\xi)$$

is Gaussian for any $t_0 < t_1 < \dots < t_n \le t$ and $n = 1, 2, \dots$. □

Note. Let $0 \le s \le t_0 < \dots < t_n \le t$. Then the conditional distribution

$$P(\theta_{t_0} \le a_0, \dots \theta_{t_n} \le a_n | \mathcal{F}_t^{\theta_s, \xi})$$

is also (P-a.s.) Gaussian; this follows from the normality of the distributions $P(\theta_s, \le a, \theta_{t_0} \le a_0 \le \dots \le a_{t_n} \le a_n | \mathcal{F}_t^\xi)$.

11.3.4. For the needs of problems of filtering, interpolation and extrapolation of the conditionally Gaussian processes, the parameters $m_t = M(\theta_t | \mathcal{F}_t^\xi)$ and $\gamma_t = M[(\theta_t - m_t)^2 | \mathcal{F}_t^\xi]$ of the conditional distribution $F_{\xi_0^t}(a) = P(\theta_t \le a | \mathcal{F}_t^\xi)$ are of special interest. They could be found if an explicit form of the elements $\tilde{\delta}_j(t,\xi)$ and $\tilde{\gamma}_{kj}(t,\xi)$ entering into (11.53) can be determined,

In the next chapter it will be shown, however, that for finding the parameters m_t and γ_t (as well as other characteristics of the conditionally Gaussian processes) it is simpler to make use of the general equations of filtering, interpolation and extrapolation developed in Chapter 8.

Notes and References. 1

11.1–11.3. The importance of distinguishing the class of conditionally Gaussian processes for effective solution of problems of optimal nonlinear filtering was noted by Liptser [194]. Conditionally Gaussian processes were discussed in Liptser and Shiryaev [205]. The proof of the theorem of conditional normality has been first given here.

Notes and References. 2

11.1–11.3. Despite the fact that the conditionally Gaussian model (11.2), (11.3) is nonlinear with respect to the observable component, its attractiveness is accounted for by many properties inherited from the Kalman–Bucy model (10.1), (10.2). For example, since the conditional distribution of an unobservable signal, given observations, is Gaussian, so that it is completely described by the conditional expectation and variance, differential equations for these parameters define the filter similar to the Kalman–Bucy one (see next section). The conditionally Gaussian model can be used as motivation for creating the so-called *extended* Kalman filter, the applicability of which is described in [253].

12. Optimal Nonlinear Filtering: Interpolation and Extrapolation of Components of Conditionally Gaussian Processes

12.1 Optimal Filtering Equations

12.1.1. Let $(\theta, \xi) = (\theta_t, \xi_t)$, $0 \le t \le T$, be a continuous random diffusion-type process with

$$d\theta_t = [a_0(t,\xi) + a_1(t,\xi)\theta_t]dt + b_1(t,\xi)dW_1(t) + b_2(t,\xi)dW_2(t), \qquad (12.1)$$

$$d\xi_t = [A_0(t,\xi) + A_1(t,\xi)\theta_t]dt + B(t,\xi)dW_2(t). \qquad (12.2)$$

Assume that the conditions given by (11.4)–(11.11) formulated in the previous chapter are satisfied. If the conditional distribution $F_{\xi_0}(a) = P(\theta_0 \le a|\xi_0)$ is (P-a.s.) Gaussian, $N(m_0, \gamma_0)$, then in accordance with Theorem 11.1 the conditional distribution $F_{\xi_0^t}(a) = P(\theta_t \le a|\mathcal{F}_t^\xi)$ will also be Gaussian, $N(m_t, \gamma_t)$. Hence if $M\theta_t^2 < \infty$, $0 \le t \le T$, then one of the moments of this distribution – the a posteriori mean $m_t = M(\theta_t|\mathcal{F}_t^\xi)$ – will be an optimal (in the mean square sense) estimate of θ_t from $\xi_0^t = \{\xi_s, s \le t\}$. The knowledge of the variance $\gamma_t = M([\theta_t - m_t]^2|\mathcal{F}_t^\xi)$ of this distribution enables us to find the filtering error

$$\Delta_t = M(\theta_t - m_t)^2 \quad (= M\gamma_t). \qquad (12.3)$$

Theorem 12.1, given below, contains equations that m_t and γ_t must satisfy. By virtue of conditional normality of the process (θ, ξ) these equations turn out to be closed ones.

It should be emphasized that Theorem 12.1 provides as a particular case the filtering equations deduced for the Kalman–Bucy scheme in Chapter 10. Whereas in the Kalman–Bucy scheme the process (θ, ξ) was Gaussian, and as a result the optimal filter was linear, in the conditionally Gaussian case the optimal filter is, generally speaking, nonlinear.

12.1.2. The deduction of equations for m_t and γ_t based on the use of the fundamental theorem of filtering (Theorem 8.1) is carried out according to the following scheme.

According to (12.1)

$$\theta_t = \theta_0 + \int_0^t [a_0(s,\xi) + a_1(s,\xi)\theta_s]ds + x_t, \qquad (12.4)$$

where

$$x_t = \int_0^t [b_1(s,\xi)dW_1(s) + b_2(s,\xi)dW_2(s)]. \tag{12.5}$$

From (12.4) and (12.5) with the help of the Itô formula we find that

$$\theta_t^2 = \theta_0^2 + \int_0^t (2\theta_s[a_0(s,\xi) + a_1(s,\xi)\theta_s] + [b_1^2(s,\xi) + b_2^2(s,\xi)]ds + \tilde{x}_t, \tag{12.6}$$

where

$$\tilde{x}_t = \int_0^t 2\theta_s[b_1(s,\xi)dW_1(s) + b_2(s,\xi)dW_2(s)]. \tag{12.7}$$

Denote

$$h_t = \theta_t, \quad H_t = a_0(t,\xi) + a_1(t,\xi)\theta_t \tag{12.8}$$

and

$$\tilde{h}_t = \theta_t^2, \quad \tilde{H}_t = 2\theta_t[a_0(t,\xi) + a_1(t,\xi)\theta_t] + [b_1^2(t,\xi) + b_2^2(t,\xi)]. \tag{12.9}$$

Then Equations (12.4) and (12.6) can be written as follows:

$$h_t = h_0 + \int_0^t H_s ds + x_t, \tag{12.10}$$

$$\tilde{h}_t = \tilde{h}_0 + \int_0^t \tilde{H}_s ds + \tilde{x}_t. \tag{12.11}$$

Therefore, the unobservable processes h_t and $\tilde{h}_t$ have the form which was assumed in Theorem 8.1.

In order to take advantage of this theorem, we need to find conditions under which the assumptions given by (8.6)–(8.8) involved in the formulation of the theorem are satisfied (other assumptions are satisfied due to (11.4)–(11.11)). In our case, (8.6)–(8.8) are reduced to the following

$$\sup_{0\le t\le T} M\theta_t^4 < \infty; \tag{12.12}$$

$$\int_0^T M[a_0(s,\xi) + a_1(s,\xi)\theta_s]^2 ds < \infty; \tag{12.13}$$

$$\int_0^T M\{2\theta_s[a_0(s,\xi) + a_1(s,\xi)\theta_s] + [b_1^2(s,\xi) + b_2^2(s,\xi)]\}^2 ds < \infty; \tag{12.14}$$

$$\int_0^T M\{A_0(s,\xi) + A_1(s,\xi)\theta_s\}^2 ds < \infty. \tag{12.15}$$

In order to have these conditions satisfied, and also to be able to assert that $X = (x_t, \mathcal{F}_t)$ and $\tilde{X} = (\tilde{x}_t, \mathcal{F}_t)$, $0 \le t \le T$, are square integrable martingales, we shall require the following conditions to be satisfied.

For all $x \in C_T$, $0 \le t \le T$:

$$|a_1(t,x)| \le L, \quad |A_1(t,x)| \le L; \tag{12.16}$$

$$\int_0^T M[a_0^4(t,\xi) + b_1^4(t,\xi) + b_2^4(t,\xi)]dt < \infty; \tag{12.17}$$

$$M\theta_0^4 < \infty. \tag{12.18}$$

In order to prove the sufficiency of these conditions, as a preliminary we shall prove the following lemma.

Lemma 12.1. *In the assumptions given by (12.16)–(12.18),*

$$M\left[\sup_{0\le t\le T} \theta_t^4\right] < \infty. \tag{12.19}$$

PROOF. Put

$$\tau_N = \inf\left\{t : \sup_{s\le t} \theta_s^4 \ge N\right\},$$

taking $\tau_N = T$ if $\sup_{s\le T} \theta_s^4 < N$. Then, by virtue of Hölder's inequality,

$$\begin{aligned}
\theta_{t\wedge\tau_N}^4 &= \left[\theta_0 + \int_0^{t\wedge\tau_N} a_0(s,\xi)ds + \int_0^{t\wedge\tau_N} a_1(s,\xi)\theta_s ds \right.\\
&\quad \left. + \sum_{i=1}^2 \int_0^{t\wedge\tau_N} b_i(s,\xi)dW_i(s)\right]^4 \\
&\le 125\left[\theta_0^4 + \left(\int_0^{t\wedge\tau_N} a_0(s,\xi)ds\right)^4\right.\\
&\quad \left. + \left(\int_0^{t\wedge\tau_N} a_1(s,\xi)\theta_s ds\right)^4 + \sum_{i=1}^2\left(\int_0^{t\wedge\tau_N} b_i(s,\xi)dW_i(s)\right)^4\right] \\
&\le 125\left[\theta_0^4 + (t\wedge\tau_N)^3\int_0^{t\wedge\tau_N} a_0^4(s,\xi)ds\right.\\
&\quad \left. + (t\wedge\tau_N)^3\int_0^{t\wedge\tau_N} a_1^4(s,\xi)\theta_s^4 ds + \sum_{i=1}^2\left(\int_0^{t\wedge\tau_N} b_i(s,\xi)dW_i(s)\right)^4\right].
\end{aligned} \tag{12.20}$$

According to Lemma 4.12,

$$M\left(\int_0^{t\wedge\tau_N} b_i(s,\xi)dW_i(s)\right)^4 \le 36T\int_0^T Mb_i^4(s,\xi)ds, \quad i = 1, 2.$$

Hence, since $\theta_s = \theta_{s\wedge\tau_N}$ for $s \le t\wedge\tau_N$,

$$M\theta^4_{t\wedge\tau_N} \le 125\left[M\theta_0^4 + T^3\int_0^T Ma_0^4(s,\xi)ds \right.$$
$$\left. + T^3L^4\int_0^t M\theta^4_{s\wedge\tau_N}ds + 36T\sum_{i=1}^2\int_0^T Mb_i^4(s,\xi)ds\right],$$

i.e.,

$$M\theta^4_{t\wedge\tau_N} \le C_1 + C_2\int_0^t M\theta^4_{s\wedge\tau_N}ds, \tag{12.21}$$

where C_1, C_2 are constants. Therefore, by Lemma 4.13,

$$M\theta^4_{t\wedge\tau_N} \le C_1e^{C_2t} \le C_1e^{C_2T},$$

and, by the Fatou lemma,

$$M\theta_t^4 \le \underline{\lim}_{N\to\infty} M\theta^4_{t\wedge\tau_N} \le C_1e^{C_2T}.$$

Thus

$$\sup_{0\le t\le T} M\theta_t^4 < \infty. \tag{12.22}$$

We shall show now that $M[\sup_{0\le t\le T}\theta_t^4] < \infty$. Substituting $t\wedge\tau_N$ for t we obtain from (12.20)

$$\sup_{0\le t\le T}\theta_t^4 \le 125\left[\theta_0^4 + T^3\int_0^T a_0^4(s,\xi)ds + T^3L^4\int_0^T \theta_s^4 ds\right.$$
$$\left. + \sum_{i=1}^2 \sup_{0\le t\le T}\left|\int_0^t b_i(s,\xi)dW_i(s)\right|^4\right].$$

According to (3.8) and Lemma 4.12,

$$M\sup_{0\le t\le T}\left|\int_0^t b_i(s,\xi)dW_i(s)\right|^4 \le \left(\frac{4}{3}\right)^4 36T\int_0^T Mb_i^4(s,\xi)ds, \quad i=1,2.$$

Hence, due to (12.22) and (12.16)–(12.18),

$$M\left[\sup_{0\le t\le T}\theta_t^4\right] \le 125\left[M\theta_0^4 + T^3\int_0^T Ma_0^4(s,\xi)ds + T^4L^4\sup_{0\le t\le T}M\theta_s^4\right.$$
$$\left. + \left(\frac{4}{3}\right)^4 36T\sum_{i=1}^2\int_0^T Mb_i^4(s,\xi)ds\right] < \infty.$$

□

Thus the conditions given by (12.12) follow from the assumptions given by (12.16)–(12.18). It is verified in an obvious manner that these assumptions guarantee the validity of (12.13)–(12.15). From the explicit form of the processes x_t and $\tilde{x}_t$ and the assumptions given by (12.16)–(12.18) it can be easily deduced that $X = (x_t, \mathcal{F}_t)$ and $\tilde{X}_t = (\tilde{x}_t, \mathcal{F}_t)$, $0 \le t \le T$, are square integrable martingales. Therefore, the conditions of Theorem 8.1 in the case considered are satisfied.

Taking into account that $\langle x, W_2 \rangle = \int_0^t b_2(s,\xi)ds$, we find from (8.9) that

$$
\begin{aligned}
m_t &= m_0 + \int_0^t [a_0(s,\xi) + a_1(s,\xi)m_s]ds \\
&\quad + \int_0^t \left\{ b_2(s,\xi) + \frac{A_1(s,\xi)[M(\theta_s^2|\mathcal{F}_s^\xi) - m_s^2]}{B(s,\xi)} \right\} d\overline{W}_s, \qquad (12.23)
\end{aligned}
$$

where

$$
\overline{W}_t = \int_0^t \frac{d\xi_s - (A_0(s,\xi) + A_1(s,\xi)m_s)ds}{B(s,\xi)}.
$$

Next, $M(\theta_s^2|\mathcal{F}_s^\xi) - m_s^2 = \gamma_s$. Hence, it follows from (12.23) that

$$
\begin{aligned}
dm_t &= [a_0(t,\xi) + a_1(t,\xi)m_t]dt \\
&\quad + \frac{b_2(t,\xi)B(t,\xi) + \gamma_t A_1(t,\xi)}{B^2(t,\xi)}[d\xi_t - (A_0(t,\xi) + A_1(t,\xi)m_t)dt].
\end{aligned} \qquad (12.24)
$$

Denote now $\delta_t = M(\theta_t^2|\mathcal{F}_t^\xi)$, so that $\delta_t - m_t^2 = \gamma_t$. Then, taking into account the equality $\langle \tilde{x}, W_2 \rangle_t = \int_0^t 2\theta_s b_2(s,\xi)ds$, again from (8.9) we obtain

$$
\begin{aligned}
\delta_t &= \delta_0 + \int_0^t [2a_0(s,\xi)m_s + 2a_1(s,\xi)\delta_s + b_1^2(s,\xi) + b_2^2(s,\xi)]ds \\
&\quad + \int_0^t \Big\{ 2m_s b_2(s,\xi) + B^{-1}(s,\xi)[A_0(s,\xi)\delta_s + A_1(s,\xi)M(\theta_s^3|\mathcal{F}_s^\xi) \\
&\quad - \delta_s(A_0(s,\xi) + A_1(s,\xi)m_s)] \Big\} d\overline{W}_s,
\end{aligned}
$$

or

$$
\begin{aligned}
\delta_t &= \delta_0 + \int_0^t [2a_0(s,\xi)m_s + 2a_1(s,\xi)\delta_s + b_1^2(s,\xi) + b_2^2(s,\xi)]ds \\
&\quad + \int_0^t B^{-1}(s,\xi)\left\{2m_s b_2(s,\xi)B(s,\xi) + A_1(s,\xi)[M(\theta_s^3|\mathcal{F}_s^\xi) - \delta_s m_s]\right\} d\overline{W}_s.
\end{aligned} \qquad (12.25)
$$

From the Itô formula and (12.24) we find that

$$\begin{aligned} m_t^2 &= m_0^2 + \int_0^t \Bigg(2m_s[a_0(s,\xi) + a_1(s,\xi)m_s] \\ &\quad + \left[\frac{b_2(s,\xi)B(s,\xi) + \gamma_s A_1(s,\xi)}{B(s,\xi)} \right]^2 \Bigg) ds \\ &\quad + \int_0^t 2m_s \frac{b_2(s,\xi)B(s,\xi) + \gamma_s A_1(s,\xi)}{B(s,\xi)} d\overline{W}_s, \end{aligned} \tag{12.26}$$

which together with (12.25) yields the following representation for $\gamma_t = \delta_t - m_t^2$:

$$\begin{aligned} \gamma_t &= \gamma_0 + \int_0^t \Bigg[2a_1(s,\xi)\gamma_s + b_1^2(s,\xi) + b_2^2(s,\xi) \\ &\quad - \left(\frac{b_2(s,\xi)B(s,\xi) + \gamma_t A_1(s,\xi)}{B(s,\xi)} \right)^2 \Bigg] ds \\ &\quad + \int_0^t \frac{A_1(s,\xi)}{B(s,\xi)} \{ M(\theta_s^3 | \mathcal{F}_s^\xi) - \delta_s m_s - 2m_s\gamma_s \} d\overline{W}_s. \end{aligned} \tag{12.27}$$

Since the conditional distribution $P(\theta_s \le a | \mathcal{F}_s^\xi)$ is Gaussian, then (see (11.1))

$$M(\theta_s^3 | \mathcal{F}_s^\xi) = 3m_s\delta_s - 2m_s^3 \quad (= \delta_s m_s + 2m_s\gamma_s).$$

Hence in (12.27) the stochastic integral is equal to zero and therefore

$$\begin{aligned} \gamma_t &= \gamma_0 + \int_0^t \Bigg[2a_1(s,\xi)\gamma_s + b_1^2(s,\xi) + b_2^2(s,\xi) \\ &\quad - \left(\frac{b_2(s,\xi)B(s,\xi) + \gamma_s A_1(s,\xi)}{B(s,\xi)} \right)^2 \Bigg] ds. \end{aligned} \tag{12.28}$$

Thus we have proved:

Theorem 12.1. *Let (θ, ξ) be a random process with differentials given by (12.1) and (12.2). If (11.4)–(11.8) and (12.16)–(12.18) are satisfied and the conditional distribution $P(\theta_0 \le a | \xi_0)$ is Gaussian, $N(m_0, \gamma_0)$, then m_t and γ_t satisfy equations*

$$\begin{aligned} dm_t &= [a_0(t,\xi) + a_1(t,\xi)m_t]dt \\ &\quad + \frac{b_2(t,\xi)B(t,\xi) + \gamma_t A_1(t,\xi)}{B^2(t,\xi)} [d\xi_t - (A_0(t,\xi) + A_1(t,\xi)m_t)dt], \end{aligned} \tag{12.29}$$

$$\dot{\gamma}_t = 2a_1(t,\xi)\gamma_t + b_1^2(t,\xi) + b_2^2(t,\xi) - \left(\frac{b_2(t,\xi)B(t,\xi) + \gamma_t A_1(t,\xi)}{B(t,\xi)}\right)^2, \tag{12.30}$$

subject to the conditions $m_0 = M(\theta_0|\xi_0)$, $\gamma_0 = M[(\theta_0 - m_0)^2|\xi_0]$.

Note 1. From (12.29) and (12.30) it follows that the a posteriori moments m_t and γ_t are continuous in t (P-a.s.).

Note 2. Let $A_1(s,x) \equiv 0$, i.e., let the observable process $\xi = (\xi_t)$, $0 \le t \le T$, have the differential

$$d\xi_t = A_0(t,\xi)dt + B(t,\xi)dW_2(t), \tag{12.31}$$

and let the observable process $\theta = (\theta_t)$, $0 \le t \le T$, satisfy the equation

$$d\theta_t = [a_0(t,\xi) + a_1(t,\xi)\theta_t]dt + b_1(t,\xi)dW_1(t) + b_2(t,\xi)dW_2(t).$$

From the proof carried out above (see (12.27)) it is seen that even without the assumption of normality of the conditional distribution $P(\theta_0 \le a|\xi_0)$ the parameters m_t and γ_t satisfy the equations

$$dm_t = [a_0(t,\xi) + a_1(t,\xi)m_t]dt + \frac{b_2(t,\xi)}{B(t,\xi)}[d\xi_t - A_0(t,\xi)dt], \tag{12.32}$$

$$\dot{\gamma}_t = 2a_1(t,\xi)\gamma_t + b_1^2(t,\xi). \tag{12.33}$$

Note 3. Let $m_{\theta_s}(t,s) = M[\theta_t|\mathcal{F}_t^{\theta_s,\xi}]$ for $s \le t$ and

$$\gamma_{\theta_s}(t,s) = M[(\theta_s - m_{\theta_s}(t,s))^2|\mathcal{F}_t^{\theta_s,\xi}].$$

Then $m_{\theta_s}(t,s)$ and $\gamma_{\theta_s}(t,s)$ satisfy (at $t > s$) Equations (12.19) and (12.30), solved under the conditions $m_{\theta_s}(s,s) = \theta_s$, $\gamma_{\theta_s}(s,s) = 0$. The proof is similar to the deduction of the equations for m_t and γ_t and exploits the fact that the conditional distribution $P(\theta_t \le a|\mathcal{F}_s^{\theta_s,\xi})$ is Gaussian (see the note to Theorem 11.1). From Equation (12.30) and the condition $\gamma_{\theta_s}(s,s) = 0$ it follows that $\gamma_{\theta_s}(t,s)$ does not actually depend on θ_s.

12.1.3. We shall discuss now one particular case of the system of equations given by (12.1) and (12.2) for which the filtering equations given by (12.29) and (12.30) permit an explicit solution.

Theorem 12.2. *Let $\theta = \theta(\omega)$ be a random variable with $M\theta^4 < \infty$. Assume that the observable process $\xi = (\xi_t)$, $0 \le t \le T$, permits a differential*

$$d\xi_t = [A_0(t,\xi) + A_1(t,\xi)\theta]dt + B(t,\xi)dW_2(t),$$

where the coefficients A_0, A_1, B satisfy the conditions of Theorem 12.1, and the conditional distribution $P(\theta \le a|\xi_0)$ is Gaussian. Then $m_t = M(\theta|\mathcal{F}_t^\xi)$ and $\gamma_t = M[(\theta_t - m_t)^2|\mathcal{F}_t^\xi]$ are given by the formulae

$$m_t = \frac{m_0 + \gamma_0 + \int_0^t \frac{A_1(s,\xi)}{B^2(s,\xi)}[d\xi_s - A_0(s,\xi)ds]}{1 + \gamma_0 \int_0^t \left(\frac{A_1(s,\xi)}{B(s,\xi)}\right)^2 ds}, \tag{12.34}$$

$$\gamma_t = \frac{\gamma_0}{1 + \gamma_0 \int_0^t \left(\frac{A_1(s,\xi)}{B(s,\xi)}\right)^2 ds}. \tag{12.35}$$

PROOF. Due to (12.29) and (12.30), m_t and γ_t satisfy the equations

$$dm_t = \frac{\gamma_t A_1(s,\xi)}{B^2(t,\xi)}[d\xi_t - (A_0(t,\xi) + A_1(t,\xi)m_t)dt], \tag{12.36}$$

$$\dot{\gamma}_t = -\left(\frac{\gamma_t A_1(t,\xi)}{B(t,\xi)}\right)^2, \tag{12.37}$$

solutions of which, as it is easy to verify, are determined by (12.34) and (12.35).

In the case considered, (12.34) and (12.35) can be obtained immediately from the Bayes formula, (11.35), without using general filtering equations for conditionally Gaussian random processes[1].

Indeed, if $\gamma_0 > 0$, then, due to (11.35),

$$\begin{aligned} P(\theta \le a|\mathcal{F}_t^\xi) &= M\{\chi_{[\theta \le a]}|\mathcal{F}_t^\xi\} \\ &= \int_{-\infty}^a \frac{1}{\sqrt{2\pi\gamma_0}} \exp\left\{-\frac{(\alpha - m_0)^2}{2\gamma_0}\right. \\ &\quad + \int_0^t \frac{A_1(s,\xi)}{B(s,\xi)}(\alpha - m_s(\xi))d\overline{W}_s \\ &\quad \left. -\frac{1}{2}\int_0^t \left[\frac{A_1(s,\xi)}{B(s,\xi)}(\alpha - m_s(\xi))\right]^2 ds\right\} d\alpha. \end{aligned}$$

From this it follows that the conditional distribution $P(\theta \le a|\mathcal{F}_t^\xi)$ has the density

$$\begin{aligned} \frac{dP(\theta \le a|\mathcal{F}_t^\xi)}{da} &= \frac{1}{\sqrt{2\pi\gamma_0}} \exp\left\{-\frac{(a - m_0)^2}{2\gamma_0}\right. \\ &\quad + \int_0^t \frac{A_1(s,\xi)}{B(s,\xi)}(a - m_s(\xi))d\overline{W}_s \\ &\quad \left. -\frac{1}{2}\int_0^t \left[\frac{A_1(s,\xi)}{B(s,\xi)}(a - m_s(\xi))\right]^2 ds\right\}. \end{aligned} \tag{12.38}$$

On the other hand, the conditional distribution $P(\theta_t \le a|\mathcal{F}_t^\xi)$ is Gaussian:

[1] In order to deduce equations for m_t and γ_t one can drop the assumptions given by (12.15) and (12.16).

$$\frac{dP(\theta \le a|\mathcal{F}_t^\xi)}{da} = \frac{1}{\sqrt{2\pi\gamma_t}} \exp\left\{-\frac{(a-m_t)^2}{2\gamma_t}\right\}. \tag{12.39}$$

Equating the terms in (12.38) and (12.39) with a and a^2, we obtain

$$-\frac{1}{2\gamma_0} - \frac{1}{2}\int_0^t \left(\frac{A_1(s,\xi)}{B(s,\xi)}\right)^2 ds = -\frac{1}{2\gamma_t} \quad (P\text{-a.s.}) \tag{12.40}$$

$$\frac{m_0}{\gamma_0} + \int_0^t \frac{A_1(s,\xi)}{B(s,\xi)} d\overline{W}_s + \int_0^t \left(\frac{A_1(s,\xi)m_s(\xi)}{B(s,\xi)}\right)^2 ds = \frac{m_t}{\gamma_t} \quad (P\text{-a.s.}). \tag{12.41}$$

(12.35) follows immediately from (12.40). If we take into account now that

$$d\overline{W}_t = \frac{d\xi_s - [A_0(s,\xi) + A_1(s,\xi)m_s(\xi)]ds}{B(s,\xi)}$$

then we obtain the required representation, (12.34), from (12.41).

If $P(\gamma_0 = 0) > 0$, then in order to deduce (12.34) and (12.35) one should consider a Gaussian distribution $P^\varepsilon(\theta \le a|\xi_0)$, with parameters $m_0^\varepsilon = m_0$, $\gamma_0^\varepsilon = \gamma_0 + \varepsilon$, $\varepsilon > 0$, instead of the distribution $P(\theta \le a|\xi_0)$. Then the associated values m_t^ε and γ_t^ε will be given by (12.34) and (12.35) with the substitution of $\gamma_0^\varepsilon = \gamma_0 + \varepsilon$ for γ_0, in which the passage to the limit should be carried out with $\varepsilon \downarrow 0$. □

12.2 Uniqueness of Solutions of Filtering Equations: Equivalence of σ-Algebras $\mathcal{F}_t^\xi$ and $\mathcal{F}_t^{\xi_0,\overline{W}}$

12.2.1. For a conditionally Gaussian process (θ, ξ) the a posteriori moments $m_t = M(\theta_t|\mathcal{F}_t^\xi)$ and $\gamma_t = M[(\theta_t - m_t)^2|\mathcal{F}_t^\xi]$ satisfy, according to Theorem 12.1, Equations (12.29) and (12.30). Therefore, this system of equations has the F^ξ-adapted solution ($F^\xi = (\mathcal{F}_t^\xi)$, $0 \le t \le T$). In this section we show that any continuous solution of this system is unique. Thus, solving this system of equations, we shall obtain moments m_t and γ_t of the conditional distribution θ_t.

Theorem 12.3. *Let the conditions of Theorem 12.1 be satisfied. Then the system of equations*

$$\begin{aligned} dx(t) &= [a_0(t,\xi) + a_1(t,\xi)x(t)]dt + \frac{b_2(t,\xi)B(t,\xi) + y(t)A_1(t,\xi)}{B^2(t,\xi)} \\ &\quad \times [d\xi_t - (A_0(t,\xi) + A_1(t,\xi)x(t))dt], \end{aligned} \tag{12.42}$$

$$\begin{aligned}\dot{y}(t) &= 2a_1(t,\xi)y(t) + b_1^2(t,\xi) + b_2^2(t,\xi) \\ &\quad - \left(\frac{b_2(t,\xi)B(t,\xi) + y(t)A_1(t,\xi)}{B(t,\xi)}\right)^2,\end{aligned} \tag{12.43}$$

subject to the initial conditions

$$x(0) = m_0, \quad y(0) = \gamma_0 \quad (|m_0| < \infty, 0 \le \gamma_0 < \infty)$$

has a unique, continuous, $\mathcal{F}_t^\xi$-measurable solution for any t, $0 \le t \le T$.

PROOF. Let $y_1(t)$ and $y_2(t)$, $0 \le t \le T$, be two nonnegative continuous solutions of Equation (12.43). Then

$$\begin{aligned}|y_1(t) - y_2(t)| &\le 2\int_0^t \left(|a_1(s,\xi)| + \left|\frac{b_2(s,\xi)}{B(s,\xi)}A_1(s,\xi)\right|\right)|y_1(s) - y_2(s)|ds \\ &\quad + \int_0^t \frac{A_1^2(s,\xi)}{B^2(s,\xi)}[y_1(s) + y_2(s)]|y_1(s) - y_2(s)|ds.\end{aligned} \tag{12.44}$$

Denote

$$r_1(s,\xi) = 2\left(|a_1(s,\xi)| + \left|\frac{b_2(s,\xi)}{B(s,\xi)}A_1(s,\xi)\right|\right) + \frac{A_1^2(s,\xi)}{B^2(s,\xi)}[y_1(s) + y_2(s)].$$

Then (12.44) can be rewritten as follows:

$$|y_1(t) - y_2(t)| \le \int_0^t r_1(s,\xi)|y_1(s) - y_2(s)|ds.$$

Hence, due to Lemma 4.13,

$$P\{y_1(t) = y_2(t)\} = 1, \quad 0 \le t \le T,$$

and, by virtue of the continuity of the solutions of $y_1(t)$ and $y_2(t)$,

$$P\left\{\sup_{0\le t\le T}|y_1(t) - y_2(t)| = 0\right\} = 1,$$

which proves the uniqueness of continuous solutions of Equation (12.43).

Let now $x_1(t)$ and $x_2(t)$ be two continuous solutions of Equation (12.42). Then

$$\begin{aligned}x_1(t) - x_2(t) &= \int_0^t \left[a_1(s,\xi) + \frac{b_2(s,\xi)}{B(s,\xi)}A_1(s,\xi)\right. \\ &\quad \left. + \frac{y(s)A_1^2(s,\xi)}{B^2(s,\xi)}\right][x_1(s) - x_2(s)]ds\end{aligned}$$

and therefore

$$|x_1(t) - x_2(t)| \le \int_0^t r_2(s,\xi)|x_1(s) - x_2(s)|ds, \tag{12.45}$$

where

$$r_2(s,\xi) = |a_1(s,\xi)| + \left|\frac{b_2(s,\xi)}{B(s,\xi)}A_1(s,\xi)\right| + \frac{y(s)A_1^2(s,\xi)}{B^2(s,\xi)}.$$

Hence, again applying Lemma 4.13 to (12.45), we find that $x_1(t) = x_2(t)$ (P-a.s.) for any t, $0 \le t \le T$. From this we obtain:

$$P\left\{\sup_{0\le t\le T} |x_1(t) - x_2(t)| = 0\right\} = 1. \qquad \square$$

Note. As proved above, γ_t, $0 \le t \le T$, is the unique continuous solution of Equation (12.43). Let us show that if $P(\gamma_0 > 0) = 1$, then $P\{\inf_{t\le T}\gamma_t > 0\} = 1$.

Indeed, by virtue of continuity, γ_t is greater than 0 for sufficiently small $t > 0$. Set $\tau = \inf(t \le T : \gamma_t = 0)$, taking $\tau = \infty$ if $\inf_{t\le T}\gamma_t > 0$. Then, for $t < \tau \wedge T$, the values $\delta_t = \gamma_t^{-1}$ are defined which satisfy the equation

$$\dot{\delta}_t = -2\tilde{a}_1(t,\xi)\delta_t + \left(\frac{A_1(t,\xi)}{B(t,\xi)}\right)^2 - \delta_t^2 b_1^2(t,\xi), \quad \delta_0 = \gamma_0^{-1}, \tag{12.46}$$

where

$$\tilde{a}_1(t,x) = a_1(t,x) - \frac{b_2(t,x)}{B(t,x)}A_1(t,x).$$

On the set $\{\omega : \tau \le T\}$, $\lim_{t\uparrow\tau}\delta_t = \infty$ (P-a.s.). However, according to (12.46),

$$\begin{aligned}
\delta_t &= \exp\left\{-2\int_0^t \tilde{a}_1(s,\xi)ds\right\}\left\{\delta_0 + \int_0^t \exp\left[2\int_0^s \tilde{a}_1(u,\zeta)du\right]\right. \\
&\quad \times \left.\left(\frac{A_1^2(s,\xi)}{B^2(s,\xi)} - \delta_s^2 b_1^2(s,\xi)\right)ds\right\} \\
&\le \exp\left\{2\int_0^T |\tilde{a}_1(s,\xi)|ds\right\}\left[\delta_0 + \int_0^T \frac{A_1^2(s,\xi)}{B^2(s,\xi)}ds\right] < \infty.
\end{aligned}$$

Therefore, $P\{\tau \le T\} = 0$. In other words,

$$\inf_{t\le T}\gamma_t = \left(\sup_{t\le T}\delta_t\right)^{-1} > 0 \quad (P\text{-a.s.}).$$

12.2.2. In deducing filtering equations for a process (θ, ξ) it was assumed that this process was a solution to Equations (12.1) and (12.2) for some Wiener processes W_1 and W_2. It was not, however, assumed that the process $(\theta, \xi) = (\theta_t, \xi_t)$, $0 \le t \le T$, was a strong solution (i.e., $\mathcal{F}_t^{\theta_0, \xi_0, W_1, W_2}$-measurable at any t) of this system of equations.

It is easy to bring out the conditions under which this system has a unique continuous strong solution.

Theorem 12.4. *Let $g(t, x)$ denote any of the nonanticipative functionals $a_i(t, x), A_i(t, x), b_j(t, x), B(t, x)$, $i = 0, 1$, $j = 1, 2$, $0 \le t \le T$, $x \in C_T$. Assume that:*

(1) for any $x, y \in C_T$,

$$|g(t, x) - g(t, y)|^2 \le L_1 \int_0^t (x_s - y_s)^2 dK(s) + L_2 (x_t - y_t)^2;$$

(2)

$$g^2(t, x) \le L_1 \int_0^t (1 + x_s^2) dK(s) + L_2 (1 + x_t^2),$$

where $K(s)$ is some nondecreasing right continuous function, $0 \le K(s) \le 1$, and L_1, L_2 are constants;

(3)

$$|a_1(t, x)| \le L_1, \quad |A_1(t, x)| \le L_2;$$

(4) $M(\theta_0^{2n} + \xi_0^{2n}) < \infty$ for some integer $n \ge 1$.

Then the system of equations given by (12.1) and (12.2) has a continuous strong solution. This solution is unique, and $\sup_{0 \le t \le T} M(\theta_t^{2n} + \xi_t^{2n}) < \infty$.

PROOF. The theorem can be proved in the same way as in the one- dimensional case (Theorem 4.9). □

12.2.3. We shall discuss now the question of the equivalence of σ-algebras $\mathcal{F}_t^{\xi}$ and $\mathcal{F}_t^{\xi_0, \overline{W}}$, $0 \le t \le T$, where $\overline{W} = (\overline{W}_t, \mathcal{F}_t^{\xi})$ is a Wiener process with the differential (see (11.27))

$$d\overline{W}_t = B^{-1}(t, \xi)[d\xi_t - (A_0(t, \xi) + A_1(t, \xi) m_t) dt], \quad \overline{W}_0 = 0. \tag{12.47}$$

According to (12.29), (12.30) and (12.47), the processes m_t, ξ_t, γ_t, $0 \le t \le T$, form a weak solution of the system of equations

$$
\begin{aligned}
dm_t &= [a_0(t,\xi) + a_1(t,\xi)m_t]dt + \left[b_2(t,\xi) + \frac{\gamma_t A_1(t,\xi)}{B(t,\xi)}\right] d\overline{W}_t, \\
d\xi_t &= [A_0(t,\xi) + A_1(t,\xi)m_t]dt + B(t,\xi)d\overline{W}_t, \qquad (12.48) \\
\dot{\gamma}_t &= 2\left[a_1(t,\xi) - \frac{b_2(t,\xi)}{B(t,\xi)}A_1(t,\xi)\right]\gamma_t + b_1^2(t,\xi) - \frac{A_1^2(t,\xi)}{B^2(t,\xi)}\gamma_t^2,
\end{aligned}
$$

for given $m_0 = M(\theta_0|\xi_0)$, ξ_0, and $\gamma_0 = M[(\theta_0 - m_0)^2|\xi_0]$.

Let us investigate the problem of the existence of a strong solution to this system of equations. A positive answer to this problem will enable us to establish the equivalence of σ-algebras $\mathcal{F}_t^{\xi}$ and $\mathcal{F}_t^{\xi_0,\overline{W}}$, $0 \le t \le T$, which, in its turn, will imply that the (innovation) processes $\overline{W}$ and ξ_0 contain the same information as the observable process ξ.

Theorem 12.5. *Let the functionals $a_i(t,x), A_i(t,x), b_j(t,x), B(t,x)$, $i = 0,1$, $j = 1,2$, satisfy (1) and (2) of Theorem 12.4. Let also $\gamma_0 = \gamma_0(x)$, $a_i(t,x), A_i(t,x), b_j(t,x)$ and $B^{-1}(t,x)$ $(i = 0,1;\ j = 1,2)$ be uniformly bounded. Then the system of equations given by (12.48) has a unique strong (i.e., $\mathcal{F}_t^{m_0,\gamma_0,\xi_0,\overline{W}}$-measurable for each t) solution. In this case*

$$
\mathcal{F}_t^{\xi} = \mathcal{F}_t^{\xi_0,\overline{W}}, \quad 0 \le t \le T. \qquad (12.49)
$$

PROOF. Let $x \in C_T$. Let $\gamma_t = \gamma_t(x)$ satisfy the equation

$$
\gamma_t(x) = \gamma_0(x) + \int_0^t \left[2\tilde{a}_1(s,x)\gamma_s(x) + b_1^2(s,x) - \frac{A_1^2(s,x)}{B^2(s,x)}\gamma_s^2(x)\right] ds. \qquad (12.50)
$$

Equation (12.50) is a Ricatti equation and its (nonnegative continuous) solution exists and is unique for each $x \in C_T$ (compare with the proof of Theorem 12.3). It is not difficult to deduce from (12.50) that

$$
\gamma_t(x) \le \exp\left\{\int_0^t 2\tilde{a}_1(s,x)ds\right\}\left\{\gamma_0(x) + \int_0^t \exp\left[-2\int_0^s \tilde{a}_1(u,x)du\right] b_1^2(s,x)ds\right\}.
$$

By virtue of the assumptions made above it follows that the $\gamma_t(x)$ are uniformly bounded over x.

We shall show that the function $\gamma_t(x)$ satisfies the Lipschitz condition

$$
|\gamma_t(x) - \gamma_t(y)|^2 \le L\int_0^t |x_s - y_s| d\tilde{K}(s), \quad x_0 = y_0
$$

for a certain nondecreasing right continuous function $\tilde{K}(s)$, $0 \le \tilde{K}(s) \le 1$.

From (12.50) we obtain

$$\gamma_t(x)-\gamma_t(y) = \int_0^t \Bigg\{ 2[\tilde{a}_1(s,x)\gamma_s(x)-\tilde{a}_1(s,y)\gamma_s(y)] + [b_1^2(s,x)-b_1^2(s,y)] - \left[\frac{A_1^2(s,x)}{B^2(s,x)}\gamma_s^2(x)-\frac{A_1^2(s,y)}{B^2(s,y)}\gamma_s^2(y)\right]\Bigg\}ds. \tag{12.51}$$

Due to (1) of Theorem 12.4

$$\begin{aligned} &|\tilde{a}_1(t,x)\gamma_t(x)-\tilde{a}_1(t,y)\gamma_t(y)|^2 \\ &\le 2\gamma_t^2(x)|\tilde{a}_1(t,x)-\tilde{a}_1(t,y)|^2 + 2|\tilde{a}_1(t,x)|^2|\gamma_t(x)-\gamma_t(y)|^2 \\ &\le d_0\int_0^t |x_s-y_s|^2 dK(s) + d_1|x_t-y_t|^2 + d_2|\gamma_t(x)-\gamma_t(y)|^2, \end{aligned} \tag{12.52}$$

where d_0, d_1 and d_2 are constants whose existence is guaranteed by uniform boundedness of the functions $\tilde{a}_1(t,x)$ and $\gamma_t(x)$, $x \in C_T$.

Similarly,

$$|b_1^2(t,x)-b_1^2(t,y)|^2 \le d_3\int_0^t |x_s-y_s|^2 dK(s) + d_4|x_t-y_t|^2 \tag{12.53}$$

and

$$\begin{aligned} &\left|\frac{A_1^2(t,x)}{B^2(t,x)}\gamma_t^2(x)-\frac{A_1^2(t,y)}{B^2(t,y)}\gamma_t^2(y)\right| \\ &\le d_5\int_0^t |x_s-y_s|^2 dK(s) + d_6|x_t-y_t|^2 + d_7|\gamma_t(x)-\gamma_t(y)|^2. \end{aligned} \tag{12.54}$$

From (12.51)–(12.54) we find that

$$\begin{aligned} |\gamma_t(x)-\gamma_t(y)|^2 &\le d_8\int_0^t \left[\int_0^s (x_u-y_u)^2 dK(u)\right] ds \\ &\quad + d_9\int_0^t (x_s-y_s)^2 ds + d_{10}\int_0^t |\gamma_s(x)-\gamma_s(y)|^2 ds \\ &\le d_8 T\int_0^t (x_s-y_s)^2 dK(s) + d_9\int_0^t (x_s-y_s)^2 ds \\ &\quad + d_{10}\int_0^t |\gamma_s(x)-\gamma_s(y)|^2 ds. \end{aligned}$$

Hence, by Lemma 4.13,

$$\begin{aligned} |\gamma_t(x)-\gamma_t(y)|^2 &\le \left[d_8 T\int_0^t (x_s-y_s)^2 dK(s) + d_9\int_0^t (x_s-y_s)^2 ds\right] e^{d_{10}t} \\ &\le d_{11}\int_0^t (x_s-y_s)^2 d\tilde{K}(s), \end{aligned} \tag{12.55}$$

where

$$\tilde{K}(s) = \frac{K(s)+s}{K(T)+T}, \quad d_{11} = e^{d_{10}T}[d_8 T + d_9](K(T)+T).$$

Let us consider now the first two equations of the system of equations given by (12.48), with $\gamma_t = \gamma_t(\xi)$ substituted, being, as it was shown above, a continuous uniformly bounded solution of the third equation of this system:

$$\begin{aligned} dm_t &= [a_0(t,\xi) + a_1(t,\xi)m_t]dt + \left[b_1(t,\xi) + \frac{A_1(t,\xi)}{B(t,\xi)}\gamma_t(\xi)\right] d\overline{W}_t \\ d\xi_t &= [A_0(t,\xi) + A_1(t,\xi)m_t]dt + B(t,\xi)d\overline{W}_t. \end{aligned} \tag{12.56}$$

According to the assumptions of the theorem and the properties of the functional $\gamma_t(x)$ established above, the system of equations given by (12.56) has a unique strong (i.e., $\mathcal{F}_t^{m_0,\xi_0,\overline{W}}$-measurable for any t) solution (see the note to Theorem 4.6). But $m_0 = M(\theta_0|\xi_0)$ is $\mathcal{F}_0^{\xi}$-measurable. Hence $\mathcal{F}_t^{m_0,\xi_0,\overline{W}} = \mathcal{F}_t^{\xi_0,\overline{W}}$, $0 \le t \le T$. Therefore, ξ_t is $\mathcal{F}_t^{\xi_0,\overline{W}}$-measurable, for any t.

Thus $\mathcal{F}_t^{\xi} \subseteq \mathcal{F}_t^{\xi_0,\overline{W}}$. The correctness of the reverse inclusion $\mathcal{F}_t^{\xi} \supseteq \mathcal{F}_t^{\xi_0,\overline{W}}$, follows from the construction of the innovation process $\overline{W}$ (see (12.47)). □

Note 1. Note that in the Kalman–Bucy scheme

$$\begin{aligned} a_0(t,x) &= a_0(t) + a_2(t)x_t, \quad a_1(t,x) = a_1(t), \\ A_0(t,x) &= A_0(t) + A_2(t)x_t, \quad A_1(t,x) = A_1(t), \\ B(t,x) &= B(t), \quad b_i(t,x) = b_i(t), \quad i = 1,2. \end{aligned} \tag{12.57}$$

In this case the coefficients in the equation determining γ_t are deterministic functions, and the equations for m_t and ξ_t have the following form

$$\begin{aligned} dm_t &= [a_0(t) + a_1(t)m_t + a_2(t)\xi_t]dt + \left[b_1(t) + \frac{A_1(t)\gamma_t}{B(t)}\right] d\overline{W}_t, \\ d\xi_t &= [A_0(t) + A_1(t)m_t + A_2(t)\xi_t]dt + B(t)d\overline{W}_t. \end{aligned} \tag{12.58}$$

This system has a unique strong solution under the same assumptions under which the Kalman–Bucy filtering equations were deduced (see (10.10), (10.11)). Hence, in this case $\mathcal{F}_t^{\xi} = \mathcal{F}_t^{\xi_0,\overline{W}}$, $0 \le t \le T$.

Note 2. The equality $\mathcal{F}_t^{\xi} = \mathcal{F}^{\xi_0,\overline{W}}$ remains valid also in the case of multidimensional processes θ and ξ (with explicit modifications in the conditions of Theorem 12.5 due to the multidimensionality). These matters will be discussed in the next section.

12.3 Optimal Filtering Equations in Several Dimensions

Let us extend the results of the previous sections to the case where each of the processes θ and ξ is vectorial.

12.3.1. Assume again that we a given a certain (complete) probability space $(\Omega, \mathcal{F}, P)$ with a nondecreasing right continuous family of sub-σ-algebras $(\mathcal{F}_t)$, $0 \le t \le T$. Let $W_1 = (W_1(t), \mathcal{F}_t)$ and $W_2 = (W_2(t), \mathcal{F}_t)$ be two mutually independent Wiener processes, where $W_1(t) = [W_{11}(t), \dots, W_{1k}(t)]$ and $W_2(t) = [W_{21}(t), \dots, W_{2l}(t)]$.

The partially observable random process

$$(\theta, \xi) = [(\theta_1(t), \dots, \theta_k(t)), (\xi_1(t), \dots, \xi_l(t)), \mathcal{F}_t], \quad 0 \le t \le T,$$

will be assumed to be a diffusion-type process with the differential

$$d\theta_t = [a_0(t, \xi) + a_1(t, \xi)\theta_t]dt + \sum_{i=1}^{2} b_i(t, \xi) dW_i(t), \tag{12.59}$$

$$d\xi_t = [A_0(t, \xi) + A_1(t, \xi)\theta_t]dt + \sum_{i=1}^{2} B_i(t, \xi) dW_i(t). \tag{12.60}$$

Here elements of the vector functions (columns)

$$\begin{aligned} a_0(t, x) &= (a_{01}(t, x), \dots, a_{0k}(t, x)), \\ A_0(t, x) &= (A_{01}(t, x), \dots, A_{0l}(t, x)) \end{aligned}$$

and matrices

$$\begin{aligned} a_1(t, x) &= \|a_{ij}^{(1)}(t, x)\|_{(k \times k)}, \quad A_1(t, x) = \|A_{ij}^{(1)}(t, x)\|_{(l \times k)}, \\ b_1(t, x) &= \|b_{ij}^{(1)}(t, x)\|_{(k \times k)}, \quad b_2(t, x) = \|b_{ij}^{(2)}(t, x)\|_{(k \times l)}, \\ B_1(t, x) &= \|B_{ij}^{(1)}(t, x)\|_{(l \times k)}, \quad B_2(t, x) = \|B_{ij}^{(2)}(t, x)\|_{(l \times l)} \end{aligned}$$

are assumed to be measurable nonanticipative functionals on

$$\{[0, T] \times C_T^l, \mathcal{B}_{[0,T]} \times \mathcal{B}_T^l\}, \quad x = (x_1, \dots, x_l) \in C_T^l.$$

The following conditions (1)–(7) are the multidimensional analog of (11.4)–(11.11), essentially used in proving Theorems 11.1 and 12.1 ($x \in C_T^l$, indices i and j take all admissible values):

(1)

$$\int_0^T [|a_{0i}(t, x)| + |a_{ij}^{(1)}(t, x)| + (b_{ij}^{(1)}(t, x))^2 + (b_{ij}^{(2)}(t, x))^2 + (B_{ij}^{(1)}(t, x))^2 + (B_{ij}^{(2)}(t, x))^2] dt < \infty;$$

(2)

$$\int_0^T [(A_{0i}(t,x))^2 + (A_{ij}^{(1)}(t,x))^2]dt < \infty;$$

(3) the matrix $B \circ B(t,x) \equiv B_1(t,x)B_1^*(t,x) + B_2(t,x)B_2^*(t,x)$ is uniformly nonsingular, i.e., the elements of the reciprocal matrix are uniformly bounded;

(4) if $g(t,x)$ denotes any element of the matrices $B_1(t,x)$ and $B_2(t,x)$, then, for $x, y \in C_T^l$,

$$|g(t,x) - g(t,y)|^2 \le L_1 \int_0^T |x_s - y_s|^2 dK(s) + L_2|x_t - y_t|^2,$$

$$g^2(t,x) \le L_1 \int_0^t (1 + |x_s|^2)dK(s) + L_2(1 + |x_t|^2),$$

where $|x_t|^2 = x_1^2(t) + \cdots x_l^2(t)$ and $K(s)$ is a nondecreasing right continuous function, $0 \le K(s) \le 1$;

(5)

$$\int_0^T M|A_{ij}^{(1)}(t,\xi)\theta_j(t)|dt < \infty;$$

(6)

$$M|\theta_j(t)| < \infty, \quad 0 \le t \le T;$$

(7)

$$P\left\{\int_0^T (A_{ij}^{(1)}(t,\xi)m_j(t))^2 dt < \infty\right\} = 1,$$

where $m_j(t) = M[\theta_j(t)|\mathcal{F}_t^\xi]$.

12.3.2. A generalization of Theorem 11.1 to the multidimensional case is the following.

Theorem 12.6. *Let conditions (1)–(7) be satisfied and, with probability one, let the conditional distribution*[2] $F_{\xi_0}(a_0) = P(\theta_0 \le a_0|\xi_0)$ *be (P-a.s.) Gaussian,* $N(m_0, \gamma_0)$, *where the vector* $m_0 = M(\theta_0|\mathcal{F}_0^\xi)$ *and the matrix* $\gamma_0 = M[(\theta_0 - m_0)(\theta_0 - m_0)^*|\mathcal{F}_0^\xi]$ *is such that* $\mathrm{Tr}\,\gamma_0 < \infty$ *(P-a.s.). Then a random process* $(\theta, \xi) = [(\theta_1(t), \dots, \theta_k(t)), (\xi_1(t), \dots, \xi_l(t))]$ *satisfying the system of equations given by (12.59) and (12.60) is conditionally Gaussian, i.e., for any* t_j, $0 \le t_0 < t_1 < \cdots < t_n \le t$, *the conditional distribution*

[2] For $\theta_0 = \{\theta_1(0), \dots, \theta_k(0)\}$ and $a_0 = (a_{01}, \dots, a_{0k})$, $\{\theta_0 \le a_0\}$ is understood as the event $\{\theta_1(0) \le a_{01}, \dots, \theta_k(0) \le a_{0k}\}$.

$$F_{\xi_0^t}(a_0, \dots, a_n) = P\{\theta_{t_0} \le a_0, \dots \theta_{t_n} \le a_n | \mathcal{F}_t^{\xi}\}$$

is (P-a.s.) Gaussian.

The proof of this theorem is analogous to the proof of Theorem 11.1. Hence we shall discuss only particular details of the proof which can present difficulties due to the multidimensionality of these processes.

First of all note that we can consider in (12.60) $B_1(t, x) \equiv 0$ and $B_2(t, x) \equiv B(t, x)$ since, due to Lemma 10.4, there exist mutually independent Wiener processes

$$\tilde{W}_1(t) = [\tilde{W}_{11}(t), \dots, \tilde{W}_{1k}(t)], \quad \tilde{W}_2(t) = \tilde{W}_{21}(t), \dots, \tilde{W}_{2l}(t)],$$

such that

$$\begin{aligned} \int_0^t \sum_{i=1}^2 b_i(s, \xi) dW_i(s) &= \int_0^t \sum_{i=1}^2 d_i(s, \xi) d\tilde{W}_i(s), \\ \int_0^t \sum_{i=1}^2 B_i(s, \xi) dW_i(s) &= \int_0^t D(s, \xi) d\tilde{W}_2(s), \end{aligned} \tag{12.61}$$

where

$$\begin{aligned} D(t, x) &= \sqrt{(B \circ B)(t, x)}, \\ d_2(t, x) &= (b \circ B)(t, x)(B \circ B)^{-1/2}(t, x), \\ d_1(t, x) &= [(b \circ b)(t, x) - (b \circ B)(t, x)(B \circ B)^{-1}(t, x)(b \circ B)^*(t, x)]^{1/2} \end{aligned} \tag{12.62}$$

with

$$B \circ B = B_1 B_1^* + B_2 B_2^*, \quad b \circ B = b_1 B_1^* + b_2 B_2^*, \quad b \circ b = b_1 b_1^* + b_2 b_2^*.$$

Next, if $f_t(\theta_0, W_1, \xi)$ is a (scalar) $\mathcal{F}_t^{\theta_0, W_1, \xi}$-measurable function with $M|f_t(\theta_0, W_1, \xi)| < \infty$, then there is a Bayes formula (compare with (11.35))

$$M(f_t(\theta_0, W_1, \xi) | \mathcal{F}_t^{\xi}) = \int_{\mathbb{R}^k} \int_{C_T^k} f_t(a, c, \xi) \rho_t(a, c, \xi) d\mu_W(c) dF_{\xi_0}(a), \tag{12.63}$$

where $a \in \mathbb{R}^k$, $c \in C_T^k$, μ_W is a Wiener measures in (C_T^k, B_T^k) and

$$\begin{aligned} \rho_t(a, c, \xi) &= \exp\left\{ \int_0^t [A_1(s, \xi)(Q_s(a, c, \xi) - m_s(\xi))]^* (B^*(s.\xi))^{-1} d\overline{W}_s \right. \\ &\quad - \frac{1}{2} \int_0^t [A_1(s, \xi)(Q_s(a, c, \xi) - m_s(\xi))]^* (B(s, \xi) B^*(s, \xi))^{-1} \\ &\quad \left. \times [A_1(s, \xi)(Q_s(a, c, \xi) - m_s(\xi))] ds \right\}. \end{aligned} \tag{12.64}$$

Here:

$$m_t(\xi) = M(\theta_t|\mathcal{F}_t^\xi);$$

$$\overline{W}_t = \int_0^t B^{-1}(s,\xi)d\xi_s - \int_0^t B^{-1}(s,\xi)[A_0(s,\xi) + A_1(s,\xi)m_s(\xi)]ds \tag{12.65}$$

is a Wiener process (with respect to $(\mathcal{F}_t^\xi)$, $0 \le t \le T$);

$$\begin{aligned} Q_t(a, W_1, \xi) &= \Phi_t(\xi)\Bigg[a + \int_0^t \Phi_s^{-1}(\xi)\tilde{a}_0(s,\xi)ds \\ &\quad + \int_0^t \Phi_s^{-1}(\xi)b_1(s,\xi)dW_1(s) + \int_0^t \Phi_s^{-1}(\xi)b_2(s,\xi)B^{-1}(s,\xi)d\xi_s\Bigg] \\ \frac{d\Phi_t(\xi)}{dt} &= \tilde{a}_1(t,\xi)\Phi_t(\xi), \quad \Phi_0(\xi) = E_{(k\times k)}; \end{aligned}$$

and

$$\begin{aligned} \tilde{a}_0(t,x) &= a_0(t,x) - b_2(t,x)B^{-1}(t,x)A_0(t,x), \\ \tilde{a}_1(t,x) &= a_1(t,x) - b_2(t,x)B^{-1}(t,x)A_1(t,x). \end{aligned}$$

With the help of (12.63), and in the same way as in the case of the one-dimensional processes θ and ξ, first we verify normality of the conditional distributions

$$P(\theta_0 \le a_0, W_1(t_1) \le y_1, \dots, W_1(t_n) \le y_n|\mathcal{F}_t^\xi),$$

$0 \le t_0 \le \dots \le t_n \le t$, and second we establish normality of the distributions

$$P(\theta_{t_0} \le a_0, \dots, \theta_{t_n} \le a_n|\mathcal{F}_t^\xi).$$

12.3.3. Assume also that in addition to (1)–(7), the following conditions are satisfied

(8)

$$|a_{ij}^{(1)}(t,x)| \le L, \quad |A_{ij}^{(1)}(t,x)\| \le L;$$

(9)

$$\int_0^T M[a_{0i}^4(t,\xi) + (b_{ij}^{(1)}(t,\xi))^4 + (b_{ij}^{(2)}(t,\xi))^4]dt < \infty;$$

(10)

$$M\sum_{i=1}^k \theta_i^4(0) < \infty.$$

The following result is the multidimensional analog of Theorems 12.1 and 12.3

Theorem 12.7. *Let conditions (1)–(10) be satisfied. Then the vector* $m_t = M(\theta_t|\mathcal{F}_t^\xi)$ *and the matrix* $\gamma_t = M\{(\theta_t - m_t)(\theta_t - m_t)^*|\mathcal{F}_t^\xi\}$ *are unique continuous* $\mathcal{F}_t^\xi$*-measurable for any* t *solutions of the system of equations*

$$\begin{aligned} dm_t &= [a_0(t,\xi) + a_1(t,\xi)m_t]dt + [(b\circ B)(t,\xi) + \gamma_t A_1^*(t,\xi)](B\circ B)^{-1}(t,\xi) \\ &\quad \times [d\xi_t - (A_0(t,\xi) + A_1(t,\xi)m_t)dt], \end{aligned} \tag{12.66}$$

$$\begin{aligned} \dot{\gamma}_t &= a_1(t,\xi)\gamma_t + \gamma_t a_1^*(t,\xi) + (b\circ b)(t,\xi) - [(b\circ B)(t,\xi) + \gamma_t A_1^*(t,\xi)] \\ &\quad \times (B\circ B)^{-1}(t,\xi)[(b\circ B)(t,\xi) + \gamma_t A_1^*(t,\xi)]^* \end{aligned} \tag{12.67}$$

with initial conditions $m_0 = M(\theta_0|\xi_0)$, $\gamma_0 = M\{(\theta_0 - m_0)\times(\theta_0 - m_0)^*|\xi_0\}$. *If in this case the matrix* γ_0 *is positive definite, then the matrices* γ_t, $0 \le t \le T$, *will have the same property.*

PROOF. In this theorem the proof of the deduction of Equations (12.66) and (12.67) corresponds to the pertinent proof of Theorem 12.1 carried out for the components

$$m_j(t) = M(\theta_j(t)|\mathcal{F}_t^\xi)$$

and

$$\gamma_{ij}(t) = M\{[\theta_i(t) - m_i(t)][\theta_j(t) - m_j(t)]|\mathcal{F}_t^\xi\}.$$

The uniqueness of solutions of the system of equations given by (12.66) and (12.67) is proved as in Theorem 12.3. □

Let us discuss the proof of the last statement of the theorem. We shall show that the matrices γ_t have inverses $\delta_t = \gamma_t^{-1}$, $0 \le t \le T$. It is seen that for sufficiently small $t = t(\omega)$ such matrices exist due to the nonsingularity of the matrix γ_0 and the continuity (P-a.s.) of the elements of the matrix γ_t in t.

Let $\tau = \inf\{t \le T : \det \gamma_t = 0\}$, with $\tau = \infty$ if $\inf_{0\le t\le T} \det\gamma_t > 0$. Then for $t < \tau \wedge T$ the matrices $\delta_t = \gamma_t^{-1}$ are defined. Note now that for $t < \tau\wedge T$,

$$0 = \frac{d}{dt}E = \frac{d}{dt}(\gamma_t\delta_t) = \dot{\gamma}_t\delta_t + \gamma_t\dot{\delta}_t = \dot{\gamma}_t\delta_t + \delta_t^{-1}\dot{\delta}_t.$$

Hence

$$\dot{\delta}_t = -\delta_t\dot{\gamma}_t\delta_t. \tag{12.68}$$

Taking into account Equation (12.67), we obtain from this that for $t < \tau \wedge T$

$$\begin{aligned}\dot{\delta}_t &= -\tilde{a}_1^*(t,\xi)\delta_t - \delta_t\tilde{a}_1(t,\xi) + A_1^*(t,\xi)(B\circ B)^{-1}(t,\xi)A_1(t,\xi)\\ &\quad -\delta_t[(b\circ b)(t,\xi) - (b\circ B)(t,\xi)(B\circ B)^{-1}(t,\xi)(b\circ B)^*(t,\xi)]\delta_t,\end{aligned} \tag{12.69}$$

where

$$\tilde{a}_1(t,x) = a_1(t,x) - (b\circ B)(t,x)(B\circ B)^{-1}(t,x)A_1(t,x).$$

On the set $\{\omega : \tau \le T\}$ the elements of the matrix δ_t must increase as $t \uparrow \tau$. We shall show that actually all the elements of the matrix δ_t are bounded.

Denote by $G_t(\xi)$ a solution of the matrix differential equation

$$\frac{dG_t(\xi)}{dt} = \tilde{a}_1(t,\xi)G_t(\xi), \quad G_0(\xi) = E_{(k\times k)}. \tag{12.70}$$

The matrix $G_t(\xi)$ being a fundamental matrix, it is, as is well known, nonsingular.

Let $V_t = G_t(\xi)\delta_t G_t^*(\xi)$. Then from (12.69) and (12.70), for $t < \tau\wedge T$ we find

$$\begin{aligned}\dot{V}_t &= \tilde{a}_1(t,\xi)V_t + V_t\tilde{a}_1^*(t,\xi) + G_t(\xi)\{-\tilde{a}_1^*(t,\xi)\delta_t - \delta_t\tilde{a}_1(t,\xi)\\ &\quad +A_1^*(t,\xi)(B\circ B)^{-1}(t,\xi)A_1(t,\xi)\\ &\quad -\delta_t[(b\circ b)(t,\xi) - (b\circ B)^*(t,\xi)(B\circ B)^{-1}(t,\xi)(b\circ B)^*(t,\xi)]\delta_t\}G_t^*(\xi).\end{aligned} \tag{12.71}$$

Since the matrix $b\circ b-(b\circ B)(B\circ B)^{-1}(b\circ B)^*$ is symmetric and nonnegative definite, we obtain from (12.71)

$$\operatorname{Tr} V_t \le \operatorname{Tr} V_0 + \int_0^T \operatorname{Tr}\{G_s(\xi)A_1^*(s,\xi)(B\circ B)^{-1}(s,\xi)A_1(s,\xi)G_s^*(\xi)\}ds,$$

which together with the nonsingularity of the matrix $G_t(\xi)$ proves the boundedness (P-a.s.) of elements of the matrix δ_t. Therefore $P(\tau \le T) = 0$.

12.3.4. We shall present, finally, the multidimensional analog of Theorem 12.2.

Theorem 12.8. *Let $\theta = (\theta_1,\dots,\theta_k)$ be a k-dimensional random variable with $\sum_{i=1}^k M\theta_i^4 < \infty$. Assume that the observable process $\xi_t = (\xi_1(t),\dots,\xi_l(t))$, $0 \le t \le T$, has the differential*

$$d\xi_l = [A_0(t,\xi) + A_1(t,\xi)\theta]dt + B(t,\xi)dW_2(t),$$

where the coefficients A_0, A_1, B satisfy the conditions of Theorem 12.6 and the conditional distribution $P(\theta \le a|\xi_0)$ is Gaussian, $N(m_0,\gamma_0)$. Then $m_t = M(\theta_t|\mathcal{F}_t^\xi)$ and $\gamma_t = M[(\theta - m_t)(\theta - m_t)^*|\mathcal{F}_t^\xi]$ are given by the formulae

$$m_t = \left[E + \gamma_0 \int_0^t A_1^*(s,\xi)(B(s,\xi)B^*(s,\xi))^{-1} A_1(s,\xi)ds\right]^{-1} \tag{12.72}$$
$$\times \left[m_0 + \gamma_0 \int_0^t A_1^*(s,\xi)(B(s,\xi)B^*(s,\xi))^{-1}(d\xi_s - A_0(s,\xi)ds)\right],$$
$$\gamma_t = \left[E + \gamma_0 \int_0^t A_1^*(s,\xi)(B(s,\xi)B^*(s,\xi))^{-1} A_1(s,\xi)ds\right]^{-1} \gamma_0. \tag{12.73}$$

The proof is similar to the pertinent proof of Theorem 12.2.

12.4 Interpolation of Conditionally Gaussian Processes

12.4.1. We shall consider the $(k+l)$-dimensional random process $(\theta, \xi) = [(\theta_1(t), \ldots, \theta_k(t)), (\xi_1(t), \ldots, \xi_l(t))]$, governed by the system of stochastic differential equations (12.59) and (12.60), and satisfying (1)–(10). Let the conditional distribution $P(\theta_0 \le a|\xi_0)$ be Gaussian, $N(m_0, \gamma_0)$. Then, due to Theorem 12.6, the conditional distribution $P(\theta_s \le a|\mathcal{F}_t^\xi)$, $s \le t$, is (P-a.s.) Gaussian with parameters

$$m(s,t) = M(\theta_s|\mathcal{F}_t^\xi),$$
$$\gamma(s,t) = M[(\theta_s - m(s,t))(\theta_s - m(s,t))^*|\mathcal{F}_t^\xi].$$

It is clear that the components $m_i(s,t) = M[\theta_i(s)|\mathcal{F}_t^\xi]$ of the vector $m(s,t) = [m_1(s,t), \ldots, m_k(s,t)]$ are the best (in the mean square sense) estimates of the components $\theta_i(s)$, $i = 1, \ldots, k$ of the vector $\theta_s = [\theta_1(s), \ldots, \theta_k(s)]$ from the observations $\xi_0^t = \{\xi_s, s \le t\}$.

In this section we shall deduce forward (over t at fixed s) and backward (over s at fixed t) equations (of interpolation) for $m(s,t)$ and $\gamma(s,t)$. Let $m_{\theta_s}(t,s) = M(\theta_t|\mathcal{F}_t^{\theta_s,\xi})$ and

$$\gamma(t,s) = M[(\theta_t - m_{\theta_s}(t,s))(\theta_t - m_{\theta_s}(t,s))^*|\mathcal{F}_t^{\theta_s,\xi}].$$

According to the multidimensional analog of Note 3 to Theorem 12.1, $m_{\theta_s}(t,s)$ and $\gamma(t,s)$ satisfy for $t \ge s$ the system of equations (compare with (12.66), (12.67))

$$d_t m_{\theta_s}(t,s) = [a_0(t,\xi) + (a(t,\xi) - \gamma(t,s)c(t,\xi))m_{\theta_s}(t,s)]dt$$
$$+[(b\circ B)(t,\xi) + \gamma(t,s)A_1^*(t,\xi)](B\circ B)^{-1}(t,\xi)[d\xi_t - A_0(t,\xi)dt], \tag{12.74}$$

$$\frac{d\gamma(t,s)}{dt} = a(t,\xi)\gamma(t,s) + \gamma(t,s)a^*(t,\xi) + b(t,\xi) - \gamma(t,\xi)c(t,\xi)\gamma(t,\xi), \tag{12.75}$$

where

$$\begin{aligned} a(t,x) &= a_1(t,x) - (b \circ B)(t,x)(B \circ B)^{-1}(t,x)A_1(t,x), \\ b(t,x) &= b{\circ}b(t,x) - (b{\circ}B)(t,x)(B{\circ}B)^{-1}(t,x)(b{\circ}B)^*(t,x), \\ c(t,x) &= A_1^*(t,x)(B \circ B)^{-1}(t,x)A_1(t,x). \end{aligned} \tag{12.76}$$

The system of equations given by (12.74) and (12.75) can be solved under the conditions $m_{\theta_s}(s,s) = \theta_s$, $\gamma(s,s) = 0$ (zero matrix of the order $(k \times k)$) and has, as has the system of equations given by (12.66) and (12.67), a unique continuous solution. From this it follows, in particular, that $\gamma(t,s)$, as the solution of Equation (12.75) with $\gamma(s,s) = 0$, does not depend on θ_s.

12.4.2. In deducing equations for $m(s,t)$ and $\gamma(s,t)$ the following two lemmas will be employed.

Lemma 12.2. *Let the matrix $\varphi_s^t(\xi)$, $t \geq s$, be a solution of the differential equation*

$$\frac{d\varphi_s^t(\xi)}{dt} = [a(t,\xi) - \gamma(t,s)c(t,\xi)]\varphi_s^t(\xi) \tag{12.77}$$

with $\varphi_s^s(\xi) = E_{(k\times k)}$,

$$\begin{aligned} q_s^t(\xi) = \int_s^t & (\varphi_s^u(\xi))^{-1}[a_0(u,\xi)du + \{(b \circ B)(u,\xi) + \gamma(u,s)A_1^*(u,\xi)\} \\ & \times (B \circ B)^{-1}(u,\xi)\{d\xi_u - A_0(u,\xi)du\}]. \end{aligned} \tag{12.78}$$

Then

$$m_{\theta_s}(t,s) = \varphi_s^t(\xi)[\theta_s + q_s^t(\xi)] \quad (P\text{-a.s.}). \tag{12.79}$$

PROOF. It is easy to convince oneself of the validity of (12.79) if one applies the Itô formula. □

Lemma 12.3. *Let $0 < s < t \leq T$. Then*

$$m_t = \varphi_s^t(\xi)[m(s,t) + q_s^t(\xi)] \quad (P\text{-a.s.}), \tag{12.80}$$

$$\gamma_t = \gamma(t,s) + \varphi_s^t(\xi)\gamma(s,t)(\varphi_s^t(\xi))^* \quad (P\text{-a.s.}) \tag{12.81}$$

PROOF. Since $\mathcal{F}_t^\xi \subseteq \mathcal{F}_t^{\theta_s,\xi}$, then

$$m_t = M(\theta_t|\mathcal{F}_t^\xi) = M[M(\theta_t|\mathcal{F}_t^{\theta_s,\xi})|\mathcal{F}_t^\xi] = M(m_{\theta_s}(t,s)|\mathcal{F}_t^\xi). \tag{12.82}$$

Note that the elements of the vector $\chi_N\varphi_s^t(\xi)\theta_s$, where

$$\chi_N = \chi_{\{\|\varphi_s^t(\xi)q_s^t(\xi)\| \leq N\}},$$

are integrable. Hence

$$\begin{aligned} \chi_N M[m_{\theta_s}(t,s)|\mathcal{F}_t^\xi] &= M[\chi_N\varphi_s^t(\xi)(\theta_s + q_s^t(\xi))|\mathcal{F}_t^\xi] \\ &= \chi_N\varphi_s^t(\xi)[m(s,t) + q_t^t(\xi)], \end{aligned}$$

which together with (12.82) proves the representation given by (12.80).

Next, since

$$M[(\theta_t - m_{\theta_s}(t,s))(m_{\theta_s}(t,s) - m_t)^* | \mathcal{F}_t^{\theta_s,\xi}] = 0 \quad (P\text{-a.s.})$$

it follows that

$$\begin{aligned}
\gamma_t &= M[(\theta_t - m_t)(\theta_t - m_t)^* | \mathcal{F}_t^\xi] \\
&= M\Big\{ [(\theta_t - m_{\theta_s}(t,s)) + (m_{\theta_s}(t,s) - m_t)] \\
&\qquad \times [(\theta_t - m_{\theta_s}(t,s)) + (m_{\theta_s}(t,s) - m_t)]^* | \mathcal{F}_t^\xi \Big\} \\
&= M\{M[(\theta_T - m_{\theta_s}(t,s))(\theta_t - m_{\theta_s}(t,s))^* | \mathcal{F}_t^{\theta_s,\xi}] | \mathcal{F}_t^\xi\} \\
&\quad + M\{(m_{\theta_s}(t,s) - m_t)(m_{\theta_s}(t,s) - m_t)^* | \mathcal{F}_t^\xi\} \\
&= \gamma(t,s) + M\{(m_{\theta_s}(t,s) - m_t)(m_{\theta_s}(t,s) - m_t)^* | \mathcal{F}_t^\xi\}. \qquad (12.83)
\end{aligned}$$

Noting that

$$\begin{aligned}
m_{\theta_s}(t,s) - m_t &= \varphi_s^t(\xi)[\theta_s + q_s^t(\xi)] - \varphi_s^t(\xi)[m(s,t) + q_s^t(\xi)] \\
&= \varphi_s^t(\xi)[\theta_s - m(s,t)],
\end{aligned}$$

we find

$$\begin{aligned}
& M\{(m_{\theta_s}(t,s) - m_t)(m_{\theta_s}(t,s) - m_t)^* | \mathcal{F}_t^\xi\} \\
&= \varphi_s^t(\xi) M[(\theta_s - m(s,t))((\theta_s - m(s,t))^* | \mathcal{F}_t^\xi](\varphi_s^t(\xi))^* \\
&= \varphi_s^t(\xi)\gamma(s,t)(\varphi_s^t(\xi))^*.
\end{aligned}$$

Together with (12.83) this proves (12.81). □

12.4.3. From (12.80) and (12.81) it is easy to obtain representations for $m(s,t)$ and $\gamma(s,t)$ which illustrate how these interpolation characteristics change with the change of t.

Theorem 12.9. *Let (1)–(10) be satisfied, and let the conditional distribution $P(\theta_0 \le a | \xi_0)$ be Gaussian. Then $m(s,t)$ and $\gamma(s,t)$ permit representations*

$$\begin{aligned}
m(s,t) &= m_s + \int_s^t \gamma(s,u)(\varphi_s^u(\xi))^* A_1^*(u,\xi)(B \circ B)^{-1}(u,\xi) \\
&\qquad \times [d\xi_u - (A_0(u,\xi) + A_1(u,\xi)m_u)du], \qquad (12.84)
\end{aligned}$$

$$\gamma(s,t) = \left(E + \gamma_s \int_s^t (\varphi_s^u(\xi))^* A_1^*(u,\xi)(B \circ B)^{-1}(u,\xi) A_1(u,\xi)\varphi_s^u(\xi)du\right)^{-1} \gamma_s. \qquad (12.85)$$

PROOF. From (12.80) we find

$$m(s,t) = (\varphi_s^t(\xi))^{-1}m_t - q_s^t(\xi). \tag{12.86}$$

The matrix $\varphi_s^t(\xi)$ is fundamental. Hence an inverse matrix $(\varphi_s^t(\xi))^{-1}$ exists and, according to (12.77), at $t \geq s$

$$\frac{d(\varphi_s^t(\xi))^{-1}}{dt} = -(\varphi_s^t(\xi))^{-1}[a(t,\xi) - \gamma(t,s)c(t,\xi)] \tag{12.87}$$

with $(\varphi_s^s(\xi))^{-1} = E_{(k\times k)}$.

From (12.86), (12,.87) and (12.29) we find by the Itô formula

$$\begin{aligned} m(s,t) &= m_s + \int_s^t (\varphi_s^u(\xi))^{-1}[\gamma_u - \gamma(u,s)]A_1^*(u,\xi)(B\circ B)^{-1}(u,\xi) \\ &\quad \times [d\xi_u - (A_0(u,\xi) + A_0(u,\xi)m_u)du]. \end{aligned} \tag{12.88}$$

But, due to (12.81),

$$(\varphi_s^u(\xi))^{-1}[\gamma_u - \gamma(u,s)] = \gamma(s,u)(\varphi_s^u(\xi))^*.$$

Substituting this expression into (12.88) we arrive at the representation sought, i.e., (12.84).

We shall now prove (12.85). From (12.81) we obtain

$$\gamma(s,t) = (\varphi_s^t(\xi))^{-1}[\gamma_t - \gamma(t,s)][(\varphi_s^t(\xi))^*]^{-1}. \tag{12.89}$$

Differentiating the right-hand side in (12.89) and taking into account (12.30), (12.87) and (12.75), after simple transformations we find that

$$\frac{d\gamma(s,t)}{dt} = -\gamma(s,t)(\varphi_s^t(\xi))^*c(t,\xi)\varphi_s^t(\xi)\gamma(s,t). \tag{12.90}$$

Equation (12.90) is a Ricatti equation, a solution of which exists and is unique. In order to solve it, let the matrices U_t, $t \geq s$, be given by the formulae

$$U_t = E + \gamma_s \int_s^t (\varphi_s^u(\xi))^*c(u,\xi)\varphi_s^u(\xi)du.$$

These matrices are nonsingular, and

$$\frac{dU_t^{-1}}{dt} = -U_t^{-1}\gamma_s(\varphi_s^t(\xi))^*c(t,\xi)\varphi_s^t(\xi)U_t^{-1}, \quad U_s^{-1} = E.$$

From this we obtain

$$\frac{d(U_t^{-1}\gamma_s)}{dt} = -(U_t^{-1}\gamma_s)(\varphi_s^t(\xi))^*c(t,\xi)\varphi_s^t(\xi)(U_t^{-1}\gamma_s), \tag{12.91}$$

where $U_s^{-1}\gamma_s = \gamma_s$.

Comparing (12.90) and (12.91) we find

$$\gamma(s,t) = U_t^{-1}\gamma_s,$$

which proves the required representation given by (12.85). □

Note. Together with (12.90) the equation for $m(s,t)$ obtained from (12.84), is called the forward equation of optimal nonlinear interpolation.

12.4.4. Let us deduce now for $m(s,t)$ and $\gamma(s,t)$ representations indicating how they must be changed for $s \uparrow t$.

Theorem 12.10. *Let (1)–(10), be satisfied and let the conditional distribution $P(\theta_0 \le a|\xi_0)$ be Gaussian, $N(m_0, \gamma_0)$. In addition, let*

$$P\left\{\inf_{0\le t\le T} \det \gamma_t > 0\right\} = 1.$$

Then

$$\begin{aligned} m(s,t) &= m_t - \int_s^t [a_0(u,\xi) + a_1(u,\xi)m(u,t) + b(u,\xi)\gamma_u^{-1}(m(u,t) - m_u)]du \\ &\quad - \int_s^t (b\circ B)(u,\xi)(B\circ B)^{-1}(u,\xi)[d\xi_u - (A_0(u,\xi) \\ &\quad + A_1(u,\xi)m(u,t))du], \end{aligned} \tag{12.92}$$

$$\begin{aligned} \gamma(s,t) &= \gamma_t - \int_s^t \Big\{[a(u,\xi) + b(u,\xi)\gamma_u^{-1}]\gamma(u,t) \\ &\quad + \gamma(u,t)[a(u,\xi) + b(u,\xi)\gamma_u^{-1}]^* - b(u,\xi)\Big\}du, \end{aligned} \tag{12.93}$$

where $a(u,x)$ and $b(u,x)$ are given by (12.76).

In order to prove this theorem we shall establish as a preliminary the following two lemmas.

Lemma 12.4. *Let $P\{\inf_{t\le T} \det \gamma_t > 0\} = 1$, and let the matrix $R_s^t(\xi)$ be a solution of the system of differential equations*

$$\frac{dR_s^t(\xi)}{dt} = [a(t,\xi) + b(t,\xi)\gamma_t^{-1}]R_s^t(\xi), \quad R_s^s(\xi) = E_{(k\times k)}. \tag{12.94}$$

Then

$$\gamma(s,t)(\varphi_s^t(\xi))^* = (R_s^t(\xi))^{-1}\gamma_t. \tag{12.95}$$

PROOF. Let $U_s^t = \gamma(s,t)(\varphi_s^t(\xi))^*$. Then, due to (12.90), and (12.77),

$$\begin{aligned}\frac{dU_s^t}{dt} &= -\gamma(s,t)(\varphi_s^t(\xi))^* c(t,\xi)\varphi_s^t(\xi)\gamma(s,t)(\varphi_s^t(\xi))^* \\ &\quad +\gamma(s,t)(\varphi_s^t(\xi))^*[a(t,\xi)-\gamma(t,s)c(t,\xi)]^* \\ &= U_s^t a^*(t,\xi) - U_s^t c(t,\xi)[\varphi_s^t(\xi)\gamma(s,t)(\varphi_s^t(\xi))^* + \gamma(t,s)].\end{aligned}$$

But, according to (12.81),

$$\varphi_s^t(\xi)\gamma(s,t)(\varphi_s^t(\xi))^* + \gamma(t,s) = \gamma_t.$$

Hence

$$\frac{dU_s^t}{dt} = U_s^t[a^*(t,\xi) - c(t,\xi)\gamma_t]. \tag{12.96}$$

Let V_s^t be a fundamental matrix solution of (12.96), i.e., let

$$\frac{dV_s^t}{dt} = V_s^t[a^*(t,\xi) - c(t,\xi)\gamma_t], \quad V_s^s = E_{(k\times k)}. \tag{12.97}$$

Since $V_s^t = V_0^t(V_0^s)^{-1}$ and the matrix $(V_0^s)^{-1}$ is a solution of the system of equations

$$\frac{d(V_0^s)^{-1}}{ds} = -[a^*(s,\xi) - c(s,\xi)\gamma_t](V_0^s)^{-1}, \quad (V_0^0)^{-1} = E_{(k\times k)},$$

the matrix V_s^t is differentiable in s and, for $s<t$,

$$\frac{dV_s^t}{ds} = -[a^*(s,\xi) - c(s,\xi)\gamma_s]V_s^t, \quad V_t^t = E_{(k\times k)}. \tag{12.98}$$

But

$$U_s^t = U_s^s V_s^t = \gamma_s V_s^t,$$

where γ_s and V_s^t are differentiable in s. Hence, the matrix U_s^t is also differentiable in s and

$$\frac{dU_s^t}{ds} = \frac{d\gamma_s}{ds}V_s^t + \gamma_s\frac{dV_s^t}{ds}.$$

From (12.30), in the notation of (12.76) we have

$$\frac{d\gamma_s}{ds} = a(s,\xi)\gamma_s + \gamma_s a^*(s,\xi) + b(s,\xi) - \gamma_s c(s,\xi)\gamma_s, \tag{12.99}$$

which, together with (12.97), yields

$$\begin{aligned}\frac{dU_s^t}{ds} &= [a(s,\xi)\gamma_s + \gamma_s a^*(s,\zeta) + b(s,\xi) - \gamma_s c(s,\xi)\gamma_s]V_s^t \\ &\quad -\gamma_s(a^*(s,\xi) - c(s,\xi)\gamma_s)V_s^t \\ &= [a(s,\xi) + b(s,\xi)\gamma_s^{-1}]U_s^t. \end{aligned} \tag{12.100}$$

From (12.94) and (12.100) it follows that $U_t^t = R_s^t U_s^t$. But $U_t^t = \gamma_t$; hence $U_s^t = (R_s^t)^{-1}\gamma_t$, which proves (12.95). □

Lemma 12.5. *Let $(\mathcal{F}_t)$, $0 \le t \le T$, be a nondecreasing family of σ-algebras, let $W = (W_t, \mathcal{F}_t)$ be a Wiener process, and let $a = (a_t, \mathcal{F}_t)$ and $b = (b_t, \mathcal{F}_t)$ be random processes with $\int_0^T |a_t|dt < \infty$, $\int_0^T b_t^2 dt < \infty$ (P-a.s.). Then, for $0 \le s \le t \le T$,*

$$\int_0^s a_u du \int_s^t b_u dW_u = \int_0^s \left[\int_u^t b_v dW_v\right] du - \int_0^s \left[\int_0^u a_v dv\right] b_u dW_u. \quad (12.101)$$

PROOF. It is obvious that

$$\int_0^s a_u du \int_s^t b_u dW_u = \int_0^s a_u du \int_0^t b_u dW_u - \int_0^s a_u du \int_0^s b_u dW_u. \quad (12.102)$$

By the Itô formula,

$$\int_0^s a_u du \int_0^s b_u dW_u = \int_0^s \left[\int_0^u b_v dW_v\right] a_u du + \int_0^s \left[\int_0^u a_v dv\right] b_u dW_u,$$

hence the right-hand side in (12.102) is equal to

$$\begin{aligned} &\int_0^s a_u du \int_0^t b_u dW_u - \int_0^s \left[\int_0^u b_v dW_v\right] a_u du - \int_0^s \left[\int_0^u a_v dv\right] b_u dW_u \\ &= \int_0^s a_u \left[\int_u^t b_v dW_v\right] du - \int_0^s \left[\int_0^u a_v dv\right] b_u dW_u, \end{aligned}$$

which proves (12.101). □

12.4.5.

PROOF OF THEOREM 12.10. According to (12.84) and (12.95),

$$m(s,t) = m_s + \int_s^t [R_s^u(\xi)]^{-1} \gamma_u A_1^*(u,\xi)(B \circ B)^{-1/2}(u,\xi) d\overline{W}_u, \quad (12.103)$$

where

$$d\overline{W}_u = (B \circ B)^{-1/2}(u,\xi)[d\xi_u - (A_0(u,\xi) + A_1(u,\xi)m_u)du].$$

The matrix $R_s^u(\xi)$ is fundamental. Hence $R_0^u(\xi) = R_0^s(\xi) R_s^u(\xi)$, and, therefore,

$$[R_s^u(\xi)]^{-1} = R_0^s(\xi)[R_0^u(\xi)]^{-1}. \quad (12.104)$$

From (12.103) and (12.104) we find

$$m(s,t) = m_s + R_0^s(\xi) \int_s^t [R_0^u(\xi)]^{-1} \gamma_u A_1^*(u,\xi)(B \circ B)^{-1/2}(u,\xi) d\overline{W}_u. \quad (12.105)$$

Next, from (12.94) and Lemma 12.5 we obtain

$$\begin{aligned} & d_s \left[R_0^s(\xi) \int_s^t (R_0^u(\xi))^{-1} \gamma_u A_1^*(u,\xi)(B \circ B)^{-1/2}(u,\xi) d\overline{W}_u \right] \\ = & \ [a(s,\xi) + b(s,\xi)\gamma_s^{-1}] \\ & \times \left[R_0^s(\xi) \int_s^t (R_0^u(\xi))^{-1} \gamma_u A_1^*(u,\xi)(B \circ B)^{-1/2}(u,\xi) d\overline{W}_u \right] ds \\ & - \gamma_s A_1^*(s,\xi)(B \circ B)^{-1/2}(s,\xi) d\overline{W}_s \\ = & \ [a(s,\xi) + b(s,\xi)\gamma_s^{-1}][m(s,t) - m_s] ds - \gamma_s A_1^*(s,\xi)(B \circ B)^{-1/2}(s,\xi) d\overline{W}_s. \end{aligned}$$

But (see (12.29))

$$\begin{aligned} dm_s = & \ [a_0(s,\xi) + a_1(s,\xi) m_s] ds + (b \circ B)(s,\xi)(B \circ B)^{-1/2}(s,\xi) d\overline{W}_s \\ & + \gamma_s A_1^*(s,\xi)(B \circ B)^{-1/2}(s,\xi) d\overline{W}_s. \end{aligned}$$

Consequently,

$$\begin{aligned} d_s m(s,t) = & \ [a_0(s,\xi) + a_1(s,\xi) m_s] ds + (b \circ B)(s,\xi)(B \circ B)^{-1/2}(s,\xi) d\overline{W}_s \\ & + [a(s,\xi) + b(s,\xi)\gamma_s^{-1}][m(s,t) - m_s] ds \\ = & \ [a_0(s,\xi) + a_1(s,\xi) m(s,t)] ds \\ & + (b \circ B)(s,\xi)(B \circ B)^{-1}(s,\xi) \\ & \times [d\xi_s - (A_0(s,\xi) + A_1(s,\xi) m(s,t)) ds] \\ & + [a(s,\xi) + b(s,\xi)\gamma_s^{-1}][m(s,t) - m_s] ds \\ & - a_1(s,\xi)[m(s,t) - m_s] ds \\ & + (b \circ B)(s,\xi)(B \circ B)^{-1}(s,\xi) A_1(s,\xi)[m_s - m(s,t)] ds. \end{aligned}$$

According to (12.76)

$$\begin{aligned} & [a(s,\xi) + b(s,\xi)\gamma_s^{-1}] - a_1(s,\xi) \\ & \quad - (b \circ B)(s,\xi)(B \circ B)^{-1}(s,\xi) A_1(s,\xi) = b(s,\xi)\gamma_s^{-1}. \end{aligned}$$

Therefore,

$$\begin{aligned} d_s m(s,t) = & \ [a_0(s,\xi) + a_1(s,\xi) m(s,t)] ds + b(s,\xi)\gamma_s^{-1}[m_s - m(s,t)] ds \\ & + (b \circ B)(s,\xi)(B \circ B)^{-1}(s,\xi) \\ & \times [d\xi_s - (A_0(s,\xi) + A_1(s,\xi) m(s,t)) ds], \end{aligned}$$

which proves (12.92).

Next let us deduce Equation (12.93) for $\gamma(s,t)$. From (12.95) and (12.90) we obtain

$$\gamma(s,t) = \gamma_s - R_0^s(\xi) \int_s^t [R_0^u(\xi)]^{-1} \gamma_u c(u,\xi) \gamma_u [(R_0^u(\xi))^*]^{-1} du (R_0^s(\xi))^*. \tag{12.106}$$

Differentiating (12.106) with respect to s, and taking into account (12.99) and (12.94), we find that

$$\begin{aligned}\frac{d\gamma(s,t)}{ds} &= a(s,\xi)\gamma_s + \gamma_s a^*(s,\xi) + b(s,\xi)\\ &\quad -\gamma_s c(s,\xi)\gamma_s - [a(s,\xi) + b(s,\xi)\gamma_s^{-1}][\gamma_s - \gamma(s,t)]\\ &\quad -[\gamma_s - \gamma(s,t)][a(s,\xi) + b(s,\xi)\gamma_s^{-1}]^* + \gamma_s c(s,\xi)\gamma_s\\ &= [a(s,\xi) + b(s,\xi)\gamma_s^{-1}]\gamma(s,t) + \gamma(s,t)[a(s,\xi) + b(s,\xi)\gamma_s^{-1}]^*\\ &\quad -b(s,\xi).\end{aligned}$$

□

Note 1. Equations (12.92) and (12.93) are linear with respect to $m(s,t)$ and $\gamma(s,t)$. Hence uniqueness of continuous solutions can be established in a standard manner.

Note 2. If $(b \circ B)(t,x) \equiv 0$, then Equations (12.92) and (12.93) become essentially simpler:

$$\begin{aligned}m(s,t) &= m_t - \int_s^t \Big\{a_0(u,\xi) + a_1(u,\xi)m(u,t)\\ &\qquad -(b\circ b)(u,\xi)\gamma_u^{-1}[m(u,t) - m_u]\Big\}du, \qquad (12.107)\end{aligned}$$

$$\begin{aligned}\gamma(s,t) &= \gamma_t - \int_s^t \Big\{[a_1(u,\xi) + (b\circ b)(u,\xi)\gamma_u^{-1}]\gamma(u,t) \qquad (12.108)\\ &\qquad +\gamma(u,t)[a_1(u,\xi)+(b\circ b)(u,\xi)\gamma_u^{-1}]^* - (b\circ b)(u,\xi)\Big\}du.\end{aligned}$$

Note 3. The Kalman–Bucy scheme discussed in Chapter 10 is a particular case of the estimation problems for conditionally Gaussian processes. Hence in this scheme the equations for $m(s,t)$ and $\gamma(s,t)$ also hold true. Note that taking into consideration the specific character of the Kalman–Bucy scheme, these equations can be deduced under the same assumptions as those for m_t and γ_t (see Theorem 10.3), requiring, in addition, nonsingularity of the matrices γ_t, $0 \le t \le T$, in deducing backward equations.

12.4.6. Let us discuss one more class of interpolation estimates for conditionally Gaussian processes.

Since the conditional distributions $P(\theta_s \le a, \theta_t \le b|\mathcal{F}_t^\xi)$ for $s \le t$ are (P-a.s.) Gaussian, then the conditional distribution $P(\theta_s \le a|\mathcal{F}_t^\xi, \theta_t)$ will also be Gaussian.

Let

$$\begin{aligned}\tilde{m}_\beta(s,t) &= M(\theta_s|\mathcal{F}_t^\xi, \theta_t = \beta),\\ \tilde{\gamma}_\beta(s,t) &= M\{(\theta_s - m_\beta(s,t))(\theta_s - m_\beta(s,t))^*|\mathcal{F}_t^\xi, \theta_t = \beta\}.\end{aligned}$$

Theorem 12.11. *If the conditions given by (1)–(10) are satisfied and the conditional distribution $P(\theta_0 \leq a|\xi_0)$ is (P-a.s.) Gaussian, then*

$$\begin{aligned}\tilde{m}_\beta(s,t) &= m(s,t) + \gamma(s,t)[\varphi_s^t(\xi)]^*\gamma_t^+(\beta - m_t), &(12.109)\\ \tilde{\gamma}_\beta(s,t) &= \gamma(s,t) - \gamma(s,t)[\varphi_s^t(\xi)]^*\gamma_t^+\varphi_s^t(\xi)\gamma(s,t), &(12.110)\end{aligned}$$

where γ_t^+ is the pseudo-inverse of the matrix γ_t, and $\varphi_s^t(\xi)$ is defined in (12.77).

PROOF. Since

$$\begin{aligned}M(\theta_t|\mathcal{F}_t^\xi) &= m_t, \quad M(\theta_s|\mathcal{F}_t^\xi) = m(s,t),\\ \mathrm{cov}\,(\theta_t,\theta_t|\mathcal{F}_t^\xi) &= \gamma_t, \quad \mathrm{cov}\,(\theta_s,\theta_s|\mathcal{F}_t^\xi) = \gamma(s,t),\\ \mathrm{cov}\,(\theta_s,\theta_t|\mathcal{F}_t^\xi) &= M[(\theta_s - m(s,t))(\theta_t - m_t)^*|\mathcal{F}_t],\end{aligned}$$

then, by the theorem on normal correlation (Theorem 13.1),

$$\begin{aligned}\tilde{m}_\beta(s,t) &= m(s,t) + \mathrm{cov}\,(\theta_s,\theta_t|\mathcal{F}_t^\xi)\gamma_t^+(\beta - m_t), &(12.111)\\ \tilde{\gamma}_\beta(s,t) &= \gamma(s,t) - \mathrm{cov}\,(\theta_s,\theta_t|\mathcal{F}_t^\xi)\gamma_t[\mathrm{cov}\,(\theta_s,\theta_t|\mathcal{F}_t^\xi)]^*. &(12.112)\end{aligned}$$

We shall show that (P-a.s.)

$$\mathrm{cov}\,(\theta_s,\theta_t|\mathcal{F}_t^\xi) = \gamma(s,t)(\varphi_s^t(\xi))^*. \tag{12.113}$$

Indeed, since

$$\mathrm{cov}\,(\theta_s,\theta_t|\mathcal{F}_t^\xi) - M[(\theta_s \quad m(s,t))M\{(\theta_t - m_t)^*|\mathcal{F}_t^{\theta_s,\xi}\}|\mathcal{F}_t^\xi]$$

and, according to (12.79) and (12.81),

$$\begin{aligned}M\{(\theta_t - m_t)^*|\mathcal{F}_t^{\theta_s,\xi}\} &= \{M[(\theta_t - m_t)|\mathcal{F}_t^{\theta_s,\xi}]\}^* = \{m_{\theta_s}(t,s) - m_t\}^*\\ &= \{\varphi_s^t(\xi)[\theta_s + q_s^t(\xi)] - \varphi_s^t(\xi)[m(s,t) + q_s^t(\xi)]\}^*\\ &= [\theta_s - m(s,t)]^*(\varphi_s^t(\xi))^*,\end{aligned}$$

then

$$\mathrm{cov}\,(\theta_s,\theta_t|\mathcal{F}_t^\xi) = M[(\theta_s - m(s,t))(\theta_s - m(s,t))^*|\mathcal{F}_t^\xi](\varphi_s^t(\xi))^*,$$

which proves (12.113).

We obtain (12.109) and (12.110) from (12.111)–(12.113). □

Note 1. If in addition to the conditions of Theorem 12.11 it is assumed that $P(\inf_{0\le t\le T}\det\gamma_t > 0) = 1$, then, differentiating (12.109) and (12.110) with respect to s, we find that

$$\begin{aligned}\tilde{m}_\beta(s,t) &= \beta - \int_s^t [a_0(u,\xi)+a_1(u,\xi)\tilde{m}_\beta(u,t)+b(u,\xi)\gamma_u^{-1}(\tilde{m}_\beta(u,t)-m_u)]du \\ &\quad -\int_s^t (b\circ B)(u,\xi)(B\circ B)^{-1}(u,\xi) \\ &\quad \times[d\xi_u - (A_0(u,\xi)+A_1(u,\xi)\tilde{m}_\beta(u,t))du], \end{aligned} \tag{12.114}$$

$$\begin{aligned}\tilde{\gamma}_\beta(s,t) &= -\int_s^t \Big\{[a(u,\xi)+b(u,\xi)\gamma_u^{-1}]\tilde{\gamma}_\beta(u,t) \\ &\quad +\tilde{\gamma}_\beta(u,t)[a(u,\xi)+b(u,\xi)\gamma_u^{-1}]^* - b(u,\xi)\Big\}du. \end{aligned} \tag{12.115}$$

Note 2. From (12.110) it follows that $\tilde{\gamma}_\beta(s,t)$ does not actually depend on β.

Note 3. Consider the Gaussian Markov process (θ_t), $0\le t\le T$, with the differential

$$d\theta_t = [a_0(t)+a_1(t)\theta_t]dt + b(t)dW(t) \tag{12.116}$$

and a given Gaussian random variable θ_0. Assume that the deterministic functions $a_0(t), a_1(t)$ and $b(t)$ are such that

$$\int_0^T |a_i(t)|dt < \infty, \quad i=0,1; \qquad \int_0^T b^2(t)dt < \infty.$$

Take, for $0\le s\le t\le T$,

$$r(t) = M\theta_t, \quad r_\beta(s,t) = M(\theta_s|\theta_t=\beta),$$

$$R(t) = M[\theta_t - r(t)]^2, \quad R_\beta(s,t) = M[(\theta_s - r_\beta(s,t))^2|\theta_t=\beta].$$

If we assume in (12.60) that $A_1(t,x)\equiv 0$ and $B_2(t,x)\equiv 0$ and observe that ξ_0 does not depend on θ_0, then it is not difficult to see that

$$r(t) = m_t, \quad R(t) = \gamma_t$$

and

$$r_\beta(s,t) = \tilde{m}_\beta(s,t), \quad R_\beta(s,t) = \tilde{\gamma}_\beta(s,t).$$

Therefore, according to (12.29) and (12.30)[3],

[3] See also Note 3 in Subsection 12.4.5.

$$r(t) = r(0) + \int_0^t [a_0(s) + a_1(s)r(s)]ds \tag{12.117}$$

and

$$R(t) = R(0) + +2\int_0^t a_1(s)R(s)ds + \int_0^t b_1^2(s)ds. \tag{12.118}$$

For $r_\beta(s,t)$ and $R_\beta(s,t)$, from (12.114) and (12.115) (on the assumption that $\inf_{0\le t\le T} R(t) > 0$) we find that

$$r_\beta(s,t) = \beta - \int_s^t \left[a_0(u) + a_1(u)r_\beta(u,t) + \frac{b^2(u)}{R(u)}(r_\beta(u,t) - r(u))\right] du. \tag{12.119}$$

$$R_\beta(s,t) = -2\int_s^t \left\{\left[a_1(u) + \frac{b^2(u)}{R(u)}\right] R_\beta(u,t) - \frac{1}{2}b^2(u)\right\} du. \tag{12.120}$$

The analogs of (12.109) and (12.110) are the formulae:

$$r_\beta(s,t) = r(s) + R(s)\exp\left(\int_s^t a_1(u)du\right) R^+(t)(\beta - r(t)), \tag{12.121}$$

$$R_\beta(s,t) = R(s) - R^2(s)\exp\left(2\int_s^t a_1(u)du\right) R^+(t). \tag{12.122}$$

12.5 Optimal Extrapolation Equations

12.5.1. In this section extrapolation equations for conditionally Gaussian processes are deduced which enable us to compute optimal (in the mean square sense) estimates of variables θ_t, from the observations $\xi_0^s = \{\xi_u, u \le s\}$, $s \le t \le T$. Unlike the problems of filtering and interpolation considered above, these equations will be deduced not for a general process (θ, ξ) given by Equations (12.1) and (12.2) but only for two particular cases given below. The restriction of the class of processes (θ, ξ) considered arises from the fact that the conditional distributions $P(\theta_t \le a|\mathcal{F}_s^\xi)$ for $t > s$ are not, generally speaking, Gaussian.

12.5.2. For $t \ge s$, let

$$n_1(t,s) = M(\theta_t|\mathcal{F}_s^\xi), \quad n_2(t,s) = M(\xi_t|\mathcal{F}_s^\xi). \tag{12.123}$$

As in the case of interpolation, equations of two types can be deduced for these characteristics: forward equations (in t for fixed s) and backward (in $s \uparrow t$ for fixed t). We can see from the forward equations how the prediction of values of θ_t deteriorates as t increases. The backward equations allow us to establish a degree of improvement for prediction of values of θ_t with 'the increase of data', i.e., with the increase of s. Note that the backward equations of extrapolation could be deduced from the general equations of extrapolation obtained in Chapter 8. We shall present here another and, we think, more natural development.

Assume that $(\theta, \xi) = (\theta_t, \xi_t)$, $0 \leq t \leq T$, is a $(k+l)$-dimensional diffusion process with

$$d\theta_t = [a_0(t) + a_1(t)\theta_t]dt + \sum_{i=1}^{2} b_i(t,\xi)dW_i(t), \tag{12.124}$$

$$d\xi_t = [A_0(t,\xi) + A_1(t,\xi)\theta_t]dt + \sum_{i=1}^{2} B_i(t,\xi)dW_i(t), \tag{12.125}$$

where the coefficients satisfy the conditions given by (1)–(10) with the elements of the vector $a_0(t)$ and the matrix $a_1(t)$ being deterministic time functions and the conditional distribution $P(\theta_0 \leq a|\xi_0)$ being Gaussian.

Next let φ_s^t be the fundamental matrix solution of the equation

$$\frac{d\varphi_s^t}{dt} = a_1(t)\varphi_s^t, \quad t \geq s, \tag{12.126}$$

with $\varphi_s^s = E_{(k\times k)}$. Under these assumptions we have the following.

Theorem 12.12. *Let the process (θ, ξ) be governed by the system of equations given by (12.124) and (12.125). Then for each fixed s, $0 \leq s \leq t \leq T$, $n_1(t,s)$ satisfies the equation*

$$\frac{dn_1(t,s)}{dt} = a_0(t) + a_1(t)n_1(t,s) \tag{12.127}$$

with $n_1(s,s) = m_s$, where m_s is defined by Equations (12.66) and (12.67). For fixed t,

$$\begin{aligned} n_1(t,s) &= n_1(t,0) + \int_0^s \varphi_u^t[(b\circ B)(u,\xi) + \gamma_u A_1^*(u,\xi)](B\circ B)^{-1}(u,\xi) \\ &\quad \times [d\xi_u - (A_0(u,\xi) + A_1(u,\xi)m_u)du], \end{aligned} \tag{12.128}$$

where m_u and γ_u can be found from Equations (12.66) and (12.67), and

$$n_1(t,0) = \varphi_0^t\left[m_0 + \int_0^t (\varphi_0^s)^{-1}a_0(s)ds\right]. \tag{12.129}$$

PROOF. Let us note that

$$n_1(t,s) = M(\theta_t|\mathcal{F}_s^\xi) = M[M(\theta_t|\mathcal{F}_t^\xi)|\mathcal{F}_s^\xi] = M(m_t|\mathcal{F}_s^\xi),$$

where, according to (12.66), m_t can be represented as follows:

$$\begin{aligned} m_t &= m_s + \int_s^t [a_0(u) + a_1(u)m_u]du \\ &+ \int_s^t (b\circ B)(u,\xi) + \gamma_u A_1^*(u,\xi)](B\circ B)^{-1/2}(u,\xi)d\overline{W}_u. \quad (12.130) \end{aligned}$$

But

$$M\left(\int_s^t [(b\circ B)(u,\xi) + \gamma_u A_1^*(u,\xi)](B\circ B)^{-1/2}(u,\xi)d\overline{W}_u \middle| \mathcal{F}_s^\xi\right) = 0;$$

hence, taking the conditional expectation $M(\cdot|\mathcal{F}_s^\xi)$ on both sides of (12.130), we arrive at Equation (12.127).

In order to deduce (12.128), we take $s = 0$ in (12.130). With the help of the Itô formula it is not difficult to convince oneself that the (unique) continuous solution m_t of Equation (12.130) with $s = 0$ can be expressed as follows:

$$\begin{aligned} m_t &= \varphi_0^t \Bigg[m_0 + \int_0^t (\varphi_0^u)^{-1} a_0(u)du \\ &+ \int_0^t (\varphi_0^u)^{-1}[(b\circ B)(u,\xi) + \gamma_u A_1^*(u,\xi)](B\circ B)^{-1/2}(u,\xi)d\overline{W}_u \Bigg]. \end{aligned}$$

From this we find that

$$\begin{aligned} m_t &= n_1(t,0) + \int_0^s \varphi_u^t[(b\circ B)(u,\xi) + \gamma_u A_1^*(u,\xi)](B\circ B)^{-1/2}(u,\xi)d\overline{W}_u \\ &+ \int_s^t \varphi_u^t[(b\circ B)(u,\xi) + \gamma_u A_1^*(u,\xi)](B\circ B)^{-1/2}(u,\xi)d\overline{W}_u. \quad (12.131) \end{aligned}$$

Subtracting the conditional expectation $M(\cdot|\mathcal{F}_s^\xi)$ from both sides of (12.131), we obtain the desired representation, (12.128). □

12.5.3. Let it be required to extrapolate the values of ξ_t from $\xi_0^s = \{\xi_u, u \le s\}$, $s \le t$, along with predicting the values of θ_t.

We shall again assume that the conditional distribution $P(\theta_0 \le a|\xi_0)$ is Gaussian and (1)–(10) are satisfied, and

$$A_0(t,x) = A_0(t) + A_2(t)x_t, \quad a_1(t,x) = a_1(t), \quad A_1(t,x) = A_1(t),$$

where the elements of the vectors and the matrices $a_i(t)$ and $A_i(t)$, $i = 0,1,2$, are deterministic functions. In other words, let

$$d\theta_t = [a_0(t) + a_1(t)\theta_t + a_2(t)\xi_t]dt + \sum_{i=1}^2 b_i(t,\xi)dW_i(t), \quad (12.132)$$

$$d\xi_t = [A_0(t) + A_1(t)\theta_t + A_2(t)\xi_t]dt + \sum_{i=1}^2 B_i(t,\xi)dW_i(t). \quad (12.133)$$

Next, let Φ_s^t be the fundamental matrix of the system ($t > s$)

$$\frac{d\Phi_s^t}{dt} = \begin{pmatrix} a_1(t) & a_2(t) \\ A_1(t) & A_2(t) \end{pmatrix} \Phi_s^t,$$

where

$$\Phi_s^s = E_{((k+l)\times(k+l))}.$$

Theorem 12.13. *Under the assumptions made, $n_1(t,s)$ and $n_2(t,s)$ (for each s) are solutions of the system of equations*

$$\begin{pmatrix} \frac{dn_1(t,s)}{dt} \\ \frac{dn_2(t,s)}{dt} \end{pmatrix} = \begin{pmatrix} a_0(t) \\ A_0(t) \end{pmatrix} + \begin{pmatrix} a_1(t) & a_2(t) \\ A_1(t) & A_2(t) \end{pmatrix} \begin{pmatrix} n_1(t,s) \\ n_2(t,s) \end{pmatrix} \qquad (12.134)$$

with $n_1(s,s) = m_2$, $n_s(s,s) = \xi_s$.

For fixed t,

$$\begin{pmatrix} n_1(t,s) \\ n_2(t,s) \end{pmatrix} = \begin{pmatrix} n_1(t,0) \\ n_2(t,0) \end{pmatrix} + \int_0^s \Phi_u^s \begin{pmatrix} [(b\circ B)(u,\xi)+\gamma_u A_1^*(u,\xi)](B\circ B)^{-1/2}(u,\xi) \\ (B\circ B)^{1/2}(u,\xi) \end{pmatrix} d\overline{W}_u, \qquad (12.135)$$

and

$$\begin{pmatrix} n_1(t,0) \\ n_2(t,0) \end{pmatrix} = \Phi_0^t \begin{pmatrix} m_0 \\ \xi_0 \end{pmatrix} + \int_0^t \Phi_s^t \begin{pmatrix} a_0(s) \\ A_0(s) \end{pmatrix} ds. \qquad (12.136)$$

PROOF. Taking into consideration the assumptions on the coefficients of the system from (12.66) and (12.133) we find that

$$\begin{pmatrix} m_t \\ \xi_t \end{pmatrix} = \begin{pmatrix} m_s \\ \xi_s \end{pmatrix} + \int_0^t \begin{pmatrix} a_0(u) \\ A_0(u) \end{pmatrix} du + \int_s^t \begin{pmatrix} a_1(u) & a_2(u) \\ A_1(u) & A_2(u) \end{pmatrix} \begin{pmatrix} m_u \\ \xi_u \end{pmatrix} du + \int_s^t \begin{pmatrix} [(b\circ B)(u,\xi)+\gamma_u A_1^*(u,\xi)](B\circ B)^{-1/2}(u,\xi) \\ (B\circ B)^{1/2}(u,\xi) \end{pmatrix} d\overline{W}_u.$$

From this (as in proving the preceding theorem) (12.134) and (12.135) can easily be deduced. □

Note. For the particular case of Equations (12.132) and (12.133) corresponding to the Kalman–Bucy scheme (see Chapter 10) the forward and backward equations of extrapolation hold true only under the assumptions of Theorem 10.3

Notes and References. 1

12.1–12.5. The results related to this chapter are due to the authors. They have been partially published in [205, 207–209].

Notes and References. 2

12.1–12.5. References to other filtering models for which filtering equations have a 'closed form' can be found in Benesh [16], Daum [47], Pardoux [252] and Yashin [321, 323].

13. Conditionally Gaussian Sequences: Filtering and Related Problems

13.1 Theorem on Normal Correlation

13.1.1. The two previous chapters dealt with problems of filtering, interpolation and extrapolation for the conditionally Gaussian processes (θ, ξ) in continuous time $t \geq 0$. In the present chapter these problems will be investigated for random sequences with discrete time $t = 0, \Delta, 2\Delta, \dots$, having the property of 'conditional normality' as well.

It should be emphasized that the complex tools of the theory of martingales, taken advantage of in the case of continuous time, will not be used in this chapter. In essence, all the results of this chapter can be deduced from the theorem on normal correlation (Theorem 13.1). Hence, the reader who wishes to become acquainted with the theory of filtering and related problems for the case of discrete time can start reading this chapter without studying the previous chapters.

The comparison of the results for discrete time and continuous time shows that there is a great similarity between them, at least formally. Moreover, a formal passage to the limit (with $\Delta \to 0$) enables us to obtain the pertinent results for the case of continuous time from the results of this chapter. However, rigorous justification is not easy and requires, in fact, all the tools employed in the two previous chapters.

13.1.2. For the formulation and proof of the main result of this section – a theorem on normal correlation – we need some properties of pseudo-inverses of matrices.

Consider a matrix equation

$$AXA = A. \tag{13.1}$$

If A is a square nonsingular matrix, then this equation has a unique solution $X = A^{-1}$. If the matrix A is singular, or even rectangular, then a solution of Equation (13.1), even if it exists, cannot be defined uniquely. Nevertheless, there exists (as will be proved below), in this case also (for a certain class of matrices), a single-valued matrix satisfying Equation (13.1). From now on this matrix will be denoted by A^+ and called a *pseudo-inverse* matrix.

Definition. A matrix A^+ (of the order $n \times m$) is called the *pseudo-inverse with respect to the matrix* $A = A_{(m\times n)}$, if the following two conditions are satisfied:

$$AA^+A = A, \tag{13.2}$$

$$A^+ = UA^* = A^*V, \tag{13.3}$$

where U and V are matrices.

It follows from (13.3) that rows and columns of the matrix A^+ are, respectively, linear combinations of rows and columns of the matrix A^*.

Lemma 13.1. *The matrix A^+ satisfying (13.2) and (13.3) exists and is unique.*

PROOF. Let us start by proving the uniqueness. Let A_1^+ and A_2^+ be two different pseudo-inverse matrices.

Then

$$AA_1^+A = A, \quad A_1^+ = U_1A^* = A^*V_1,$$

and

$$AA_2^+A = A, \quad A_2^+ = U_2A^* = A^*V_2,$$

for some matrices U_1, V_1, U_2, and V_2. Let $D = A_1^+ - A_2^+$, $U = U_1 - U_2$, $V = V_1 - V_2$, Then[1]

$$ADA = 0, \quad D = UA^* = A^*V.$$

But $D^* = V^*A$; hence,

$$(DA)^*(DA) = A^*D^*DA = A^*V^*ADA = 0,$$

and therefore $DA = 0$.

Making use of the formula $D^* = AU^*$ we find that

$$DD^* = DAU^* = 0.$$

Therefore $A_1^+ - A_2^+ = D = 0$.

In order to prove the existence of the matrix A^+, assume first that the rank of the matrix A (of the order $m \times n$ with $m \geq n$) is equal to n.

We shall show that in this case the matrix

$$A^+ = (A^*A)^{-1}A^* \tag{13.4}$$

satisfies (13.2) and (13.3).

[1] 0 denotes the zero matrix.

(13.2) is obviously satisfied since

$$AA^+A = A(A^*A)^{-1}(A^*A) = A,$$

where A^*A is a nonsingular matrix of the order $n \times n$. The equality $A^+ = UA^*$ is satisfied with $U = (A^*A)^{-1}$. The equality $A^+ = A^*V$ can be satisfied as is easy to verify, if it is assumed that $V = A(A^*A)^{-2}A^*$.

Similarly it can be shown that if the rank of the matrix A (of the order $m \times n$ with $m \leq n$) is equal to m, then the matrix

$$A^+ = A^*(AA^*)^{-1} \tag{13.5}$$

is the pseudo-inverse with respect to the matrix A.

In order to prove the existence of a pseudo-inverse matrix in the general case we shall make use of the fact that any matrix A of the order $m \times n$ of rank r can be represented as a product

$$A = B \cdot C, \tag{13.6}$$

with matrices $B_{(m\times r)}$ and $C_{(r\times n)}$ of rank $r \leq m \wedge n$.

Indeed, let us construct a matrix B having r independent columns of the matrix A. Then all the columns of the matrix A can be expressed in terms of columns of the matrix B, which is justified because (13.6) determines a 'skeleton' decomposition of the matrix A.

Now set

$$A^+ = C^+B^+, \tag{13.7}$$

where, according to (13.4) and (13.5),

$$C^+ = C^*(CC^*)^{-1}, \tag{13.8}$$

$$B^+ = (B^*B)^{-1}B^*. \tag{13.9}$$

Then

$$AA^+A = BCC^*(CC^*)^{-1}(B^*B)^{-1}B^*BC = BC = A.$$

Next, if it is assumed that $U = C^*(CC^*)^{-1}(B^*B)^{-1}(CC^*)^{-1}C$, it can be easily checked that:

$$UA^* = A^+.$$

Analogously, if $V = B(B^*B)^{-1}(CC^*)^{-1}(B^*B)^{-1}B^*$, then $A^+ = A^*V$. □

13.1.3. We shall present a number of properties of pseudo-inverse matrices to be used further on:

(1°) $AA^+A = A$, $A^+AA^+ = A^+$;
(2°) $(A^*)^+ = (A^+)^*$;
(3°) $(A^+)^+ = A$;
(4°) $(A^+A)^2 = A^+A$, $(A^+A)^* = A^+A$, $(AA^+)^2 = AA^+$, $(AA^+)^* = AA^+$;
(5°) $(A^*A)^+ = A^+(A^*)^+ = A^+(A^+)^*$;
(6°) $A^+ = (A^*A)^+A^* = A^*(AA^*)^+$;
(7°) $A^+AA^* = A^*AA^+ = A^*$;
(8°) if S is an orthogonal matrix, then $(SAS^*)^+ = SA^+S^*$;
(9°) if A is a symmetric nonnegative definite matrix of order $n \times n$ of rank $r < n$, then

$$A^+ = T^*(TT^*)^{-2}T, \tag{13.10}$$

where the matrix $T_{(r\times n)}$ of rank r is defined by the decomposition

$$A = T^*T; \tag{13.11}$$

(10°) if the matrix A is nonsingular, then $A^+ = A^{-1}$.

The properties given above can be verified by immediate calculation. Thus (1°) and (2°) follow from (13.2) and (13.6)–(13.9). The equalities

$$A^+ = C^+B^+ = C^*(CC^*)^{-1}(B^*B)^{-1}B^* = \tilde{B}\tilde{C},$$

where

$$\tilde{B} = C^*(CC^*)^{-1}, \quad \tilde{C} = (B^*B)^{-1}B^*,$$

give a skeleton decomposition of the matrix A^+ from which (3°) follows. (4°) follows from (1°), (2°) and (13.7)–(13.9). In order to prove (5°), one should make a skeleton decomposition $A = BC$ and represent the matrix A^*A as a product $\tilde{B}\tilde{C}$ where $\tilde{B} = C^*$ and $\tilde{C} = B^*BC$. (6°) and (7°) follow from (1°)–(5°).

In order to prove (8°) it suffices to note that, by virtue of the orthogonality ($SS^* = E$) of the matrix S,

$$(SAS^*)(SA^+S^*)(SAS^*) = SAA^+AS^* = SAS^*. \tag{13.12}$$

Next, if $A^+ = UA^* = A^*V$, then

$$SA^+S^* = S(UA^*)S = SU(S^*S)A^*S = \tilde{U}(SA^*S) = \tilde{U}(SAS^*)^*, \tag{13.13}$$

with $\tilde{U} = SUS^*$.

Similarly, it is established that

$$SA^+S^* = (SAS^*)^*\tilde{V} \tag{13.14}$$

with $\tilde{V} = SVS^*$.

It follows from (13.12)–(13.14) that $(SAS^*)^+ = SA^+S^*$.

Finally, (9°) follows from the skeleton decomposition $A = T^*T$ and (13.7)–(13.9).

Note. According to (9°), in the case of symmetric nonnegative definite matrices A the pseudo-inverse matrix A^+ can be defined by (13.10) where the matrix T is defined from the decomposition $A = T^*T$. This decomposition is not, in general, unique. The pseudo-inverse matrix $A^+ = T^*(TT^*)^{-2}T$ is, however, defined uniquely regardless of the way of decomposing A as T^*T. Therefore, in the case of symmetric nonnegative definite matrices A, the pseudo-inverse matrix

$$A^+ = \begin{cases} A^{-1} & \text{if the matrix } A \text{ is nonsingular,} \\ T^*(TT^*)^{-2}T, & \text{if the matrix } A \text{ is singular.} \end{cases} \tag{13.15}$$

13.1.4. We recall that the random vector $\xi = (\xi_1, \dots, \xi_n)$ is called Gaussian (normal), if its characteristic function [2]

$$\varphi_\xi(z) = M\exp[iz^*\xi], \quad z = (z_1, \dots, z_n), \quad z^*\xi = \sum_{i=1}^n z_i\xi_i,$$

is given by the formula

$$\varphi_\xi(z) = \exp\left[iz^*m - \frac{1}{2}z^*Rz\right], \tag{13.16}$$

where $m = (m_1, \dots, m_n)$ and $R = \|R_{ij}\|$ is a nonnegative definite symmetric matrix of the order $(n \times n)$. The parameters m and R have a simple meaning. The vector m is a vector of the mean values, $m = M\xi$, and the matrix R is a matrix of covariances

$$R \equiv \operatorname{cov}(\xi, \xi) = M(\xi - m)(\xi - m)^*.$$

Let us note a number of simple properties of Gaussian vectors.

(1) If $\xi = (\xi_1, \dots, \xi_n)$ is a Gaussian vector, $A_{(m\times n)}$ a matrix and $a = (a_1, \dots, a_m)$ a vector, then the random vector $\eta = A\xi + a$ is Gaussian with

$$\varphi_n(z) = \exp\left\{iz^*(a + Am) - \frac{1}{2}z^*(ARA^*)z\right\} \tag{13.17}$$

and

$$M\eta = a + Am, \quad \operatorname{cov}(\eta, \eta) = A\operatorname{cov}(\xi, \xi)A^*. \tag{13.18}$$

[2] In algebraic operations, vectors a are regarded as columns, and vectors a^* are regarded as rows.

(2) Let $(\theta, \xi) = [(\theta_1, \dots, \theta_k), (\xi_1, \dots, \xi_l)]$ be a Gaussian vector with $m_\theta = M\theta$, $m_\xi = M\xi$, $D_{\theta\theta} = \operatorname{cov}(\theta, \theta) = M(\theta - m_\theta)(\theta - m_\theta)^*$, $D_{\xi\xi} = \operatorname{cov}(\xi, \xi) = M(\xi - m_\xi)(\xi - m_\xi)^*$ and $D_{\theta\xi} = \operatorname{cov}(\theta, \xi) = M(\theta - m_\theta)(\xi - m_\xi)$.
If $D_{\theta\xi} = 0$, then the (Gaussian) vectors θ and ξ are independent and

$$\varphi_{(\theta,\xi)}(z_1, z_2) = \varphi_\theta(z_1)\varphi_\xi(z_2),$$

where $z_1 = (z_{11}, \dots, z_{1k})$, $z_2 = (z_{21}, \dots, z_{2l})$ and

$$\begin{aligned} \varphi_0(z_1) &= \exp\left[iz_1^* m_\theta - \frac{1}{2} z_1^* D_{\theta\theta} z_1\right], \\ \varphi_\xi(z_2) &= \exp\left[iz_2^* m_\xi - \frac{1}{2} z_2^* D_{\xi\xi} z_2\right]. \end{aligned}$$

(3) Let $\xi = (\xi_1, \dots, \xi_n)$ be a Gaussian vector with $m = M\xi$ and $R = \operatorname{cov}(\xi, \xi)$. Then there exists a Gaussian vector $\varepsilon = (\varepsilon_1, \dots, \varepsilon_n)$ with independent components, $M\varepsilon = 0$ and $\operatorname{cov}(\varepsilon, \varepsilon) = E_{(n \times n)}$, such that

$$\xi = R^{1/2}\varepsilon + m. \tag{13.19}$$

For this purpose let us introduce a Gaussian vector[3] $\nu = (\nu_1, \dots, \nu_n)$ independent of ξ, with $M\nu = 0$, $\operatorname{cov}(\nu, \nu) = E$. Assume $T = R^{1/2}$;

$$\varepsilon = (T^+)^*(\xi - m) + (E - TT^+)\nu. \tag{13.20}$$

Since the vectors ξ and ν are independent, then the vector ε is also Gaussian. It is seen that $M\varepsilon = 0$. Compute now the covariance $\operatorname{cov}(\varepsilon, \varepsilon)$. We have

$$\operatorname{cov}(\varepsilon, \varepsilon) = M\varepsilon\varepsilon^* = (T^+)^* R T^+ + (E - TT^+)(E - TT^+)^*.$$

But by property (4°) of pseudo-inverse matrices

$$(E - TT^+)^* = E - TT^+, \quad (E - TT^+)^2 = E - TT^+$$

and

$$(T^+)^* R T^+ = (T^+)^* T^* T T^+ = [(T^+)^* T^*][TT^+] = TT^+.$$

Hence, $\operatorname{cov}(\xi, \xi) = E$, which proves the independence of the components of the vector ε.
Next we obtain from (13.20)

$$\begin{aligned} T^*\varepsilon &= T^*(T^+)^*(\xi - m) + (T^* - T^*TT^+)\nu \\ &= (\xi - m) - (E - T^*(T^+)^*)(\xi - m) + (T^* - T^*TT^+)\nu. \end{aligned}$$

But $T^* = T^*TT^+$ (from (7°)), $T^*(T^+)^* = (T^+T)^* = T^+T$ (from (4°)), and $(E - T^+T)\operatorname{cov}(\xi, \xi)(E - T^+T)^* = (E - T^+T)(T^*T) \times (E - T^+T) = 0$, which proves the equality $R^{1/2}\varepsilon = \xi - m$.

[3] Here we assume that the initial probability space is sufficiently 'rich'.

(4) Let ξ_n, $n = 1, 2, \ldots$, be a sequence of Gaussian vectors converging in probability to a vector ξ. Then ξ is also a Gaussian vector.
Indeed, let $m_n = M\xi_n$ and $R_n = \operatorname{cov}(\xi_n, \xi_n)$. Then, since $P\text{-}\lim_{n\to\infty} \xi_n = \xi$ and $|\exp[iz^*\xi_n]| \le 1$, by the Lebesgue dominated convergence theorem

$$\lim_{n\to\infty} \exp\left[iz^* m_n - \frac{1}{2} z^* R_n z\right] = \lim_{n\to\infty} M \exp[iz^*\xi_n] = M \exp[iz^*\xi].$$

From this, by virtue of the arbitrariness of z, there exist a vector m and a nonnegative definite matrix R such that

$$m = \lim_n m_n, \quad R = \lim_n R_n.$$

Therefore,

$$M \exp[iz^*\xi] = \exp\left[iz^* m - \frac{1}{2} z^* R z\right],$$

which proves the normality of the vector ξ.

13.1.5.

Theorem 13.1 (Theorem on normal correlation). *Let* $(\theta, \xi) = ([\theta_1, \ldots, \theta_k], [\xi_1, \ldots, \xi_l])$ *be a Gaussian vector with*

$$m_\theta = M\theta, \quad m_\xi = M\xi,$$

$$D_{\theta\theta} = \operatorname{cov}(\theta, \theta), \quad D_{\theta\xi} = \operatorname{cov}(\theta, \xi), \quad D_{\xi\xi} = \operatorname{cov}(\xi, \xi).$$

Then the conditional expectation $M(\theta|\xi)$ *and the conditional covariance*

$$\operatorname{cov}(\theta, \theta|\xi) = M\{[\theta - M(\theta|\xi)][\theta - M(\theta|\xi)]^*|\xi\}$$

are given by the formulae

$$M(\theta|\xi) = m_\theta + D_{\theta\xi} D_{\xi\xi}^{+}(\xi - m_\xi), \tag{13.21}$$

$$\operatorname{cov}(\theta, \theta|\xi) = D_{\theta\theta} - D_{\theta\xi} D_{\xi\xi}^{+}(D_{\theta\xi})^*. \tag{13.22}$$

PROOF. Set

$$\eta = (\theta - m_\theta) + C(\xi - m_\xi), \tag{13.23}$$

where we want to select the matrix $C_{(k\times l)}$ such that $M\eta(\xi - m_\xi)^* = 0$.

If such a matrix exists, then it is a solution of the linear system

$$D_{\theta\xi} + C D_{\xi\xi} = 0. \tag{13.24}$$

If $D_{\xi\xi}$ is a positive definite matrix, then

$$C = -D_{\theta\xi} D_{\xi\xi}^{-1}. \tag{13.25}$$

Otherwise it can be assumed that

$$C = -D_{\theta\xi} D^{+}_{\xi\xi}. \tag{13.26}$$

According to (3), the property given in Subsection 13.1.4, there exists a Gaussian vector ε with $M\varepsilon = 0$, $M\varepsilon\varepsilon^* = E$, such that

$$\xi - m_\xi = D^{1/2}_{\xi\xi} \varepsilon.$$

Then, setting $T = D^{1/2}_{\xi\xi}$, we obtain

$$D_{\theta,\xi} = M[(\theta - m_\theta)(\xi - m_\xi)^*] = M(\theta_t - m_\theta)\varepsilon^* T = d_{\theta\varepsilon} T,$$

where $d_{\theta\varepsilon} = M(\theta - m_\theta)\varepsilon^*$. Therefore,

$$D_{\theta\xi} = d_{\theta\varepsilon} T, \quad D_{\theta\xi} D^{+}_{\xi\xi} D_{\xi\xi} = d_{\theta\varepsilon} T(TT)^{+} TT = d_{\theta\varepsilon} T,$$

where we take advantage of (1°), (4°) and (5°), according to which it follows that

$$\begin{aligned} D^{+}_{\xi\xi} = (TT)^{+} = T^{+}T^{+}, \quad T(TT)^{+}TT &= TT^{+}T^{+}TT = TT^{+}(T^{+}T)^{*}T \\ &= (TT^{+})^{2}T = TT^{+}T = T, \end{aligned}$$

i.e.,

$$D_{\theta\xi} = D_{\theta\xi} D^{+}_{\xi\xi} D_{\xi\xi},$$

which proves (13.24) with $C = -D_{\theta\xi} D^{+}_{\xi\xi}$.

Therefore, the vector

$$\eta = (\theta - m_\theta) - D_{\theta\xi} D^{+}_{\xi\xi} (\xi - m_\xi) \tag{13.27}$$

has the property that $M\eta(\xi - m_\xi)^* = 0$.

Since (θ, ξ) is Gaussian, so is η. Moreover, the vector (η, ξ) will also be Gaussian, since the characteristic function

$$\begin{aligned} \varphi_{(\eta,\xi)}(z_1, z_2) &= M \exp[i z_1^* \eta + i z_2^* \xi] \\ &= M \exp\{i z_1^* [(\theta - m_0) + C(\xi - m_\xi)] + i z_2^* \xi\} \end{aligned}$$

can be written in the same form as (13.16) due to the normality of the vector (θ, ξ). Next, $M\eta = 0$ and $M\eta(\xi - m_\xi)^* = 0$.

Hence, according to (2), given in Subsection 13.1.4, the Gaussian vectors η and ξ are independent.

Therefore,

$$M(\eta|\xi) = M\eta = 0 \quad (P\text{-a.s.}),$$

which, together with (13.27), yields (13.21).

In order to prove (13.22), note that $\theta - M(\theta|\xi) = \eta$ and, due to the independence of ξ and η,

$$\operatorname{cov}(\theta, \theta|\xi) = M(\eta\eta^*|\xi) = M\eta\eta^* \quad (P\text{-a.s.}). \tag{13.28}$$

But, according to (13.27),

$$\begin{aligned} M\eta\eta^* &= D_{\theta\theta} + D_{\theta\xi}D_{\xi\xi}^+ D_{\xi\xi}D_{\xi\xi}^+ D_{\theta\xi}^* - 2D_{\theta\xi}D_{\xi\xi}^+ D_{\xi\xi}D_{\xi\xi}^+ D_{\theta\xi}^* \\ &= D_{\theta\theta} - D_{\theta\xi}D_{\xi\xi}^+ D_{\xi\xi}D_{\xi\xi}^+ D_{\theta\xi}^* = D_{\theta\theta} - D_{\theta\xi}D_{\xi\xi}^+ D_{\theta\xi}^*, \end{aligned} \tag{13.29}$$

where we take advantage of the fact that, according to (1°), $D_{\xi\xi}^+ D_{\xi\xi} D_{\xi\xi}^+ = D_{\xi\xi}^+$.

From (13.28) and (13.29) we obtain (13.22) for $\operatorname{cov}(\theta, \theta|\xi)$. □

13.1.6.

Corollary 1. *If $k = l = 1$ and $D\xi > 0$, then*

$$M(\theta|\xi) = M\theta + \frac{\operatorname{cov}(\theta, \xi)}{D\xi}(\xi - M\xi), \tag{13.30}$$

$$D(\theta|\xi) = D\theta - \frac{\operatorname{cov}^2(\theta, \xi)}{D\xi}, \tag{13.31}$$

where $D(\theta|\xi) = M\{[\theta - M(\theta|\xi)]^2|\xi\}$.

Assuming $\sigma_\theta = +\sqrt{D\theta}$, $\sigma_\xi = +\sqrt{D\xi}$ and introducing the correlation coefficient

$$\rho = \frac{\operatorname{cov}(\theta, \xi)}{\sigma_\theta \sigma_\xi},$$

(13.30) and (13.31) can be rewritten as follows:

$$M(\theta|\xi) = M\theta + \rho\frac{\sigma_\theta}{\sigma_\xi}(\xi - M\xi), \tag{13.32}$$

$$D(\theta|\xi) = \sigma_\theta^2(1 - \rho^2). \tag{13.33}$$

Corollary 2. *If*

$$\theta = b_1\varepsilon_1 + b_2\varepsilon_2, \quad \xi = B_1\varepsilon_1 + B_2\varepsilon_2,$$

where $\varepsilon_1, \varepsilon_2$ are independent Gaussian variables with $M\varepsilon_i = 0$, $D\varepsilon_i = 1$, $i = 1, 2$, and $B_1^2 + B_2^2 > 0$, then

$$M(\theta|\xi) = \frac{b_1B_1 + b_2B_2}{B_1^2 + B_2^2}\xi, \tag{13.34}$$

$$D(\theta|\xi) = \frac{(B_1b_2 - b_1B_2)^2}{B_1^2 + B_2^2}. \tag{13.35}$$

Corollary 3. *Let the random variables $(\theta, \xi_1, \ldots, \xi_l)$ form a Gaussian vector where $\xi_1, \ldots, \xi_l$ are independent and $D\xi_i > 0$. Then*

$$M(\theta|\xi_1,\dots,\xi_l) = M\theta + \sum_{i=1}^{l} \frac{\operatorname{cov}(\theta,\xi_i)}{D\xi_i}(\xi_i - M\xi_i).$$

In particular, if $M\theta = M\xi_i = 0$, *then*

$$M(\theta|\xi_1,\dots,\xi_l) = \sum_{i=1}^{l} \frac{\operatorname{cov}(\theta,\xi_i)}{D\xi_i}\xi_i.$$

Note. Let $[\theta,\xi] = [(\theta_1,\dots,\theta_k),(\xi_1,\dots,\xi_l)]$ be a random vector specified on a probability space $(\Omega,\mathcal{F},P)$. Let $\mathcal{G}$ be a certain sub-σ-algebra of $\mathcal{F}$ ($\mathcal{G}\subseteq\mathcal{F}$). Assume that ($P$-a.s.) the conditional (with respect to $\mathcal{G}$) distribution of a vector (θ,ξ) is Gaussian with means $M(\theta|\mathcal{G})$ and $M(\xi|\mathcal{G})$, and covariances $d_{11} = \operatorname{cov}(\theta,\theta|\mathcal{G})$, $d_{12} = \operatorname{cov}(\theta,\xi|\mathcal{G})$, and $d_{22} = \operatorname{cov}(\xi,\xi|\mathcal{G})$. Then the vector of conditional expectations $M(\theta|\xi,\mathcal{G})$ and the conditional matrix of covariances $\operatorname{cov}(\theta,\theta|\xi,\mathcal{G})$ are given (P-a.s.) by the formulae

$$M(\theta|\xi,\mathcal{G}) = M(\theta|\mathcal{G}) + d_{12}d_{22}^{+}[\xi - M(\xi|\mathcal{G})], \tag{13.36}$$

$$\operatorname{cov}(\theta,\theta|\xi,\mathcal{G}) = d_{11} - d_{12}d_{22}^{+}d_{12}^{*}. \tag{13.37}$$

This result is proved in the same ways as in the case $\mathcal{G} = \{\emptyset,\Omega\}$ and will be used frequently from now on.

13.1.7.

Theorem 13.2. *Under the assumptions of Theorem 13.1, the conditional distribution*[4] $P(\theta\le x|\xi)$ *is Gaussian with parameters* $M(\theta|\xi)$ *and* $\operatorname{cov}(\theta,\theta|\xi)$ *given (respectively) by (13.21), and (13.22).*

PROOF. It suffices to show that the conditional characteristic function

$$M\left(\exp[iz^*\theta]|\xi\right) = \exp\left(iz^*M(\theta|\xi) - \frac{1}{2}z^*\operatorname{cov}(\theta,\theta)z\right). \tag{13.38}$$

According to (13.27) and (13.21),

$$\theta = m_\theta + D_{\theta\xi}D_{\xi\xi}^{+}(\xi - M\xi) + \eta = M(\theta|\xi) + \eta,$$

where the Gaussian vectors ξ and η are independent. Hence

$$\begin{aligned} M(\exp[iz^*\theta]|\xi) &= \exp[iz^*M(\theta|\xi)]M(\exp[iz^*\eta]|\xi) \\ &= \exp[iz^*M(\theta|\xi)]M\exp[iz^*\eta] \\ &= \exp\left[iz^*M(\theta|\xi) - \frac{1}{2}z^*\operatorname{cov}(\theta,\theta|\xi)z\right]. \end{aligned}$$

[4] By $\{\theta\le x\}$ we mean the event $\{\theta_1\le x_1,\dots,\theta_k\le x_k\}$.

Note. Let the matrix $\operatorname{cov}(\theta,\theta|\xi) = D_{\theta\theta} - D_{\theta\xi}D^{+}_{\xi\xi}D^{*}_{\theta\xi}$ be positive definite. Then the distribution function $P(\theta \le x|\xi) = P(\theta_1 \le x_1, \dots, \theta_k \le x_k|\xi)$ has (P-a.s.) the density

$$\begin{aligned} P(x_1,\dots,x_k|\xi) &= \frac{[\det(D_{\theta\theta} - D_{\theta\xi}D^{+}_{\xi\xi}D_{\theta\xi})]^{-1/2}}{(2\pi)^{k/2}} \\ &\quad \times \exp\left\{ -\frac{1}{2}(x - M(\theta|\xi))^{*}[D_{\theta\theta} - D_{\theta\xi}D^{+}_{\xi\xi}D^{*}_{\theta\xi}]^{-1} \right. \\ &\quad \left. \times (x - M(\theta|\xi)) \right\}. \end{aligned} \tag{13.39}$$

□

13.1.8. The theorem on normal correlation allows us to establish easily the following auxiliary results.

Lemma 13.2. *Let b_1, b_2, B_1, B_2, be matrices of the orders $k \times k$, $k \times l$, $l \times k$, $l \times l$, respectively, and let*

$$\begin{aligned} b \circ b &= b_1 b_1^{*} + b_2 b_2^{*}, \\ b \circ B &= b_1 B_1^{*} + b_2 B_2^{*}, \\ B \circ B &= B_1 B_1^{*} + B_2 B_2^{*}. \end{aligned} \tag{13.40}$$

Then the symmetric matrix

$$b \circ b - (b \circ B)(B \circ B)^{+}(b \circ B)^{*} \tag{13.41}$$

is nonnegative definite.

PROOF. Let $\varepsilon_1 = [\varepsilon_{11}, \dots, \varepsilon_{1k}]$, $\varepsilon_2 = [\varepsilon_{21}, \dots, \varepsilon_{2l}]$ be independent Gaussian vectors with independent components, $M\varepsilon_{ij} = 0$, $D\varepsilon_{ij} = 1$.

Set

$$\begin{aligned} \theta &= b_1 \varepsilon_1 + b_2 \varepsilon_2, \\ \xi &= B_1 \varepsilon_1 + B_2 \varepsilon_2. \end{aligned}$$

Then, according to (13.22),

$$b \circ b - (b \circ B)(B \circ B)^{+}(B \circ B)^{*} = \operatorname{cov}(\theta, \theta|\xi),$$

which proves the lemma since the matrix of covariances $\operatorname{cov}(\theta,\theta|\xi)$ is nonnegative definite. □

Lemma 13.3. Let $R_{(n\times n)}$, $P_{(m\times m)}$ *be nonnegative definite symmetric matrices, and let $Q_{(m\times n)}$ be an arbitrary matrix. Then the system of linear algebraic equations*

$$(R + Q^*PQ)x = Q^*Py \tag{13.42}$$

is solvable (for x) given any vector $y = (y_1, \dots, y_m)$, and one solution is given by

$$\tilde{x} = (R + Q^*PQ)^+Q^*Py. \tag{13.43}$$

PROOF. Let $\theta = (\theta_1, \dots, \theta_m)$, $\varepsilon = (\varepsilon_1, \dots, \varepsilon_n)$ be independent Gaussian vectors with $M\theta = 0$, $\operatorname{cov}(\theta, \theta) = P$, $\operatorname{cov}(\varepsilon, \varepsilon) = E$. Set $\xi = Q^*\theta + R^{1/2}\varepsilon$. Then, in this case, $D_{\theta\xi} = \operatorname{cov}(\theta, \xi) = PQ$, $D_{\xi\xi} = \operatorname{cov}(\xi, \xi) = R + Q^*PQ$, since it was proved in Theorem 13.1 that the system $D_{\theta\xi} + CD_{\xi\xi} = 0$ is solvable with respect to C and that $C = -D_{\theta\xi}D_{\xi\xi}^+$. As applied to the present situation, this implies that the system

$$PQ + C(R + Q^*PQ) = 0 \tag{13.44}$$

is solvable with respect to C and that $C = -PQ[R + Q^*PQ]^+$.

From the solvability of the system given by (13.44) follows the solvability (with respect to C^*) of the adjoint system

$$Q^*P + [R + Q^*PQ]C^* = 0. \tag{13.45}$$

Now consider an arbitrary vector y. Assume $\tilde{x} = -C^*y$. Then, multiplying (13.45) by $(-y)$, we obtain $(R + Q^*PQ)\tilde{x} = Q^*Py$, which proves the lemma. □

Lemma 13.4. *Let $\theta_t = (\theta_1(t), \dots, \theta_n(t))$, $t = 0, 1, \dots$, be a Gaussian Markov process with mean $r(t) = M\theta_t$ and correlation*

$$R(t, s) = M[(\theta_t - r(t))(\theta_s - r(s))^*], \quad t, s = 0, 1, \dots.$$

Then we can find a sequence of independent Gaussian vectors

$$\varepsilon(t) = (\varepsilon_1(t), \dots, \varepsilon_n(t)), \quad t \geq 1,$$

with $M\varepsilon(t) \equiv 0$ and $M\varepsilon(t)\varepsilon^*(t) \equiv E_{(n\times n)}$, such that

$$\begin{aligned}\theta_{t+1} &= [r(t+1) - R(t+1, t)R^+(t, t)r(t)] + R(t+1, t)R^+(t, t)\theta_t \\ &\quad + [R(t+1, t+1) - R(t+1, t)R^+(t, t)R^*(t+1, t)]^{1/2}\varepsilon(t+1).\end{aligned}$$

PROOF. Put $V_{t+1} = \theta_{t+1} - M(\theta_{t+1}|\theta_t)$. By the theorem on normal correlation,

$$M[\theta_{t+1}|\theta_t] = r(t+1) + R(t+1, t)R^+(t, t)(\theta_t - r(t)).$$

From this it follows that the vectors V_t, $t \geq 1$, are independent Gaussian. Indeed, for $t > s$, because of the Markovian nature of the process (θ_t), $t = 0, 1, \dots$,

$$M[\theta_t - M(\theta_t|\theta_{t-1})|\theta_s, \theta_{s-1}] = M[\theta_t|\theta_s] - M[\theta_t|\theta_s] = 0,$$

and therefore

$$\begin{aligned} MV_tV_s^* &= M[\theta_t - M(\theta_t|\theta_{t-1}))(\theta_s - M(\theta_s|\theta_{s-1}))^*] \\ &= M\{M[\theta_t - M(\theta_t|\theta_{t-1})|\theta_s, \theta_{s-1}][\theta_s - M(\theta_s|\theta_{s-1})]^*\} = 0. \end{aligned}$$

The equality $MV_tV_s^* = 0$ for $t < s$ is verified in a similar way.

Next we find from (13.22), that

$$MV_{t+1}V_{t+1}^* = R(t+1, t+1) - R(t+1, t)R^+(t, t)R^*(t+1, t).$$

Therefore by (3) we can find a Gaussian vector ε_{t+1} such that (see (13.19))

$$V_{t+1} = [R(t+1, t+1) - R(t+1, t)R^+(t, t)R^*(t+1, t)]^{1/2}\varepsilon(t+1),$$

$$M\varepsilon_{t+1} = 0, \quad \operatorname{cov}(\varepsilon_{t+1}, \varepsilon_{t+1}) = E.$$

The independence of the Gaussian vectors ε_t, $t = 1, 2, \ldots$, follows from the independence of the vectors V_t, $t = 1, 2, \ldots$, and from the method of construction of vectors ε_t according to (13.20).

The required recursive equation for θ_t follows now from the formulae for V_{t+1} and the representation for the conditional expectation $M(\theta_{t+1}|\theta_t)$. □

13.2 Recursive Filtering Equations for Conditionally Gaussian Sequences

13.2.1. On a probability space $(\Omega, \mathcal{F}, P)$, let there be given a partially observable random sequence $(\theta, \xi) = (\theta_t, \xi_t)$, $t = 0, 1, \ldots$, where

$$\theta_t = (\theta_1(t), \ldots, \theta_k(t)), \quad \xi_t = (\xi_1(t), \ldots, \xi_l(t)),$$

defined by recursive equations

$$\begin{aligned} \theta_{t+1} &= a_0(t,\xi) + a_1(t,\xi)\theta_t + b_1(t,\xi)\varepsilon_1(t+1) \\ &\quad + b_2(t,\xi)\varepsilon_2(t+1), \qquad (13.46) \\ \xi_{t+1} &= A_0(t,\xi) + A_1(t,\xi)\theta_t + B_1(t,\xi)\varepsilon_1(t+1) \\ &\quad + B_2(t,\xi)\varepsilon_2(t+1). \qquad (13.47) \end{aligned}$$

Here, $\varepsilon_1(t) = (\varepsilon_{11}(t), \ldots, \varepsilon_{1k}(t))$ and $\varepsilon_2(t) = (\varepsilon_{21}(t), \ldots, \varepsilon_{2l}(t))$ are independent Gaussian vectors with independent components, each of which is normally distributed, $N(0, 1)$, while

$$\begin{aligned} a_0(t,\xi) &= (a_{01}(t,\xi), \ldots, a_{0k}(t,\xi)), \\ A_0(t,\xi) &= (A_{01}(t,\xi), \ldots, A_{0l}(t,\xi)) \end{aligned}$$

are vector functions and

$$\begin{aligned} b_1(t,\xi) &= \|b_{ij}^{(1)}(t,\xi)\|, \quad b_2(t,\xi) = \|b_{ij}^{(2)}(t,\xi)\|, \quad B_1(t,\xi) = \|B_{ij}^{(1)}(t,\xi)\|, \\ B_2(t,\xi) &= \|B_{ij}^{(2)}(t,\xi)\|, \quad a_1(t,\xi) = \|a_{ij}^{(1)}(t,\xi)\|, \quad A_1(t,\xi) = \|A_{ij}^{(1)}(t,\xi)\|, \end{aligned}$$

are matrix functions having (respectively) the orders $k \times k$, $k \times l$, $l \times k$, $l \times l$, $k \times k$, $l \times k$.

Any element of these vector functions and matrices is assumed to be nonanticipative, i.e., $\mathcal{F}_t^\xi$-measurable where $\mathcal{F}_t^\xi = \sigma\{\xi_0, \dots, \xi_t\}$ for any $t = 0, 1, \dots$.

The system of equations given by (13.46) and (13.47) can be solved under the initial conditions of (θ_0, ξ_0), where the random vector (θ_0, ξ_0) is assumed to be independent of sequences $(\varepsilon_1, \varepsilon_2) = [\varepsilon_1(t), \varepsilon_2(t)]$, $t = 1, 2, \dots$. As to the coefficients of the system of equations given by (13.46) and (13.47) and the initial conditions of (θ_0, ξ_0), the following assumptions will be adopted throughout the chapter.

(1) If $g(t,\xi)$ is any of the functions[5] $a_{0i}, A_{0j}, b_{ij}^{(1)}, b_{ij}^{(2)}, B_{ij}^{(1)}, B_{ij}^{(2)}$, then

$$M|g(t,\xi)|^2 < \infty, \quad t = 0, 1 \dots. \tag{13.48}$$

(2) With probability one

$$|a_{ij}^{(1)}(t,\xi)| \le c, \quad |A_{ij}^{(1)}(t,\xi)| \le c.$$

(3) $M(\|\theta_0\|^2 + \|\xi_0\|^2) < \infty$, where, for

$$x = (x_1, \dots, x_n), \quad \|x\|^2 = \sum_{i=1}^n x_i^2.$$

(4) The conditional distribution $P(\theta_0 \le a|\xi_0)$ is (P-a.s.) Gaussian.

It follows from (1)–(3) that, at any time $t < \infty$,

$$M(\|\theta_t\|^2 + \|\xi_t\|^2) < \infty. \tag{13.49}$$

13.2.2. If the sequence (θ, ξ) is assumed to be partially observable, the problem of filtering involves the construction of an estimate for the unobservable variables θ_t from the observations $\xi_0^t = (\xi_0, \dots, \xi_t)$. Let $F_{\xi_0^t}(a) = P(\theta_t \le a|\mathcal{F}_t^\xi)$,

$$m_t = M(\theta_t|\mathcal{F}_t^\xi), \quad \gamma_t = M[(\theta_t - m_t)(\theta_t - m_t)^*|\mathcal{F}_t^\xi].$$

It is obvious that, due to (13.49) the a posteriori mean

$$m_t = (m_1(t), \dots, m_k(t))$$

is the optimal estimate (in the mean square sense) of the vector θ_t based on the variables $\xi_0^t = \{\xi_0, \dots, \xi_t\}$, and

[5] For simplicity, arguments in the functions considered are sometimes omitted.

$$\operatorname{Tr} M\gamma_t = \sum_{i=1}^{k} M[\theta_i(t) - m_i(t)]^2$$

yields the estimation error.

In the case of an arbitrary partially observable sequence (θ, ξ) it is difficult to find the form of the distribution $F_{\xi_0^t}(a)$ and its parameters m_t, γ_t. For the sequences (θ, ξ) governed by the system of equations given by (13.46) and (13.47) with the additional assumption of normality of the conditional distribution $P(\theta_0 \leq a|\xi_0)$ the solution of the problem of filtering (i.e., finding m_t and γ_t) becomes possible. The following result, analogous to Theorem 11.1 for the case of continuous time, is the basis for the method of solution.

Theorem 13.3. *Let (1)–(4) be satisfied. Then the sequence (θ, ξ) governed by (13.46) and (13.47) is conditionally Gaussian, i.e., the conditional distributions*

$$P(\theta_0 \leq a_0, \ldots, \theta_t \leq a_t|\mathcal{F}_t^\xi)$$

are (P-a.s.) Gaussian for any $t = 0, 1, \ldots$.

PROOF. Let us establish the normality of the conditional distribution $P(\theta_t < a|\mathcal{F}_t^\xi)$. This suffices for our present purposes; the proof for the general case will be given in Subsection 13.3.6.

The proof will be carried out by induction. Assume that the distribution $F_{\xi_0^t} = P(\theta_t \leq a|\mathcal{F}_t^\xi)$ is normal, $N(m_t, \gamma_t)$.

Because of (13.46) and (13.47), the conditional distribution

$$P(\theta_{t+1} \leq a, \xi_{t+1} \leq x|\mathcal{F}_t^\xi, \theta_t = b)$$

is Gaussian with vector of mathematical expectations

$$\mathbf{A}_0 + \mathbf{A}_1 b = \begin{pmatrix} a_0 + a_1 b \\ A_0 + A_1 b \end{pmatrix}, \tag{13.50}$$

and with covariance matrix

$$\mathbf{B} = \begin{pmatrix} b \circ b & b \circ B \\ (b \circ B)^* & B \circ B \end{pmatrix}, \tag{13.51}$$

where $b \circ b = b_1 b_1^* + b_2 b_2^*$, $b \circ B = b_1 B_1^* + b_2 B_2^*$, and $B \circ B = B_1 B_1^* + B_2 B_2^*$.

Let $\nu_t = (\theta_t, \xi_t)$, $z = (z_1, \ldots, z_{k+l})$. Then the conditional characteristic function of the vector ν_{t+1} is given by the formula

$$M(\exp[iz^*\nu_{t+1}]|\mathcal{F}_t^\xi, \theta_t) = \exp\left[iz^*(\mathbf{A}_0(t,\xi)+\mathbf{A}_1(t,\xi)\theta_t) - \frac{1}{2}z^*\mathbf{B}(t,\xi)z\right]. \tag{13.52}$$

Assuming that, for some t,

$$\begin{aligned} &M(\exp[iz^*(\mathbf{A}_1(t,\xi)\theta_t)]|\mathcal{F}_t^\xi) \\ &= \exp\left[iz^*\left(\mathbf{A}_1(t,\xi)m_t - \frac{1}{2}z^*(\mathbf{A}_1(t,\xi)\gamma_t\mathbf{A}_1^*(t,\xi))z\right)\right], \end{aligned} \tag{13.53}$$

we obtain from (13.52) and (13.53)

$$\begin{aligned} M(\exp[iz^*\nu_{t+1}]|\mathcal{F}_t^\xi) = \exp\Big[&iz^*(\mathbf{A}_0(t,\xi) + \mathbf{A}_1(t,\xi)m_t) - \frac{1}{2}z^*\mathbf{B}(t,\xi)z \\ &-\frac{1}{2}z^*(\mathbf{A}_1(t,\xi)\gamma_t\mathbf{A}_1^*(t,\xi))z\Big]. \end{aligned}$$

Therefore, by induction, the conditional distributions

$$P(\theta_{t+1} \le a, \xi_{t+1} \le x|\mathcal{F}_t^\xi) \tag{13.54}$$

are Gaussian.

Consider now the vector

$$\eta = [\theta_{t+1} - M(\theta_{t+1}|\mathcal{F}_t^\xi)] - C[\xi_{t+1} - M(\xi_{t+1}|\mathcal{F}_t^\xi)].$$

By virtue of the theorem on normal correlation (and its accompanying note) there exists a matrix C such that

$$M[\eta(\xi_{t+1} - M(\xi_{t+1}|\mathcal{F}_t^\xi))^*|\mathcal{F}_t^\xi] = 0 \quad (P\text{-a.s.}).$$

It follows from this that the conditionally Gaussian vectors η and ξ_{t+1} (under the condition $\mathcal{F}_t^\xi$) are independent. Hence ($z = (z_1, \dots, z_k)$)

$$\begin{aligned} &M[\exp(iz^*\theta_{t+1})|\mathcal{F}_t^\xi, \xi_{t+1}] \\ &= M\{\exp(iz^*[M(\theta_{t+1}|\mathcal{F}_t^\xi) + C(\xi_{t+1} - M(\xi_{t+1}|\mathcal{F}_t^\xi)) + \eta]|\mathcal{F}_t^\xi, \xi_{t+1}\} \\ &= \exp(iz^*[M(\theta_{t+1}|\mathcal{F}_t^\xi) + C(\xi_{t+1} - M(\xi_{t+1}|\mathcal{F}_t^\xi))])M\{\exp(iz^*\eta)|\mathcal{F}_t^\xi, \xi_{t+1}\} \\ &= \exp(iz^*[M(\theta_{t+1}|\mathcal{F}_t^\xi) + C(\xi_{t+1} - M(\xi_{t+1}|\mathcal{F}_t^\xi))])M\{\exp(iz^*\eta)|\mathcal{F}_t^\xi, \xi_{t+1}\}. \end{aligned} \tag{13.55}$$

Due to (13.54), the conditional distribution $P(\eta \le y|\mathcal{F}_t^\xi)$ is Gaussian. Together with (13.55) this proves the normality of the conditional distribution $P(\theta_{t+1} \le a|\mathcal{F}_{t+1}^\xi)$.

Thus, for all t, $t = 0, 1, \dots$, the conditional distributions $P(\theta_t \le a|\mathcal{F}_t^\xi)$ are Gaussian. □

Note. It can be shown in like fashion that if, at some s, the distribution $P(\theta_s \le a|\mathcal{F}_s^\xi)$ is Gaussian, then the conditional distributions $P(\theta_t \le a|\mathcal{F}_t^\xi)$ will be the same for all $t \ge s$.

13.2.3. Conditional normality of the sequence (θ, ξ) enables us to deduce a closed system (compare with Section 12.1) of recursive equations for the parameters m_t, γ_t.

Theorem 13.4 *In (1)–(4) the parameters m_t and γ_t can be defined by the recursive equations*[6]

$$\begin{aligned} m_{t+1} &= [a_0 + a_1 m_t] \\ &\quad + [b \circ B + a_1 \gamma_t A_1^*][B \circ B + A_1 \gamma_t A_1^*]^+ [\xi_{t+1} - A_0 - A_1 m_t], \end{aligned} \quad (13.56)$$

$$\begin{aligned} \gamma_{t+1} &= [a_1 \gamma_t a_1^* + b \circ b] \\ &\quad - [b \circ B + a_1 \gamma_t A_1^*][B \circ B + A_1 \gamma_t A_1^*]^+ [b \circ B + a_1 \gamma_t A_1^*]^*. \end{aligned} \quad (13.57)$$

PROOF. Let us find first the parameters of the conditional Gaussian distribution

$$P(\theta_{t+1} \le a, \xi_{t+1} \le x | \mathcal{F}_t^\xi) = M[P(\theta_{t+1} \le a, \xi_{t+1} \le x | \theta_t, \mathcal{F}_t^\xi) | \mathcal{F}_t^\xi].$$

Due to (13.50),

$$\begin{aligned} M(\theta_{t+1} | \mathcal{F}_t^\xi) &= a_0(t,\xi) + a_1(t,\xi) m_t, \\ M(\xi_{t+1} | \mathcal{F}_t^\xi) &= A_0(t,\xi) + A_1(t,\xi) m_t. \end{aligned} \quad (13.58)$$

In order to find the matrices of covariances, let us take advantage of the fact that, according to (13.56)–(13.58),

$$\begin{aligned} \theta_{t+1} - M(\theta_{t+1} | \mathcal{F}_t^\xi) &= a_1(t,\xi)[\theta_t - m_t] \\ &\quad + b_1(t,\xi)\varepsilon_1(t+1) + b_2(t,\xi)\varepsilon_2(t+1), \\ \xi_{t+1} - M(\xi_{t+1} | \mathcal{F}_t^\xi) &= A_1(t,\xi)[\theta_t - m_t] \\ &\quad + B_1(t,\xi)\varepsilon_1(t+1) + B_2(t,\xi)\varepsilon_2(t+1). \end{aligned} \quad (13.59)$$

We obtain from this

$$\begin{aligned} d_{11} &= \operatorname{cov}(\theta_{t+1}, \theta_{t+1} | \mathcal{F}_t^\xi) - a_1(t,\xi)\gamma_t a_1^*(t,\xi) + (b \circ b)(t,\xi), \\ d_{12} &= \operatorname{cov}(\theta_{t+1}, \xi_{t+1} | \mathcal{F}_t^\xi) = a_1(t,\xi)\gamma_t A_1^*(t,\xi) + (b \circ B)(t,\xi), \\ d_{22} &= \operatorname{cov}(\xi_{t+1}, \xi_{t+1} | \mathcal{F}_t^\xi) = A_1(t,\xi)\gamma_t A_1^*(t,\xi) + (B \circ B)(t,\xi). \end{aligned}$$

Since the conditional (under the condition $\mathcal{F}_t^\xi$) distribution of the vector $(\theta_{t+1}, \xi_{t+1})$ is normal, by virtue of the theorem on normal correlation (and its accompanying note)

$$M(\theta_{t+1} | \mathcal{F}_t^\xi, \xi_t + 1) = M(\theta_{t+1} | \mathcal{F}_t^\xi) + d_{12} d_{22}^+ (\xi_{t+1} - M(\xi_{t+1} | \mathcal{F}_t^\xi)) \quad (13.60)$$

and

[6] In the coefficients $a_0, A_0, \dots, b \circ b$, the arguments (t, ξ) are omitted.

$$\operatorname{cov}(\theta_{t+1}, \theta_{t+1} | \mathcal{F}_t^\xi, \xi_{t+1}) = d_{11} - d_{12} d_{22}^+ d_{12}^*. \tag{13.61}$$

Substituting here the expressions for $M(\theta_{t+1}|\mathcal{F}_t^\xi)$, $M(\xi_{t+1}|\mathcal{F}_t^\xi)$, d_{11}, d_{12} and d_{22}, we obtain recursive equations (13.56) and (13.57) from (13.58) and (13.59). □

Corollary 1. *Let*

$$\begin{aligned} a_0(t,\xi) &= a_0(t) + a_2(t)\xi_t, & A_0(t,\xi) &= A_0(t) + A_2(t)\xi_t, \\ a_1(t,\xi) &= a_1(t), & A_1(t,\xi) &= A_1(t), \\ b_i(t,\xi) &= b_i(t), & B_i(t,\xi) &= B_i(t), \quad i = 1,2, \end{aligned}$$

where all the functions $a_j(t), A_j(t), b_i(t), B_i(t)$, $j = 0,1,2$, and $i = 1,2$, are functions only of t. If the vector (θ_0, ξ_0) is Gaussian, then the process (θ_t, ξ_t) $t = 0,1,2,\ldots$, will also be Gaussian. In this case the covariance γ_t does not depend on 'chance' and, therefore, $\operatorname{Tr}\gamma_t$ determines the mean square estimation error corresponding to θ_t based on the observations $\xi_0^t = (\xi_0, \ldots, \xi_t)$.

Corollary 2. *Let a partially observable sequence $(\theta, \xi) = (\theta_t, \xi_t)$, $t = 0,1,\ldots$, satisfy for $t \geq 1$ the system of equations*

$$\theta_{t+1} = a_0(t,\xi) + a_1(t,\xi)\theta_t + b_1(t,\xi)\varepsilon_1(t+1) + b_2(t,\xi)\varepsilon_2(t+1), \tag{13.62}$$

$$\xi_t = \tilde{A}_0(t-1,\xi) + \tilde{A}_1(t-1,\xi)\theta_t + \tilde{B}_1(t-1,\xi)\varepsilon_1(t) + \tilde{B}_2(t-1,\xi)\varepsilon_2(t), \tag{13.63}$$

with $P(\theta_1 \leq a|\xi_1) \sim N(m_1, \gamma_1)$.

Although the system of equations for θ_{t+1} and ξ_t, considered in a formal way, does not fit the scheme of (13.46) and (13.47), nevertheless, in finding equations for $m_t = M(\theta_t|\mathcal{F}_t^\xi)$ and $\gamma_t = \operatorname{cov}(\theta_t, \theta_t|\mathcal{F}_t^\xi)$, one can take advantage of the results of Theorem 13.4. Indeed, we find from (13.62) and (13.63) that

$$\begin{aligned} \xi_{t+1} = {} & \tilde{A}_0(t,\xi) + \tilde{A}_1(t,\xi)[a_0(t,\xi) + a_1(t,\xi)\theta_t + b_1(t,\xi)\varepsilon_1(t+1) \\ & + b_2(t,\xi)\varepsilon_2(t+1)] + \tilde{B}_1(t,\xi)\varepsilon_1(t+1) + \tilde{B}_2(t,\xi)\varepsilon_2(t+1). \end{aligned}$$

Setting

$$\begin{aligned} A_0 &= \tilde{A}_0 + \tilde{A}_1 a_0, & A_1 &= \tilde{A}_1 a_1, \\ B_1 &= \tilde{A}_1 b_1 + \tilde{B}_1, & B_2 &= \tilde{A}_1 b_2 + \tilde{B}_2, \end{aligned} \tag{13.64}$$

we note that the sequence (θ, ξ) satisfies Equations (13.46) and (13.47), and m_t and γ_t satisfy Equations (13.56) and (13.57).

Corollary 3 (Kalman–Bucy Filter). *Let the Gaussian sequence (θ, ξ) satisfy the equations*

$$\theta_{t+1} = a_0(t) + a_1(t)\theta_t + b_1(t)\varepsilon_1(t+1) + b_2(t)\varepsilon_2(t+1), \tag{13.65}$$

$$\xi_t = A_0(t) + A_1(t)\theta_t + B_1(t)\varepsilon_1(t) + B_2(t)\varepsilon_2(t). \tag{13.66}$$

Then, due to (13.56) and (13.57) and the previous corollary, m_t and γ_t satisfy the system of equations

$$\begin{aligned} m_{t+1} &= [a_0(t) + a_1(t)m_t] + P_\gamma(t)Q_\gamma^+(t) \\ &\quad \times[\xi_{t+1} - A_0(t+1) - A_1(t+1)a_0(t) - A_1(t+1)a_1(t)m_t], \end{aligned} \tag{13.67}$$

$$\gamma_{t+1} = [a_1(t)\gamma_t a_1^*(t) + b\circ b(t)] - P_\gamma(t)Q_\gamma^+(t)P_\gamma^*(t), \tag{13.68}$$

where

$$\begin{aligned} P_\gamma(t) &= b_1(t)[A_1(t+1)b_1(t) + B_2(t+1)]^* + b_2(t)[A_1(t+1)b_2(t) \\ &\quad +B_2(t+1)]^* + a_1(t)\gamma_t a_1^*(t)A_1^*(t+1), \end{aligned} \tag{13.69}$$

$$\begin{aligned} Q_\gamma(t) &= [A_1(t+1)b_1(t) + B_1(t+1)][A_1(t+1)b_1(t) + B_1(t+1)]^* \\ &\quad +[A_1(t+1)b_2(t) + B_2(t+1)][A_1(t+1)b_2(t) + B_2(t+1)]^* \\ &\quad +A_1(t+1)a_1(t)\gamma_t a_1^*(t)A_1^*(t+1). \end{aligned} \tag{13.70}$$

With the help of the theorem on normal correlation we obtain the following expressions for $m_0 = M(\theta_0|\xi_0)$ and $\gamma_0 = \text{cov}\,(\theta_0, \theta_0|\xi_0)$:

$$\begin{aligned} m_0 &= M\theta_0 + \text{cov}\,(\theta_0,\theta_0)A_1^*(0)[A_1(0)\text{cov}\,(\theta_0,\theta_0)A_1^*(0) + B\circ B(0)]^+ \\ &\quad \times[\xi_0 - A_0(0) - A_1(0)M\theta_0], \end{aligned} \tag{13.71}$$

$$\begin{aligned} \gamma_0 &= \text{cov}\,(\theta_0,\theta_0) - \text{cov}(\theta_0,\theta_0)A_1^*(0)[A_1(0)\text{cov}(\theta_0,\theta_0)A_1^*(\theta) \\ &\quad +B\circ B(0)]^+A_1(0)\text{cov}\,(\theta_0,\theta_0). \end{aligned} \tag{13.72}$$

Note. In the assumptions of the theorem, the conditional distribution $P(\theta_t \le b|\mathcal{F}_t^\xi, \theta_s = a)$, $t \ge s$, is also Gaussian and its parameters $m_a(t,s) = M(\theta_t|\mathcal{F}_t^\xi, \theta_s = a)$ and $\gamma_a(t,s) = \text{cov}\,(\theta_t,\theta_t|\mathcal{F}_t^\xi, \theta_s = a)$ satisfy, for $t \ge s$ the system of equations

$$\begin{aligned} m_a(t+1,s) &= [a_0(t,\xi) + a_1(t,\xi)m_a(t,s)] \\ &\quad +[b\circ B(t,\xi) + a_1(t,\xi)\gamma_a(t,s)A_1^*(t,\xi)] \\ &\quad \times[B\circ B(t,\xi) + A_1(t,\xi)\gamma_a(t,s)A_1^*(t,\xi)]^+ \\ &\quad \times[\xi_{t+1} - A_0(t,\xi) - A_1(t,\xi)m_a(t,s)], \end{aligned} \tag{13.73}$$

$$\begin{aligned} \gamma_a(t+1,s) &= [a_1(t,\xi)\gamma_a(t,s)a_1^*(t,\xi) + b\circ b(t,\xi)] \\ &\quad -[b\circ B(t,\xi) + a_1(t,\xi)\gamma_a(t,s)A_1^*(t,\xi)] \\ &\quad \times[B\circ B(t,\xi) + A_1(t,\xi)\gamma_a(t,s)A_1^*(t,\xi)]^+ \\ &\quad \times[b\circ B(t,\xi) + a_1(t,\xi)\gamma_a(t,s)A_1^*(t,\xi)]^*, \end{aligned} \tag{13.74}$$

with $m_a(s,s) = a$, $\gamma_a(s,s) = 0$.

It follows from (13.74) that $\gamma_a(t,s)$, for $t \ge s$, does not depend on a.

13.2.4. Notice a number of useful properties of the processes m_t and γ_t, $t = 0, 1, \ldots$, assuming the conditions of Theorem 13.4 to be satisfied.

Property 1. For any $t = 0, 1, \ldots$, the values of m_t and $(\theta_t - m_t)$ are uncorrelated, i.e.,

$$M\{m_t^*(\theta_t - m_t)\} = M\{(\theta_t - m_t)^* m_t\} = 0,$$

and, therefore,

$$M\theta_t^*\theta_t = Mm_t^* m_t + M(\theta_t - m_t)^*(\theta_t - m_t). \tag{13.75}$$

Property 2. The conditional covariance γ_t does not depend explicitly on the coefficients $a_0(t, \xi)$ and $A_0(t, \xi)$.

Property 3. Let γ_0 and all the coefficients of the system of equations given by (13.46) and (13.47), possibly with the exception of the coefficients $a_0(t, \xi)$ and $A_0(t, \xi)$, be independent of ξ. Then the conditional covariance γ_t is a function of time t alone and $\gamma_t = M\{(\theta_t - m_t)(\theta_t - m_t)^*\}$. In this case the distribution of the value of $\Delta_t = \theta_t - m_t$ is normal, $N(0, \gamma_t)$.

Property 4. The estimate of m_t is unbiased:

$$Mm_t = M\theta_t, \quad t = 0, 1, \ldots. \tag{13.76}$$

13.2.5. In the following theorem a special representation is given for the sequence ξ_t, $t = 0, 1, \ldots$, (compare with Theorem 7.12), which will be used frequently further on.

Theorem 13.5. *Let (1)–(4) be satisfied. Then there exist Gaussian vectors* $\bar{\varepsilon}(t) = (\bar{\varepsilon}_1, \ldots, \bar{\varepsilon}_l(t))$ *with independent coordinates and with*

$$M\bar{\varepsilon}(t) = 0, \quad M\bar{\varepsilon}(t)\bar{\varepsilon}^*(s) = \delta(t - s)E_{(l\times l)} \tag{13.77}$$

such that (P-a.s.)

$$\begin{aligned}\xi_{t+1} &= A_0(t, \xi) + A_1(t, \xi)m_t \\ &\quad + [(B \circ B)(t, \xi) + A_1(t, \xi)\gamma_t A_1^*(t, \xi)]^{1/2}\bar{\varepsilon}(t + 1).\end{aligned} \tag{13.78}$$

If, in addition, the matrices $(B \circ B)(t, \xi) + A_1(t, \xi)\gamma_t A_1^*(t, \xi)$ *are nonsingular (P-a.s.),* $t = 0, \ldots,$ *then*[7]

$$\mathcal{F}_t^{\xi} = \mathcal{F}_t^{(\xi_0, \bar{\varepsilon})}, \quad t = 1, 2, \ldots. \tag{13.79}$$

PROOF. Assume first that for all $t = 0, 1, \ldots$, the matrices $(B \circ B)(t, \xi)$ are positive definite. Then, since the matrices $A_1(t, \xi)\gamma_t A_1^*(t, \xi)$ are at least nonnegative definite, the matrices $[(B \circ B)(t, \xi) + A_1(t, \xi) \times \gamma_t A_1^*(t, \xi)]^{1/2}$ are positive definite and, therefore, the following is a random vector

[7] All the σ-algebras considered here are assumed to be augmented by sets of $\mathcal{F}$-measure zero.

$$\begin{aligned}\overline{\varepsilon}(t+1) &= [(B\circ B)(t,\xi)+A_1(t,\xi)\gamma_t A_1^*(t,\xi)]^{-1/2}[A_1(t,\xi)(\theta_t-m_t)\\&\quad+B_1(t,\xi)\varepsilon_1(t+1)+B_2(t,\xi)\varepsilon_2(t+1)].\end{aligned} \tag{13.80}$$

The conditional (conditioned on $\mathcal{F}_t^\xi$) distribution of the vector θ_t is Gaussian according to Theorem 13.3, and the random vectors $\varepsilon_1(t+1)$ and $\varepsilon_2(t+1)$ do not depend on $\xi_0^t = (\xi_0,\dots,\xi_t)$. Hence it follows from (13.80) that the conditional distribution $P(\overline{\varepsilon}(t+1)\le x|\mathcal{F}_t^\xi)$ is Gaussian and it is not difficult to compute

$$M[\overline{\varepsilon}(t+1)|\mathcal{F}_t^\xi]=0, \tag{13.81}$$

$$\operatorname{cov}(\overline{\varepsilon}(t+1),\overline{\varepsilon}(t+1)|\mathcal{F}_t^\xi)=E_{(l\times l)}. \tag{13.82}$$

From this it is seen that the parameters of the conditional distribution of the vector $\overline{\varepsilon}(t+1)$ do not depend on the condition and, therefore, the (unconditional) distribution of the vector $\overline{\varepsilon}(t+1)$ is also Gaussian. Here

$$M\overline{\varepsilon}(t+1)=0,\quad \operatorname{cov}(\overline{\varepsilon}(t+1),\overline{\varepsilon}(t+1))=E_{(l\times l)}.$$

In a similar way it can be shown, employing Theorem 13.3, that at any t the joint distribution of the vectors $(\overline{\varepsilon}(1),\dots,\overline{\varepsilon}(t))$ is also Gaussian with $\operatorname{cov}(\overline{\varepsilon}(u),\overline{\varepsilon}(v))=\delta(u-v)E$. From this follows the independence of the vectors $\overline{\varepsilon}(1),\dots,\overline{\varepsilon}(t)$. The required representation, (13.78), follows explicitly from (13.80) and (13.47).

In order to prove (13.79) note first of all that, according to (13.78),

$$\mathcal{F}_t^\xi\subseteq\mathcal{F}_t^{(\xi_0,\overline{\varepsilon})}. \tag{13.83}$$

If the matrix $(B\circ B)(t,\xi)+A_1(t,\xi)\gamma_t A_1^*(t,\xi)$ is nonsingular, then, due to (13.78),

$$\begin{aligned}\overline{\varepsilon}(t) &= [(B\circ B)(t-1,\xi)+A_1(t-1,\xi)\gamma_{t-1}A_1^*(t-1,\xi)]^{-1/2}\\&\quad\times[\xi_t-A_0(t-1,\xi)-A_1(t-1,\xi)m_{t-1}].\end{aligned}$$

Hence $\mathcal{F}_t^\xi\supseteq\mathcal{F}_t^{(\xi_0,\overline{\varepsilon})}$, which, together with (13.83), proves the coincidence of the σ-algebras $\mathcal{F}_t^\xi$ and $\mathcal{F}_t^{(\xi_0,\overline{\varepsilon})}$, $t=1,2,\dots$.

Assume now that at some time t the matrix $(B\circ B)(t,\xi)+A_1(t,\xi)\gamma_t A_1^*(t,\xi)$ is singular (with positive probability).

Let us construct (at the expense of extending the main probability space) a sequence of independent Gaussian random vectors $z(t)=(z_1(t),\dots,z_l(t))$, $Mz(t)=0$, $Mz(t)z^*(t)=E_{(l\times l)}$, independent of the processes $\varepsilon_1(t),\varepsilon_2(t)$, $t\ge 0$, and the vectors (θ_0,ξ_0) as well. Set

$$\begin{aligned}\overline{\varepsilon}(t+1) &= D^+(t,\xi)[A_1(t,\xi)(\theta_t-m_t)+B_1(t,\xi)\varepsilon_1(t+1)\\&\quad+B_2(t,\xi)\varepsilon_2(t+1)]+(E-D^+(t,\xi)D(t,\xi))z(t+1),\end{aligned} \tag{13.84}$$

where $D(t,\xi)=[(B\circ B)(t,\xi)+A_1(t,\xi)\gamma_t A_1^*(t,\xi)]^{1/2}$. It is easy to convince oneself that the sequence $\overline{\varepsilon}(1),\overline{\varepsilon}(2),\dots$, of the vectors thus defined has the properties given in the formulation of the theorem. In order to prove (13.78) it obviously suffices to show that

$$D(t,\xi)\overline{\varepsilon}(t+1) = A_1(t,\xi)[\theta_t - m_t] + B_1(t,\xi)\varepsilon_1(t+1) + B_2(t,\xi)\varepsilon_2(t+1). \quad (13.85)$$

Multiplying the left- and right-hand sides in (13.84) by $D(t,\xi)$, we obtain

$$\begin{aligned} D(t,\xi)\overline{\varepsilon}(t+1) &= [A_1(t,\xi)(\theta_t - m_t) + B_1(t,\xi)\varepsilon_1(t+1) + B_2(t,\xi)\varepsilon_2(t+1)] \\ &\quad -[E - D(t,\xi)D^+(t,\xi)][A_1(t,\xi)(\theta_t - m_t) \\ &\quad +B_1(t,\xi)\varepsilon_1(t+1) + B_2(t,\xi)\varepsilon_2(t+1)] \\ &\quad +D(t,\xi)[E - D^+(t,\xi)D(t,\xi)]z(t+1). \end{aligned} \quad (13.86)$$

By the first property of pseudo-inverse matrices, $D[E - D^+D] = D - DD^+D = 0$, and, therefore, ($P$-a.s.)

$$D(t,\xi)[E - D^+(t,\xi)D(t,\xi)]z(t+1) = 0. \quad (13.87)$$

Write

$$\begin{aligned} \zeta(t+1) &= [E - D(t,\xi)D^+(t,\xi)][A_1(t,\xi)(\theta_t - m_t) + B_1(t,\xi)\varepsilon_1(t+1) \\ &\quad +B_2(t,\xi)\varepsilon_2(t+1)]. \end{aligned}$$

Then

$$\begin{aligned} M\zeta(t+1)\zeta^*(t+1) &= M\{M(\zeta(t+1)\zeta^*(t+1)|\mathcal{F}_t^\xi)\} \\ &= M\{(E - DD^+)DD^*(E - DD^+\} \\ &= M\{(DD^* - DD^+DD^*)(E - DD^+)\} \\ &= M[(DD^* - DD^*)(E - DD^+)] = 0. \end{aligned}$$

Consequently, $\zeta(t+1) = 0$ (P-a.s.), which, together with (13.86) and (13.87), proves (13.85). □

Note. When the matrices $B \circ B(t,\xi) + A_1(t,\xi)$ and $\gamma_t A_1^*(t,\xi)$, $t \geq 0$, are nonsingular:

$$\mathcal{F}_t^\xi = \mathcal{F}_t^{\xi_0,\overline{\varepsilon}}, \quad t = 1, 2, \ldots;$$

hence, the sequence $\overline{\varepsilon} = (\overline{\varepsilon}(1), \overline{\varepsilon}(2), \ldots)$ (by analogy with the definition given in Subsection 7.4.2) is naturally called an *innovation sequence.*

13.3 Forward and Backward Interpolation Equations

13.3.1. For the random sequence $(\theta,\xi)=(\theta_t,\xi_t)$, $t=0,1,\ldots$, governed by Equations (13.46) and (13.47), interpolation is understood as a problem of constructing an optimal (in the mean square sense) estimate of the vector θ_s from the observations $\xi_0^t=\{\xi_0,\ldots,\xi_t\}$, $t\geq s$.

For $t\geq s$, let

$$m(s,t)=M(\theta_s|\mathcal{F}_t^\xi),\quad \gamma(s,t)=\operatorname{cov}(\theta_s,\theta_s|\mathcal{F}_t^\xi)$$

denote (respectively) the vector of mean values and the matrix of covariances of the conditional distribution $\Pi_a(s,t)=P(\theta_s\leq a|\mathcal{F}_t^\xi)$. It is seen that $m(s,t)$ is an optimal estimate of θ_s from ξ_0^t. For this estimation both forward equations (over t at fixed s) and backward equations (over s at fixed t) can be deduced. The forward equations demonstrate how much the interpolation improves with the increase of the data, i.e., with the increase of t. The backward equations are of interest in those statistical problems where the vector $\xi_0^t=\{\xi_0,\ldots,\xi_t\}$ is known and by means of which the unobservable component θ_s for all $s=0,\ldots,t$ has to be estimated. The backward equations provide a convenient recursive technique for calculating the estimates $m(t-1,t)$ from $m(t,t)=m_t$ and ξ_t, $m(t-2,t)$ from $m(t-1,t)$, $m(t,t)$, ξ_{t-1}, and ξ_t, etc.

13.3.2. (1)–(3) in Section 13.2 will be assumed to be satisfied

For the deduction of forward equations of interpolation the following theorem is useful.

Theorem 13.6. *If the conditional distribution $\Pi_a(s,s)=P(\theta_s\leq a|\mathcal{F}_s^\xi)$ is normal (P-a.s.), then the distributions $\Pi_a(s,t)=P(\theta_s\leq a|\mathcal{F}_t^\xi)$ at $t\geq s$ are also normal.*

In order to prove this we shall need:

Lemma 13.5. *If the conditional distribution $\Pi_a(s,s)=P(\theta_s\leq a|\mathcal{F}_s^\xi)$ is normal, then the conditional expectation*

$$m_\alpha(t,s)=M(\theta_t|\mathcal{F}_t^\xi,\theta_s=\alpha),\quad t\geq s,$$

permits the representation

$$m_\alpha(t,s)=\varphi_s^t\alpha+\psi_s^t,\tag{13.88}$$

where the matrices[8]

[8] $\prod_{u=s}^{t-1}A_u$ denotes the product of the matrices $A_{t-1},\ldots,A_s$.

$$\begin{aligned}\varphi_s^s &= E_{(k\times k)},\\ \varphi_s^t &= \prod_{u=s}^{t-1}\Big\{a_1(u,\xi)-[b\circ B(u,\xi)+a_1(u,\xi)\gamma(u,s)A_1^*(u,\xi)]\\ &\quad\times[(B\circ B)(u,\xi)+A_1(u,\xi)\gamma(u,s)A_1^*(u,\xi)]^+A_1(u,\xi)\Big\}\end{aligned} \tag{13.89}$$

and the vectors

$$\begin{aligned}\psi_s^t &= \sum_{u=s}^{t-1}\varphi_u^{t-1}\Big\{a_0(u,\xi)+[(b\circ B)(u,\xi)+a_1(u,\xi)\gamma(u,s)A_1^*(u,\xi)] \\ &\quad\times[(B\circ B)(u,\xi)+A_1(u,\xi)\gamma(u,s)A_1^*(u,\xi)]^+(\xi_{u+1}-A_0(u,\xi))\Big\}\end{aligned} \tag{13.90}$$

do not depend on α. The matrices $\gamma(u,s)$, $u\geq s$, can be defined from the equations

$$\begin{aligned}\gamma(u,s) &= [a_1(u-1,\xi)\gamma(u-1,s)a_1^*(u-1,\xi)+(b\circ b)(u-1,\xi)]\\ &\quad-[(b\circ B)(u-1,\xi)+a_1(u-1,\xi)\gamma(u-1,s)A_1^*(u-1,\xi)]\\ &\quad\times[(B\circ B)(u-1,\xi)+A_1(u-1,\xi)\gamma(u-1,s)A_1^*(u-1,\xi)]^+\\ &\quad\times[(b\circ B)(u-1,\xi)+a_1(u-1,\xi)\gamma(u-1,s)A_1^*(u-1,\xi)]^*\end{aligned} \tag{13.91}$$

with an initial condition $\gamma(s,s)=0$.

PROOF. Note first that the pertinent analog of (13.88) was given in Lemma 12.2 (compare (13.88) with (12.79)).

According to the note to Theorem 13.4, $m_\alpha(t,s)$ and $\gamma_\alpha(t,s)=\operatorname{cov}(\theta_t,\theta_t|\mathcal{F}_t^\xi,\theta=\alpha)$ satisfy Equations (13.73), (13.74) with an initial condition $m_\alpha(s,s)=\alpha$, $\gamma_\alpha(s,s)=0$. Since $\gamma_\alpha(t,s)$ does not depend on α we shall write $\gamma(t,s)=\gamma_\alpha(t,s)$. (13.88) can be deduced from (13.73) by induction. □

PROOF OF THEOREM 13.6. Let us first show that the conditional distribution $P(\theta_s\leq a,\xi_t\leq x|\mathcal{F}_{t-1}^\xi)$ is Gaussian. For this purpose we compute the conditional characteristic function

$$\begin{aligned}&M(\exp i[z_1^*\theta_s+z_2^*\xi_t]|\mathcal{F}_{t-1}^\xi)\\ &= M(\exp i[z_1^*\theta_s]M\{\exp i[z_2^*\xi_t]|\mathcal{F}_{t-1}^\xi,\theta_s\}|\mathcal{F}_{t-1}^\xi).\end{aligned} \tag{13.92}$$

It is obvious that

$$\begin{aligned}M(\exp i[z_2^*\xi_t]|\mathcal{F}_{t-1}^\xi,\theta_{t-1},\theta_s) &= \exp\Big\{iz_2^*(A_0(t-1,\xi)+A_1(t-1,\xi)\theta_{t-1})\\ &\quad-\frac{1}{2}z_2^*(B\circ B)(t-1,\xi)z_2\Big\}.\end{aligned} \tag{13.93}$$

Next,

$$P\{\theta_{t-1} \le b | \mathcal{F}_{t-1}^{\xi}, \theta_s\} \sim N(m_{\theta_s}(t-1,s), \gamma(t-1,s))$$

and, due to (13.93),

$$\begin{aligned}
&M\{M[\exp i(z_2^*\xi_t) | \mathcal{F}_{t-1}^{\xi}, \theta_{t-1}, \theta_s] | \mathcal{F}_{t-1}^{\xi}, \theta_s\} \\
&= \exp\left\{ i[z_2^* A_0(t-1,\xi)] - \frac{1}{2} z_2^*(B \circ B)(t-1,\xi) z_2 \right\} \\
&\times M\{\exp i[z_2^* A_1(t-1,\xi)\theta_{t-1}] | \mathcal{F}_{t-1}^{\xi}, \theta_s\} \\
&= \exp\left\{ i z_2^* A_0(t-1,\xi) - \frac{1}{2} z_2^*(B \circ B)(t-1,\xi) z_2 \right\} \\
&\times \exp\Big\{ i[z_2^* A_1(t-1,\xi) m_{\theta_s}(t-1,s)] \\
&-\frac{1}{2} z_2^* A_1(t-1,\xi)\gamma(t-1,s) A_1^*(t-1,\xi) z_2) \Big\}.
\end{aligned}$$

By Lemma 13.5.

$$m_{\theta_s}(t-1,s) = \varphi_s^{t-1}\theta_s + \psi_s^{t-1} \quad (P\text{-a.s.}).$$

Hence

$$\begin{aligned}
M\{\exp i[z_2^*\xi_t] | \mathcal{F}_{t-1}^{\xi}, \theta_s\} \;=\; & \exp\Big\{ i z_2^*(A_0(t-1,\xi) + A_1(t-1,\xi)\psi_s^{t-1}) \\
& -\frac{1}{2} z_2^*((B \circ B)(t-1,\xi) + A_1(t-1,\xi)\gamma(t-1,s) \\
& \times A_1^*(t-1,\xi)) z_2 + i z_2^* A_1(t-1,\xi)\varphi_s^{t-1}\theta_s \Big\},
\end{aligned}$$

which, together with (13.92), leads to the equality

$$\begin{aligned}
M(\exp i[z_1^*\theta_s + z_2^*\xi_t] | \mathcal{F}_{t-1}^{\xi}) \;=\; & \exp\Big\{ i z_2^*(A_0(t-1,\xi) + A_1(t-1,\xi)\psi_s^{t-1}) \\
& -\frac{1}{2} z_2^*((B \circ B)(t-1,\xi) + A_1(t-1,\xi) \\
& \times \gamma(t-1,s) A_1^*(t-1,\xi)) z_2 \Big\} M\{\exp i[z_1^*\theta_s \\
& + z_2^*(A_1(t-1,\xi)\varphi_s^{t-1}\theta_s)] | \mathcal{F}_{t-1}^{\xi}\}. \qquad (13.94)
\end{aligned}$$

Let $t = s+1$. Since the distribution $\Pi_a(s,s) = P(\theta_s \le a | \mathcal{F}_s^{\xi}) \sim N(m_s, \gamma_s)$, it follows from (13.94) that the distribution $P(\theta_s \le a, \xi_{s+1} \le x | \mathcal{F}_s^{\xi})$ is also Gaussian. It is not difficult to deduce from this that the distribution $\Pi_a(s, s+1)$ is Gaussian. It can be proved by induction from (13.94) that for any $t > s$ the conditional distribution $\Pi_a(s,t)$ is also Gaussian. □

Note. Normality of the conditional distributions $P\{\theta_s \le a | \mathcal{F}_t^{\xi}, \theta_u = b\}$ for $u < s \le t$ can be proved in the same way.

13.3.3. Therefore, according to Theorem 13.6 the distribution $\Pi_a(s,t) = P(\theta_s \le a|\mathcal{F}_t^\xi) \sim N(m(s,t), \gamma(s,t))$, if the distribution $\Pi_a(s,s)$ is Gaussian. Let us find forward equations (of interpolation) for $m(s,t)$ and $\gamma(s,t)$.

Theorem 13.7. *If $\Pi_a(s,s) \sim N(m_s, \gamma_s)$, then $m(s,t)$ and $\gamma(s,t)$ for $t > s$ satisfy the equations*

$$\begin{aligned} m(s,t+1) &= m(s,t) + \gamma(s,t)(\varphi_s^t)^* A_1^*(t,\xi)[(B\circ B)(t,\xi) + A_1(t,\xi)\gamma_t A_1^*(t,\xi)]^+ \\ &\quad \times[\xi_{t+1} - A_0(t,\xi) - A_1(t,\xi)m_t], \end{aligned} \tag{13.95}$$

$$\begin{aligned} \gamma(s,t+1) &= \gamma(s,t) - \gamma(s,t)(\varphi_s^t)^* A_1^*(t,\xi) \\ &\quad \times[(B\circ B)(t,\xi) + A_1(t,\xi)\gamma_t A_1^*(t,\xi)]^+ \\ &\quad \times A_1(t,\xi)\varphi_s^t\gamma(s,t), \end{aligned} \tag{13.96}$$

where $m(t,t) = m_t$, $\gamma(t,t) = \gamma_t$, and the matrices φ_s^t are defined from (13.89).

PROOF. From Theorem 13.6 it follows that the conditional distribution $P(\theta_s \le a, \xi_t \le x|\mathcal{F}_{t-1}^\xi)$ is normal. Parameters of this distribution could be obtained from (13.94), but it is easier to find them by taking advantage of the theorem on normal correlation.

According to the note to this theorem,

$$M(\theta_s|\xi_t, \mathcal{F}_{t-1}^\xi) = M(\theta_s|\mathcal{F}_{t-1}^\xi) + d_{12}d_{22}^+[\xi_t - M(\xi_t|\mathcal{F}_{t-1}^\xi)], \tag{13.97}$$

where

$$d_{12} = \operatorname{cov}(\theta_s, \xi_t|\mathcal{F}_{t-1}^\xi), \tag{13.98}$$

$$d_{22} = \operatorname{cov}(\xi_t, \xi_t|\mathcal{F}_{t-1}^\xi) = A_1(t-1,\xi)\gamma_{t-1}A_1^*(t-1,\xi) + (B\circ B)(t-1,\xi). \tag{13.99}$$

In order to find d_{12}, note that, due to Lemma 13.5.

$$\begin{aligned} m_{t-1} &= M(\theta_{t-1}|\mathcal{F}_{t-1}^\xi) = M[M(\theta_{t-1}|\mathcal{F}_{t-1}^\xi, \theta_s)|\mathcal{F}_{t-1}^\xi] \\ &= M[\varphi_s^{t-1}\theta_s + \psi_s^{t-1}|\mathcal{F}_{t-1}^\xi] \\ &= \varphi_s^{t-1}m(s,t-1) + \psi_s^{t-1}. \end{aligned} \tag{13.100}$$

Next,

$$\begin{aligned} M[\theta_{t-1} - m_{t-1}|\mathcal{F}_{t-1}^\xi, \theta_s] &= \varphi_s^{t-1}\theta_s + \psi_s^{t-1} - [\varphi_s^{t-1}m(s,t-1) + \psi_s^{t-1}] \\ &= \varphi_s^{t-1}[\theta_s - m(s,t-1)], \end{aligned} \tag{13.101}$$

$$M[\xi_t|\mathcal{F}_{t-1}^\xi] = A_0(t-1,\xi) + A_1(t-1,\xi)m_{t-1}, \tag{13.102}$$

and, by Lemma 13.5,

$$\begin{aligned}
&M\{[\xi - M(\xi_t|\mathcal{F}_{t-1}^{\xi})]^*|\mathcal{F}_{t-1}^{\xi},\theta_s\}\\
&= M\{[A_1(t-1,\xi)(\theta_{t-1}-m_{t-1})\\
&\quad +B_1(t-1,\xi)\varepsilon_1(t)+B_2(t-1,\xi)\varepsilon_2(t)]^*|\mathcal{F}_{t-1}^{\xi},\theta_s\}\\
&= M\{[A_1(t-1,\xi)(\theta_{t-1},-m_{t-1})]^*|\mathcal{F}_{t-1}^{\xi},\theta_s\}\\
&= [\theta_s - m(s,t-1)]^*(\varphi_s^{t-1})^*A_1^*(t-1,\xi).
\end{aligned}\tag{13.103}$$

Hence, from (13.100)–(13.103), we find that

$$\begin{aligned}
d_{12} &= \mathrm{cov}\,(\theta_s,\xi_t|\mathcal{F}_{t-1}^{\xi})\\
&= M\{[\theta_s - m(s,t-1)][\xi_t - M(\xi_t|\mathcal{F}_{t-1}^{\xi})]^*|\mathcal{F}_{t-1}^{\xi}\}\\
&= M\{[\theta_s - m(s,t-1)][\theta_s - m(s,t-1)]^*(\varphi_s^{t-1})^*A_1^*(t-1,\xi)|\mathcal{F}_{t-1}^{\xi}\}\\
&= \gamma(s,t-1)(\varphi_s^{t-1})^*A_1^*(t-1,\xi).
\end{aligned}\tag{13.104}$$

We obtain (13.95) from (13.97), (13.98), (13.102) and (13.104).

In order to deduce Equation (13.96), it should be noted that, according to the note to the theorem on normal correlation

$$\gamma(s,t) = \mathrm{cov}\,(\theta_s,\theta_s|\mathcal{F}_{t-1}^{\xi},\xi_t) = d_{11} - d_{12}d_{22}^{+}d_{12}^{+},\tag{13.105}$$

where

$$d_{11} = \mathrm{cov}\,(\theta_s,\theta_s|\mathcal{F}_{t-1}^{\xi}) = \gamma(s,t-1).\tag{13.106}$$

We obtain the required equation, (13.96), for $\gamma(s,t)$ from (13.105), (13.106), (13.104) and (13.99). □

13.3.4.

Theorem 13.8. *If the matrices $(B\circ B)(u,\xi)$, $u=0,1,\ldots$, are nonsingular, then solutions $m(s,t)$ and $\gamma(s,t)$ of Equations (13.95) and (13.96) are given by the formulae*

$$\begin{aligned}
m(s,t) &= \Bigg[E + \gamma_s\sum_{u=s}^{t-1}(\varphi_s^u)^*A_1^*(u,\xi)((B\circ B)(u,\xi)\\
&\quad +A_1(u,\xi)\gamma(u,s)A_1^*(u,\xi))^{-1}A_1(u,\xi)\varphi_s^u\Bigg]^{-1}\\
&\quad\times\Bigg[m_s + \gamma_s\sum_{u=s}^{t-1}(\varphi_s^u)^*A_1^*(u,\xi)((B\circ B)(u,\xi)+A_1(u,\xi)\gamma(u,s)\\
&\quad\times A_1^*(u,\xi))^{-1}(\xi_{u+1}-A_0(u,\xi)-A_1(u,\xi)\psi_s^u)\Bigg],
\end{aligned}\tag{13.107}$$

$$\gamma(s,t) = \left[E + \gamma_s \sum_{u=s}^{t-1} (\varphi_s^u)^* A_1^*(u,\xi)((B \circ B)(u,\xi) + A_1(u,\xi)\gamma(u,s)A_1^*(u,\xi))^{-1} A_1(u,\xi)\varphi_s^u\right]^{-1} \gamma_s, \quad (13.108)$$

where φ_s^u, ψ_s^u and $\gamma(u,s)$ are defined by (13.89), (13.90) and (13.91).

PROOF. Let us show first that at all $t > s$,

$$\gamma_{t-1} = \gamma(t-1,s) + \varphi_s^{t-1}\gamma(s,t-1)(\varphi_s^{t-1})^*. \quad (13.109)$$

Indeed,

$$\begin{aligned}
\gamma_{t-1} &= \operatorname{cov}(\theta_{t-1}, \theta_{t-1} | \mathcal{F}_{t-1}^{\xi}) = M\{[\theta_{t-1} - m_{t-1}][\theta_{t-1} - m_{t-1}]^* | \mathcal{F}_{t-1}^{\xi}\} \\
&= M\{[\theta_{t-1} - m_{\theta_s}(t-1,s) + m_{\theta_s}(t-1,s) - m_{t-1}] \\
&\quad \times [\theta_{t-1} - m_{\theta_s}(t-1,s) + m_{\theta_s}(t-1,s) - m_{t-1}]^* | \mathcal{F}_{t-1}^{\xi}\} \\
&= M\{M[(\theta_{t-1} - m_{\theta_s}(t-1,s))(\theta_{t-1} - m_{\theta_s}(t-1,s))^* | \mathcal{F}_{t-1}^{\xi}, \theta_s] | \mathcal{F}_{t-1}^{\xi}\} \\
&\quad + M\{(m_{\theta_s}(t-1,s) - m_{t-1})(m_{\theta_s}(t-1,s) - m_{t-1})^* | \mathcal{F}_{t-1}^{\xi}\} \\
&= M\{\gamma(t-1,s) | \mathcal{F}_{t-1}^{\xi})\} + M\{\varphi_s^{t-1}(\theta_s - m(s,t-1)) \\
&\quad \times (\theta_s - m(s,t-1))^* (\varphi_s^{t-1})^* | \mathcal{F}_{t-1}^{\xi}\} \\
&= \gamma(t-1,s) + \varphi_s^{t-1}\gamma(s,t-1)(\varphi_s^{t-1})^*,
\end{aligned}$$

where (13.100) is used:

$$m_{t-1} = \varphi_s^{t-1} m(s,t-1) + \psi_s^{t-1}.$$

We obtain from (13.96) and (13.109)

$$\begin{aligned}
\gamma(s,t) &= \gamma(s,t-1) - \gamma(s,t-1)(\varphi_s^{t-1}(\xi))^* A_1^*(t-1,\xi) \\
&\quad \times [(B \circ B)(t-1,\xi) + A_1(t-1,\xi)\gamma(t-1,s)A_1^*(t-1,\xi) \\
&\quad + A_1(t-1,\xi)\varphi_s^{t-1}\gamma(s,t-1)(\varphi_s^{t-1})^* A_1^*(t-1,\xi)]^{-1} \\
&\quad \times A_1(t-1,\xi)\varphi_s^{t-1}\gamma(s,t-1).
\end{aligned} \quad (13.110)$$

For $t > s$, define

$$\tilde{A}_1(t-1,\xi) = A_1(t-1,\xi)\varphi_s^{t-1},$$

$$(\widetilde{B \circ B})(t-1,\xi) = (B \circ B)(t-1,\xi) + A_1(t-1,\xi)\gamma(t-1,s)A_1^* t-1,\xi). \quad (13.111)$$

Then $\gamma(s,t)$ will satisfy (over $t > s$) the equation

$$\begin{aligned}
\gamma(s,t) &= \gamma(s,t-1) - \gamma(s,t-1)\tilde{A}_1^*(t-1,\xi) \\
&\quad \times [(\widetilde{B \circ B})(t-1,\xi) + \tilde{A}_1(t-1,\xi)\gamma(s,t-1)\tilde{A}_1^*(t-1,\xi)]^{-1} \\
&\quad \times \tilde{A}_1(t-1,\xi)\gamma(s,t-1).
\end{aligned}$$

Along with (13.111), let $\tilde{A}_0(t-1,\xi) = A_0(t-1,\xi) + A_1(t-1,\xi)\psi_s^{t-1}$. Then Equation (13.95) can be rewritten as follows:

$$\begin{aligned} m(s,t) &= m(s,t-1) + \gamma(s,t-1)\tilde{A}_1^*(t-1,\xi) \\ &\times[\widetilde{(B\circ B)}(t-1,\xi) + \tilde{A}_1(t-1,\xi)\gamma(s,t-1)\tilde{A}_1^*(t-1,\xi)]^{-1} \\ &\times[\xi_t - \tilde{A}_0(t-1,\xi) - \tilde{A}_1(t-1,\xi)m(s,t-1)]. \end{aligned}$$

Solutions of this equation (see also Theorem 13.15) can be defined by (13.107) and (13.108). □

13.3.5. We shall discuss one more class of interpolation problems involving the construction of the optimal (in the mean square sense) estimates of a vector θ_s from the observations $\xi_0^t = \{\xi_0,\dots,\xi_t\}$ and the known value of $\theta_t = \beta$ (compare with Subsection 12.4.6).

Write

$$\Pi_{\alpha\beta}(s,t) = P(\theta_s \le \alpha|\mathcal{F}_t^\xi, \theta_t = \beta), \quad t \ge s,$$

and

$$\tilde{m}_\beta(s,t) = M(\theta_s|\mathcal{F}_t^\xi, \theta_t = \beta), \quad \tilde{\gamma}_\beta(s,t) = \operatorname{cov}(\theta_s,\theta_s|\mathcal{F}_t^\xi, \theta_t = \beta).$$

Theorem 13.9. *If the conditional distribution $\Pi_\alpha(s) = P(\theta_s \le \alpha|\mathcal{F}_s^\xi)$ is normal, then the a posteriori distribution $\Pi_{\alpha\beta}(s,t)$ at all $t \ge s$ is also normal.*

PROOF. Let us calculate the conditional characteristic function

$$M\{\exp i[z^*\theta_s + \tilde{z}^*\theta_t]|\mathcal{F}_t^\xi\} = M\{\exp i[z^*\theta_s]M(\exp i[\tilde{z}\theta_t]|\mathcal{F}_t^\xi,\theta_s)|\mathcal{F}_t^\xi\}.$$

where $z = (z_1,\dots,z_k)$ and $\tilde{z} = (\tilde{z}_1,\dots,\tilde{z}_k)$. According to the note to Theorem 13.4, the distribution $P(\theta_t \le \beta|\theta_s,\mathcal{F}_t^\xi)$ is Gaussian, $N(m_{\theta_s}(t,s),\gamma_{\theta_s}(t,s))$. By Lemma 13.5, $m_{\theta_s}(t,s) = \varphi_s^t\theta_s + \psi_s^t$, and the covariance $\gamma_{\theta_s}(t,s)$ does not depend on θ_s: $\gamma_{\theta_s}(t,s) = \gamma(t,s)$. Hence

$$M\{\exp[i\tilde{z}^*\theta_t]|\mathcal{F}_t^\xi,\theta_s\} = \exp\left[i\tilde{z}^*(\varphi_s^t\theta_s + \psi_s^t) - \frac{1}{2}\tilde{z}^*\gamma(t,s)\tilde{z}\right]$$

and

$$\begin{aligned} M(\exp i[z^*\theta_s + \tilde{z}^*\theta_t]|\mathcal{F}_t^\xi) &= \exp\left[i(\tilde{z}^*\psi_s^t) - \frac{1}{2}\tilde{z}^*\gamma(t,s)\tilde{z}\right] \\ &\times M(\exp i[z^*\theta_s + \tilde{z}^*\varphi_s^t\theta_s]|\mathcal{F}_t^\xi). \quad (13.112) \end{aligned}$$

However, the conditional distribution $P(\theta_s \le \alpha|\mathcal{F}_t^\xi)$ is Gaussian (Theorem 13.6). Hence, it follows from (13.112) that the distribution $P(\theta_s \le \alpha, \theta_t \le \beta|\mathcal{F}_t^\xi)$ will also be Gaussian; this, along with the normality of the distribution $P(\theta_t \le \beta|\mathcal{F}_t^\xi)$ (see the note to Theorem 13.3) proves the normality of the a posteriori distribution $\Pi_{\alpha\beta}(s,t) = P(\theta_s \le \alpha|\mathcal{F}_t^\xi, \theta_t = \beta)$.

13.3.6. The techniques applied in proving Theorem 13.9 enable us to complete the proof of Theorem 13.3.

PROOF OF THEOREM 13.3. We have

$$
\begin{aligned}
& M\left(\exp i\left[\sum_{s=0}^{t} z_s^*\theta_s\right]\Big|\mathcal{F}_t^\xi\right) \\
&= M\left\{\left(\exp i\left[\sum_{s=0}^{t-1} z_s^*\theta_s\right]\right) M(\exp[iz_t^*\theta_t]|\mathcal{F}_t^\xi,\theta_0,\dots,\theta_{t-1})\Big|\mathcal{F}_t^\xi\right\} \\
&= M\left\{\left(\exp i\left[\sum_{s=0}^{t-1} z_s^*\theta_s\right]\right) M(\exp[iz^*\theta_t]|\mathcal{F}_t^\xi,\theta_{t-1})\Big|\mathcal{F}_t^\xi\right\} \\
&= M\left\{\exp i\left[\sum_{s=0}^{t-1} z_s^*\theta_s + z_{t-1}^*\theta_{t-1} + z_t^*(\varphi_{t-1}^t\theta_{t-1}+\psi_{t-1}^t)\right]\Big|\mathcal{F}_t^\xi\right\} \\
&\quad \times \exp\left\{-\frac{1}{2}z_t^*\gamma(t,t-1)z_t\right\} = \exp\left\{i[z_t^*\psi_{t-1}^t] - \frac{1}{2}z_t^*\gamma(t,t-1)z_t\right\} \\
&= M\left\{\left(\exp i\left[\sum_{s=0}^{t-2} z_s^*\theta_s\right]\right)\right. \\
&\quad \left.\times M[\exp i(z_{t-1}+(\varphi_{t-1}^t)^* z_t)^*\theta_{t-1}|\mathcal{F}_t^\xi,\theta_{t-2}\Big|\mathcal{F}_t^\xi\right\}.
\end{aligned}
\tag{13.113}
$$

The distribution $P(\theta_{t-1}\le\beta|\mathcal{F}_t^\xi,\theta_{t-2})$ is normal (see the note to Theorem 13.6); its a posteriori mean depends linearly on θ_{t-2}, and the covariance does not depend on θ_{t-2} at all, since equations analogous to Equations (13.95) and (13.96) hold for them. Hence,

$$
\begin{aligned}
& M\{\exp i[z_{t-1}+(\varphi_{t-1}^t)^* z_t]\theta_{t-1}|\mathcal{F}_t^\xi,\theta_{t-2}\} \\
&= \exp\left[i(z_{t-1}+(\varphi_{t-1}^t)^* z_t)(a(t-1,t-2)\theta_{t-2}+b(t-1,t-2))\right. \\
&\quad \left. -\frac{1}{2}(z_{t-1}+(\varphi_{t-1}^t)^* z_t)^* c(t-1,t-2)(z_{t-1}+(\varphi_t^{t-1})^* z_t)\right],
\end{aligned}
\tag{13.114}
$$

where $a(\cdot), b(\cdot)$ and $c(\cdot)$ are matrix functions (their explicit forms are of no consequence now), dependent only on time and ξ_0^t. It follows from this that θ_{t-2} enters into the exponent of the right-hand side of (13.114) linearly, and the variables z_t, z_{t-1} quadratically.

Therefore,

$$
M\left[\exp\left(i\sum_{s=0}^{t} z_s^*\theta_s\right)\Big|\mathcal{F}_t^\xi\right]
$$

$$= \exp\Big\{ i[z_t^* \psi_{t-1}^t + (z_{t-1} + (\varphi_{t-1}^t)^* z_t)^* b(t-1, t-2)] - \frac{1}{2} z_t^* \gamma(t, t-1) z_t$$
$$- \frac{1}{2}(z_{t-1} + (\varphi_{t-1}^t)^* z_t)^* c(t-1, t-2)(z_{t-1} - (\varphi_{t-1}^t)^* z_t) \Big\}$$
$$\times M \Bigg(\exp i \Bigg[\sum_{s=0}^{t-3} z_s^* \theta_s + \{z_{t-2} + (z_{t-1} + (\varphi_{t-1}^t)^* z_t)$$
$$\times a(t-1, t-2)\} \theta_{t-2} \Bigg] \Bigg| \mathcal{F}_t^\xi \Bigg). \tag{13.115}$$

Extending the techniques of 'splitting off' variables given above we can see that the characteristic function

$$M \left[\exp \left(i \sum_{s=0}^{t} z_s^* \theta_s \right) \Bigg| \mathcal{F}_t^\xi \right]$$

too is (negative) exponential in the nonnegative definite quadratic form of the variables $z_0, \dots, z_t$, which proves the conditional normality of the sequence (θ, ξ) governed by Equations (13.46) and (13.47).

13.3.7. Let us continue our study of the interpolation problem discussed in Subsection 13.3.5.

Theorem 13.10. *If the conditional distribution $\Pi_\alpha(s) = P(\theta_s \le \alpha | \mathcal{F}_s^\xi)$ is normal, then the parameters $\tilde{m}_\beta(s,t)$ and $\tilde{\gamma}_\beta(s,t)$ of the distribution $\Pi_{\alpha,\beta}(s,t) = P(\theta_s \le \alpha | \mathcal{F}_t^\xi, \theta_t = \beta)$ for all $t > s$ can be defined by the relations (compare with (12.109) and (12.110))*

$$\tilde{m}_\beta(s,t) = m(s,t) + \gamma(s,t)(\varphi_s^t)^* \gamma_t^+ (\beta - m_t), \tag{13.116}$$

$$\tilde{\gamma}_\beta(s,t) = \gamma(s,t) - \gamma(s,t)(\varphi_s^t)^* \gamma_t^+ \varphi_s^t \gamma(s,t) \tag{13.117}$$

with $\tilde{m}_\beta(s,s) = \beta$, $\tilde{\gamma}_\beta(s,s) = 0$.

PROOF. The conditional distribution $P(\theta_s \le \alpha, \theta_t \le \beta | \mathcal{F}_t^\xi)$ is normal. Hence, according to the note to the theorem on normal correlation,

$$\tilde{m}_\beta(s,t) = M(\theta_s | \mathcal{F}_t^\xi, \theta_t = \beta) = M(\theta_s | \mathcal{F}_t^\xi) + d_{12} d_{22}^+ (\beta - M(\theta_t | \mathcal{F}_t^\xi)) \tag{13.118}$$

and

$$\tilde{\gamma}_\beta(s,t) = d_{11} - d_{12} d_{22}^+ d_{12}^*, \tag{13.119}$$

where

$$\begin{aligned} d_{11} &= \operatorname{cov}(\theta_s, \theta_s | \mathcal{F}_t^\xi) = \gamma(s,t), \\ d_{12} &= \operatorname{cov}(\theta_s, \theta_t | \mathcal{F}_t^\xi), \\ d_{22} &\quad \operatorname{cov}(\theta_t, \theta_t | \mathcal{F}_t^\xi). \end{aligned} \tag{13.120}$$

According to (13.100) and Lemma 13.5,

$$\begin{aligned} M[(\theta_t - m_t)^*|\mathcal{F}_t^\xi, \theta_s] &= \theta_s^*(\varphi_s^t)^* + (\psi_s^t)^* - (m^*(s,t)(\varphi_s^t)^*) + (\psi_s^t)^* \\ &= (\theta_s - m(s,t))^*(\varphi_s^t)^*. \end{aligned}$$

Hence,

$$\begin{aligned} d_{12} &= \operatorname{cov}(\theta_s, \theta_t|\mathcal{F}_t^\xi) = M[(\theta_s - m(s,t))(\theta_t - m_t)^*|\mathcal{F}_t^\xi] \\ &= M\{(\theta_s - m(s,t))M[(\theta_t - m_t)^*|\mathcal{F}_t^\xi, \theta_s]|\mathcal{F}_t^\xi\} \\ &= d_{11}(\varphi_s^t)^* = \gamma(s,t)(\varphi_s)^*. \end{aligned} \tag{13.121}$$

We obtain (13.116) and (13.117) from (13.118)–(13.121). □

Note. It follows from (13.117) that the covariance $\tilde{\gamma}_\beta(s,t)$ does not depend on β.

13.3.8. We shall deal now with the deduction of backward interpolation equations (over s at fixed t) for $m(s,t)$, $\gamma(s,t)$ and $\tilde{m}_\beta(s,t)$, $\tilde{\gamma}_\beta(s,t)$.

Theorem 13.11. *Let (1)–(4) be satisfied. Then the moments $\tilde{m}_\beta(s,t)$ and $\tilde{\gamma}_\beta(s,t)$ satisfy the equations (over $s < t$)*

$$\begin{aligned} \tilde{m}_\beta(s,t) &= m(s,s+1) + \gamma(s,s+1)(\varphi_s^{s+1})^*\gamma_{s+1}^+ \\ &\quad \times[\tilde{m}_\beta(s+1,t) - m_{s+1}], \end{aligned} \tag{13.122}$$

$$\begin{aligned} \tilde{\gamma}_\beta(s,t) &= \tilde{\gamma}_\beta(s,s+1) + \gamma(s,s+1)(\varphi_s^{s+1})^*\gamma_{s+1}^+\tilde{\gamma}_\beta(s+1,t) \\ &\quad \times\gamma_{s+1}^+\varphi_s^{s+1}\gamma(s,s+1) \end{aligned} \tag{13.123}$$

with $\tilde{m}_\beta(t,t) = \beta$, $\tilde{\gamma}_\beta(t,t) = 0$.

PROOF. We obtain from (13.116) and (13.117) the following:

$$\tilde{m}_\beta(s,s+1) = m(s,s+1) + \gamma(s,s+1)(\varphi_s^{s+1})^*\gamma_{s+1}^+(\beta - m_{s+1}), \tag{13.124}$$

$$\tilde{\gamma}_\beta(s,s+1) = \gamma(s,s+1) - \gamma(s,s+1)(\varphi_s^{s+1})^*\gamma_{s+1}^+\varphi_s^{s+1}\gamma(s,s+1). \tag{13.125}$$

Let us show that, for the process (θ, ξ) governed by Equations (13.46) and (13.47) and for all $s < u \le t$,

$$P(\theta_s \le \alpha|\mathcal{F}_t^\xi, \theta_u, \ldots, \theta_t) = P(\theta_s \le a|\mathcal{F}_u^\xi, \theta_u). \tag{13.126}$$

For this purpose we shall consider the arbitrary measurable bounded functions $f(\theta_s)$, $\chi_{u+1}^t(\theta,\xi)$, $g_0^u(\xi)$, $\lambda(\theta_u)$ of θ_s, $(\theta_{u+1}, \ldots, \theta_t, \xi_{u+1}, \ldots, \xi_t)$, $(\xi_0, \ldots, \xi_u)$, θ_u, respectively, and note that for $s < u$,

$$M\{\chi_{u+1}^t(\theta,\xi)|\mathcal{F}_u^\xi, \theta_s, \ldots, \theta_u\} = M\{\chi_{u+1}^t(\theta,\xi)|\mathcal{F}_u^\xi, \theta_u\}$$

and

$$\begin{aligned}
&M\{\lambda(\theta_u)g_0^u(\xi)M[f(\theta_s)\chi_{u+1}^t(\theta,\xi)|\mathcal{F}_u^\xi,\theta_u]\}\\
&= M\{\lambda(\theta_u)g_0^u(\xi)f(\theta_s)\chi_{u+1}^t(\theta,\xi)\}\\
&= M\{\lambda(\theta_u)g_0^u(\xi)f(\theta_s)M[\chi_{u+1}^t(\theta,\xi)|\mathcal{F}_u^\xi,\theta_s,\dots,\theta_u]\}\\
&= M\{\lambda(\theta_u)g_0^u(\xi)f(\theta_s)M[\chi_{u+1}^t(\theta,\xi)|\mathcal{F}_u^\xi,\theta_u]\}\\
&= M\{\lambda(\theta_u)g_0^u(\xi)M[f(\theta_s)|\mathcal{F}_u^\xi,\theta_u]M[\chi_{u+1}^t(\theta,\xi)|\mathcal{F}_u^\xi,\theta_u]\}.
\end{aligned}$$

Therefore, by virtue of the arbitrariness of the functions $\lambda(\theta_s)$ and $g_0^u(\xi)$,

$$\begin{aligned}
&M[f(\theta_s)|\mathcal{F}_u^\xi,\theta_u]M[\chi_{u+1}^t(\theta,\xi)|\mathcal{F}_u^\xi,\theta_u]\\
&= M[f(\theta_s)\chi_{u+1}^t(\theta,\xi)|\mathcal{F}_u^\xi,\theta_u]\\
&= M\{M[f(\theta_s)\chi_{u+1}^t(\theta,\xi)|\mathcal{F}_t^\xi,\theta_u,\dots,\theta_t]|\mathcal{F}_u^\xi,\theta_u\}\\
&= M\{\chi_{u+1}^t(\theta,\xi)M[f(\theta_s)|\mathcal{F}_t^\xi,\theta_u,\dots,\theta_t]|\mathcal{F}_u^\xi,\theta_u\}.
\end{aligned}$$

Because of the arbitrariness of $\chi_{u+1}^t(\theta,\xi)$, the required equality, (13.126), follows.

Taking into account (13.126), we find that

$$\Pi_{\alpha\beta}(s,t) = M[\Pi_{\alpha,\theta_{s+1}}(s,s+1)|\mathcal{F}_t^\xi,\theta_t=\beta]. \tag{13.127}$$

It follows from this formula that

$$\tilde{m}_\beta(s,t) = M[\tilde{m}_{\theta_{s+1}}(s,s+1)|\mathcal{F}_t^\xi,\theta_t=\beta],$$

which, together with (13.124), leads to Equation (13.122).

We shall employ the following known formula to compute the conditional covariances: if $\xi,\tilde{\xi}$ are random vectors such that $M\xi^*\xi<\infty$, and if $\mathcal{G}$ is a certain σ-algebra, then

$$\operatorname{cov}(\xi,\xi|\mathcal{G}) = M[\operatorname{cov}(\xi,\xi|\mathcal{G},\tilde{\xi})|\mathcal{G}] + \operatorname{cov}[M(\xi|\mathcal{G},\tilde{\xi}),M(\xi|\mathcal{G},\tilde{\xi})|\mathcal{G}]. \tag{13.128}$$

According to this formula and (13.127)

$$\begin{aligned}
\tilde{\gamma}_\beta(s,t) &= \operatorname{cov}(\theta_s,\theta_s|\mathcal{F}_t^\xi,\theta_t=\beta)\\
&= M[\operatorname{cov}(\theta_s,\theta_s|\mathcal{F}_t^\xi,\theta_t,\theta_{s+1})|\mathcal{F}_t^\xi,\theta_t=\beta]\\
&\quad+\operatorname{cov}[M(\theta_s|\mathcal{F}_t^\xi,\theta_t,\theta_{s+1}),M(\theta_s|\mathcal{F}_t^\xi,\theta_t,\theta_{s+1})|\mathcal{F}_t^\xi,\theta_t=\beta]\\
&= M[\operatorname{cov}(\theta_s,\theta_s|\mathcal{F}_{s+1}^\xi,\theta_{s+1})|\mathcal{F}_t^\xi,\theta_t=\beta]\\
&\quad+\operatorname{cov}[M(\theta_s|\mathcal{F}_{s+1}^\xi,\theta_{s+1}),M(\theta_s|\mathcal{F}_{s+1}^\xi,\theta_{s+1})|\mathcal{F}_t^\xi,\theta_t=\beta]\\
&= M[\tilde{\gamma}_{\theta_{s+1}}(s,s+1)|\mathcal{F}_t^\xi,\theta_t=\beta]\\
&\quad+\operatorname{cov}[\tilde{m}_{\theta_{s+1}}(s,s+1),\tilde{m}_{\theta_{s+1}}(s,s+1)|\mathcal{F}_t^\xi,\theta_t=\beta]\\
&= \tilde{\gamma}_\beta(s,s+1) + M[(\tilde{m}_{\theta_{s+1}}(s,s+1)-\tilde{m}_\beta(s,t))\\
&\quad\times\tilde{m}_{\theta_{s+1}}(s,s+1)-\tilde{m}_\beta(s,t))^*|\mathcal{F}_t^\xi,\theta_t=\beta].
\end{aligned} \tag{13.129}$$

But it follows from (13.122) and (13.124) that

$$\tilde{m}_{\theta_{s+1}}(s,s+1)-\tilde{m}_{\beta}(s,t)=\gamma(s,s+1)(\varphi_s^{s+1})^*\gamma_{s+1}^{+}[\theta_{s+1}-\tilde{m}_{\beta}(s+1,t)], \tag{13.130}$$

which together with (13.129) leads to Equation (13.123).

Theorem 13.12. *Let (1)–(4) be satisfied. Then the moments $m(s,t)$ and $\gamma(s,t)$ of the conditional distribution $\Pi_\alpha(s,t) = P(\theta_s \le \alpha|\mathcal{F}_t^\xi)$ satisfy, for $s<t$, the (backward) equations*

$$m(s,t)=m(s,s+1)+\gamma(s,s+1)(\varphi_s^{s+1})^*\gamma_{s+1}^{+}[m(s+1,t)-m_{s+1}], \tag{13.131}$$

$$\gamma(s,t)=\tilde{\gamma}(s,s+1)+\gamma(s,s+1)(\varphi_s^{s+1})^*\gamma_{s+1}^{+}\gamma(s+1,t)\gamma_{s+1}^{+}\varphi_s^{s+1}\gamma(s,s+1) \tag{13.132}$$

with $m(t,t)=m_t$, $\gamma(t,t)=\gamma_t$, $\tilde{\gamma}(s,s+1)\equiv\tilde{\gamma}_\beta(s,s+1)$.

PROOF. Equation (13.131) can be deduced immediately from (13.122). In order to deduce (13.132) let us make use of (13.127) and (13.128). We obtain

$$\begin{aligned}\gamma(s,t) &= \operatorname{cov}(\theta_s,\theta_s|\mathcal{F}_t^\xi)\\ &= M[\operatorname{cov}(\theta_s,\theta_s|\mathcal{F}_t^\xi,\theta_{s+1})|\mathcal{F}_t^\xi]\\ &\quad +\operatorname{cov}[M(\theta_s|\mathcal{F}_t^\xi,\theta_{s+1}),M(\theta_s|\mathcal{F}_t^\xi,\theta_{s+1})|\mathcal{F}_t^\xi]\\ &= M[\operatorname{cov}(\theta_s,\theta_s|\mathcal{F}_{s+1}^\xi,\theta_{s+1})|\mathcal{F}_t^\xi]\\ &\quad +\operatorname{cov}[M(\theta_s|\mathcal{F}_{s+1}^\xi,\theta_{s+1}),M(\theta_s|\mathcal{F}_{s+1}^\xi,\theta_{s+1})|\mathcal{F}_t^\xi]\\ &= \tilde{\gamma}(s,s+1)+M\{[\tilde{m}_{\theta_{s+1}}(s,s+1)-m(s,t)]\\ &\quad \times[\tilde{m}_{\theta_{s+1}}(s,s+1)-m(s,t)]^*|\mathcal{F}_t^\xi\}. \end{aligned}\tag{13.133}$$

But, according to (13.122) and (13.131),

$$\tilde{m}_{\theta_{s+1}}(s,s+1)-m(s,t)=\gamma(s,s+1)(\varphi_s^{s+1})^*\gamma_{s+1}^{+}[\theta_{s+1}-m(s+1,t)],$$

which, together with (13.133), yields Equation (13.132). □

13.4 Recursive Equations of Optimal Extrapolation

13.4.1. Extrapolation is understood as estimation of vectors θ_t, ξ_t from the observations $\xi_0^s=\{\xi_0,\dots,\xi_s\}$, where $t>s$. As in the case of continuous time (Section 12.5) equations of extrapolation will be deduced only in two particular cases due to the fact that the conditional distributions

$$P(\theta_t\le a,\xi_t\le b|\mathcal{F}_s^\xi)$$

are, generally speaking, no longer Gaussian.

Before formulating the theorems, we shall set forth the way of identifying the cases for which extrapolation estimates can be constructed.

Due to (13.56) and (13.78),

$$\begin{aligned} m_{t+1} &= [a_0(t,\xi) + a_1(t,\xi)m_t] \\ &\quad +[(b \circ B)(t,\xi) + a_1(t,\xi)\gamma_t A_1^*(t,\xi)] \\ &\quad \times[(B \circ B)(t,\xi) + A_1(t,\xi)\gamma_y A_1^*(t,\xi)]^+ \\ &\quad \times[(B \circ B)(t,\xi) + A_1(t,\xi)\gamma_t A_1^*(t,\xi)]^{1/2}\overline{\varepsilon}(t+1), \quad (13.134) \end{aligned}$$

$$\begin{aligned} \xi_{t+1} &= [A_0(t,\xi) + A_1(t,\xi)m_t] \\ &\quad +[(B \circ B)(t,\xi) + A_1(t,\xi)\gamma_t A_1^*(t,\xi)]^{1/2}\overline{\varepsilon}(t+1). \quad (13.135) \end{aligned}$$

Denote by

$$n_1(t,s) = M(\theta_t|\mathcal{F}_s^\xi), \quad n_2(t,s) = M(\xi_t|\mathcal{F}_s^\xi)$$

the optimal (in the mean square sense) estimates θ_t and ξ_t from $\xi_0^s = \{\xi_0, \dots, \xi_{s0}\}$. Since $n_1(t,s) = M[M(\theta_t|\mathcal{F}_t^\xi)|\mathcal{F}_s^\xi] = M[m_t|\mathcal{F}_s^\xi)$ and

$$M(\overline{\varepsilon}(t+1)|\mathcal{F}_s^\xi) = 0$$

for all $t+1 > s$, then equations for $n_1(t,s)$ and $n_2(t,s)$ can be found by taking $M(\cdot|\mathcal{F}_s^\xi)$ on both sides in (13.134) and (13.135).

It is easy to see from this that the simultaneous determination of $n_1(t,s)$ and $n_2(t,s)$ becomes possible if

$$a_0(t,\xi) = a_0(t) + a_2(t)\xi_t, \quad a_1(t,\xi) = a_1(t), \qquad (13.136)$$

$$A_0(t,\xi) = A_0(t) + A_2(t)\xi_t, \quad A_1(t,\xi) = A_1(t) \qquad (13.137)$$

where the matrix functions $a_i(t)$ and $A_i(t)$, $i = 1, 2$, and the vectors $a_0(t)$ and $A_0(t)$ depend only on time.

If we are interested in nothing but estimation of variables θ_t, then determination of $n_1(t,s)$ becomes possible if we require (13.136) with $a_2(t) \equiv 0$ to be satisfied.

13.4.2.

Theorem 13.13. *Let (1)-(4), (13.136) and (13.137) be satisfied. Then the moments $n_1(t,s)$ and $n_2(t,s)$ satisfy the equations*

$$n_1(t+1,s) = a_0(t) + a_1(t)n_1(t,s) + a_2(t)n_2(t,s), \qquad (13.138)$$

$$n_2(t+1,s) = A_0(t) + A_1(t)n_1(t,s) + A_2(t)n_2(t,s), \qquad (13.139)$$

with $n_1(s,s) = m_s$, $n_2(s,s) = \xi_s$.

If (13.136) is satisfied and, in addition, $a_2(t) \equiv 0$, then

$$n_1(t+1,s) = a_0(t) + a_1(t)n_1(t,s), \quad n_1(s,s) = m_s. \qquad (13.140)$$

Proof is immediate by taking $M(\cdot|\mathcal{F}_s^\xi)$ on both sides of (13.134) and (13.135).

Let us now consider the backward equations for $n_1(t,s)$ and $n_2(t,s)$.

Theorem 13.14. *Let (1)–(4), (13.136) and (13.137) be satisfied. Then*

$$\begin{aligned}\begin{pmatrix} n_1(t,s+1) \\ n_2(t,s+1)\end{pmatrix} &= \begin{pmatrix} n_1(t,s) \\ \times n_2(t,s)\end{pmatrix} + \Phi_{s+1}^{t-1}\begin{pmatrix} D_1(s,\xi)\cdot D_2^+(s,\xi) \\ E\end{pmatrix} \\ &\quad \times[\xi_{s+1} - A_0(s) - A_1(s)m_s - A_2(s)\xi_s], \qquad (13.141)\end{aligned}$$

where

$$\begin{aligned} D_1(s,\xi) &= (b\circ B)(s,\xi) + a_1(s,\xi)\gamma_s A_1^*(s,\xi), \\ D_2(s,\xi) &= (B\circ B)(s,\xi) + A_1(s,\xi)\gamma_s A_1^*(s,\xi).\end{aligned}$$

$E = E_{(l\times l)}$, *the matrix* Φ_s^t *can be defined by the recursive equations*

$$\Phi_s^t = \begin{pmatrix} a_1(t-1) & a_2(t-1) \\ A_1(t-1) & A_2(t-1)\end{pmatrix}\Phi_s^{t-1}, \quad \Phi_s^s = E_{(k\times l)\times(k\times l)}, \qquad (13.142)$$

and

$$\begin{pmatrix} n_1(t,0) \\ n_2(t,0)\end{pmatrix} = \Phi_0^t\begin{pmatrix} m_0 \\ \xi_0\end{pmatrix} + \sum_{u=0}^{t-1}\Phi_u^{t-1}\begin{pmatrix} a_0(u) \\ A_0(u)\end{pmatrix}. \qquad (13.143)$$

If (13.136) is satisfied, and, in addition, $a_2(t)\equiv 0$, *then*

$$\begin{aligned} n_1(t,s+1) &= n_1(t,s) + \psi_{s+1}^{t-1}[(b\circ B)(s,\xi) + a_1(s)\gamma_s A_1^*(s,\xi)] \\ &\quad \times[(B\circ B)(s,\xi) + A_1(s,\xi)\gamma_s A_1^*(s,\xi)]^+ \\ &\quad \times[\xi_{s+1} - A_0(s,\xi) + A_1(s,\xi)m_s], \qquad (13.144)\end{aligned}$$

where the matrix ψ_s^t *can be defined by the equations*

$$\psi_s^t = a_1(t-1)\psi_s^{t-1}, \quad \psi_s^s = E_{(k\times k)}, \qquad (13.145)$$

and

$$n_1(t,0) = \psi_0^t m_0 + \sum_{u=0}^{t-1}\psi_u^{t-1}a_0(u). \qquad (13.146)$$

PROOF. By induction we obtain from (13.134) and (13.135) the following:

$$\begin{aligned}\begin{pmatrix} m_t \\ \xi_t\end{pmatrix} &= \Phi_0^t\,(m_0\xi_0) + \sum_{u=0}^{t-1}\Phi_u^{t-1}\begin{pmatrix} a_0(u) \\ A_0(u)\end{pmatrix} \\ &\quad + \sum_{u=0}^{t-1}\Phi_u^{t-1}\begin{pmatrix} D_1(u,\xi)D_2^+(u,\xi)D_2^{1/2}(u,\xi) \\ D_2^{1/2}(u,\xi)\end{pmatrix}\overline{\varepsilon}(u+1). \end{aligned} \qquad (13.147)$$

Take the conditional expectation $M(\cdot|\mathcal{F}_{s+1}^{\xi})$ on both sides in (13.147). Then, taking into account that $M[\overline{\varepsilon}(u+1)|\mathcal{F}_{s+1}^{\xi}]=0$, $u>s$, we easily find from (13.147) that

$$\begin{pmatrix} n_1(t,s+1) \\ n_2(t,s+1) \end{pmatrix} = \begin{pmatrix} n_1(t,s) \\ n_2(t,s) \end{pmatrix} + \Phi_{s+1}^{t-1} \begin{pmatrix} D_1(s,\xi)D_2^{+}(s,\xi)D_2^{1/2}(s,\xi) \\ D_2^{1/2}(s,\xi) \end{pmatrix} \overline{\varepsilon}(s+1);$$

this, together with (13.135), leads to the system of equations given by (13.141).

(13.144) can be deduced in a similar way. □

13.5 Examples

13.5.1. We shall present here examples which illustrate the potential applications for the equations of filtering, interpolation and extrapolation deduced above.

EXAMPLE 1 (Parameter Estimation). Let $\theta=(\theta_1,\dots,\theta_k)$ be a Gaussian vector with $M\theta=m$ and $\operatorname{cov}(\theta,\theta)=\gamma$. It is required to estimate θ from the observation of the l-dimensional process ξ_t, $t=0,1,\dots$, satisfying the recursive equations

$$\xi_{t+1}=A_0(t,\xi)+A_1(t,\xi)\theta+B_1(t,\xi)\varepsilon_1(t+1) \tag{13.148}$$

with $\xi_0=0$.

Assuming[9] (1)–(4) we obtain, from (13.56) and (13.57), for $m_t=M(\theta|\mathcal{F}_t^{\xi})$ and $\gamma_t=\operatorname{cov}(\theta,\theta|\mathcal{F}_t^{\xi})$, recursive equations

$$\begin{aligned} m_{t+1} &= m_t+\gamma_t A_1^*(t,\xi)[(B_1B_1^*)(t,\xi)+A_1(t,\xi)\gamma_t A_1^*(t,\xi)]^{+} \\ &\quad \times[\xi_{t+1}-A_0(t,\xi)-A_1(t,\xi)m_t], \end{aligned} \tag{13.149}$$

$$\gamma_{t+1}=\gamma_t-\gamma_t A_1^*(t,\xi)[(B_1B_1^*)(t,\xi)+A_1(t,\xi)\gamma_t A_1^*(t,\xi)]^{+}A_1(t,\xi)\gamma_t, \tag{13.150}$$

with $m_0=m$, $\gamma_0=\gamma$.

Theorem 13.15. *If the matrices $(B_1B_1^*)(t,\xi)$ are nonsingular (P-a.s.), $t=0,1,\dots$, then solutions of Equations (13.149) and (13.150) are given by the formulae*[10]

[9] Assumption (2) in this case can be replaced by the condition $M\operatorname{Tr}A_1(t,\xi)A_1^*(t,\xi)<\infty$.

[10] Compare with Theorems 12.2 and 12.8.

$$m_{t+1} = \left[E + \gamma \sum_{s=0}^{t} A_1^*(s,\xi)(B_1B_1^*)^{-1}(s,\xi)A_1^*(s,\xi)\right]^{-1} \times \left[m + \gamma \sum_{s=0}^{t} A_1^*(s,\xi)(B_1B_1^*)^{-1}(s,\xi)(\xi_{s+1} - A_0(s,\xi))\right], \tag{13.151}$$

$$\gamma_{t+1} = \left[E + \gamma \sum_{s=0}^{t} A_1^*(s,\xi)(B_1B_1^*)^{-1}(s,\xi)A_1(s,\xi)\right]^{-1} \gamma. \tag{13.152}$$

PROOF. By Theorem 13.3, the conditional distribution $P\{\theta \le a|\mathcal{F}_t^\xi\}$ is Gaussian (P-a.s.) with the parameters (m_t, ξ_t).

Assume that the matrix γ_t is positive definite. Then the conditional distribution $P\{\theta \le a|\mathcal{F}_t^\xi\}$ has the density

$$f_\theta(a|\xi_0^t) = \frac{dP\{\theta \le a|\mathcal{F}_t^\xi\}}{da}.$$

The conditional distribution $P\{\xi_{t+1} \le b|\mathcal{F}_t^\xi, \theta\}$ is also (P-a.s.) Gaussian with the parameters $\{(A_0(t,\xi) + A_1(t,\xi)\theta), (B_1B_1^*)(t,\xi)\}$. Since the matrices $(B_1B_1^*)(t,\xi)$, $t = 0, 1, \ldots$, are nonsingular (P-a.s.), the distribution $P\{\xi_{t+1} \le b|\mathcal{F}_t^\xi, \theta\}$ has the density

$$f_{\xi_{t+1}}(b|\xi_0^{t+1}, 0) = \frac{(dP\{\xi_{t+1} \le b|\mathcal{F}_t^\xi, \theta\}}{db}.$$

But according to the Bayes formula, there exists a density

$$f_\theta(a|\xi_0^{t+1}) = \frac{dP\{\theta \le a|\mathcal{F}_{t+1}^\xi\}}{da},$$

given by the formula

$$f_\theta(a|\xi_0^{t+1}) = \frac{f_\theta(a|\xi_0^t) f_{\xi_{t+1}}(\xi_{t=1}|\xi_0^t, a)}{\int_{\mathbb{R}^k} f_{\xi_{t+1}}(\xi_{t+1}|\xi_0^t, x) f_\theta(x|\xi_0^t)dx} \quad (P\text{-a.s.}). \tag{13.153}$$

Let us write

$$g_1(t+1,\xi) = (2\pi)^{k/2}\sqrt{\det \gamma_{t+1}}, \tag{13.154}$$

$$g_2(t+1,\xi) = (2\pi)^{(k+l)/2}\sqrt{\det \gamma_t \cdot \det(B_1B_1^*)(t,\xi)} \times \int_{\mathbb{R}^k} f_{\xi_{t+1}}(\xi_{t+1}|\xi_0^t, x) f_\theta(x|\xi_0^t)dx. \tag{13.155}$$

By Theorem 13.3, the density $f_\theta(a|\xi_0^{t+1})$ is (P-a.s.) Gaussian with the parameters (m_{t+1}, γ_{t+1}), where γ_{t+1} is a positive definite matrix. Taking this fact as well as (13.154) and (13.155) into account, we find from (13.153) that (P-a.s.)

$$
\begin{aligned}
&[g_1(t+1,\xi)]^{-1}\exp\left\{-\frac{1}{2}(a-m_{t+1})^*\gamma_{t+1}^{-1}(a-m_{t+1})\right\}\\
=\ &[g_2(t+1,\xi)]^{-1}\exp\left\{-\frac{1}{2}(a-m_t)^*\gamma_t^{-1}(a-m_t)\right.\\
&-\frac{1}{2}(\xi_{t+1}-A_0(t,\xi)-A_1(t,\xi)a)^*(B_1B_1^*)^{-1}(t,\xi)\\
&\left.\times(\xi_{t+1}-A_0(t,\xi)-A_1(t,\xi)a)\right\}. \qquad (13.156)
\end{aligned}
$$

Equating now the square and linear forms over a in the left- and right-hand sides of (13.156), respectively, we obtain, by virtue of the arbitrariness of the vector a, the recursive equations

$$\gamma_{t+1}^{-1}=\gamma_t^{-1}+A_1^*(t,\xi)(B_1B_1^*)^{-1}(t,\xi)A_1(t,\xi), \qquad (13.157)$$

$$\gamma_{t+1}^{-1}m_{t+1}=\gamma_t^{-1}m_t+A_1^*(t,\xi)(B_1B_1^*)^{-1}(t,\xi)[\xi_{t+1}-A_0(t,\xi)]. \qquad (13.158)$$

If the matrix $\gamma_0=\gamma$ is positive definite then by induction it follows that recursive equations (13.157) and (13.158) hold true for all t. Hence, in the case where γ is nonsingular, (13.151) and (13.152) for m_{t+1},γ_{t+1}, $t\ge 0$, follow from (13.157) and (13.158).

If the matrix γ is singular, then, assuming $\gamma_0^\varepsilon=\gamma_0+\varepsilon E$, $\varepsilon>0$, we find γ_{t+1}^ε and m_{t+1}^ε from (13.151) and (13.152) with the substitution of $\gamma+\varepsilon E$ for γ. In particular,

$$\gamma_{t+1}^\varepsilon=\left\{E+(\gamma_\varepsilon E)\sum_{s=0}^{t}A_1^*(s,\xi)(B_1B_1^*)^{-1}(s,\xi)A_1(s,\xi)\right\}^{-1}[\gamma+\varepsilon E].$$

After a passage to the limit $\varepsilon\downarrow 0$, we obtain the required representations for m_{t+1} and γ_{t+1} for any symmetric nonnegative definite matrix γ. □

Note. Let $m_t^{(n)}$ and $\gamma_t^{(n)}$ be parameters of the a posteriori distributions $P(\theta\le u|\mathcal{F}_t^\xi)$, corresponding to the a priori distributions $P(\theta\le a)\sim N(m^{(n)},\gamma^{(n)})$.

Let $0<\gamma^{(n)}$, $\operatorname{Tr}\gamma^{(n)}<\infty$. Then, if $\lim_{n\to\infty}(\gamma^{(n)})^{-1}=0$ and the matrices $\sum_{s=0}^{t}A_1^*(s,\xi)(B_1B_1^*)^{-1}(s,\xi)A_1(s,\xi)$ are nonsingular (P-a.s.), it is not difficult to prove that there exist

$$\tilde m_t=\lim_{n\to\infty}m_t^{(n)},\quad \tilde\gamma_t=\lim_{n\to\infty}\gamma_t^{(n)}$$

and

$$\tilde\gamma_{t+1}=\left[\sum_{s=0}^{t}A_1^*(s,\xi)(B_1B_1^*)^{-1}(s,\xi)A_1(s,\xi)\right]^{-1},$$

$$\tilde{m}_{t+1} = \tilde{\gamma}_{t+1}\left[\sum_{s=0}^{t} A_1^*(s,\xi)(B_1B_1^*)^{-1}(s,\xi)(s,\xi)(\xi_{s+1} - A_0(s,\xi))\right]. \quad (13.159)$$

Note that the estimate given by (13.159) coincides with the maximum likelihood estimate for the vector θ from the observations $\xi_0^{t+1} = \{\xi_0, \dots, \xi_{t+1}\}$.

13.5.2.

EXAMPLE 2 (Interpolation of a Gaussian Markov Chain). Let

$$\theta_t = (\theta_1(t), \dots, \theta_k(t)), \quad t = 0, 1, \dots$$

be the Markov chain defined by the recursive equations

$$\theta_{t+1} = a_0(t) + a_1(t)\theta_t + b(t)\varepsilon_1(t+1), \quad (13.160)$$

where $a_0(t), a_1(t)$ and $b(t)$ depend only on t, and the random vector $\theta_0 \sim N(m, \gamma)$,

Let us discuss the problem of estimating the variables θ_s on the assumption that $\theta_t = \beta$, $s < t$.

Let

$$\begin{aligned} \tilde{m}_\beta(s,t) &= M(\theta_s | \theta_t = \beta), \\ \tilde{\gamma}(s,t) &\equiv \tilde{\gamma}_\beta(s,t) = M[(\theta_s - \tilde{m}_\beta(s,t)(\theta_s - \tilde{m}_\beta(s,t))^* | \theta_t = \beta], \\ m_t &= M\theta_t, \quad \gamma_t - = \operatorname{cov}(\theta_t, \theta_t). \end{aligned}$$

Then, according to Theorems 13.4 and 13.10,

$$m_{t+1} = a_0(t) + a_1(t)m_t, \quad \gamma_{t+1} = a_1(t)\gamma_t a_1^*(t) + b(t)b^*(t) \quad (13.161)$$

and

$$\tilde{m}_\beta(s,t) = m_t + \gamma_t(\varphi_s^t)^*\gamma_t^+(\beta - m_t), \quad \tilde{\gamma}(s,t) = \gamma_s - \gamma_s(\varphi_s^t)^*\gamma_t^+\varphi_s^t\gamma_s, \quad (13.162)$$

where $\varphi_s^t = a_1(t-1)\cdots a_1(s)$. In particular, if $\theta_{t+1} = \theta_t + \varepsilon_1(t+1)$, then

$$\tilde{m}_\beta(s,t) = m + \frac{s+\gamma}{t+\gamma}(\beta - m), \quad \tilde{\gamma}(s,t) = (s+\gamma)\left[1 - \frac{s+\gamma}{t+\gamma}\right]. \quad (13.163)$$

EXAMPLE 3 (Interpolation with Fixed Delay). Let us discuss the problem of estimating the variables θ_s from the observations $\xi_0^{s+h} = \{\xi_0, \dots, \xi_{s+h}\}$ where h is a fixed value., Let $m_h(s) = m(s, s+h)$, $\gamma_h(s) = \gamma(s, s+h)$, and assume that for all s, $s = 0, 1, \dots$, the matrices $\gamma(s, s+1)(\varphi_s^{s+1})^*\gamma_{s+1}^{-1}$ are nonsingular.

Then the forward equation given by (13.95) yields

$$\begin{aligned} m_h(s+1) &= m(s+1, s+h) + \gamma(s+1, s+h)(\varphi_{s+1}^{s+h})^* A_1^*(s+h, \xi) \\ &\quad \times [(B \circ B)(s+h, \xi) + A_1(s+h, \xi)\gamma_{s+h}A_1^*(s+h, \xi)]^+ \\ &\quad \times [\xi_{s+h+1} - A_0(s+h, \xi) - A_1(s+h, \xi)m_{s+h}]. \quad (13.164) \end{aligned}$$

From the backward equation given by (13.131), on the assumption of nonsingularity of the matrices $\gamma(s,s+1)(\varphi_s^{s+h})^*\gamma_{s+1}^{-1}$, we obtain

$$m(s+1,s+h) = m_{s+1} + [\gamma(s,s+1)(\varphi_s^{s+1})^*\gamma_{s+1}^{-1}]^{-1}[m_h(s) - m(s,s+1)], \tag{13.165}$$

which, together with (13.164), yields the following equation for $m_h(s)$:

$$\begin{aligned} m_h(s+1) &= m_{s+1} + [\gamma(s,s+1)(\varphi_s^{s+1})^*\gamma_{s+1}^{-1}]^{-1}[m_h(s) - m(s,s+1)] \\ &\quad +\gamma(s+1,s+h)(\varphi_{s+1}^{s+h})^* A_1^*(s+h,\xi)[(B\circ B)(s+h,\xi) \\ &\quad +A_1(s+h,\xi)\gamma_{s+h}A_1^*(s+h,\xi)]^+ \\ &\quad \times[\xi_{s+h+1} - A_0(s+h,\xi) + A_1(s+h,\xi)m_{s+h}]. \end{aligned} \tag{13.166}$$

Similarly, from the forward equation given by (13.96) we find for $\gamma_h(s+1) = \gamma(s+1,s+h+1)$ that

$$\begin{aligned} \gamma_h(s+1) &= \gamma(s+1,s+h) - \gamma(s+1,s+h)(\varphi_{s+1}^{s+h})^* A_1^*(s+h,\xi) \\ &\quad \times[(B\circ B)(s+h,\xi) + A_1(s+h,\xi)\gamma_{s+h}A_1^*(s+h,\xi)]^+ \\ &\quad \times A_1(s,h,\xi)\varphi_{s+1}^{s+h}(s+1,s+h). \end{aligned} \tag{13.167}$$

From (13.132) we obtain

$$\begin{aligned} \gamma(s+1,s+h) &= [\gamma(s,s+1)(\varphi_s^{s+1})^*\gamma_{s+1}^{-1}]^{-1}[\gamma_h(s) - \tilde{\gamma}(s,s+1)] \\ &\quad \times[\gamma_{s+1}^{-1}\varphi_s^{s+1}\gamma(s,s+1)]^{-1}. \end{aligned}$$

Substituting this expression for $\gamma(s+1,s+h)$ in (13.166) and (13.167) we obtain equations describing the evolution of $m_h(s)$ and $\gamma_h(s)$. In this case $m_h(0) = m(0,h)$ and $\gamma_h(0) = \gamma(0,h)$ are defined from the forward equations given by (13.95) and (13.96).

In the particular case $h = 1$,

$$\begin{aligned} m_1(s+1) &= m_{s+1} + \gamma_{s+1}A_1^*(s+1,\xi)[(B\circ B)(s+1,\xi) \\ &\quad +A_1(s+1,\xi)\gamma_{s+1}A_1^*(s+1,\xi)]^+ \\ &\quad \times[\xi_{s+2} - A_0(s+1,\xi) - A_1(s+1,\xi)m_{s+1}]. \end{aligned} \tag{13.168}$$

13.5.3.
EXAMPLE 4 (Linear Prediction of Stationary Sequences). Let $\tilde{\xi}_t$, $t = 0,\pm1,\pm2$, be a stationary wide-sense process with $M\tilde{\xi}_t \equiv 0$ and the spectral density

$$\tilde{f}(\lambda) = \frac{|e^{i\lambda}+1|^2}{|e^{2i\lambda} + \frac{1}{2}e^{i\lambda} + \frac{1}{2}|^2}. \tag{13.169}$$

Let it be required to construct an optimal (in the mean square sense) linear estimate of the variables $\tilde{\xi}_t$ from $\tilde{\xi}_0^s = \{\tilde{\xi}_0,\dots,\tilde{\xi}_s\}$, $s \le t$.

We shall construct the Gaussian process ξ_t, $t = 0,\pm1,\dots$, with $M\xi_t \equiv 0$ and the spectral density $f(\lambda) \equiv \tilde{f}(\lambda)$. Such a process can be obtained by solving the equation

$$\xi_{t+1} + \frac{1}{2}(\xi_{t+1} + \xi_t) = \varepsilon(t+2) + \varepsilon(t+1),$$

where $\varepsilon(t)$, $t = 0, \pm 1, \ldots$, is a sequence of Gaussian random variables with

$$M\varepsilon(t) = 0, \quad M\varepsilon(t)\varepsilon(s) = \delta(t,s) = \begin{cases} 1, & t = s \\ 0, & t \neq s \end{cases}.$$

Set $\theta_t = \xi_{t+1} - \varepsilon(t+1)$. Then for (θ_t, ξ_t), $t = 0, \pm 1, \ldots$, we obtain the system of equations

$$\begin{aligned} \theta_{t+1} &= -\frac{1}{2}\theta_t - \frac{1}{2}\xi_t + \frac{1}{2}\varepsilon(t+1), \\ \xi_{t+1} &= \theta_t + \varepsilon(t+1). \end{aligned} \tag{13.170}$$

According to Theorem 13.13, $n_1(t,s) = M(\theta_t | \mathcal{F}_s^\xi)$ and $n_2(t,s) = M(\xi_t | \mathcal{F}_s^\xi)$ can be defined from Equations (13.138) and (13.139):

$$\begin{aligned} n_1(t+1,s) &= -\frac{1}{2}n_1(t,s) - \frac{1}{2}n_2(t,s), \\ n_2(t+1,s) &= n_1(t,s), \end{aligned} \tag{13.171}$$

with $n_1(s,s) = m_s$, $n_2(s,s) = \xi_s$.

γ_s and the initial condition $m_s = M(\theta_s | \mathcal{F}_s^\xi)$ entering into (13.171) can be defined by the equations (see (13.56) and (13.57))

$$m_{s+1} = -\frac{1}{2}m_s - \frac{1}{2}\xi_s + \frac{1 - \gamma_s}{2(1+\gamma_s)}(\xi_{s+1} - m_s), \tag{13.172}$$

$$\gamma_{s+1} = \frac{\gamma_s}{1 + \gamma_s}. \tag{13.173}$$

Note here that $m_0 = 0$, $\gamma_0 = 1$.

Indeed, by virtue of the stationarity of the process (θ_t, ξ_t), $t = 0, \pm 1, \ldots$, parameters $d_{11} = M\theta_t^2$, $d_{12} = M\theta_t\xi_t$, and $d_{22} = M\xi_t^2$ are easily found from the following system obtained from (13.170):

$$\begin{aligned} d_{11} &= \frac{1}{4}d_{11} + \frac{1}{4}d_{22} + \frac{1}{2}d_{12} + \frac{1}{4}, \\ d_{12} &= -\frac{1}{2}d_{11} - \frac{1}{2}d_{12} + \frac{1}{2}, \\ d_{22} &= d_{11} + 1. \end{aligned}$$

Thus, $d_{11} = 1$, $d_{12} = 0$, $d_{22} = 2$, and, by the theorem on normal correlation, $m_0 = 0$, $\gamma_0 = 1$.

Returning to the initial process $\tilde{\xi}_t$, $t = 0, \pm 1, \ldots$, we find that optimal linear prediction can be defined from (13.171)–(13.173) where (in (13.172)) ξ_t should be substituted for $\tilde{\xi}_t$ (see Lemma 14.1).

Notes and References. 1

13.1. The theorem on normal correlation (Theorem 13.1), proved in general by Marsaglia [225] (see also Anderson [3]), has been repeatedly used in several chapters of this book. The authors owe the proof of Theorem 13.2 to Kitsul. For the properties of pseudo-inverse matrices see also Gantmacher [69]. Lemma 13.3 was proved by Pjatetsky (Masters Thesis).

13.2–3.5. The material of these sections is based on the papers of Liptser and Shiryaev [213], and Glonti [77–79].

Notes and References. 2

13.1. The theorem of normal correlation in a general setting uses the Moore–Penrose pseudo-inverse matrix. This matrix is also extensively employed in statistical applications and in Kalman filtering for the discrete-time case. The main properties of the pseudo-inverse matrix and its various applications can be found in Albert [1].

14. Application of Filtering Equations to Problems of Statistics of Random Sequences

14.1 Optimal Linear Filtering of Stationary Sequences with Rational Spectra

14.1.1. The objective of this chapter is to show how the equations of optimal nonlinear filtering obtained for conditionally Gaussian random sequences can be applied to solving various problems of mathematical statistics. In particular, the present section deals with the problem of linear estimation of unobservable components of a multidimensional stationary wide-sense process (discrete time) with rational spectral density in the components accessible for observation.

The possibility of applying the filtering equations obtained above to this problem is based on the fact (Theorem 14.1) that any stationary sequence with a rational spectrum is a component of a multidimensional process satisfying a system of recursive equations of the type given by (13.46) and (13.47).

More precisely, let $\eta(t)$, $t = 0, \pm 1, \pm 2, \ldots$, be a (real or complex) stationary wide-sense random process permitting the spectral representation

$$\eta(t) = \int_{-\pi}^{\pi} e^{i\lambda t} \frac{P_{n-1}(e^{i\lambda})}{Q_n(e^{i\lambda})} \Phi(d\lambda), \tag{14.1}$$

where $\Phi(d\lambda)$ is an orthogonal (random) measure with

$$M\Phi(d\lambda) = 0, \quad M|\Phi(d\lambda)|^2 = \frac{d\lambda}{2\pi},$$

$$P_{n-1}(z) = \sum_{k=0}^{n-1} b_k z^k, \quad Q_n(z) = \sum_{k=0}^{n} a_k z^k, \quad a_n = 1, \quad a_k, b_k \in \mathbb{R}^1.$$

Assume that all the roots of the equation $Q_n(z) = 0$ lie within the unit circle.

It follows from (14.1) that the process $\eta(t)$ has the rational spectral density

$$f_n(\lambda) = \left| \frac{P_{n-1}(e^{i\lambda})}{Q_n(e^{i\lambda})} \right|^2. \tag{14.2}$$

Construct from the measure $\Phi(d\lambda)$ the process

$$\varepsilon(t) = \int_{-\pi}^{\pi} e^{i\lambda(t-1)} \Phi(d\lambda). \tag{14.3}$$

It is clear that

$$M\varepsilon(t) = 0, \quad M|\varepsilon(t)|^2 = \int_{-\pi}^{\pi} \frac{d\lambda}{2\pi} = 1,$$

and

$$M\varepsilon(t)\overline{\varepsilon}(s) = \int_{-\pi}^{\pi} e^{i\lambda(t-s)} \frac{d\lambda}{2\pi} = \delta(t,s), \tag{14.4}$$

where $\delta(t,s)$ denotes the Kronecker function.

It follows from (14.4) that the sequence of values of $\varepsilon(t)$, $t = 0, \pm 1, \ldots,$ is a sequence of uncorrelated variables.

In addition to the process $\eta(t)$, permitting the spectral representation given by (14.1), we shall define new processes $\eta_1(t), \ldots, \eta_n(t)$ by the formulae

$$\eta_j(t) = \int_{-\pi}^{\pi} e^{i\lambda t} W_j(e^{i\lambda}) \Phi(d\lambda), \quad j = 1, \ldots, n, \tag{14.5}$$

where the frequency characteristics $W_j(z)$, $j = 1, \ldots, n$, are selected in the following specific manner:

$$W_j(z) = z^{-(n-j)} W_n(z) + \sum_{k=j}^{n-1} \beta_k z^{-(k-j+1)}, \quad j = 1, \ldots, n-1; \tag{14.6}$$

$$W_n(z) = -z^{-1} \sum_{k=0}^{n-1} a_k W_{k+1}(z) + z^{-1}\beta_n; \tag{14.7}$$

$$\beta_1 = b_{n-1}, \quad \beta_j = b_{n-j} - \sum_{i=1}^{j-1} \beta_i a_{n-j+1}, \quad j = 2, \ldots, n. \tag{14.8}$$

It follows from (14.6) and (14.7) that

$$W_j(z) = z^{-1}[W_{j+1}(z) + \beta_j] \tag{14.9}$$

and

$$W_n(z) = z^{-1} \left[-\sum_{k=0}^{n-1} a_k W_{k+1}(z) + \beta_n \right]. \tag{14.10}$$

It is not difficult to deduce from this that

$$W_n(z) = z^{-1} \left[-\sum_{k=0}^{n-1} a_k z^{-(n-k-1)} W_n(z) + \sum_{j=k+1}^{n-1} \beta_j z^{-(j-k)} + \beta_n \right], \tag{14.11}$$

and, therefore,

$$W_n(z) = \frac{P_{n-1}^{(n)}(z)}{Q_n(z)}, \tag{14.12}$$

where $P_{n-1}^{(n)}(z)$ is a polynomial of degree less than or equal to $n-1$.

Next, due to (14.9)–(14.12),

$$W_j(z) = \frac{P_{n-1}^{(j)}(z)}{Q_n(z)}, \tag{14.13}$$

where the polynomials $P_{n-1}^{(j)}(z)$ also have degree less than or equal to $n-1$ and, due to (14.8),

$$P_{n-1}^{(1)}(z) \equiv P_{n-1}(z). \tag{14.14}$$

Thus $\eta_1(t) = \eta(t)$.

Theorem 14.1. *The stationary (wide-sense) process $\eta(t)$, $t = 0, \pm 1, \ldots,$ with spectral representation given by (14.1) is one component of the n-dimensional stationary (wide-sense) process $(\eta_1(t), \ldots, \eta_n(t))$, $\eta_1(t) = \eta(t)$, obeying the system of recurrent equations*

$$\begin{aligned} \eta_j(t+1) &= \eta_{j+1}(t) + \beta_j \varepsilon(t+1), \quad j = 1, \ldots, n-1, \\ \eta_n(t+1) &= -\sum_{j=0}^{n-1} a_j \eta_{j+1}(t) + \beta_n \varepsilon(t+1). \end{aligned} \tag{14.15}$$

The process $\varepsilon(t)$, $t = 0, \pm 1, \ldots,$ permits the representation given by (14.3)

$$M\eta_j(s)\overline{\varepsilon}(t) = 0, \quad s < t, \quad j = 1, \ldots, n, \tag{14.16}$$

and the coefficients $\beta_1, \ldots, \beta_n$ are given by (14.8).

PROOF. Note first that from (14.12) and (14.13) it follows that all the poles of the functions $W_j(z)$ lie within the unit circle.

Taking advantage of (14.6), (14.7) and (14.5), we find easily that the process $(\eta_1(t), \ldots, \eta_n(t))$ satisfies the system of recursive equations given by (14.15).

Let us establish now the validity of (14.16). Let[1]

$$A = \begin{pmatrix} 0 & 1 & 0 & \ldots & 0 \\ 0 & 0 & 1 & \ldots & 0 \\ \ldots & \ldots & \ldots & \ldots & \ldots \\ 0 & 0 & 0 & \ldots & 1 \\ -a_0 & -a_1 & -a_2 & \ldots & -a_{n-1} \end{pmatrix}, \quad B = \begin{pmatrix} \beta_1 \\ \beta_2 \\ \ldots \\ \beta_n \end{pmatrix}. \tag{14.17}$$

[1] In algebraic operations Y_t is regarded as a column vector.

Then in matrix notation the system of equations given by (14.15) permits the representation

$$Y_t = AY_{t-1} + B\varepsilon_t. \tag{14.18}$$

Let $t > s$. Then, due to (14.18) and (14.4),

$$MY_s\overline{\varepsilon}(t) = AMY_{s-1}\overline{\varepsilon}(t) = A^2MY_{s-2}\overline{\varepsilon}(t) = \cdots = A^N MY_{s-N}\overline{\varepsilon}(t);$$

in this case, for each j, $j = 1, \ldots, n$,

$$|M\eta_j(s-N)\overline{\varepsilon}(t)| \le (M|\eta_j(s-N)|^2)^{1/2} = \left(\int_{-\pi}^{\pi}\left|\frac{P^{(j)}_{n-1}(e^{i\lambda})}{Q_n(e^{i\lambda})}\right|^2\frac{d\lambda}{2\pi}\right)^{1/2} < \infty.$$

Therefore, in order to prove (14.16), it suffices to show that

$$\lim_{N\to\infty} A^N = 0 \tag{14.19}$$

(0 is the zero matrix).

The eigenvalues of the matrix A are the roots of the equation $Q_n(z) = 0$ and, therefore, they lie within the unit circle. Transform the matrix A into a Jordan form

$$A = CJC^{-1},$$

where the eigenvalues of the matrix A are on the main diagonal of the matrix J. Let $\tilde{\lambda}$ be a maximal eigenvalue of the matrix A. Then, since $|\tilde{\lambda}| < 1$, no element of the matrix J^N exceeds in magnitude the values of $N|\tilde{\lambda}|^{N-1}$. But $A^N = CJ^NC^{-1}$ and $N|\tilde{\lambda}|^{N-1} \to 0$, $N \to \infty$, which proves (14.19). □

Note 1. If $\eta(t)$, $t = 0, \pm 1, \ldots$, is a real process, then each of the processes $\eta_1(t)$, $\eta_2(t)$, $\ldots$, $\eta_n(t)$ is also real. Here the covariance matrix $\Gamma = MY_tY_t^*$ satisfies the equation

$$\Gamma = A\Gamma A^* + BB^*. \tag{14.20}$$

If $t > s$, then

$$\operatorname{cov}(Y_t, Y_s) = MY_tY_s^* = A^{t-s}\Gamma, \tag{14.21}$$

which follows from the equalities

$$\begin{aligned} Y_t = AY_{t-1} + B\varepsilon(t) &= A^2Y_{t-2} + AB\varepsilon(t-1) + B\varepsilon(t) \\ &= A^{t-s}Y_s + \sum_{j=2}^{t-1} A^{t-1-j}B\varepsilon(j+1). \end{aligned} \tag{14.22}$$

Similarly, at $t < s$,

$$\operatorname{cov}(Y_t, Y_s) = \Gamma(A^*)^{s-t}.$$

Note 2. If $\eta(t)$, $t = 0, \pm 1, \ldots$, is a Gaussian process, then $\varepsilon(t)$, $t = 0, \pm 1, \ldots$, is a Gaussian sequence of independent random variables.

14.1.2. We can take advantage of (14.15) in order to deduce filtering equations of stationary sequence components with rational spectra.

Let $v_t = [\theta_t, \xi_t] = [(\theta_1(t), \dots, \theta_k(t)), (\xi_1(t), \dots, \xi_l(t))]$, $t = 0, \pm 1, \dots$, be a real stationary (wide-sense) $(k+l)$-dimensional process permitting the representation

$$v_t = \int_{-\pi}^{\pi} e^{i\lambda t} W(e^{i\lambda}) \Phi(d\lambda), \tag{14.23}$$

where $W(z) = \|W_{r,q}(z)\|$ is the matrix of order $N \times m$, $N = k + l$, with the rational elements

$$W_{r,q}(z) = \frac{P^{(r,q)}_{n_{r,q}-1}}{Q^{(r,q)}_{n_{r,q}}}, \tag{14.24}$$

and $\Phi(d\lambda) = [\Phi_1(d\lambda), \dots, \Phi_m(d\lambda)]$ is the random vector measure with uncorrelated components $M\Phi_j(d\lambda) = 0$, $M|\Phi_j(d\lambda)|^2 = d\lambda/2\pi$. Assume as well that the roots of the equations $Q^{(r,q)}_{n_{r,q}}(z) = 0$ lie within the unit circle.

Applying Theorem 14.1 to each of the processes

$$v_{p,r,q}(t) = \int_{-\pi}^{\pi} e^{i\lambda t} W_{r,q}(e^{i\lambda}) \Phi_p(d\lambda), \tag{14.25}$$

after simple transformations for the vector $\xi_t = (\xi_1(t), \dots, \xi_l(t))$ and the vector $\hat{\theta}_t$ (composed of the vector $\theta_t = (\theta_1(t), \dots, \theta_k(t))$ and all those additional components of the type $\eta_2(t), \dots, \eta_n(t)$, which arise by Theorem 14.1 in the system of equations given by (14.15)), we obtain the system of recursive equations

$$\begin{aligned} \hat{\theta}_{t+1} &= a_1 \hat{\theta}_t + a_2 \xi_t + b\varepsilon(t+1), \\ \xi_{t+1} &= A_1 \hat{\theta}_t + A_2 \xi_t + B\varepsilon(t+1), \end{aligned} \tag{14.26}$$

where $\varepsilon(t) = (\varepsilon_1(t), \dots, \varepsilon_m(t))$ is the sequence of uncorrelated vectors with uncorrelated components, $M\varepsilon_j(t) = 0$, $M\varepsilon_l^2(t) = 1$,

$$\varepsilon_j(t) = \int_{-\pi}^{\pi} e^{i\lambda(t-1)} \Phi_j(d\lambda). \tag{14.27}$$

The matrices a_i, A_i, b and B, $i = 1, 2$, in (14.26), can be found by immediate computation.

Assume now that in the vector $v_t = (\theta_t, \xi_t)$ the first component is unobservable. Consider the problem of constructing for each t, $t = 0, 1, \dots$, the linear optimal (in the mean square sense) estimate for θ_t from the observations $(\xi_0, \dots, \xi_t)$.

If v_t, $t = 0, 1, \dots$, is a Gaussian process, then by Theorem 13.4 and Corollary 1 of the thcorem, $\hat{m}_t = M(\hat{\theta}_t | \mathcal{F}_t^{\xi})$ and $\hat{\gamma}_t = M([\hat{\theta}_t - \hat{m}_t][\hat{\theta}_t - \hat{m}_t]^*)$ can be defined from the system of equations

$$\begin{aligned} \hat{m}_{t+1} &= a_1 \hat{m}_t + a_2 \xi_t \\ &\quad + (bB^* + a_1 \hat{\gamma}_t A_1^*)(BB^* + A_1 \hat{\gamma}_t A_1^*)^+ (\xi_{t+1} - A_1 \hat{m}_t - A_2 \xi_t), \end{aligned} \tag{14.28}$$

$$\hat{\gamma}_{t+1} = a_1\hat{\gamma}_t a_1^* + bb^* \\ -(bB^*+a_1\hat{\gamma}_t A_1^*)(BB^*+A_1\hat{\gamma}_t A_1^*)^+(bB^*+a_1\hat{\gamma}_t A_1^*)^*, \quad (14.29)$$

to be solved under the initial conditions

$$\hat{m}_0 = M(\hat{\theta}_0|\xi_0), \quad \hat{\gamma}_0 = M([\hat{\theta}_0 - \hat{m}_0][\hat{\theta}_0 - \hat{m}_0]^*).$$

According to the theorem on normal correlation (Theorem 13.1),

$$\hat{m}_0 = \mathrm{cov}\,(\hat{\theta}_0, \xi_0)\mathrm{cov}^+(\xi_0, \xi_0)\xi_0, \tag{14.30}$$

$$\hat{\gamma}_0 = \mathrm{cov}\,(\hat{\theta}_0, \hat{\theta}_0) - \mathrm{cov}\,(\hat{\theta}_0, \xi_0)\mathrm{cov}^+(\xi_0, \xi_0)\mathrm{cov}\,(\hat{\theta}_0, \xi_0). \tag{14.31}$$

Since $\hat{m}_t = M(\hat{\theta}_t|\mathcal{F}_t^\xi)$ depends linearly on $\xi_0, \dots, \xi_t$, for the Gaussian process $\gamma_t = [\theta_t, \xi_t]$ the solution of the problem of constructing the optimal linear estimate of θ_t from $\xi_0, \dots, \xi_t$ is given by Equations (14.28) and (14.29).

In the general case the optimal (in the mean square sense) linear estimate can be also defined from the same equations. This assertion is validated by the following:

Lemma 14.1. *Let (α, β) be a random vector with $M(\alpha^2 + \beta^2) < \infty$ and let $(\tilde{\alpha}, \tilde{\beta})$ be the Gaussian vector with the same two first moments as in (α, β), i.e.,*

$$M\tilde{\alpha}^i = M\alpha^i, \quad M\tilde{\beta}^i = M\beta^i, \quad i = 1, 2,$$

$$M\tilde{\alpha}\tilde{\beta} = M\alpha\beta.$$

Let $l(b)$ be the linear function of $b \in \mathbb{R}^1$, such that (P-a.s.)

$$l(\tilde{\beta}) = M(\tilde{\alpha}|\tilde{\beta}). \tag{14.32}$$

Then $l(\beta)$ is the optimal (in the mean square sense) linear estimate of the value of α from β, $Ml(\beta) = M\alpha$.

PROOF. First of all note that the existence of the linear function $l(\beta)$ with the property given by (14.32) follows from the theorem on normal correlation.

The unbiasedness ($Ml(\beta) = M\alpha$) of the linear estimate follows from the following explicit chain of equalities:

$$Ml(\beta) = Ml(\tilde{\beta}) = M[M(\tilde{\alpha}|\tilde{\beta})] = M\tilde{\alpha} = M\alpha.$$

Next, if $\tilde{l}(\beta)$ is some other linear estimate, then

$$M[\tilde{\alpha} - \tilde{l}(\tilde{\beta})]^2 \geq M[\tilde{\alpha} - l(\tilde{\beta})]^2.$$

Hence, by virtue of the linearity of the estimates $l(\beta)$ and $\tilde{l}(\beta)$,

$$M[\alpha - \tilde{l}(\beta)]^2 = M[\tilde{\alpha} - \tilde{l}(\tilde{\beta})]^2 \geq M[\tilde{\alpha} - l(\tilde{\beta})]^2 = M[\alpha - l(\beta)]^2,$$

which proves the optimality (in the mean square sense) of $l(\beta)$ in the class of linear estimates. □

Note. The assertion of the lemma holds true if α and β are vectors, $\alpha = (\alpha_1, \dots, \alpha_k)$, $\beta = (\beta_1, \dots, \beta_l)$.

In order to apply Lemma 14.1 to prove that the optimal estimate of θ_t from $\xi_0, \dots, \xi_t$ is defined by the system of equations given by (14.28) and (14.29), it remains only to note that the process $(\hat{\theta}_t, \xi_t)$ satisfying (14.26) and the Gaussian process defined by the same system have the same first two moments.

14.1.3. To illustrate the approach suggested above to the problems of estimating components of stationary processes we shall discuss the following.

EXAMPLE 1. Let θ_t and ζ_t, $t = 0, \pm 1, \dots$, be mutually uncorrelated stationary (wide-sense) sequences with $M\theta_t = M\zeta_t = 0$ and the spectral densities

$$f_\theta(\lambda) = \frac{1}{|e^{i\lambda} + c_1|^2}, \quad f_\zeta(\lambda) = \frac{1}{|e^{i\lambda} + c_2|^2},$$

where $|c_i| < 1$, $i = 1, 2$.

We shall assume that θ_t is a 'useful signal', ζ_t is 'noise', and that the process

$$\xi_t = \theta_t + \zeta_t \tag{14.33}$$

is observed.

According to Theorem 14.1, we can find uncorrelated sequences $\varepsilon_1(t)$ and $\varepsilon_2(t)$, $t = 0, \pm 1, \dots$, with $M\varepsilon_i(t) = 0$, $M\varepsilon_i(t)\varepsilon_i(s) = \delta(t, s)$, $i = 1, 2$, such that

$$\theta_{t+1} = c\theta_t + \varepsilon_1(t+1), \quad \xi_{t+1} = c_2\zeta_t + \varepsilon_2(t+1). \tag{14.34}$$

Taking into account (14.33) and (14.34), we obtain

$$\xi_{t+1} = \theta_{t+1} + \zeta_{t+1} = (c_1 - c_2)\theta_t + c_2\xi_t + \varepsilon_1(t+1) + \varepsilon_2(t+1).$$

Hence the 'unobservable' process θ_t and the 'observable' process ξ_t satisfy the system of equations

$$\begin{aligned} \theta_{t+1} &= c_1\theta_t + \varepsilon_1(t+1), \\ \xi_{t+1} &= (c_1 - c_2)\theta_t + c_2\xi_t + \varepsilon_1(t+1) + \varepsilon_2(t+1). \end{aligned} \tag{14.35}$$

Due to (14.28) and (14.29), the optimal linear estimate m_t, $t = 0, 1, \dots$, of the values of θ_t and the mean square filtering error $\gamma_t = M(\theta_t - m_t)^2$ satisfy the recursive equations

$$m_{t+1} = c_1 m_t + \frac{1 + c_1(c_1 - c_2)\gamma_t}{2 + (c_1 - c_2)^2\gamma_t}[\xi_{t+1} - (c_1 - c_2)m_t - c_2\xi_t], \tag{14.36}$$

$$\gamma_{t+1} = c_1^2\gamma_t + 1 - \frac{[1 + c_1(c_1 - c_2)\gamma_t]^2}{2 + (c_1 - c_2)^2\gamma_t}. \tag{14.37}$$

Let us find the initial conditions m_0, γ_0 for this system of equations.

The process (θ_t, ξ_t), $t = 0, \pm 1, \ldots$, is a stationary (wide-sense) process with $M\theta_t = M\xi_t = 0$ and the covariances $d_{11} = M\theta_t^2$, $d_{12} = M\theta_t\xi_t$, and $d_{22} = M\xi_t^2$ satisfying, due to (14.35) and (14.20), the system of equations

$$\begin{aligned} d_{11} &= c_1^2 d_{11} + 1, \\ d_{12} &= c_1(c_1 - c_2)d_{11} + c_1c_2d_{12} + 1, \\ d_{22} &= (c_1 - c_2)^2 d_{11} + c_2^2 d_{22} + 2c_2(c_1 - c_2)d_{12} + 2. \end{aligned}$$

From this we find

$$d_{11} = \frac{1}{1 - c_1^2}, \quad d_{12} = \frac{1}{1 - c_1^2}, \quad d_{22} = \frac{2 - c_1^2 - c_2^2}{(1 - c_1^2)(1 - c_2^2)},$$

which, together with (14.30) and (14.31), gives

$$m_0 = \frac{d_{12}}{d_{22}}\xi_0 = \frac{1 - c_2^2}{2 - c_1^2 - c_2^2}\xi_0,$$

$$\gamma_0 = d_{11} - \frac{d_{12}^2}{d_{22}} = \frac{1}{1 - c_1^2} - \frac{1 - c_2^2}{(1 - c_1^2)(2 - c_1^2 - c_2^2)} = \frac{1}{2 - c_1^2 - c_2^2}.$$

Thus the optimal (in the mean square sense) linear estimate m_t of the 'useful signal' θ_t from $\xi_0, \ldots, \xi_t$ and the mean square error γ_t are defined by means of the system of equations given by (14.36) and (14.37), and can be solved under the initial conditions

$$m_0 = \frac{1 - c_2^2}{2 - c_1^2 - c_2^2}\xi_0, \quad \gamma_0 = \frac{1}{2 - c_1^2 - c_2^2}.$$

In the case of estimation of the parameter θ_t from the observations $(\xi_{-N}, \ldots, \xi_0, \ldots, \xi_t)$ the system of equations given by (14.36) and (14.37) also holds true, and

$$m_{-N} = \frac{1 - c_2^2}{2 - c_1^2 - c_2^2}\xi_{-N}, \quad \gamma_{-N} = \frac{1}{2 - c_1^2 - c_2^2}.$$

14.1.4. In conclusion, we note that the optimal linear estimates of interpolation and extrapolation for a stationary sequence with a rational spectrum can be obtained (as in the case of filtering) from the results of the previous chapter if we discuss only Gaussian sequences with the same first two moments.

14.2 Maximum Likelihood Estimates for Coefficients of Linear Regression

14.2.1. At $t = 0, 1, \dots$, let the random process

$$\xi(t) = \sum_{i=1}^{N} \alpha_i(t)\theta_i + \eta(t), \tag{14.38}$$

be observed, where $\theta = (\theta_1, \dots, \theta_N)$ is the vector (column) of unknown parameters, $-\infty < \theta_i < \infty$, $i = 1, \dots, n$, $\alpha(t) = (\alpha_1(t), \dots, \alpha_N(t))$ is the known vector function (row) and $\eta(t)$, $t = 0, \pm 1, \dots$, is the Gaussian stationary random process with $M\eta(t) = 0$ and the rational spectral density

$$f_\eta(\lambda) = \left| \frac{P_{n-1}(e^{i\lambda})}{Q_n(e^{i\lambda})} \right|^2 . \tag{14.39}$$

In (14.39),

$$P_{n-1}(z) = \sum_{j=0}^{n-1} b_j z^j, \quad b_{n-1} \neq 0,$$

$$Q_n(z) = \sum_{j=0}^{n} a_j z^j, \quad a_n = 1,$$

where it is assumed that the roots of the equation $Q_n(z) = 0$ lie within the unit circle.

In order to obtain the estimates of maximal likelihood of the vector $\theta = (\theta_1, \dots, \theta_N)$ one needs to find the Radon–Nikodym derivative $d\mu_\xi^\theta / d\mu_\xi^0$ of the measure μ_ξ^θ, corresponding to the process $\xi = (\xi(t))$, $t = 0, 1, \dots$, defined in (14.38), over the measure μ_ξ^0 for the same process with $\theta = 0$ (0 is the zero vector).

According to Theorem 14.1, the process $\eta(t)$, $t = 0, \pm 1, \dots$, is a component of the process $(\eta_1(t), \dots, \eta_n(t))$ with $\eta_1(t) = \eta(t)$ defined by the equations

$$\begin{aligned} \eta_j(t+1) &= \eta_{j+1}(t) + \beta_j \varepsilon(t+1), \quad j = 1, \dots, n-1, \\ \eta_n(t+1) &= -a_0 \eta_1(t) - \sum_{j=1}^{n-1} a_j \eta_{j+1}(t) + \beta_n \varepsilon(t+1), \end{aligned} \tag{14.40}$$

where $\varepsilon(t)$, $t = 0, \pm 1, \dots$, is some sequence of independent Gaussian random variables with $M\varepsilon(t) = 0$, $M\varepsilon^2(t) = 1$, and where the numbers $\beta_1, \dots, \beta_n$ are given by (14.8).

Since $\xi(t+1) = \alpha(t+1)\theta + \eta_1(t+1)$, for the process $(\xi_1(t), \dots, \xi_n(t))$, with $\xi_1(t) = \xi(t)$, $\xi_j(t) = \eta_j(t)$, $j = 2, \dots, n$, we have the system of recursive equations

$$\begin{aligned} \xi_1(t+1) &= \alpha(t+1)\theta + \xi_2(t) + \beta_1\varepsilon(t+1), \\ \xi_k(t+1) &= \xi_{k+1}(t) + \beta_k\varepsilon(t+1), \quad 1 < k < n, \\ \xi_n(t+1) &= -a_0(\xi_1(t) - \alpha(t)\theta) - \sum_{j=1}^{n-1} a_j\xi_{j+1}(t) + \beta_n\varepsilon(t+1). \end{aligned} \tag{14.41}$$

For a fixed value of θ let us write

$$m_k^\theta(t) = M[\xi_k(t)|\mathcal{F}_t^\xi], \quad k > 1,$$

$$\gamma_{ij}^\theta(t) = M[(\xi_i(t) - m_i^\theta(t))(\xi_j(t) - m_j^\theta(t))], \quad i, j > 1.$$

The system of equations given by (14.41) is a particular case of the system of equations given by (13.46) and (13.47), and, therefore, $m_k^\theta(t)$ and $\gamma_{ij}^\theta(t)$ satisfy Equations (13.56) and (13.57). It should be noted that the coefficients of the equations from which $\gamma_{ij}^\theta(t)$ are defined do not include θ. The initial conditions $\gamma_{ij}^\theta(0)$ do not depend on θ either. Therefore, the elements of the matrix $\gamma^\theta(t) = \|\gamma_{ij}^\theta(t)\|$ do not depend on θ. Hence we shall denote it simply by $\gamma(t) = \|\gamma_{ij}(t)\|$, $i, j \geq 2$.

For fixed θ, the equations for $m_k^\theta(t)$, $k = 2, \dots, n$, according to (13.56) have the following form:

$$\begin{aligned} m_k^\theta(t+1) &= m_{k+1}^\theta(t) + \frac{\beta_1\beta_k + \gamma_{2k}(t)}{\beta_1^2 + \gamma_{22}(t)} \\ &\quad \times [\xi_{t+1} - \alpha(t+1)\theta - m_2^\theta(t)], \quad 2 \leq k \leq n-1, \end{aligned} \tag{14.42}$$

$$\begin{aligned} m_n^\theta(t+1) &= -a_0(\xi_1(t) - \alpha(t)\theta) - \sum_{j=1}^{n-1} a_j m_{j+1}^\theta(t) \\ &\quad + \frac{\beta_1\beta_n - \sum_{j=1}^{n-1} a_j\gamma_{1,j+1}(t)}{\beta_1^2 + \gamma_{22}(t)}[\xi_{t+1} - \alpha(t+1)\theta - m_2^\theta(t)]. \end{aligned} \tag{14.43}$$

In solving the linear system of equations given by (14.42) and (14.43) we establish that

$$m_2^\theta(t) = \nu_0(t, \xi) + \nu_1(t)\theta, \tag{14.44}$$

where $\nu_0(t, \xi)$ is a $\mathcal{F}_t^\xi$-measurable function linearly dependent on $\xi_0, \dots, \xi_t$, and $\nu_1(t) = (\nu_{11}(t), \dots, \nu_{1N}(t))$ is a nonrandom vector function (row).

Let us apply Theorem 13.5 to $\xi_1(t) = \xi(t)$. Then (for fixed θ) there exists a sequence of independent Gaussian random variables $\tilde{\varepsilon}(t)$, $t = 0, 1, \dots$, with

$M\tilde{\varepsilon}(t) = 0$, $M\tilde{\varepsilon}^2(t) = 1$, $\mathcal{F}_t^\xi = \sigma\{\omega : \xi(0), \dots, \xi(t)\}$-measurable for each t (since $\beta_1 = b_{n-1} \neq 0$), such that (P-a.s.)

$$\xi(t+1) = \alpha(t+1)\theta + m_2^\theta(t) + \sqrt{\beta_1^2 + \gamma_{22}(t)}\tilde{\varepsilon}(t+1). \tag{14.45}$$

Therefore, making use of (14.44), we obtain

$$\xi(t+1) = [\alpha(t+1) + \nu_1(t)]\theta + \nu_0(t,\xi) + \beta(t)\tilde{\varepsilon}(t+1), \tag{14.46}$$

where

$$\beta(t) = \sqrt{\beta_1^2 + \gamma_{22}(t)}.$$

But $\tilde{\varepsilon}(t)$, $t = 0, 1, \dots$, are independent Gaussian random variables with $M\tilde{\varepsilon}(t) = 0$, $M\tilde{\varepsilon}^2(t) = 1$. Hence, we find from (14.46) that

$$\begin{aligned}\frac{d\mu_\xi^\theta}{d\mu_\xi^0}(\xi(0), \dots, \xi(t)) = \exp\Bigg\{ &\frac{\xi(0)\alpha(0)\theta}{\delta^2} - \frac{(\alpha(0)\theta)^2}{2\delta^2} \\ &+ \sum_{s=1}^{t}\Bigg(\frac{[\xi(s) - \nu_0(s-1,\xi)][\alpha(s) + \nu_1(s-1)]\theta}{\beta^2(s-1)} \\ &- \frac{1}{2}\frac{[(\alpha(s) + \nu_1(s-1))\theta]^2}{\beta^2(s-1)}\Bigg)\Bigg\},\end{aligned} \tag{14.47}$$

where $\delta^2 = M\eta^2(0)$.

Assume that at some $t \geq N-1$ the matrix

$$D_t = \frac{\alpha^*(0)\alpha(0)}{\delta^2} + \sum_{s=1}^{t} \frac{[\alpha(s) + \nu_1(s-1)]^*[\alpha(s) + \nu_1(s-1)]}{\beta^2(s-1)} \tag{14.48}$$

is nonsingular. Then from (14.47) we obtain the maximum likelihood estimate $\hat{\theta}_t$ (which maximizes the right-hand side of (14.47)) given by the formula

$$\hat{\theta}_t = D_t^{-1}\left\{ \frac{\alpha^*(0)\xi(0)}{\delta^2} + \sum_{s=1}^{t} \frac{[\alpha(s) + \nu_1(s-1)]^*[\xi(s) - \nu_0(s-1,\xi)]}{\beta^2(s-1)} \right\}. \tag{14.49}$$

It is easy to deduce from (14.48) and (14.49) that the estimate $\hat{\theta}_t$ is unbiased ($M_\theta\hat{\theta}_t = \theta$) and that

$$M_\theta[(\hat{\theta}_t - \theta)(\hat{\theta}_t - \theta)^*] = D_t^{-1}. \tag{14.50}$$

With the help of simple transformations it follows from (14.47) and (14.49) that

$$\frac{d\mu_\xi^\theta}{d\mu\xi^0}(\xi(0), \dots, \xi(t)) = \exp\left\{\theta^* D_t\hat{\theta}_t - \frac{1}{2}\theta^* D_t\theta\right\}. \tag{14.51}$$

It is seen from this, in particular, that $\hat{\theta}_t$ is a sufficient statistic for the problem under consideration (see Section 1.5).

We shall show now that in the class of unbiased estimates

$$\tilde{\theta}_t = (\tilde{\theta}_1(t), \dots, \tilde{\theta}_k(t))$$

with $M \sum_{i=1}^N \tilde{\theta}_i^2(t) < \infty$ the estimate is $\hat{\theta}_t$ efficient, i.e.,

$$M_\theta(\tilde{\theta}_t - \theta)(\tilde{\theta}_t - \theta)^* \geq M_\theta(\hat{\theta}_t - \theta)(\hat{\theta}_t - \theta)^* = D_t^{-1}. \tag{14.52}$$

Indeed, according to the Cramer–Rao matrix inequality (1.50),

$$M(\tilde{\theta}_t - \theta)(\tilde{\theta}_t - \theta)^* \geq I^{-1}(\theta), \tag{14.53}$$

where $\tilde{\theta}_t$ is an unbiased estimate of the vector θ ($M_\theta \tilde{\theta}_t = \theta$) and $I(\theta) = \|I_{ij}(\theta)\|$ is the Fisher information matrix with the elements

$$I_{ij}(\theta) = M_\theta \left\{ \frac{\partial}{\partial \theta_i} \ln \frac{d\mu_\xi^\theta}{d\mu_\xi^0}(\xi(0), \dots, \xi(t)) \right\} \times \left\{ \frac{\partial}{\partial \theta_j} \ln \frac{d\mu_\xi^\theta}{d\mu_\xi^0}(\xi(0), \dots, \xi(t)) \right\}.$$

But in our case,

$$I(\theta) = D_t. \tag{14.54}$$

In order to prove (14.54), introducing the notation $D_{ij}(t)$ and $\tilde{D}_{ij}(t)$ for the elements of the matrices D_t and D_t^{-1} respectively, we note that

$$\ln \frac{d\mu_\xi^\theta}{d\mu_\xi^0}(\xi(0), \dots, \xi(t)) = \sum_{k,l=1}^N D_{kl}(t) \theta_k \left[\hat{\theta}_l(t) - \frac{1}{2}\theta_l \right],$$

and therefore,

$$\frac{\partial}{\partial \theta_j} \ln \frac{d\mu_\xi^\theta}{d\mu_\xi^0}(\xi(0), \dots, \xi(t)) = \sum_{l=1}^N D_{jl}(t)[\hat{\theta}_l(t) - \theta_l],$$

$$\begin{aligned} I_{ij}(\theta) &= M \left\{ \frac{\partial}{\partial \theta_i} \ln \frac{d\mu_\xi^\theta}{d\mu_\xi^0}(\xi(0), \dots, \xi(t)) \right\} \left\{ \frac{\partial}{\partial \theta_j} \ln \frac{d\mu_\xi^\theta}{d\mu_\xi^0}(\xi(0), \dots, \xi(t)) \right\} \\ &= \sum_{l,k=1}^N D_{jl}(t) D_{ik}(t) M[\hat{\theta}_l(t) - \theta_l][\hat{\theta}_k(t) - \theta_k] \\ &= \sum_{l,k=1}^N D_{jl}(t) D_{ik}(t) \tilde{D}_{lk}(t) = \sum_{l=1}^N D_{jl}(t) \left(\sum_{k=1}^N D_{ik}(t) \tilde{D}_{lk})(t) \right) \\ &= \sum_{l=1}^N D_{jl}(t) \delta(i,l) = D_{ij}(t). \end{aligned}$$

We obtain the required inequality (14.52) from (14.53) and (14.54), which proves the efficiency of the estimate θ_t.

Note. The method used above for the deduction of (14.49) and (14.51) can also be applied in the case where $b_{n-1} = b_{n-2} = \cdots = b_{n-m} = 0$, $b_{n-m-1} \neq 0$.

14.2.2.
EXAMPLE 2. Let $\xi(t) = \theta + \eta(t)$, where θ is an unknown parameter, $-\infty < \theta < \infty$, and $\eta(t)$, $t = 0, \pm 1, \ldots$, is a stationary Gaussian process with $M\eta(t) = 0$ and the spectral density

$$f(\lambda) = \left| \frac{e^{i\lambda} + 1}{e^{2(i\lambda)} + e^{i\lambda} + \frac{1}{2}} \right|^2 .$$

The maximum likelihood estimate of the unknown parameter θ can be also interpreted as an estimate of the mean $M\xi(t) = \theta$ of the process $\xi(t)$, $t = 0, \pm 1, \ldots$.

By Theorem 14.1 the process $\eta(t)$ is a component of the two-dimensional process $(\eta_1(t), \eta_2(t))$ with $\eta_1(t) = \eta(t)$ defined by the recursive equations

$$\begin{aligned} \eta_1(t+1) &= \eta_2(t) + \varepsilon(t+1), \\ \eta_2(t+1) &= -\frac{1}{2}\eta_1(t) - \eta_2(t) + \frac{1}{2}\varepsilon(t+1) \end{aligned}$$

and a sequence of the independent Gaussian random variables $\varepsilon(t)$, $t = 0, \pm 1, \ldots$, where $M\varepsilon(t) = 0$ and $M\varepsilon^2(t) = 1$. From this we see that

$$\xi(t+1) = \theta + \eta_2(t) + \varepsilon(t+1),$$

$$\eta_2(t+1) = -\frac{\theta - \xi(t)}{2} - \eta_2(t) + \frac{1}{2}\varepsilon(t+1).$$

In accordance with this, $m^\theta(t) = M(\eta_2(t) | \mathcal{F}_t^\xi)$ is a solution of the recursive equation

$$m_{t+1}^\theta = -\frac{\theta - \xi(t)}{2} - m_t^\theta + \frac{\frac{1}{2} + \gamma_t}{1 + \gamma_t}(\xi(t+1) - \theta - m^\theta(t)),$$

where (see (13.57))

$$\gamma_{t+1} = \gamma_t + \frac{1}{4} - \frac{\left(\frac{1}{2} + \gamma_t\right)^2}{1 + \gamma_t}.$$

By virtue of Equation (14.20),

$$M\eta_1^2(t) = \frac{12}{5}, \quad M\eta_2^2(t) = \frac{7}{5}, \text{ and } M\eta_1(t)\eta_2(t) = -\frac{9}{15}.$$

Hence

$$M[\xi(t)-\theta]^2 = M\eta_1^2(t) = \frac{12}{5}, \quad M[\xi(t)-\theta]\eta_2(t) = M\eta(t)\eta_2(t) = -\frac{9}{15},$$

and, therefore, by the theorem on normal correlation (Theorem 13.1),

$$m^\theta(0) = \frac{1}{4}(\theta - \xi(0)), \quad \gamma_0 = \frac{5}{4}.$$

In solving the equation for $m^\theta(t)$ with the initial condition $m^\theta(0) = \frac{1}{4}(\theta - \xi(0))$, we obtain

$$\begin{aligned} m^\theta(t) &= \frac{1}{4}\prod_{s=0}^{t-1}\left(-\frac{\frac{3}{2}+2\gamma_s}{1+\gamma_s}\right)[\theta-\xi(0)] \\ &\quad + \sum_{s=0}^{t-1}\prod_{j=s+1}^{t-1}\left(-\frac{\frac{3}{2}+2\gamma_j}{1+\gamma_j}\right)\left[\frac{1}{2}(\xi(s)-\theta) + \frac{\frac{1}{2}+\gamma_s}{1+\gamma_s}\xi(s+1)\right] \\ &= \nu_0(t,\xi) - \nu_1(t)\theta, \end{aligned}$$

where

$$\begin{aligned} \nu_0(t,\xi) &= -\frac{1}{4}\prod_{s=0}^{t-1}\left(-\frac{\frac{3}{2}+2\gamma_s}{1+\gamma_s}\right)\xi_0 \\ &\quad + \sum_{s=0}^{t-1}\prod_{j=s+1}^{t-1}\left(-\frac{\frac{3}{2}+2\gamma_j}{1+\gamma_j}\right)\left[\frac{1}{2}\xi(s) + \frac{\frac{1}{2}+\gamma_s}{1+\gamma_s}\xi(s+1)\right], \end{aligned}$$

$$\nu_1(t) = \frac{1}{4}\prod_{s=0}^{t-1}\left(-\frac{\frac{3}{2}+2\gamma_s}{1+\gamma_s}\right) - \sum_{s=0}^{t-1}\sum_{j=s+1}^{t-1}\left(-\frac{\frac{3}{2}+2\gamma_j}{1+\gamma_j}\right).$$

Now, due to (14.48) and (14.49), we have ($t \geq 1$)

$$D_t = \left[\frac{5}{12} + \sum_{s=1}^{t}\frac{1+\nu_1(s-1))^2}{1+\gamma(s-1)}\right],$$

$$\hat{\theta}_t = D_t^{-1}\left[\frac{5}{12}\xi(0) + \sum_{s=1}^{t}\frac{(1+\nu_1(s-1))(\xi(s)-\nu_0(s-1,\xi))}{1+\gamma(s-1)}\right].$$

14.3 A Control Problem with Incomplete Data (Linear System with Quadratic Performance Index)

14.3.1. In this section we shall show how the optimal nonlinear filtering equations deduced in the preceding chapter can be applied to optimal control problems.

It will be assumed that the state of some 'controlled' system is described by the process $(\theta,\xi) = [(\theta_1(t),\dots,\theta_k(t)),(\xi_1(t),\dots,\xi_l(t))]$, $t = 0,1,\dots,T < \infty$, which obeys the equations

$$\begin{aligned}\theta_{t+1} &= c(t)u_t + a(t)\theta_t + b(t)\varepsilon_1(t+1),\\ \xi_{t+1} &= A(t)\theta_t + B(t)\varepsilon_2(t+1).\end{aligned} \tag{14.55}$$

Here $c(t), a(t), b(t), A(t)$ and $B(t)$ are matrices of dimension $(k\times r)$, $(k\times k)$, $(k\times k)$, $(l\times k)$, $(l\times l)$, respectively, whose elements are deterministic bounded functions, $t = 0,1,\dots,T-1$. The mutually independent random sequences

$$\varepsilon_1(t) = (\varepsilon_{11}(t),\dots,\varepsilon_{1k}(t)),\quad \varepsilon_2(t) = (\varepsilon_{21}(t),\dots,\varepsilon_{2l}(t)),\quad t = 1,\dots,T,$$

in (14.55) are Gaussian with the independent components, $M\varepsilon_{ij}(t) = 0$, $M\varepsilon_{ij}^2(t) = 1$.

The system of equations given by (14.55) can be solved under the initial condition θ_0, where θ_0 is Gaussian,

$$M\theta_0 = m,\quad M[(\theta_0 - m_0)(\theta_0 - m_0)^*] = \gamma,$$

independent of the sequences $\varepsilon_i(t)$, $i = 1,2$, $t = 1,\dots,T$. (14.55) includes as well the vector column $u_t = (u_1(t,\xi),\dots,u_r(t,\xi))$, where at each t, $t = 0,1,\dots,T-1$, the functions $u_i(t,\xi)$, playing the role of controlling actions, are $\mathcal{F}_i^\xi = \sigma\{\omega : \xi_0,\dots,\xi_t\}$-measurable ($\xi_0 = 0$).

All the controls $u = (u_0,\dots,u_{T-1})$ discussed from now on will be assumed to satisfy

$$\sum_{i=1}^{r} Mu_i^2(t,\xi) < \infty,\quad t = 0,1,\dots,T-1. \tag{14.56}$$

Assume that the control performance of $u = (u_0,\dots,u_{T-1})$ is measured by the quadratic performance index

$$V(u) = M\left[\sum_{t=0}^{T-1}(\theta_t^* H(t)\theta_t + u_t^* R(t)u_t) + \theta_T^* H(T)\theta_T\right], \tag{14.57}$$

where $H(t)$ and $R(t)$ are deterministic, bounded, symmetric nonnegative definite matrices of the orders $(k\times k)$ and $(r\times r)$, respectively.

It is necessary to find the (optimal) control $\tilde{u} = (\tilde{u}_0,\dots,\tilde{u}_{T-1})$ for which

$$V(\tilde{u}) = \inf V(u), \tag{14.58}$$

where 'inf' is taken over all the controls satisfying (14.56).

This problem is an example of control problems with incomplete data where the control must be based on the observable part of the coordinates $(\xi_0,\xi_1,\dots)$ describing the state of the control system.

14.3.2. In searching for optimal controls (in a given problem the existence of such controls will be clear in what follows), both the ideas of dynamic programming and the results of optimal nonlinear filtering will be employed.

We shall introduce now some additional notation.

Let $P(t)$ and γ_t, $t = 0, 1, \dots, T$, be matrix functions of the order $(k \times k)$ defined as solutions of the recursive equations

$$\begin{aligned} P(t) &= H(t) + a^*(t)P(t+1)a(t) \\ &\quad - a^*(t)P(t+1)c(t)[R(t)+c^*(t)P(t+1)c(t)]^+c^*(t)P(t+1)a(t) \end{aligned} \tag{14.59}$$

with $P(T) = H(T)$ and

$$\begin{aligned} \gamma_{t+1} &= a(t)\gamma_t a^*(t) + b(t)b^*(t) \\ &\quad - a(t)\gamma_t A^*(t)[B(t)B^*(t) + A(t)\gamma_t A^*(t)]^+ A(t)\gamma_t a^*(t) \end{aligned} \tag{14.60}$$

with $\gamma_0 = \gamma$. Also let,

$$D(t) = a(t)\gamma_t A^*(t)\left\{[B(t)B^*(t) + A(t)\gamma_t A^*(t)]^{1/2}\right\}^+, \tag{14.61}$$

and let $p(t)$, $t = 0, 1, \dots, T$, be a sequence of nonnegative numbers defined in a recursive manner:

$$p(t) = p(t+1) + \operatorname{Tr} P^{1/2}(t+1)D(t)D^*(t)P^{1/2}(t+1), \quad p(T) = 0. \tag{14.62}$$

It follows that, from (14.62),

$$p(t) = \sum_{s=t}^{T-1} \operatorname{Tr} P^{1/2}(s+1)D(s)D^*(s)P^{1/2}(s+1). \tag{14.63}$$

The matrices $P(t)$ and γ_t and the numbers $p(t)$ are found from the coefficients of the system of equations given by (14.55) and the specified matrices $H(t)$ and $R(t)$. Hence they do not depend on the data and, being only functions of t, can be found a priori.

Note that the matrices $P(t)$, $t = 0, 1, \dots, T$, found from the system of recursive equations given by (14.59) are symmetric and nonnegative definite. In order to convince ourselves of this, let us consider the problem of filtering for processes

$$\begin{aligned} \tilde{\theta}_{s+1} &= a^*(T-s)\tilde{\theta}_s + H^{1/2}(T-s)\tilde{\varepsilon}_1(s+1), \\ \tilde{\xi}_{s+1} &= c^*(T-s)\tilde{\theta}_s + R^{1/2}(T-s)\tilde{\varepsilon}_2(s+1), \end{aligned}$$

where $\tilde{\varepsilon}_1(s)$ and $\tilde{\varepsilon}_2(s)$ are independent Gaussian vectors with independent components whose means are equal to zero and whose variances are equal to one. Assume that $\tilde{\theta}_0$ is a Gaussian vector, $M\tilde{\theta}_0 = 0$, $M\tilde{\theta}_0\tilde{\theta}_0^* = H(t)$, independent of $\tilde{\varepsilon}_1(s)$, $\tilde{\varepsilon}_2(s)$, $s = 0, \dots, T-1$.

If we compare (13.57) for $\tilde{\gamma}_t = M[(\tilde{\theta}_t - \tilde{m}_t)(\tilde{\theta}_t - \tilde{m}_t)^*]$, where $\tilde{m}_t = M(\tilde{\theta}_t | \tilde{\xi}_1, \dots, \tilde{\xi}_t)$, with (14.59) for $P(t)$ we shall see that $P(t) = \tilde{\gamma}_{T-t}$. Therefore, the matrices $P(t)$ are symmetric and nonnegative definite.

14.3.3.

Theorem 14.2. *In the class of controls satisfying (14.56) an optimal control $\tilde{u} = (\tilde{u}_0, \dots, \tilde{u}_{T-1})$ exists and is given by the formulae*

$$\tilde{u}(t,\xi) = -[R(t) + c^*(t)P(t+1)c(t)]^+ c^*(t)P(t+1)a(t)\tilde{m}_t, \tag{14.64}$$

where the matrices $P(t)$ can be defined from (14.59), and $\tilde{m}_t$ can be found from the recursive equations of optimal filtering

$$\begin{aligned}\tilde{m}_{t+1} &= c(t)\tilde{u}_t + a(t)\tilde{m}(t)\\ &\quad + a(t)\gamma_t A^*(t)[B(t)B^*(t) + A(t)\gamma_t A^*(t)]^+[\xi_{t+1} - A(t)\tilde{m}_t]\end{aligned} \tag{14.65}$$

with $\tilde{m}_0 = m$ and the matrices γ_t defined in (14.60).

The observable process ξ_t, $t = 1, \dots, T$, in (14.65), can be defined by the system of equations

$$\begin{aligned}\theta_{t+1} &= c(t)\tilde{u}_t + a(t)\theta_t + b(t)\varepsilon_1(t+1),\\ \xi_{t+1} &= A(t)\theta_t + B(t)\varepsilon_2(t+1),\end{aligned} \tag{14.66}$$

and

$$V(\tilde{u}) = p(0) + m^*P(0)m + \sum_{t=0}^{T} \operatorname{Tr} H^{1/2}(t)\gamma_t H^{1/2}(t). \tag{14.67}$$

PROOF. Let $u = (u_0, \dots, u_{T-1})$ be some control satisfying (14.56). Then $M\sum_{t=0}^{T} \theta_t^*\theta_t < \infty$ and

$$V(u) = M\sum_{t=0}^{T} M(\theta_t^* H(t)\theta_t | \mathcal{F}_t^\xi) + M\sum_{t=0}^{T-1} u_t^* R(t) u_t. \tag{14.68}$$

For the control $u = (u_0, \dots, u_{T-1})$, let

$$m_t^u = M(\theta_t^u | \mathcal{F}_t^{\xi^u}), \quad \gamma_t^u = M[(\theta_t^u - m_t^u)(\theta_t^u - m_t^u)^*],$$

where the corresponding controlled processes θ_t^u and ξ_t^u were defined in (14.55). It should be emphasized that for any control $u = (u_0, \dots, u_{T-1})$ subject to (14.56) the matrices γ_t^u satisfy the system of recursive equations given by (14.60) (see Theorem 13.4 and Property 3 in Subsection 13.2.4). Since neither the coefficients of these equations nor the initial conditions depend on the control, the matrices γ_t^u are the same for different u. Hence, $\gamma_t^u \equiv \gamma_t$ (see (14.60)). Let us show now that in (14.68)

$$M(\theta_t^{u*} H(t)\theta_t^u | \mathcal{F}_t^{\xi^u}) = (m_t^u)^* H(t) m_t^u + \operatorname{Tr} H^{1/2}(t)\gamma_t H^{1/2}(t). \tag{14.69}$$

We have

$$
\begin{aligned}
& M(\theta_t^{u*} H(t)\theta_t^u | \mathcal{F}_t^{\xi^u}) = M[(\theta_t^u - m_t^u + m_t^u)^* H(t)(\theta_t^u - m_t^u - m_t^u)|\mathcal{F}_t^{\xi^u}] \\
= & (m_t^u)^* H(t) m_t^u + 2M[(\theta_t^u - m_t^u)H(t)m_t^u|\mathcal{F}_t^{\xi^u}] \\
& + M[(\theta_t^u - m_t^u)^* H(t)(\theta_t^u - m_t^u)|\mathcal{F}_t^{\xi^u}] \\
= & (m_t^u)^* H(t) m_t^u + \operatorname{Tr} M[H^{1/2}(t)(\theta_t^u - m_t^u)(\theta_t^u - m_t^u)^* H^{1/2}(t)|\mathcal{F}_t^{\xi^u}] \\
= & (m_t^u)^* H(t) m_t^u + \operatorname{Tr} H^{1/2}(t) M[(\theta_t^u - m_t^u)(\theta_t^u - m_t^u)^*|\mathcal{F}_t^{\xi^u}] H^{1/2}(t).
\end{aligned}
\tag{14.70}
$$

But, according to Property 3 in Subsection 13.2.4,

$$
M[(\theta_t^u - m_t^u)(\theta_t^u - m_t^u)^*|\mathcal{F}_t^{\xi^u}] = M[(\theta_t^u - m_t^u)(\theta_t^u - m_t^u)^*] = \gamma_t,
$$

which, together with (14.70), proves (14.69).

Thus, due to (14.68) and (14.69),

$$
V(u) = \sum_{t=0}^{T} \operatorname{Tr} H^{1/2}(t)\gamma_t H^{1/2}(t) + M\sum_{t=0}^{T}(m_t^u)^* H(t) m_t^u + M\sum_{t=0}^{T-1} u_t^* R(t) u_t.
\tag{14.71}
$$

Since the functions $\operatorname{Tr} H^{1/2}(t)\gamma_t H^{1/2}(t)$ depend only on t and do not depend on either the control or the processes describing the state of the system, it is obvious that the optimal control $\tilde{u}$ in the primary problem (assuming it exists) coincides with the optimal control in the problem of minimization of the functional

$$
\overline{V}(u) = M\left(\sum_{t=0}^{T}(m_t^u)^* H(t) m_t^u + \sum_{t=0}^{T-1} u_t^* R(t) u_t\right).
\tag{14.72}
$$

The 'controlled' process m_t^u is defined by the equation

$$
\begin{aligned}
m_{t+1}^u = {} & c(t)u_t + a(t)m_t^u \\
& + a(t)\gamma_t A^*(t)[B(t)B^*(t) + A(t)\gamma_t A^*(t)]^+[\xi_{t+1}^u - A(t)m_t^u].
\end{aligned}
\tag{14.73}
$$

According to Theorem 13.5, there exists a sequence of independent Gaussian vectors $\overline{\varepsilon}^u(t) = (\overline{\varepsilon}_1^u(t), \dots, \overline{\varepsilon}_l^u(t))$, $t = 1, \dots, T$, with independent components $M\overline{\varepsilon}_i^u(t) = 0$, $M(\overline{\varepsilon}_u^i(t))^2 = 1$, $i = 1, \dots, l$, such that

$$
m_{t+1}^u = c(t)u_t + a(t)m_t^u + D(t)\overline{\varepsilon}^u(t+1).
\tag{14.74}
$$

It should be noted here that for every permissible u the values of $\overline{\varepsilon}^u(t)$ coincide ($\overline{\varepsilon}^u(t) \equiv \overline{\varepsilon}(t)$, $t = 1, \dots, T$). This follows from (13.84) and the fact that the $\theta_t^u - m_t^u$ do not depend on u (see (14.73) and (14.55))

Thus the primary problem of determining the optimal control for the system (14.55) and the functional (14.57) is reduced to a problem of finding the optimal control for the filtered system given by (14.74) with the functional (14.72) ('the separation principle' [313]).

14.3.4. In finding optimal controls in this reduced problem the following two lemmas will be useful.

Lemma 14.2. *If $u = (u_0, \dots, u_{T-1})$ is the control subject to (14.56), then for any nonnegative definite symmetric matrix $S(t+1)$*

$$\begin{aligned} & M[(m_{t+1}^u)^* S(t+1) m_{t+1}^u | \mathcal{F}_t^{\xi^u}] \\ = {} & M[(m_{t+1}^u)^* S(t+1) m_{t+1}^u | m_t^u, u_t] \\ = {} & (m_t^u)^* a^*(t) S(t+1) a(t) m_t^u + u_t^* c^*(t) S(t+1) c(t) u_t \\ & + 2u_t^* c^*(t) S(t+1) a(t) m_t^u \\ & + \operatorname{Tr} S^{1/2}(t+1) D(t) D^*(t) S^{1/2}(t+1). \end{aligned} \tag{14.75}$$

PROOF. Due to (14.74),

$$\begin{aligned} & M[(m_{t+1}^u)^* S(t+1) m_{t+1}^u | \mathcal{F}_t^{\xi^u}] \\ = {} & M\Big\{ [c(t)u_t + a(t)m_t^u - D(t)\overline{\varepsilon}(t+1)]^* \\ & \times S(t+1)[c(t)u_t + a(t)m_t^u + D(t)\overline{\varepsilon}(t+1)] | \mathcal{F}_t^{\xi^u} \Big\} \\ = {} & (m_t^u)^* a^*(t) S(t+1) a(t) m_t^u + u_t^* c^*(t) S(t+1) c(t) u_t \\ & + 2u_t^* c^*(t) S(t+1) a(t) m_t^u \\ & + 2M(\overline{\varepsilon}^*(t+1) | \mathcal{F}_t^{\xi^u}) D^*(t) S(t+1)(c(t)u_t + a(t)m_t^u) \\ & + M[\overline{\varepsilon}^*(t+1) D^*(t) S(t+1) D(t) \overline{\varepsilon}(t+1) | \mathcal{F}_t^{\xi^u}] \\ = {} & (m_t^u)^* a^*(t) S(t+1) a(t) m_t^u + u_t^* c^*(t) S(t+1) c(t) u_t \\ & + 2u_t^* c(t) S(t+1) a(t) m_t^u + \operatorname{Tr} S^{1/2}(t+1) D(t) D^*(t) S^{1/2}(t+1), \end{aligned}$$

where we took advantage of the fact that $M(\overline{\varepsilon}(t+1) | \mathcal{F}_t^{\xi^u}) = 0$ and

$$\begin{aligned} & M[\overline{\varepsilon}^*(t+1) D^*(t) S(t+1) D(t) \overline{\varepsilon}(t+1) | \mathcal{F}_t^{\xi^u}] \\ = {} & M[\overline{\varepsilon}^*(t+1) D^*(t) S(t+1) D(t) \overline{\varepsilon}(t+1)] \\ = {} & \operatorname{Tr} S^{1/2}(t+1) D(t) M\overline{\varepsilon}(t+1)\overline{\varepsilon}^*(t+1) D^*(t) S^{1/2}(t+1) \\ = {} & \operatorname{Tr} S^{1/2}(t+1) D(t) D^*(t) S^{1/2}(t). \end{aligned} \tag{14.76}$$

□

Note. Let $\overline{\delta}(1), \dots, \overline{\delta}(T)$ be a sequence of independent Gaussian vectors $(\overline{\delta}(t) = (\overline{\delta}_1(t), \dots, \overline{\delta}_l(t)))$ with the independent components having zero mean and unit variances. Consider the process m, $t = 0, \dots, T$, defined by the recursive relations

$$m_{t+1} = c(t)u_t + a(t)m_t + D(t)\overline{\delta}(t+1), \quad m_0 = m, \tag{14.77}$$

where $u_t = u_t(\omega)$ does not depend on $\overline{\delta}(t+1)$. As in the proof of (14.75), we show here that

$$\begin{aligned} & M[m_{t+1}^* S(t+1) m_{t+1} | u_t, m_t] \\ = & \; m_t^* a^*(t) S(t+1) a(t) m_t \\ & + u_t^* c^*(t) S(t+1) c(t) u_t \\ & + 2u_t^* c^*(t) S(t+1) a(t) m_t \\ & + \operatorname{Tr} S^{1/2}(t+1) D(t) D^*(t) S^{1/2}(t+1). \end{aligned} \tag{14.78}$$

Let the matrices $P(x)$ introduced above and the functions $p(t)$, $t = 0, \dots, T$, be related to the scalar functions

$$Q_t(x) = p(t) + x^* P(t) x, \tag{14.79}$$

where $x \in \mathbb{R}^k$. Since $p(T) = 0$, and $P(T) = H(T)$,

$$Q_T(x) = x^* H(T) x. \tag{14.80}$$

Lemma 14.3. *The functions $Q_t(x)$, $t = 0, 1, \dots, T$, satisfy the recursive equations*

$$Q_t(x) = \inf_V \left\{ x^* H(t) x + V^* R(t) V + M[Q_{t+1}(x_{t+1}^{x,V})] \right\}, \tag{14.81}$$

where $V \in \mathbb{R}^r$, $x \in \mathbb{R}^k$,

$$x_{t+1}^{x,V} = c(t) V + a(t) x + D(t) \overline{\delta}(t+1). \tag{14.82}$$

In this case, the inf *in (14.81) can be attained on the r-dimensional vector*

$$\tilde{V} = -[R(t) + c^*(t) P(t+1) c(t)]^+ c^*(t) P(t+1) a(t) x. \tag{14.83}$$

PROOF. Let us verify that the functions $Q_t(x) = p(t) + x^* P(t) x$ satisfy Equation (14.81), i.e., that

$$\begin{aligned} p(t) + x^* P(t) x = & \inf_V \Big\{ x^* H(t) x + V^* R(t) V + p(t+1) \\ & + M[(x_{t+1}^{x,V})^* P(t+1) x_{t+1}^{x,V}] \Big\}. \end{aligned} \tag{14.84}$$

Set

$$J(V, x) = V^*[R(t) + c^*(t) P(t+1) c(t)] V + 2V^* c^*(t) P(t+1) a(t) x. \tag{14.85}$$

Then, taking into account the note to Lemma 14.2, we find that (14.84) is equivalent to the equations

$$p(t) + x^* P(t) x = p(t+1) + \operatorname{Tr} P^{1/2}(t+1) D(t) D^*(t) P^{1/2}(t+1) + \inf_V J(V, x).$$

But, due to (14.62),

$$p(t) = p(t+1) + \operatorname{Tr} P^{1/2}(t+1)D(t)D^*(t)P^{1/2}(t+1).$$

Hence, it need only be verified that

$$x^*P(t)x = \inf_V J(V,x) \tag{14.86}$$

for any $x \in \mathbb{R}^k$.

If the matrices $R(t), t \geq 0$, in $J(V,x)$ were positive definite, then $J(V,x) > -\infty$ and $\inf_V J(V,x)$ would be attained on the vector

$$\tilde{V} = -[R(t) + c^*(t)P(t+1)c(t)]^+ c^*(t)P(t+1)a(t)x, \tag{14.87}$$

and it would be easy to check immediately that $J(\tilde{V},x) = x^*P(t)x$.

In order to prove (14.86) in the general case we shall consider the system of algebraic equations (with respect to $V = (V_1, \dots, V_r)$)

$$\frac{1}{2}\nabla J(V,x) = 0, \tag{14.88}$$

i.e., the system

$$[R(t) + c^*(t)P(t+1)c(t)V] = -c^*(t)P(t+1)a(t)x. \tag{14.89}$$

According to Lemma 13.3, this system is solvable and the vector $\tilde{V}$ defined by (14.83) is one of its solutions. Hence the minimum of the quadratic form $J(V,x)$ is attained on the vector $\tilde{V}$, and in order to verify (14.86) it remains only to establish that $x^*P(t)x = J(\tilde{V},x)$, i.e., that

$$\begin{aligned} x^*P(t)x \;=\; x^*\Big[&H(t) + a^*(t)P(t+1)a(t+1) \\ &-a^*(t)P(t+1)c(t)\,(R(t) + c^*(t)P(t)c(t))^+ \\ &\times c^*(t)P(t+1)a(t)\Big]x. \end{aligned} \tag{14.90}$$

The validity of this equality follows from the definition of the matrices $P(t)$ (see Equation (14.59)). □

14.3.5. Returning to the proof of Theorem 14.2, consider the control

$$\tilde{u} = (\tilde{u}_0, \dots, \tilde{u}_{T-1})$$

defined in (14.64). Then, due to Lemma 14.3,

$$-M[Q_{t+1}(\tilde{m}_{t+1}) - Q_t(\tilde{m}_t)] = M[\tilde{m}_t^*H(t)\tilde{m}_t + \tilde{u}_t^*R(t)\tilde{u}_t]. \tag{14.91}$$

Summing (14.91) over t from 0 to $T-1$ and taking into account that $\tilde{m}_0 = m$, we find

$$
\begin{aligned}
Q_0(m) &= MQ_T(\tilde{m}_T) + \sum_{t=0}^{T-1} M[\tilde{m}_t^* H(t)\tilde{m}_t + \tilde{u}_t^* R(t)\tilde{u}_t] \\
&= \sum_{t=0}^{T} M\tilde{m}_t^* H(t)\tilde{m}_t + \sum_{t=0}^{T-1} M\tilde{u}_t^* R(t)\tilde{u}_t.
\end{aligned} \tag{14.92}
$$

On the other hand, let $u = (u_0, \ldots, u_{T-1})$ be any of the controls satisfying (14.56). Then, due to Lemmas 14.2 and 14.3,

$$
-M[Q_{t+1}(m_{t+1}^u) - Q_t(m_t^u)] \leq M[(m_t^u)^* H(t) m_t^u + u_t^* R(t) u_t],
$$

whence it follows that

$$
Q_0(m) \leq \sum_{t=0}^{T} M(m_t^u)^* H(t) m_t^u + \sum_{t=0}^{T-1} M u_t^* R(t) u. \tag{14.93}
$$

The comparison of (14.92) with (14.93) proves the optimality of the control $\tilde{u} = (\tilde{u}_0, \ldots, \tilde{u}_{T-1})$. (14.67) follows from (14.71), (14.79) and the fact that

$$
V(\tilde{u}) = Q_0(m) + \sum_{t=0}^{T} \operatorname{Tr} H^{1/2}(t)\gamma_t H^{1/2}(t). \qquad \square
$$

Note. Let $\theta_0 = m$ be a deterministic vector, $b(t) \equiv 0$. Consider the problem (with complete data) of controlling the deterministic process θ_t, $t = 0, \ldots, T$, with

$$
\theta_{t+1} = c(t)u_t + a(t)\theta_t, \quad \theta_0 \equiv m, \tag{14.94}
$$

and the functional

$$
V(u) = \sum_{t=0}^{T} \theta_t^* H(t)\theta_t + \sum_{t=0}^{T-1} u_t^* H(t) u_t. \tag{14.95}
$$

In this particular case, the optimal control is

$$
\tilde{u}_t = -[R(t) + c^*(t)P(t+1)c(t)]^+ c^*(t)P(t+1)a(t)\tilde{\theta}_t, \tag{14.96}
$$

where

$$
\tilde{\theta}_{t+1} = c(t)\tilde{u}_t + a(t)\tilde{\theta}_t, \quad \tilde{\theta}_0 = m,
$$

and

$$
V(\tilde{u}) = m^* P(0) m. \tag{14.97}
$$

14.4 Asymptotic Properties of the Optimal Linear Filter

14.4.1. Consider the filtering problem[2] for the Gaussian process

$$(\tilde{\theta}, \tilde{\xi}) = [(\tilde{\theta}_1(t), \dots, \tilde{\theta}_k(t)), (\tilde{\xi}_1(t), \dots, \tilde{\xi}_l(t))], \quad t = 0, 1, \dots,$$

satisfying the recursive equations

$$\begin{aligned} \tilde{\theta}_{t+1} &= a_1\tilde{\theta}_t + a_2\tilde{\xi}_t + b_1\varepsilon_1(t+1) + b_2\varepsilon_2(t+1), \\ \tilde{\xi}_{t+1} &= A_1\tilde{\theta}_t + A_2\tilde{\xi}_t + B_1\varepsilon_1(t+1) + B_2\varepsilon_2(t+1) \end{aligned} \tag{14.98}$$

with the constant matrices $a_1, a_2, b_1, b_2, A_1, A_2, B_1$ and B_2 of order $(k \times k)$, $(k \times l)$, $(k \times k)$, $(k \times l)$, $(l \times k)$, $(l \times l)$, $(l \times k)$, $(l \times l)$, respectively.

Let $m_t = M(\tilde{\theta}_t | \mathcal{F}_t^{\xi})$ and $\gamma_t = M[(\tilde{\theta}_t - \tilde{m}_t)(\tilde{\theta}_t - \tilde{m}_t)^*]$. Then, according to Theorem 13.4, the error matrix γ_t satisfies the equation

$$\begin{aligned} \gamma_{t+1} &= a_1\gamma_t a_1^* + b \circ b \\ &\quad -[b\circ B + a_1\gamma_t A_1^*][B\circ B + A_1\gamma_t A_1^*]^+[b\circ B + a_1\gamma_t A_1^*]^*, \end{aligned} \tag{14.99}$$

where $b \circ b = b_1 b_1^* + b_2 b_2^*$, $b \circ B = b_1 B_1^* + b_2 B_2^*$ and $B \circ B = B_1 B_1^* + B_2 B_2^*$.

In this section we investigate the asymptotic behavior of the matrices γ_t at $t \to \infty$. Under the assumptions formulated later in Theorem 14.3 we shall show that $\lim_{t\to\infty} \gamma_t = \gamma^0$ exists and $0 < \operatorname{Tr} \gamma^0 < \infty$.

The existence of such a limit is crucial for the application since in this case the optimal mean square estimate m_t, $t \geq 0$, 'tracks' the values of $\tilde{\theta}_t$, $t \geq 0$, with finite error even when

$$\sum_{j=1}^{k} M\tilde{\theta}_j^2(t) \to \infty, \quad t \to \infty.$$

Before passing to a clarification of the conditions guaranteeing the existence of the limit $\gamma^0 = \lim_{t\to\infty} \gamma_t$, note that it is enough to consider the system of equations

$$\begin{aligned} \theta_{t+1} &= a\theta_t + b\varepsilon_1(t+1), \\ \xi_{t+1} &= A\theta_t + B\varepsilon_2(t_+1) \end{aligned} \tag{14.100}$$

instead of the system of equations given by (14.98), with $\theta_0 = \tilde{\theta}_0$, $\xi_0 = \tilde{\xi}_0$,

$$a = a_1 - (b \circ B)(B \circ B)^+ A_1, \quad A = A_1, \tag{14.101}$$

$$b = [(b \circ b) - (b \circ B)(B \circ B)^+(b \circ B)^*], \quad B = (B \circ B)^{1/2}, \tag{14.102}$$

since the equations for γ_t in (14.98) and (14.100) will coincide.

Indeed, if $m_{\tilde{\theta}_t}(t+1, t) = M(\tilde{\theta}_{t+1} | \mathcal{F}_{t+1}^{\xi}, \tilde{\theta}_t)$, then

---

[2] Regarding the notation adopted here, see Section 13.2.

$$\begin{aligned}
\gamma_{t+1} &= M[(\tilde{\theta}_{t+1} - m_{t+1})(\tilde{\theta}_{t+1} - m_{t+1})^*] \\
&= M\Big[(\tilde{\theta}_{t+1} - m_{\tilde{\theta}_t}(t+1,t) + m_{\tilde{\theta}_t}(t+1,t) - m_{t+1}) \\
&\qquad \times(\tilde{\theta}_{t+1} - m_{\tilde{\theta}_t}(t+1,t) + m_{\tilde{\theta}_t}(t+1,t) - m_{t+1})^*\Big] \\
&= M[(\tilde{\theta}_{t+1} - m_{\tilde{\theta}_t}(t+1,t))(\tilde{\theta}_{t+1} - m_{\tilde{\theta}_t}(t+1,t))^*] \\
&\quad +M[(m_{\tilde{\theta}_t}(t+1,t) - m_{t+1})(m_{\tilde{\theta}_t}(t+1,t) - m_{t+1})^*].
\end{aligned}$$

Due to (13.91),

$$\begin{aligned}
&M[(\tilde{\theta}_{t+1} - m_{\tilde{\theta}_t}(t+1,t))(\tilde{\theta}_{t+1} - m_{\tilde{\theta}_t}(t+1,t))] = \gamma(t+1,t) \\
&= b\circ b - (b\circ B)(B\circ B)^+(b\circ B)^* = bb^*,
\end{aligned}$$

and it follows from the definition of $m_{\tilde{\theta}_t}(t+1,t)$, due to the note to Theorem 13.4, that

$$m_{\tilde{\theta}_t}(t+1,t) = a_1\tilde{\theta}_t - a_2\tilde{\xi}_t + (b\circ B)(B\circ B)^+(\tilde{\xi}_{t+1} - A_1\tilde{\theta}_t - A_2\tilde{\xi}_t).$$

Since $m_{t+1} = M[m_{\tilde{\theta}_t}(t+1,t)|\mathcal{F}_{t+1}^{\tilde{\xi}}]$, we obtain from the recursive equation for $m_{\tilde{\theta}_t}(t+1,t)$ that

$$\begin{aligned}
m_{t+1} &= a_1 m(t,t+1) + a_2\tilde{\xi}_t \\
&\quad +(b\circ B)(B\circ B)^+(\xi_{t+1} - A_1 m(t,t+1) - A_2\xi_t),
\end{aligned}$$

where $m(t,t+1) = M[\tilde{\theta}_t|\mathcal{F}_{t+1}^{\tilde{\xi}}]$. Consequently,

$$\begin{aligned}
m_{\tilde{\theta}_t}(t+1) - m_{t+1} &= [a_1 - (b\circ B)(B\circ B)^+ A_1](\tilde{\theta}_t - m(t,t+1) \\
&= a(\tilde{\theta}_t - m(t,t+1)),
\end{aligned}$$

and

$$M[(m_{\tilde{\theta}_t}(t+1,t) - m_{t+1})(m_{\tilde{\theta}_t}(t+1,t) - m_{t+1})^*] = a\gamma(t,t+1)a^*,$$

where

$$\gamma(t,t+1) = M[(\tilde{\theta}_t - m(t,t+1))(\tilde{\theta}_t - m(t,t+1))^*].$$

But, according to (13.110),

$$\gamma(t,t+1) = \gamma_t - \gamma_t A_1^*[B\circ B + A_1\gamma_t A_1^*]^+ A_1\gamma_t.$$

Therefore, for γ_t, $t>0$, we have:

$$\begin{aligned}
\gamma_{t+1} &= [a_1 - (b\circ B)(B\circ B)^+ A_1]\gamma_t[a_1 - (b\circ B)(B\circ B)^+ A_1]^* \\
&\quad +[b\circ b - (b\circ B)(B\circ B)^+(b\circ B)^*] \\
&\quad -[a_1 - (b\circ B)(B\circ B)^+(b\circ B)^*]\gamma_t A_1^* \\
&\quad \times[B\circ B + A_1\gamma_t A_1^*]^+ A_1\gamma_t[a_1 - (b\circ B)(B\circ B)^+(b\circ B)^*]^*.
\end{aligned}$$

Hence, in what follows, we shall discuss only the system of equations given by (14.100) and study the asymptotic behavior of the matrices γ_t satisfying the recursive equation

$$\gamma_{t+1} = a\gamma_t a^* + bb^* - a\gamma_t A^*[BB^* + A\gamma_t A^*]^+ A\gamma_t a. \tag{14.103}$$

Theorem 14.3. *Let the following conditions be satisfied:*

(1) the rank of the block matrix

$$G_1 = \begin{pmatrix} A \\ Aa \\ \dots \\ Aa^{k-1} \end{pmatrix}$$

of dimension $(kl \times k)$ is equal to k;

(2) the rank of the block matrix $G_2 = (b\ ab \dots a^{k-1}b)$ of dimension $(k \times lk)$ is equal to k;

(3) the matrix BB^* is nonsingular.

Then $\lim_{t\to\infty} \gamma_t = \gamma^0$ exists and does not depend on γ_0. $\operatorname{Tr}\gamma^0 < \infty$ and the matrix γ^0 is the unique solution (in the class of symmetric positive definite matrices) of the matrix equation

$$\gamma = a\gamma a^* + bb^* - a\gamma A^*(BB^* + A\gamma A^*)^{-1} A\gamma a^*. \tag{14.104}$$

14.4.2. Before proving this theorem let us make some auxiliary assertions.

Lemma 14.4. *Let D and d be matrices of dimension $(l \times k)$ and $(k \times k)$, respectively, and let*

$$D_n = \begin{pmatrix} D \\ Dd \\ \dots \\ Dd^{n-1} \end{pmatrix}, \quad n \geq k,$$

be block matrices of dimension $(nl \times k)$.

Then the matrices $D_k^* D_k$ and $D_n^* D_n$, $n > k$, are either both singular or both nonsingular.

PROOF. From the rule for multiplication of block matrices it follows that

$$D_n^* D_n = D_k^* D_k + \sum_{j=k}^{n-1} (d^*)^j D^* D d^j. \tag{14.105}$$

It is seen from this that the singularity of the matrix $D_n^* D_n$ implies the singularity of the matrix $D_k^* D_k$.

Let now the matrix $D_k^* D_k$ be singular. We shall show that the matrices $D_n^* D_n$, $n > k$ are also singular.

Denote by $x = (x_1, \dots, x_k)$ a nonzero column vector such that

$$x^* D_k^* D_k x = 0. \tag{14.106}$$

We shall show that $Dd^j x = 0$ for all $j \geq k$. Since

$$D_k^* D_k = \sum_{j=0}^{k-1} (d^*)^j D^* D d^j,$$

due to (14.106) it follows that

$$Dx = 0, \quad Ddx = 0, \dots, \quad Dd^{k-1} x = 0. \tag{14.107}$$

Set

$$y_0 = x, \quad y_1 = dx = dy_0, \quad y_{j+1} = dy_j, \quad j \leq k-1.$$

Then

$$Dy_0 = 0, \quad Dy_1 = 0, \dots, \quad Dy_{k-1} = 0. \tag{14.108}$$

But the system of vectors $(y_0, y_1, \dots, y_k)$, where each vector has the dimension k, is linearly dependent. Hence, there exist numbers $c_0, \dots, c_k$, not all equal to zero, such that

$$\sum_{j=0}^{k} c_j y_j = 0. \tag{14.109}$$

Let $i = \max[j \leq k : c_j \neq 0]$. Then, from (14.109), we obtain

$$y_i = \sum_{j=0}^{i-1} c_j' y_j, \quad c_j' = -\frac{c_j}{c_i},$$

and, therefore,

$$y_k = d^{k-i} y_i = \sum_{j=0}^{i-1} c_j' d^{k-i} y_j = \sum_{j=0}^{i-1} c_j' y_{k-i+j}.$$

Hence, due to (14.108),

$$Dd^k x = Dy^k = \sum_{j=0}^{i-1} c_j' D y_{k-i+j} = 0.$$

We establish by induction from this that $Dd^j x = 0$, $j \geq k$, which, together with (14.105), proves the statement of the lemma. □

Corollary. *Let $D = D_{(k \times l)}$, $d = d_{(k \times k)}$ be some matrices and let*

$$\tilde{D}_n = (D\, dD \cdots d^{n-1} D)$$

be a block matrix of order $(k \times nl)$, $n \geq k$. *Then the matrices* $\tilde{D}_n \tilde{D}_n^*$ *and* $\tilde{D}_k \tilde{D}_k^*$ *are either both singular or both nonsingular.*

Lemma 14.5. *Let* $\theta = [\theta_1(t), \dots, \theta_k(t)]$, $t = 0, 1, \dots$, *be a Gaussian sequence satisfying the recursive equation*

$$\theta_{t+1} = a\theta_t + b\varepsilon(t+1), \quad \theta_0 = 0, \tag{14.110}$$

where a *and* b *are matrices of dimension* $(k \times k)$ *and* $\varepsilon(t)$ *is a sequence of independent Gaussian vectors* $\varepsilon(t) = (\varepsilon_1(t), \dots, \varepsilon_k(t))$ *with independent components,* $M\varepsilon_j(t) = 0$, $M\varepsilon_j^2(t) = 1$, $j = 1, \dots, k$, $t = 0, 1, \dots$.

If the matrix $G_2 = (b\, ab \dots a^{k-1} b)$ *of dimension* $(k \times lk)$ *has rank* k, *then the matrix* $\Gamma_t = M\theta_t\theta_t^*$ *at* $t \geq k$ *is positive definite.*

PROOF. We find from (14.110) that

$$\begin{aligned} \Gamma_{t+1} &= M\theta_{t+1}\theta_{t+1}^* = M[a\theta_t + b\varepsilon(t+1)][a\theta_t + b\varepsilon(t+1)]^* \\ &= aM\theta_t\theta_t^* a^* + bM\varepsilon(t+1)\varepsilon^*(t+1)b^*. \end{aligned}$$

Hence,

$$\Gamma_{t+1} = a\Gamma_t a^* + bb^*, \quad \Gamma_0 = 0. \tag{14.111}$$

Therefore

$$\begin{aligned} \Gamma_1 &= bb^*, \quad \Gamma_2 = bb^* + abb^*a^*, \dots, \\ \Gamma_t &= bb^* + abb^*a^* + \cdots + +a^{t-1}bb^*(a^*)^{t-1}. \end{aligned}$$

Let $t = k$. Then, obviously, $\Gamma_t = G_2 G_2^*$ and at $t > k$

$$\Gamma_t = G_2 G_2^* + \sum_{j=k}^{t-1} a^j bb^*(a^*)^j. \tag{14.112}$$

Since the rank of the matrix G_2 is assumed to be equal to k, then the rank of the matrix $G_2 G_2^*$ is also equal to k. It follows, therefore, from (14.112) that for $t \geq k$ the matrix Γ_t is nonsingular. □

Lemma 14.6. *Let* $(\tilde{\theta}, \tilde{\xi}) = ([\tilde{\theta}_1, \dots, \tilde{\theta}_n], [\tilde{\xi}_1, \dots, \tilde{\xi}_N])$ *be a Gaussian vector with the positive definite matrices*[3]

$$\operatorname{cov}(\tilde{\theta}, \tilde{\theta}) = M[(\tilde{\theta} - M\tilde{\theta})(\tilde{\theta} - M\tilde{\theta})^*], \tag{14.113}$$

$$\operatorname{cov}(\tilde{\xi}, \tilde{\xi}|\tilde{\theta}) = M[(\tilde{\xi} - M(\tilde{\xi}|\tilde{\theta}))(\tilde{\xi} - M(\tilde{\xi}|\tilde{\theta}))^*]. \tag{14.114}$$

[3] By the theorem on normal correlation (Theorem 13.1) the matrices $\operatorname{cov}(\tilde{\xi}, \tilde{\xi}|\tilde{\theta})$ and $\operatorname{cov}(\tilde{\theta}, \tilde{\theta}|\tilde{\xi})$ do not depend on $\tilde{\theta}$ and $\tilde{\xi}$, respectively.

Then the matrix

$$\operatorname{cov}(\tilde{\theta},\tilde{\theta}|\tilde{\xi}) = M[(\tilde{\theta} - M(\tilde{\theta}|\tilde{\xi}))(\tilde{\theta} - M(\tilde{\theta}|\tilde{\xi}))^*] \tag{14.115}$$

is also positive definite.

PROOF. Because of the nonsingularity of the matrices given by (14.113) and (14.114), the Gaussian distributions[4] $P(\tilde{\theta} \le a)$ and $P(\tilde{\xi} \le b|\tilde{\theta} = a)$ have the densities $f_{\tilde{\theta}}(a)$ and $f_{\tilde{\xi}|\tilde{\theta}}(b|a)$. It can be easily deduced from this that there exists a density $f_{\tilde{\xi}}(b)$. Hence, it follows from the Bayes formula that the distribution $P(\tilde{\theta} \le a|\tilde{\xi})$ has density $f_{\tilde{\theta}|\tilde{\xi}}(a|b)$ as well, and

$$f_{\tilde{\theta}|\tilde{\xi}}(a|b) = \frac{f_{\tilde{\xi}|\tilde{\theta}}(b|a) f_{\tilde{\theta}}(a)}{f_{\tilde{\xi}}(b)}.$$

The existence of this (Gaussian) density implies that the corresponding matrix of the covariances $\operatorname{cov}(\tilde{\theta},\tilde{\theta}|\tilde{\xi})$ is nonsingular and, therefore, positive definite. □

Lemma 14.7. *Let γ_t^0, $t = 0, 1, \ldots$, be a solution of the equation*

$$\gamma_{t+1} = a\gamma_t a^* + bb^* - a\gamma_t A^*(BB^* + A\gamma_t A^*)^+ A\gamma_t a^* \tag{14.116}$$

with the initial condition $\gamma_0^0 = 0$ (0 being the zero matrix of order $(k \times k)$). If the matrix BB^* is positive definite, and the rank of the matrix G_2 is equal to k, then the matrix γ_k^0 is positive definite.

PROOF. Let θ_t^0, $t = 0, 1, \ldots$, be a solution of the equation

$$\theta_{t+1} = a\theta_t + b\varepsilon_1(t+1) \tag{14.117}$$

(see (14.100)) with $\theta_0 = 0$. Then $\gamma_t^0 = M[(\theta_t^0 - m_t^0)(\theta_t^0 - m_t^0)^*]$ and $m_t^0 = M(\theta_t^0|\mathcal{F}_t^\xi)$ where

$$\xi_{t+1} = A\theta_t^0 + B\varepsilon_2(t+1). \tag{14.118}$$

Write: $\tilde{\theta} = \theta_k^0$; $\tilde{\xi} = (\xi_1, \ldots, \xi_k)$; $\hat{\theta} = (\theta_0^0, \ldots, \theta_{k-1})$; $\tilde{\varepsilon} = (\varepsilon_2(1), \ldots, \varepsilon_2(k))$. Also, let

$$\tilde{B} = \operatorname{diag}(B \cdots B), \quad \tilde{a} = \operatorname{diag}(a \cdots a)$$

be block diagonal matrices in which only the blocks situated on the diagonals of the matrices B and a, respectively, are different from zero. Then the system of equations given by (14.118) for $t = 0, 1, \ldots, k-1$ can be represented as $\tilde{\xi} = \tilde{a}\hat{\theta} + \tilde{B}\tilde{\varepsilon}$. The vectors $(\tilde{\theta}, \hat{\theta})$ and $\tilde{\varepsilon}$ are independent since the sequences $\varepsilon_1(t)$ and $\varepsilon_2(t)$, $t = 1, 2, \ldots$ are independent. Hence

[4] The notation $\{\tilde{\theta} \le a\}$ denotes the event$\{\tilde{\theta}_1 \le a_1, \ldots, \tilde{\theta}_n \le a_n\}$.

$$M(\tilde{\xi}|\tilde{\theta}) = \tilde{a}M(\hat{\theta}|\tilde{\theta})$$

and

$$\tilde{\xi} - M(\tilde{\xi}|\tilde{\theta}) = \tilde{a}[\hat{\theta} - M(\hat{\theta}|\tilde{\theta})] + B\tilde{\varepsilon}.$$

Because of the independence of the vectors $\tilde{\theta}$ and $\tilde{\varepsilon}$, from this it follows that

$$\operatorname{cov}(\tilde{\xi}, \tilde{\xi}|\tilde{\theta}) = \tilde{a}\operatorname{cov}(\hat{\theta}, \hat{\theta}|\tilde{\theta})\tilde{a}^* + \tilde{B}\tilde{B}^*.$$

Since the matrix BB^* is nonsingular, the matrix

$$\tilde{B}\tilde{B}^* = \operatorname{diag}(BB^* \cdots BB^*)$$

is also nonsingular. Next, the matrix $\operatorname{cov}(\tilde{\theta}, \tilde{\theta}) = M\theta_k^0(\theta_k^0)^*$ is nonsingular by Lemma 14.5. Hence, by Lemma 14.6, the matrix

$$\operatorname{cov}(\tilde{\theta}, \tilde{\theta}|\tilde{\xi}) = M[(\theta_k^0 - M(\theta_k^0|\mathcal{F}_k^\xi))(\theta_k^0 - M(\theta_k^0|\mathcal{F}_k^\xi))^*] = \gamma_k^0$$

will also be nonsingular. □

Lemma 14.8. *If the rank of the matrix G_1 is equal to k, then for any vector $x = (x_1, \ldots, x_k)$, $|x_i| < \infty$, $i = 1, \ldots, k$,*

$$\sup_{t \geq 0} x^* \gamma_t x < \infty. \tag{14.119}$$

PROOF. Let $x_t = (x_1(t), \ldots, x_k(t))$, $t = 0, 1, \ldots T > k$, be the controlled process satisfying the recursive equation $x_{t+1} = a^* x_t + A^* u_t$, $x_0 = x$, where the control $u_t = (u_1(t, x_0, \ldots, x_t), \ldots, u_l(t, x_0, \ldots, x_t))$ is chosen to minimize the functional

$$V_T(x; u) = x_T^* \gamma_0 x_T + \sum_{t=0}^{T-1} [x_t^* bb^* x_t + u_t^* BB^* u_t]. \tag{14.120}$$

According to the note to Theorem 14.2, the optimal control $\tilde{u}_t$, $t = 0, 1, \ldots, T-1$, exists and is given by the formula

$$\tilde{u}_t = -[BB^* + AP(t+1)A^*]^+ AP(t+1)a^* \tilde{x}_t,$$

where $\tilde{x}_{t+1} = a^* \tilde{x}_t + A^* \tilde{u}_t$ and

$$\begin{aligned} P(t) &= bb^* + aP(t+1)a^* - aP(t+1)A^*[BB^* + AP(t+1)A^*]^+ \\ &\quad \times AP(t+1)a^*, \quad P(T) = \gamma_0. \end{aligned} \tag{14.121}$$

Comparing this equation with Equation (14.103) we convince ourselves that

$$P(t) = \gamma_{T-t}. \tag{14.122}$$

Since (see (14.94)) for the optimal control $\tilde{u} = (\tilde{u}_0, \ldots, \tilde{u}_{T-1})$

$$V_T(x; \tilde{u}) = x^* P(0) x = x^* \gamma_T x,$$

in order to prove the lemma it suffices to show that

$$V_T(x; \tilde{u}) \le c < \infty, \tag{14.123}$$

where the constant c does not depend on T.

By the conditions of the lemma, the matrix G_1 has rank k. Hence, the matrix $G_1^* G_1$ is nonsingular.

Consider the control $\hat{u}_t = (\hat{u}_t(t, x_0), \ldots, \hat{u}_l(t, x_0))$ defined as

$$\hat{u}_t = \begin{cases} -Aa^{k-t-1}(G_1^* G_1)^{-1}(a^*)^k x_0, & t \le k \\ 0, & t > k \end{cases}.$$

The associated controlled process $\hat{x}_t$, $t = 0, 1, \ldots, \hat{x}_{t+1} = a^* \hat{x} t + A^* \hat{u}_t$, goes to zero in k steps, since

$$\begin{aligned} \hat{x}_k &= (a^*)^k x_0 + \sum_{t=0}^{k-1} (a^*)^{k-t-1} A^* \hat{u}_t \\ &= (a^*)^k \left\{ E - \left[\sum_{t=0}^{k-1} (a^*)^{k-t-1} A^* A a^{k-t-1} \right] (G_1^* G_1)^{-1} \right\} x_0 \\ &= (a^*)^k \{ E - (G_1^* G_1)(G_1^* G_1)^{-1} \} x_0 = 0. \end{aligned}$$

Consider the functional $V_T(x; \hat{u})$. Since $\hat{u}_t = 0$. $\hat{x}_t = 0$, $t > k$, we have $\sup_{T>k} V_T(x; u) < \infty$. But by virtue of the optimality of the control $\tilde{u} = (\tilde{u}_0, \ldots, \tilde{u}_{T-1})$,

$$\sup_{T \ge k} V_T(x, \tilde{u}) \le \sup_{T \ge k} V_T(x, \hat{u}).$$

Hence

$$\sup_{T \ge 0} x^* \gamma_T x = \sup_{T \ge 0} V_T(x; \tilde{u}) \le \sup_{T \ge 0} V_T(x; \hat{u}) = \max_{0 \le T \le k} V_T(x, \hat{u}) < \infty. \qquad \square$$

Lemma 14.9. *Let γ_t^0, $t = 0, 1, \ldots$, be the solution of Equation (14.116) with the initial condition $\gamma_0^0 = 0$. If the rank of the matrix G_1 is equal to k, then there exists*

$$\lim_{t \to \infty} \gamma_t^0 = \gamma^0, \tag{14.124}$$

where γ^0 is a nonnegative definite symmetric matrix with $\operatorname{Tr} \gamma^0 < \infty$. If, in addition, the rank of the matrix G_2 is equal to k and the matrix BB^* is nonsingular, then the matrix γ^0 is positive definite.

PROOF. According to Lemma 14.8, the values of $x^* \gamma_T x$ are bounded for any $T \ge 0$ ($|x_i| < \infty$, $i = 1, \ldots, k$). Let us show that these values are monotone nondecreasing functions of T.

Let $T_2 > T_1$, $\tilde{u}^0(T_1)$ and $\tilde{u}^0(T_2)$ be optimal controls corresponding to the observation durations T_1 and T_2, respectively.

If $\tilde{x}_t^0(T_1)$ and $\tilde{x}_t^0(T_2)$ are trajectories of the controlled processes for the controls $\tilde{u}^0(T_1)$ and $\tilde{u}^0(T_2)$, respectively[5], then

$$\begin{aligned} x^*\gamma_{T_2}^0 x &= V_{T_2}^0(x;\tilde{u}^0(T_2)) \\ &= \sum_{t=0}^{T_2-1}[(\tilde{x}_t^0(T_2))^* bb^*(\tilde{x}_t^0(T_2)) + (\tilde{u}_t^0(T_2))^* BB^*(\tilde{u}_t^0(T_2))] \\ &\geq \sum_{t=0}^{T_1-1}[(\tilde{x}_t^0(T_2))^* bb^*(\tilde{x}_t^0(T_2)) + (\tilde{u}_t^0(T_2))^* BB^*(\tilde{u}_t^0(T_2))] \\ &\geq V_{T_1}^0(x;\tilde{u}^0(T_1)) = x^*\gamma_{T_1}x. \end{aligned}$$

Hence, if $\tilde{u}^0(T_n)$ is an optimal control on the interval T_n, and $T_{n+1} = T_n + 1$, then

$$V_{T_1}^0(x;\tilde{u}^0(T_1)) \leq V_{T_2}^0(x;\tilde{u}^0(T_2)) \leq \cdots \leq V_{T_n}^0(x;\tilde{u}^0(T_n)),$$

and, because of the uniform (over T_n) boundedness of the values of $V_{T_n}^0(x;\tilde{u}^0(T_n))$, there exists

$$\lim_{T_n\to\infty} V_{T_n}^0(x;\tilde{u}^0(T_n)) = x^*\gamma^0 x.$$

Because of the arbitrariness of the vector x it is seen that the limit matrix γ^0 is symmetric nonnegative definite and that $\operatorname{Tr}\gamma^0 < \infty$.

If, finally, rank $G_2 = k$ and the matrix BB^* is nonsingular, then, by Lemma 14.7, we have that $x^*\gamma_k x > 0$ for any nonzero vector x. But $x^*\gamma_T x$ is monotone nondecreasing in T. Hence, for any nonzero vector x the values of $x^*\gamma_T x > 0$, $T > k$, which proves the positive definiteness of the matrix γ^0. □

14.4.3.

PROOF OF THEOREM 14.3. Take the control

$$\overline{u}_t = -[BB^* + A\gamma^0 A^*]^{-1} A\gamma^0 a^*\overline{x}_t, \tag{14.125}$$

where

$$\overline{x}_{t+1} = a^*\overline{x}_t + A^*\overline{u}_t \tag{14.126}$$

and the matrix γ^0 is defined by (14.124). We shall show that

$$\lim_{t\to\infty} \overline{x}_t^*\gamma^0\overline{x}_t = 0. \tag{14.127}$$

Due to (14.125) and (14.126),

[5] The index 0 in $V_T^0(x;\cdot)$, $\tilde{u}^0(T)$ and $\hat{x}_t^0(T)$ indicates that $\gamma_0 = 0$.

$$
\begin{aligned}
\overline{x}_{t+1}^*\gamma^0\overline{x}_{t+1} &= \{\overline{x}_t^* a + \overline{u}_t^* A\}\gamma^0\{a^*\overline{x}_t + A^*\overline{u}_t\} \\
&= \overline{x}_t^*\{a\gamma^0 a^* - 2a\gamma^0 A^*(BB^* + A\gamma^0 A^*)^{-1}A\gamma^0 a^* \\
&\quad + a\gamma^0 A(BB^* + A\gamma^0 A^*)^{-1}[BB^* + A\gamma^0 A^*] \\
&\quad \times (BB^* + A\gamma^0 A^*)^{-1}A\gamma^0 a^a\}x_t - \overline{u}_t^* BB^*\overline{u}_t \\
&= \overline{x}_t^*\{a\gamma^0 a^* - a\gamma^0 A^*(BB^* + A\gamma^0 A^*)^{-1}A\gamma^0 a^*\}\overline{x}_t - \overline{u}_t^* BB^*\overline{u}_t.
\end{aligned} \tag{14.128}
$$

γ^0 is the limit of the sequence of matrices γ_t^0 satisfying (14.116), and the matrix BB^* is nonsingular; hence it satisfies the equation

$$\gamma^0 = a\gamma^0 a^* + bb^* - a\gamma^0 A^*(BB^* + A\gamma^0 A^*)^{-1}A\gamma^0 a^*.$$

We find from this and (14.128) that

$$\overline{x}_{t+1}^*\gamma^0\overline{x}_{t+1} - \overline{x}_t^*\gamma^0\overline{x}_t = [\overline{x}_t^* bb^*\overline{x}_t + \overline{u}_t^* BB^*\overline{u}_t].$$

Therefore, according to Lemma 14.9,

$$
\begin{aligned}
0 \le \overline{x}_T^*\gamma^0\overline{x}_T &= x^*\gamma^0 x - \sum_{t=0}^{T-1}[\overline{x}_t^* bb^*\overline{x}_t - \overline{u}_t^* BB^*\overline{u}_t] \\
&\le x^*\gamma^0 x - V_T^*(x; \tilde{u}^0(T)) \to 0, \quad T \to \infty.
\end{aligned}
$$

Now it is seen that, since the matrix γ^0 is nonsingular (Lemma 14.9),

$$\lim_{T\to\infty} \overline{x}_T = 0 \tag{14.129}$$

and

$$\lim_{T\to\infty} V_T^0(x; \tilde{u}^0(T)) = x^*\gamma^0 x = \lim_{T\to\infty} \sum_{t=0}^{T-1}[\overline{x}_t^* bb^*\overline{x}_t + \overline{u}_t^* BB^*\overline{u}_t]. \tag{14.130}$$

Let γ_0 be any nonnegative definite symmetric matrix. Then, due to (14.120),

$$V_T^0(x; \tilde{u}^0(T)) \le V_T(x; \tilde{u}(T)) \le \overline{x}_T^*\gamma_0\overline{x}_T + \sum_{t=0}^{T-1}[\overline{x}_t^* bb^*\overline{x}_t + \overline{u}_t^* BB^*\overline{u}_t]. \tag{14.131}$$

Passing in these inequalities to the limit ($T \to \infty$) we find, taking into account (14.129) and (14.30), that

$$\lim_{T\to\infty} x^*\gamma_T x = \lim_{T\to\infty} V_T(x; \tilde{u}(T)) = \lim_{T\to\infty} V_T^0(x; \tilde{u}^0(T)) = x^*\gamma^0 x. \tag{14.132}$$

Therefore, because of the arbitrariness of the vector x, $\lim_{T\to\infty}\gamma_T = \gamma^0$ exists, and γ^0 does not depend on the initial matrix γ_0.

It was noted above that γ^0 is a positive definite solution of the matrix equation given by (14.104). We shall show that in the class of the positive definite symmetric matrices this solution is unique.

Indeed, let $\gamma^{(1)}$ and $\gamma^{(2)}$ be two such solutions. Denote by $\gamma_t^{(i)}$, $t \geq 0$, the solutions of (14.103) with $\gamma_0^{(1)} = \gamma^{(1)}$ and $\gamma_0^{(2)} = \gamma^{(2)}$, respectively. Then, according to what we have already proved,

$$\lim_{T\to\infty} \gamma_T^{(i)} = \gamma^0 = \gamma^{(i)}, \quad i = 1, 2. \qquad \square$$

Note 1. If $\sup_{t\geq 0} \operatorname{Tr} M\theta_t\theta_t^* < \infty$, then in the formulation of Theorem 14.3 one can discard the first assumption since $\operatorname{Tr}\gamma_t \leq \operatorname{Tr} M\theta_t\theta_t^*$.

Note 2. Let the process $(\theta_t, \xi_t) = ([\theta_1(t), \dots, \theta_k(t)], [\xi_1(t), \dots, \xi_l(t)])$ satisfy the recursive equations (Kalman–Bucy problem)

$$\begin{aligned} \theta_{t+1} &= a_1\theta_t + b_1\varepsilon_1(t+1), \\ \xi_t &= A_1\theta_t + B_1\varepsilon_2(t) \end{aligned} \tag{14.133}$$

(compare with (14.100)). In order to formulate the conditions providing the existence of the limit $\lim_{t\to\infty}\gamma_t$ in terms of the matrices a_1, b_1, A_1 and B_1, it suffices to note the following. Since

$$\xi_{t+1} = A_1a_1\theta_t + A_1b_1\varepsilon_1(t+1) + B_1\varepsilon_2(t+1),$$

assuming

$$\begin{aligned} a &= a_1 - b_1b_1^*A_1[A_1b_1b_1^*A_1^* + B_1B_1^*]^{-1}A_1a_1, \\ A &= A_1a_1, \\ b &= [b_1b_1^* - b_1b_1^*A_1^*(A_1b_1b_1^*A_1^* + B_1B_1^*)^{-1}A_1b_1b_1^*]^{1/2}, \\ B &= (A_1b_1b_1^*A_1^* + B_1B_1^*)^{1/2}, \end{aligned}$$

reduces the problem of the existence of $\lim_{t\to\infty}\gamma_t$ to the problem studied for the system given by (14.100).

14.4.4.
EXAMPLE 3. Let θ_t and ξ_t be one-dimensional processes with

$$\theta_{t+1} = a\theta_t + b\varepsilon_1(t+1), \quad \xi_{t+1} = A\theta_t + B\varepsilon_2(t+1).$$

Then, if $A \neq 0$, $b \neq 0$ and $B \neq 0$, the conditions of Theorem 14.3 are satisfied and the limiting filtering error $\gamma^0 = \lim_{t\to\infty}\gamma_t$ ($\gamma_t = M(\theta_t - m_t)^2$; $m_t = M(\theta_1|\xi_0, \dots, \xi_t)$) can be defined as the positive root of the quadratic equation

$$\gamma^2 + \left[\frac{B^2(1-a^2)}{A^2} - b^2\right]\gamma - \frac{b^2B^2}{A^2} = 0.$$

14.5 Recursive Computation of the Best Approximate Solutions (Pseudo-solutions) of Linear Algebraic Systems

14.5.1. Let the vector $y = (y_1, \dots, y_k)$ and the matrix $A = \|a_{ij}\|$ of order $(k \times n)$ and rank $A \le \min(k, n)$ be given. Then the system of linear algebraic equations

$$Ax = y \tag{14.134}$$

need have no solutions, generally speaking, and even if it has, the solution need not be unique.

The vector x^0 is said to be the best approximate solution (pseudo-solution) of the system of equations given by (14.134) if

$$|y - Ax^0|^2 = \inf_x |y - Ax|^2; \tag{14.135}$$

if, also, $|y - Ax'| = \inf_x |y - Ax|$, then

$$|x^0|^2 \le |x'|^2, \tag{14.136}$$

where

$$|y - Ax|^2 = \sum_{i=1}^{k} \left| y_i - \sum_{j=1}^{n} a_{ij}x_j \right|^2, \quad |x|^2 = \sum_{j=1}^{n} |x_j|^2.$$

In other words, the pseudo-solution is an approximate solution having the least norm.

It is well known[6] that such a solution x^0 is given by the formula

$$x^0 = A^+ y, \tag{14.137}$$

where A^+ is the matrix which is the pseudo-inverse with respect to the matrix A (see Section 13.1).

It is seen from (14.137) that in order to find pseudo-solutions it is necessary for the pseudo-inverse matrix A^+ to be found. As will be shown in this section, taking advantage of the optimal filtering equations given by (13.56) and (13.57), one can, however, offer recursive procedures for finding the pseudo-solutions which do not require the 'pseudo-inversion' of the matrix A.

[6] See, for example, Chapter 1, Section 5, in [69].

14.5.2. Let us start with the case where the system of algebraic equations $Ax = y$ is solvable ($k \le n$). In this case the pseudo-solution $x^0 = A^+y$ is distinguished among all the solutions x by the fact that its norm is the least, i.e., $|x^0| \le |x|$.

Let us introduce some additional notation. Let $t = 1, 2, \ldots, k$ be the numbers of the rows of the matrix A, let a_t be the rows of the matrix A,

$$A_t = \begin{pmatrix} a_1 \\ \cdots \\ a_t \end{pmatrix},$$

and let y_t be the elements of the vector y, $t = 1 \ldots, k$,

$$y^t = \begin{pmatrix} y_1 \\ \cdots \\ y_t \end{pmatrix}.$$

Consider for each t (solvable) systems of linear algebraic equations[7]

$$A_t x = y^t. \tag{14.138}$$

Let

$$x_t = A_t^+ y^t, \quad \gamma_t = E - A_t^+ A_t. \tag{14.139}$$

Theorem 14.4. *The vectors x_t and the matrices γ_t, $t = 1, \ldots, k$, satisfy the system of recursive equations*

$$x_{t+1} = x_t + \gamma_t a_{t+1}^*(a_{t+1}\gamma_t a_{t+1}^*)^+(y_{t+1} - a_{t+1}x_t), \quad x_0 = 0, \tag{14.140}$$

$$\gamma_{t+1} = \gamma_t - \gamma_t a_{t+1}^*(a_{t+1}\gamma_t a_{t+1}^*)^+ a_{t+1}\gamma_t, \quad \gamma_0 = E, \tag{14.141}$$

where

$$(a_{t+1}\gamma_t a_{t+1}^*)^+ = \begin{cases} [a_{t+1}\gamma_t a_{t+1}^*]^{-1}, & a_{t+1}\gamma_t a_{t+1}^* > 0, \\ 0, & a_{t+1}\gamma_t a_{t+1}^* = 0, \end{cases} \tag{14.142}$$

and the vector x_k coincides with the pseudo-solution x^0.

If the rank of A is equal to k, then $(a_{t+1}\gamma_t a_{t+1}^*)^+ = (a_{t+1}\gamma_t a_{t+1}^*)^{-1}$ for all $t = 0, \ldots, k-1$.

PROOF. Let $\theta = (\theta_1, \ldots, \theta_k)$ be a Gaussian vector with $M\theta = 0$, $M\theta\theta^* = E$, and let

$$\xi^t = A_t\theta. \tag{14.143}$$

Then, by the theorem on normal correlation (Theorem 13.1) and the fact that $M\xi^t = 0$, $M\theta(\xi^t)^* = A_t^*$, $M(\xi^t)(\xi^t)^* = A_t A_t^*$, we have

$$m_t = M(\theta|\xi^t) = A_t^*(A_t A_t^*)^+ \xi^t.$$

[7] The dimension of the vector x is equal to n for any t.

But by (6°) of Section 13.1 regarding pseudo-inverse matrices,

$$A_t^*(A_tA_t^*)^+ = A_t^+.$$

Hence

$$m_t = A_t^+\xi^t. \tag{14.144}$$

Next, again by the theorem on normal correlation,

$$\begin{aligned}\gamma_t &= E - A_t^+A_t = E - A_t^*(A_tA_t^*)^+A_t \qquad (14.145)\\ &= M\theta\theta^* - M\theta(\xi^t)^*(M(\xi^t)(\xi^t)^*)^+(M\theta(\xi^t)^*)^* = M[(\theta - m_t)(\theta - m_t)^*].\end{aligned}$$

On the other hand, the system of equations given by (14.143) can be represented in the following equivalent form adopted in the filtering scheme considered above:

$$\theta_{t+1} = \theta_t, \quad \theta_0 = \theta; \quad \xi_{t+1} = a_{t+1}\theta, \quad \xi_0 = 0 \tag{14.146}$$

(compare with the system of equations given by (13.46) and (13.47)). We find from the filtering equations given by (13.56) and (13.57), as an application of (14.146), that

$$m_{t+1} = m_t + \gamma_t a_{t+1}^*(a_{t+1}\gamma_t a_{t+1}^*)^+(\xi_{t+1} - a_{t+1}m_t), \quad m_0 = 0, \tag{14.147}$$

$$\gamma_{t+1} = \gamma_t - \gamma_t a_{t+1}^*(a_{t+1}\gamma_t a_{t+1}^*)^+a_{t+1}\gamma_t, \quad \gamma_0 = E. \tag{14.148}$$

Thus, the required recursive equation, (14.141), for γ_t is established. In order to deduce (14.140) from (14.147) we proceed as follows.

Let $z = \theta^*x$. Then

$$\begin{aligned}M\xi^t z &= MA_t\theta\theta^*x = A_tx = y^t,\\ M\xi_t z &= Ma_t\theta\theta^*x = a_tx = y_t,\\ Mm_t z &= MA_t^+\xi^t z = A_t^+M\xi^t z = A_t^+y^t = x_t.\end{aligned} \tag{14.149}$$

Multiplying the left- and right-hand sides of (14.147) by z and then taking the mathematical expectation of the expressions obtained, we find

$$Mm_{t+1}z = Mm_tz + \gamma_t a_{t+1}^*(a_{t+1}\gamma_t a_{t+1}^*)^+[M\xi_{t+1}z - a_{t+1}Mm_tz],$$

which, together with (14.149), leads to the desired equation, (14.140).

It also follows from (14.139) and (14.137) that $x_k = x^0$.

In order to prove the concluding part of the theorem, for each prescribed t, let

$$b = a_{t+1} - \sum_{s=1}^{t} c_s as_s, \tag{14.150}$$

where the numbers $c_1, \dots, c_t$ are chosen so that the value of bb^* is minimal. Denoting by c the vector row $(c_1, \dots, c_t)$, we shall write (14.150) in vectorial form

$$b = a_{t+1} - cA_t. \tag{14.151}$$

Then

$$bb^* = a_{t+1}a_{t+1}^* - 2a_{t+1}A_t^*c^* + cA_tA_t^*c.$$

From this, because of the minimality of the value of bb^*, it follows that the vector $c = (c_1, \ldots, c_t)$ satisfies the system of linear algebraic equations $c(A_tA_t^*) = a_{t+1}A_t^*$ and, therefore,

$$c = a_{t+1}A_t^*(A_tA_t^*)^+ = a_{t+1}A_t^+. \tag{14.152}$$

It follows from (14.151) and (14.152) that

$$b = a_{t+1}(E - A_t^+A_t)$$

and

$$\begin{aligned} bb^* &= a_{t+1}(E - 2A_t^+A_t + (A_t^+A_t)^2)a_{t+1}^* \\ &= a_{t+1}(E - A_t^+A_t)a_{t+1}^* = a_{t+1}\gamma_t a_{t+1}^*, \end{aligned}$$

where we have made use of (4°), one of the properties of pseudo-inverse matrices (see Section 13.1).

If the rank of the matrix A is equal to k, then the ranks of the matrices A_t, $t = 1, \ldots, k_t$, are all equal to t. Hence, for any $t = 1, \ldots, k$, the row a_{t+1} is not a linear combination of the rows $a_1, , \ldots, a_t$, and, therefore, $bb^* > 0$. But $bb^* = a_{t+1}\gamma_t a_{t+1}^*$, hence $a_{t+1}\gamma_t a_{t+1}^* > 0$. □

14.5.3. Let us discuss now the case where the system of algebraic equations $Ax = y$ is insolvable. It turns out that in this case in order to find the pseudo-solution $x^0 = A^+y$, a recursive procedure can be constructed which does not require 'pseudo-inversion' of the matrix A.

Assume that the matrix $A = \|a_{ij}\|$ has the order $(k \times n)$. In describing recursive processes it is essential to distinguish between the cases $k \leq n$ and $k > n$. Here we consider only the case $k \leq n$.

Theorem 14.5. *Let $k \leq n$ and let the rank of A equal k. Then the pseudo-solution $x^0 = A^+y$ coincides with the vector x_k obtained from the system of recursive equations (14.140) and (14.141).*

In order to prove this we need the following.

Lemma 14.10. *Let B be a matrix of order $(m \times n)$ and let E be the unit matrix of order $(n \times n)$. Then*

$$\lim_{\alpha \downarrow 0}(\alpha E + B^*B)^{-1}B^* = B^+, \tag{14.153}$$

$$\lim_{\alpha \downarrow 0} (\alpha E + B^* B)^{-1} \alpha = E - B^+ B. \tag{14.154}$$

PROOF. We have

$$\begin{aligned} \Delta(\alpha) &\equiv B^+ - (\alpha E + B^* B)^{-1} B^* = (\alpha E + B^* B)^{-1} [(\alpha E + B^* B) B^+ - B^*] \\ &= (\alpha E + B^* B)^{-1} [\alpha B^+ + B^* B B^+ - B]. \end{aligned}$$

But $B^* B B^+ = B^*$ (see (7°), Section 13.1). Hence

$$\Delta(\alpha) = \alpha (\alpha E + B^* B)^{-1} B^+$$

and

$$\Delta(\alpha)(\Delta(\alpha))^* = \alpha^2 (\alpha E + B^* B)^{-1} (B^* B)^+ (\alpha E + B^* B)^{-1}, \tag{14.155}$$

since $B^+ (B^+)^* = (B^* B)^+$ (see (5°), Section 13.1).

If $B^* B$ is a diagonal matrix, then the validity of (14.153) follows from (14.155), since the zeros on the diagonals of the matrices $B^* B$ and $(B^* B)^+$ coincide. Otherwise, with the aid of orthogonal transformation of S ($S^* = S^{-1}$), we obtain

$$S^* (B^* B) S = \operatorname{diag}(B^* B), \quad S^* (B^* B)^+ S = \operatorname{diag}(B^* B)^+$$

and

$$\begin{aligned} S^* \Delta(\alpha)(\Delta(\alpha))^* S &= \alpha [\alpha E + \operatorname{diag}(B^* B)]^{-1} \\ &\quad \times \operatorname{diag}(B^* B)^+ [\alpha E + \operatorname{diag}(B^* B)]^{-1} \to 0, \quad \alpha \downarrow 0. \end{aligned}$$

From this, because of the nonsingularity of the matrix S, we obtain

$$\Delta(\alpha)(\Delta(\alpha))^* \to 0, \quad \alpha \downarrow 0.$$

Thus (14.153) is established.

In order to prove (14.154), it remains only to note that, due to (14.153),

$$\begin{aligned} E - B^+ B &= E - \lim_{\alpha \downarrow 0} (\alpha E + B^* B)^{-1} B^* B \\ &= E - \lim_{\alpha \downarrow 0} (\alpha E + B^* B)^{-1} (B^* B + \alpha E - \alpha E) \\ &= \lim_{\alpha \downarrow 0} (\alpha E + B^* B)^{-1} \alpha. \end{aligned}$$

□

PROOF OF THEOREM 14.5. If the system $Ax = y$ is solvable, then the required statement follows from Theorem 14.4. Let us proceed to the general case.

First of all we shall show that the vector $x_t = A_t^+ y^t$ can be obtained in the following way:

$$x_t = \lim_{\alpha \downarrow 0} x_t^\alpha, \tag{14.156}$$

where x_t^α, $\alpha > 0$ is a solution of the solvable system of linear equations

$$(\alpha E - A_t^* A_t) x_t^\alpha = A_t^* y^t. \tag{14.157}$$

Indeed, let the vector $x_t^\alpha(t) = (x_1^\alpha(t), \dots, x_n^\alpha(t))$ minimize the functional

$$J(x^\alpha) = \sum_{s=1}^{t} [a_s x^\alpha - y_s]^2 + \alpha \sum_{j=1}^{m} (x_j^\alpha)^2,$$

where $x^\alpha = (x_1^\alpha, \dots, x_n^\alpha)$. Then it is not difficult to see that

$$x_t^\alpha = (\alpha E + A_t^* A_t)^{-1} A_t^* y^t. \tag{14.158}$$

It follows immediately that x_t^α is a solution of the solvable system of equations given by (14.157). But, by Lemma 14.10,

$$\lim_{\alpha \downarrow 0} (\alpha E + A_t^* A_t)^{-1} A_t^* = A_t^+,$$

which, together with (14.158), proves the equality

$$x_t = \lim_{\alpha \downarrow 0} x_t^\alpha.$$

We can deduce recursive equations for the vectors x_t^α, $t \le k$. For this purpose let us take advantage of the technique applied in proving the previous theorem.

Let $\theta = (\theta_1, \dots, \theta_n)$ be a Gaussian vector with $M\theta = 0$, $M\theta\theta^* = E$, and let ε_t, $t = 1, \dots, k$, be a Gaussian sequence of independent random variables with $M\varepsilon_t = 0$, $M\varepsilon_t^2 = 1$, independent of the vector θ.

Set

$$\xi_{t+1} = a_{t+1}\theta_t + \alpha^{1/2}\varepsilon_{t+1}, \quad \alpha > 0, \tag{14.159}$$

where $\theta_t \equiv \theta$. Then $m_t^\alpha = M(\theta_t|\xi_1, \dots, \xi_t) = M(\theta|\xi_1, \dots, \xi_t)$ and $\gamma_t^\alpha = M[(\theta - m_t^\alpha)(\theta - m_t^\alpha)^*]$, according to Theorem 13.4, satisfy the following system of equations:

$$m_{t+1}^\alpha = m_t^\alpha + \frac{\gamma_t^\alpha a_{t+1}^*}{\alpha + a_{t+1}\gamma_t^\alpha a_{t+1}^*}(\xi_{t+1} - a_{t+1} m_t^\alpha), \quad m_0^\alpha = 0, \tag{14.160}$$

$$\gamma_{t+1}^\alpha = \gamma_t^\alpha - \frac{\gamma_t^\alpha a_{t+1}^* a_{t+1} \gamma_t^\alpha}{\alpha + a_{t+1}\gamma_t^\alpha a_{t+1}^*}, \quad \gamma_0^\alpha = E. \tag{14.161}$$

According to Theorem 13.15, the solutions m_t^α and γ_t^α of these equations are given by the formulae

$$m_t^\alpha = (\alpha E + A_t^* A_t)^{-1} \sum_{s=0}^{t-1} a_{s+1}^* \xi_{s+1} = (\alpha E + A_t^* A_t)^{-1} A_t^* \xi^t, \tag{14.162}$$

$$\gamma_t^\alpha = \alpha \left(\alpha E + \sum_{s=0}^{t-1} a_{s+1}^* a_{s+1} \right)^{-1} = \alpha(\alpha E + A_t^* A_t)^{-1}. \tag{14.163}$$

Let $\Delta_t^\alpha = y_t - a_t x_t^\alpha$, $\Delta^\alpha = (\Delta_1^\alpha, \dots, \Delta_k^\alpha)$, $\varepsilon = (\varepsilon_1, \dots, \varepsilon_k)$ and $E_t \Delta^\alpha = y^t - A_t x^\alpha$, where E_t is the matrix formed by the first t rows of the unit matrix E of dimension $(k \times k)$. Set $z = \theta^* x^\alpha + \alpha^{-1/2} \varepsilon^* \Delta^\alpha$. Then

$$\begin{aligned} M\xi_t z &= M[a_t\theta + \alpha^{1/2}\varepsilon_t][\theta^* x^\alpha + \alpha^{-1/2}\varepsilon^*\Delta^\alpha] = a_t x^\alpha + \Delta_t^\alpha = y_t. \\ M\xi^t z &= M[A_t\theta + \alpha^{1/2} E_t \Delta^\alpha][\theta^* x^\alpha + \alpha^{-1/2}\varepsilon^*\Delta^\alpha] \\ &= A_t x^\alpha + E_t\Delta^\alpha = y^t, \end{aligned} \tag{14.164}$$

$$Mm_t^\alpha z = (\alpha E + A_t^* A_t)^{-1} A_t^* M\xi^t z = (\alpha E + A_t^* A_t)^{-1} A_t^* y^t = x_t^\alpha. \tag{14.165}$$

Multiplying (on the right) the left- and right-hand sides of (14.160) by z, then taking the mathematical expectation and taking into account relations (14.164) and (14.165), we find that

$$x_{t+1}^\alpha = x_t^\alpha + \frac{\gamma_t^\alpha a_{t+1}^*}{\alpha + a_{t+1}\gamma_t^\alpha a_{t+1}^*}(y_{t+1} - a_{t+1}x_t^*), \quad x_0^\alpha = 0. \tag{14.166}$$

From Lemma 14.10 we have

$$\lim_{\alpha \downarrow 0} \gamma_t^\alpha = E - A_t^+ A_t \quad (= \gamma_t).$$

Since the rank of A is equal to k, $\alpha_{t+1}\gamma_t^\alpha a_{t+1}^* > 0$ for all $\alpha \geq 0$, which follows from (14.163) and Theorem 14.4. Hence in (14.161) we may take the limit as $\alpha \downarrow 0$, yielding for $\gamma_t = \lim_{\alpha\downarrow 0} \gamma_t^\alpha$ the equation

$$\gamma_{t+1} = \gamma_t - \gamma_t a_{t+1}^* (a_{t+1}\gamma_t a_{t+1}^*)^{-1} a_{t+1}\gamma_t, \quad \gamma_0 = E.$$

Finally, taking the limit as $\alpha \downarrow 0$ in (14.166), we obtain from (14.156) the required equation, (14.140). □

Note. The system of recursive relations given by (14.166) and (14.161) for $\alpha > 0$ holds true for the case $k > n$, rank $A \leq n$ as well. Thus, with the aid of this system the vectors $x_k^\alpha = (\alpha E + A^*A)^{-1}A^*y \to A^+ y$ for the matrix A $(k \times n)$ of rank $r \leq \min(k, n)$ can be found (see Lemma 14.10).

14.6 Kalman Filter under Wrong Initial Conditions

Here, we consider a Kalman filtering model with a vector signal θ_t (of size k) and a vector observation ξ_t (of size ℓ) defined by recursions

$$\begin{aligned} \theta_{t+1} &= a\theta_t + b\varepsilon_1(t+1) \\ \xi_{t+1} &= A\xi_t + B\varepsilon_2(t+1), \end{aligned} \tag{14.167}$$

where a, A, b, and B are matrices of dimension $(k \times k)$, $(\ell \times k)$, $(k \times k)$, $(\ell \times \ell)$ respectively and $((\varepsilon_1(t))_{t\geq 1}, (\varepsilon_2(t))_{t\geq 1}$ are zero-mean vector white noises (of sizes k, ℓ respectively) with unit covariance matrices. White noises $((\varepsilon_1(t))_{t\geq 1}$ and $(\varepsilon_2(t))_{t\geq 1}$ are assumed to be orthogonal to each other. The recursion for θ_t is subject to a random initial condition θ_0 with $M\theta_0^*\theta_0 < \infty$ orthogonal to $((\varepsilon_1(t))_{t\geq 1}, (\varepsilon_2(t))_{t\geq 1}$. Assume that $\xi_0 = 0$ and that BB^* is a nonsingular matrix. If

$$m_0 = M\theta_0 \quad \text{and} \quad \gamma_0 = M(\theta_0 - m)(\theta_0 - m)^*$$

are known parameters, then the Kalman filter is defined as (see Corollary 3 to Theorem 13.4)

$$\begin{aligned} m_t &= am_{t-1} + a\gamma_{t-1}A^*[A\gamma_{t-1}A^* + BB^*]^{-1}(\xi_t - Am_{t-1}) \qquad (14.168)\\ \gamma_t &= a\gamma_{t-1}a^* + bb^* - a\gamma_{t-1}A^*[A\gamma_{t-1}A^* + BB^*]^{-1}A\gamma_{t-1}a^*, \end{aligned}$$

subject to the initial conditions m_0 and γ_0. For every fixed t, m_t is the orthogonal projection $\widehat{M}(\theta_t|\xi_{[1,t]})$ of θ_t on the linear space generated by $(1, \xi_1, \cdots, \xi_t)$ while

$$\gamma_t = M(\theta_t - m_t)(\theta_t - m_t)^*,$$

i.e., m_t is a linear optimal (in the mean square sense) filtering estimate for θ_t, given the observation $\xi_{[1,t]} = \{\xi_s, 1 \leq s \leq t\}$. If $(\theta_0, (\varepsilon_1(t), \varepsilon_2(t))_{t\geq 0})$ forms the Gaussian object, then m_t is the conditional expectation for θ_t given the σ-algebra generated by $\xi_{[1,t]}$. If only the noises $(\varepsilon_1(t), \varepsilon_2(t))_{t\geq 0}$ are Gaussian but θ_0 is not, the Kalman filter (14.168) creates only the linear optimal estimate, that is for any t

$$\gamma_t \geq M(X_t - \pi_t)(X_t - \pi_t)^*, \tag{14.169}$$

where π_t is the conditional expectation $M(\theta_t|\xi_{[1,t]})$ defined by a nonlinear filter of more sophisticated structure than the Kalman filter. Nevertheless, the use of the Kalman filter instead of the nonlinear one makes sense, if m_t 'forgets' the distribution of θ_0 in the sense that ($\|\cdot\|^2$ is the Euclidean norm)

$$\lim_{t\to\infty} M\|m_t - \pi_t\|^2 = 0. \tag{14.170}$$

Assume now that even the parameters m_0 and γ_0 of the distribution θ_0 are unknown. Then one can apply wrong m_0' and γ_0' (nonnegative definite matrix). In this case, m_t', γ_t', being defined by the same Kalman filter, are neither the orthogonal projection nor the matrix of filtering errors. Moreover, the use of such a filter makes sense provided that

$$\lim_{t\to\infty} M\|m_t' - \pi_t\|^2 = \lim_{t\to\infty} M\|m_t' - m_t\|^2 = 0. \tag{14.171}$$

We establish below the validity of (14.171) under the assumptions of Theorem 14.3 on matrices

$$G_1 = \begin{pmatrix} A \\ Aa \\ \vdots \\ Aa^{k-1} \end{pmatrix} \quad \text{and} \quad G_2 = \begin{pmatrix} b & ab & \cdots & a^{k-1}b \end{pmatrix}.$$

In what follows, we assume that matrices G_1, G_2 have rank k. Then by Theorem 14.3 $\lim_{t\to\infty} \gamma_t = \gamma$ exists and is independent of γ_0. Moreover, γ is a positive definite matrix, being the unique solution (in the class of symmetric positive definite matrices) of the algebraic equation

$$\gamma = a\gamma a^* + bb^* - a\gamma A^*[A\gamma A^* + BB^*]^{-1}A\gamma a^* \tag{14.172}$$

and

$$\lim_{t\to\infty} M(\theta_t - m_t)(\theta_t - m_t)^* = \gamma.$$

For further analysis, introduce, the so-called Kalman gain

$$K_t = a\gamma_t A^*(A\gamma_t A^* + BB^*)^{-1}$$

and note that there exists a limit matrix

$$K := a\gamma A^*(A\gamma A^* + BB^*)^{-1} = \lim_{t\to\infty} K_t. \tag{14.173}$$

Lemma 4.11. *Let the assumptions of Theorem 14.3 be fulfilled. Then the eigenvalues of the matrix $a - KA$ lie inside the unit circle.*

PROOF. With K_{t-1} defined as above, the recursion for γ_t can be rewritten as

$$\gamma_t = (a - K_{t-1}A)\gamma_{t-1}(a - K_{t-1}A)^* + bb^* + K_{t-1}BB^*K_{t-1}^*$$

and, therefore, passing to limit with $t \to \infty$, we find

$$\gamma = (a - KA)\gamma(a - KA)^* + bb^* + KBB^*K^*. \tag{14.174}$$

Let φ be a left eigenvector of the matrix $a - KA$ corresponding to eigenvalue λ (λ^*). Since γ is a positive definite matrix, $\varphi\gamma\varphi^* = c > 0$. Then, multiplying the right-hand side of (14.174) from the left by φ and from the right by φ^*, we arrive at a linear equation with respect to $|\lambda|^2$

$$c = |\lambda|^2 c + \varphi[bb^* + KBB^*K^*]\varphi^*$$

which implies $|\lambda| \leq 1$. If simultaneously $\varphi K \neq 0$, $\varphi b \neq 0$, then, by virtue of the assumption that BB^* is positive definite, we have $\varphi[bb^* + KBB^*K^*]\varphi^* = c_1 > 0$ and so, $|\lambda| < 1$. Hence, it remains to show only that '$\varphi K = 0, \varphi b = 0$' contradicts the assumptions of the lemma. Assume '$\varphi K = 0, \varphi b = 0$' holds. Then φ is the left eigenvector of the matrix a with the eigenvalue λ. In this case, the $(k \times k)$-matrix $G_2G_2^* = bb^* + abb^*a^* + \cdots + a^{k-1}bb^*(a^{s-1})^*$ is singular:

$$\varphi G_2 G_2^* \varphi^* = 0$$

and at the same time the rank of G_2 is k. The contradiction obtained confirms the statement of the lemma. □

14.6.1 Asymptotically Optimal Kalman Filter. Assume the rank of both G_1 and G_2 is k. Consider a linear Kalman type filter in which the limiting matrix γ involves γ instead of γ_t

$$\widetilde{m}_t = a\widetilde{m}_{t-1} + a\gamma A^*[A\gamma A^* + BB^*]^{-1}(\xi_t - A\widetilde{m}_{t-1}) \tag{14.175}$$

subject to a reasonable initial condition $\widetilde{m}_0$.

Theorem 14.6. *Assume θ_0 is an arbitrarily distributed random vector with $M\|\theta_0\|^2 < \infty$. Denote $m_t = \widehat{M}(\theta_t|\xi_{[1,t]})$ and $\pi_t = M(\theta_t|\xi_{[0,t]})$. Then*

1. $\lim_{t\to\infty} M(\theta_t - m_t)(\theta_t - m_t)^* = \lim_{t\to\infty} M(\theta_t - \widetilde{m}_t)(\theta_t - \widetilde{m}_t)^* = \gamma$;
2. *with Gaussian noises $(\varepsilon_1(t), \varepsilon_2(t))_{t\geq 1}$,*
$$\lim_{t\to\infty} M(\theta_t - \pi_t)(\theta_t - \pi_t)^* = \lim_{t\to\infty} M(\theta_t - \widetilde{m}_t)(\theta_t - \widetilde{m}_t)^* = \gamma;$$
3. [8] *with Gaussian noises $(\varepsilon_1(t), \varepsilon_2(t))_{t\geq 1}$, $\theta_t - \pi_t, t \to \infty$ converges in distribution to a zero-mean Gaussian vector with covariance γ.*

PROOF. 1. Let us note that

$$\begin{aligned} M(\theta_t - \widetilde{m}_t)(\theta_t - \widetilde{m}_t)^* &= M(\theta_t - m_t)(\theta_t - m_t)^* \\ &\quad + M(m_t - \widetilde{m}_t)(m_t - \widetilde{m}_t)^*. \end{aligned}$$

It suffices therefore to show that

$$\lim_{t\to\infty} M\|m_t - \widetilde{m}_t\|^2 = 0. \tag{14.176}$$

We now convert the recursion for m_t to a form more relevant for verifying the validity of (14.176). Introduce an innovation difference (see Theorem 13.5)

$$\overline{\varepsilon}_t = [A\gamma_{t-1}A^* + BB^*]^{-1/2}(\xi_t - Am_{t-1})$$

which forms white noise $(\overline{\varepsilon}_t)_{t\geq 1}$ with a unit covariance matrix. Then, by the definition of the matrices K_{t-1} and K (see (14.173))

$$m_t = am_{t-1} + K(\xi_t - Am_{t-1}) + (K_{t-1} - K)[A\gamma_{t-1}A^* + BB^*]^{1/2}\overline{\varepsilon}_t \tag{14.177}$$

and at the same time

$$\widetilde{m}_t = A\widetilde{m}_{t-1} + K(\xi_t - A\widetilde{m}_{t-1}). \tag{14.178}$$

(14.177) and (14.178) imply for $\Delta_t = m_t - \widetilde{m}_t$ the recursion

$$\Delta_t = (a - KA)\Delta_{t-1} + (K_{t-1} - K)[A\gamma_{t-1}A^* + BB^*]^{1/2}\overline{\varepsilon}_t. \tag{14.179}$$

Denote by $V_t = M\Delta_t\Delta_t^*$ and note that (14.179) implies

[8] Here, we give a simple proof of the result from [224].

$$V_t = (a - KA)V_{t-1}(a - KA)^* + (K_{t-1} - K)[A\gamma_{t-1}A^* + BB^*](K_{t-1} - K)^*.$$

Since by Lemma 4.11 the eigenvalues of the matrix $(a - KA)$ lie within the unit circle and $K_{t-1} - K \to 0$, $t \to \infty$, it follows that $\lim_{t\to\infty} V_t = 0$.

2. We use the upper bound $M(\theta_t - \pi_t)(\theta_t - \pi_t)^* \le M(\theta_t - \widetilde{m}_t)(\theta_t - \widetilde{m}_t)^*$ and the lower bound $M(\theta_t - \pi_t)(\theta_t - \pi_t)^* \ge M(\theta_t - \pi_t^\circ)(\theta_t - \pi_t^\circ)^*$, where $\pi_t^\circ = M(\theta_t|\theta_0, \xi_{[1,t]})$. Under the assumption made, the conditional distributions $P(\theta_t \le x|\theta_0, \xi_{[1,t]})$, $t \ge 0$ are Gaussian with probability one (see Chapter 13), and moreover π_t° is defined as:

$$\begin{aligned}
\pi_t^\circ &= a\pi_{t-1}^\circ + a\gamma_{t-1}^\circ A^*[A\gamma_{t-1}^\circ A^* + BB^*]^{-1}(\xi_t - A\pi_{t-1}^\circ) \\
\gamma_t^\circ &= a\gamma_{t-1}^\circ a^* + bb^* - a\gamma_{t-1}^\circ A^*[A\gamma_{t-1}^\circ A^* + BB^*]^{-1}A\gamma_{t-1}^\circ a
\end{aligned}$$

subject to $\pi_0^\circ = \theta_0, \gamma_0^\circ = 0$. By Theorem 14.3 $\lim_{t\to\infty} \gamma_t^\circ = \gamma$, that is

$$\lim_{t\to 0} M(\theta_t - \pi_t^\circ)(\theta_t - \pi_t^\circ)^* = \gamma.$$

Coupled with the first statement of the theorem, this implies the validity of the second statement as well.

3. Since for any t, $\theta_t - \pi_t^\circ$ is a zero-mean Gaussian vector with covariance γ_t°, its distribution converges weakly to the distribution of a zero-mean Gaussian vector with covariance γ. Therefore (see, for example, Theorem 4.1 in [19]) the required result holds provided that $\pi_t - \pi_t^\circ \to 0$, $t \to 0$ in probability. To verify this, note that

$$M(\pi_t - \pi_t^\circ)(\pi_t - \pi_t^\circ)^* = M(\theta_t - \pi_t)(\theta_t - \pi_t)^* - M(\theta_t - \pi_t^\circ)(\theta_t - \pi_t^\circ)^*.$$

Thus 3. is implied by 2. □

Notes and References. 1

14.1–14.2. In these sections we have systematically used the fact that a stationary sequence with rational spectrum is a component of a multidimensional stationary process obeying the system of recursive equations given by (14.15) (see also Section 15.3). The idea of deducing recursive equations has been borrowed from Laning and Battin [184].

14.3. The optimal control problem for a linear system with a quadratic performance index has been studied by Krasovsky and Lidsky [162], Letov [188] and Kalman, Falb and Arbib [141]. The same control problem with incomplete data has been presented in Aoki [5], Meditch [227] and Wonham [311].

14.4. Theorem 14.3 is analogous to the similar result due to Kalman [139] for the case of continuous time, see also Section 16.2.

14.5. The results obtained in this section have been obtained by Albert and Sittler [2] and also Zhukovsky and Liptser [334].

Notes and References. 2

14.1. The application of the Kalman filter of the minimal dimension for a homogeneous finite-state-space Markov process as an unobservable signal can be found in [200, 201].

14.6. For the continuous time case, statements similar to Theorem 14.6 can be found in Ocone and Pardoux [249]. Statement 3. of Theorem 14.6 is proved differently in Makowski and Sowers [224]. Different approaches to the analysis for discrete time filters can be found in Budhiraja and Ocone [33].

15. Linear Estimation of Random Processes

15.1 Wide-Sense Wiener Processes

15.1.1. In the previous chapter the interrelation between properties in the 'wide' and in the 'strict' sense, which is frequently applied in probability theory, was used in finding optimal linear estimates for stationary sequences with rational spectra. Thus it was enough for our purposes to consider the case of Gaussian sequences (Lemma 14.1) for the construction of the optimal mean square linear estimate. This technique will now be used in problems of linear estimation of processes with continuous time. Here the consideration of the concept of a wide-sense Wiener process turns out to be useful.

15.1.2.
Definition. The measurable random process $W = (W_t)$, $t \geq 0$, given on a probability space $(\Omega, \mathcal{F}, P)$ is called a *wide-sense Wiener process* if

$$\begin{aligned} W_0 &= 0 \quad (P\text{-a.s.}), \\ MW_t &= 0, \quad t \geq 0, \\ MW_tW_s &= t \wedge s. \end{aligned} \tag{15.1}$$

It is clear that any Wiener process is a wide-sense Wiener process at the same time. Another example of a wide-sense Wiener process is the process

$$W_t = \pi_t - t, \tag{15.2}$$

where $\Pi = (\pi_t)$, $t \geq 0$, is a Poisson process with $P(\pi_0 = 0) = 1$ and $P(\pi_t = k) = e^{-t}(t^k/k!)$.

Let $\mathcal{F}_t$, $t \geq 0$, be a nondecreasing family of sub-σ-algebras of $\mathcal{F}$, let $z = (z_t, \mathcal{F}_t)$, $t \geq 0$, be a Wiener process, and let $a = (a_t(\omega), \mathcal{F}_t)$, $t \geq 0$, be some process with $Ma_t^2(\omega) > 0$, $0 < t < T$. Then the process

$$W_t = \int_0^t \frac{a_s(\omega)}{\sqrt{Ma_s^2(\omega)}} dz_s, \quad 0 \leq t \leq T, \tag{15.3}$$

is another example of a wide-sense Wiener process. Note that this process has (P-a.s.) a continuous modification.

It is seen from the definition that a wide-sense Wiener process is a 'process with orthogonal increments', i.e.,

$$M[W_{t_2} - W_{t_1}]W_{s_2} - W_{s_1}] = 0,$$

if $s_1 < s_2 < t_1 < t_2$.

Let $\Phi(d\lambda)$, $-\infty < \lambda < \infty$, be the orthogonal spectral measure with $M\Phi(d\lambda) = 0$, $M|\Phi(d\lambda)|^2 = d\lambda/2\pi$. It is known from the spectral theory of stationary processes that for any measurable function $\varphi(\lambda)$, such that

$$\int_{-\infty}^{\infty} |\varphi(\lambda)|^2 d\lambda < \infty,$$

one can define the stochastic integral[1]

$$I(\varphi, \Phi) = \int_{-\infty}^{\infty} \varphi(\lambda)\Phi(d\lambda),$$

having the following two essential properties:

$$M\int_{-\infty}^{\infty} \varphi(\lambda)\Phi(d\lambda) = 0, \tag{15.4}$$

$$M\int_{-\infty}^{\infty} \varphi_1(\lambda)\Phi(d\lambda)\overline{\int_{-\infty}^{\infty} \varphi_2(\lambda)\Phi(d\lambda)} = \frac{1}{2\pi}\int_{-\infty}^{\infty} \varphi_1(\lambda)\overline{\varphi}_2(\lambda)d\lambda. \tag{15.5}$$

Lemma 15.1. *The random process*

$$W_t = \int_{-\infty}^{\infty} \frac{e^{i\lambda t} - 1}{i\lambda}\Phi(d\lambda) \tag{15.6}$$

is a wide-sense Wiener process.

PROOF. Only the property $MW_sW_t = s \wedge t$ is not obvious. In order to verify it we shall denote by $\Delta = (t_1, t_2)$ and $\Delta' = (s_1, s_2)$ two nonintersecting intervals. Then

$$\begin{aligned} M[W_{t_2} - W_{t_1}][W_{s_2} - W_{s_1}] &= M[W_{t_2} - W_{t_1}]\overline{[W_{s_2} - W_{s_1}]} \\ &= \frac{1}{2\pi}\int_{-\infty}^{\infty} (e^{i\lambda t_2} - e^{i\lambda t_1})(e^{-i\lambda s_2} - e^{-i\lambda s_1})\frac{d\lambda}{\lambda^2}. \end{aligned}$$

But if

$$\chi_\Delta(t) = \begin{cases} 1, & t \in \Delta \\ 0, & t \notin \Delta, \end{cases}$$

then, by virtue of Parseval's theorem,

[1] This integral is the limit (in the mean square) of the explicitly defined integrals $I(\varphi_n, \Phi)$ of the simple functions $\varphi_n(\lambda)$, $n = 1, 2, \ldots$, such that $\int_{-\infty}^{\infty} |\varphi(\lambda) - \varphi_n(\lambda)|^2 d\lambda \to 0$, $n \to \infty$ (compare with the construction of the Itô integral in Section 4.2).

$$\frac{1}{2\pi}\int_{-\infty}^{\infty}(e^{i\lambda t_2}-e^{i\lambda t_1})(e^{-i\lambda s_2}-e^{-i\lambda s_1})\frac{d\lambda}{\lambda^2}=\int_{-\infty}^{\infty}\chi_\Delta(t)\chi_{\Delta'}(t)dt=0.$$

Hence

$$M[W_{t_2}-W_{t_1}][W_{s_2}-W_{s_1}]=0. \tag{15.7}$$

Similarly, it can be shown that

$$M[W_{t_2}-W_{t_1}]^2=\int_{-\infty}^{\infty}(\chi_\Delta(t))^2dt=t_2-t_1. \tag{15.8}$$

It follows from (15.7) and (15.8) that this process is a process with uncorrelated increments and with $MW_t^2=t$. Hence, if $t>s$, then

$$MW_tW_s=M[W_t-W_s+W_s]W_s=MW_s^2=s=t\wedge s.$$

Similarly, at $t<s$

$$MW_tW_s=t\wedge s. \qquad \square$$

It is useful to note that if the wide-sense Wiener process W_t, $t\geq 0$, is Gaussian, then it has a continuous modification that is a Brownian motion process. Indeed, because of the normality, $M[W_t-W_s]^4=3(M[W_t-W_s]^2)^2=3|t-s|^2$. Hence, by the Kolmogorov criterion (Theorem 1.10) the process considered has a continuous modification that by definition (see Section 1.4) is a Brownian motion process.

15.1.3. Let $f(\cdot)\in L_2[0,T]$. Using the wide-sense Wiener process $W=(W_t)$, $t\geq 0$, one can define the Itô stochastic integral (in a wide sense)

$$I_T(f)=\int_0^T f(s)dW_s, \tag{15.9}$$

by defining

$$I_T(f)=\text{l.i.m.}_n\sum_k f_n(t_k^{(n)})[W_{t_{k+1}^{(n)}}-W_{t_k^{(n)}}] \tag{15.10}$$

(where $f_n(t)$ is an array of simple functions ($f_n(t)=f_n(t_k^{(n)})$ for $t_k^{(n)}<t\leq t_{k+1}^{(n)}$, $0=t_0^{(n)}<t_1^{(n)}<\cdots<t_n^{(n)}=T$), having the property that

$$\lim_n\int_0^T[f(t)-f_n(t)]^2dt=0. \tag{15.11}$$

The integral thus defined has the following properties (compare with Subsection 4.2.5):

$$I_T(af_1+bf_2)=aI_T(f_1)+bI_T(f_2),\quad a,b=\text{constant},\ f_i\in L_2[0,T], \tag{15.12}$$

$$\int_0^t f(s)dW_s=\int_0^u f(s)dW_s+\int_u^t f(s)dW_s, \tag{15.13}$$

where

$$\int_u^t f(s)dW_s = \int_0^T f(s)\chi_{(u,t]}(s)dW_s, \tag{15.14}$$

and $\chi_{(ut]}(s)$ is the characteristic function of the set $u < s \le t$. The process $I_t(t) = \int_0^t f(s)dW_s$ is continuous over t in the mean square

$$M\int_0^t f(s)dW_s = 0, \tag{15.15}$$

$$M\int_0^t f_1(s)dW_s \int_0^t f_2(s)dW_s = \int_0^t f_1(s)f_2(s)ds, \quad f_i \in L_2[0,T]. \tag{15.16}$$

If[2]

$$\int_0^T |g(s)|ds < \infty, \quad \int_0^T f^2(s)ds < \infty,$$

then

$$\begin{aligned}\int_0^t g(s)ds \int_0^t f(s)dW_s &= \int_0^t \left(\int_0^s g(u)du\right) f(s)dW_s \\ &+ \int_0^t \left(\int_0^s f(u)dW_u\right) g(s)ds. \end{aligned}\tag{15.17}$$

The existence of the integral in (15.9) and the properties formulated can be verified in the same way as in the case of the Itô stochastic integral for a Wiener process (see Section 4.2).

15.1.4. Let $a(t), b(t), f(t)$, $t \le T$, be measurable (deterministic) functions such that

$$\int_0^T |a(t)|dt < \infty, \quad \int_0^T b^2(t)dt < \infty, \tag{15.18}$$

$$\int_0^T \left(|f(t)a(t)| + |f(t)b(t)|^2\right) dt < \infty. \tag{15.19}$$

Set

$$\xi_t = \int_0^t a(s)ds + \int_0^t b(s)dW_s, \tag{15.20}$$

where W_s, $s \ge 0$, is a wide-sense Wiener process.

[2] The last integral in (15.17) exists due to the Fubini theorem and the inequality

$$\begin{aligned}M\int_0^T \left|\int_0^s f(u)dW_u\right| |g(s)|ds &\le \int_0^T \left(M\left[\int_0^s f(u)dW_u\right]^2\right)^{1/2} |g(s)|ds \\ &= \int_0^T \left(\int_0^s f^2(u)du\right)^{1/2} |g(s)|ds < \infty.\end{aligned}$$

With this process the integral $\int_0^t f(s)d\xi_s$ can be defined by setting

$$\int_0^t f(s)d\xi_s = \text{l.i.m.}_n \sum_k f_n(t_k^{(n)})[\xi_{t_{k+1}^{(n)}} - \xi_{t_k^{(n)}}], \tag{15.21}$$

where $f_n(t)$ is a sequence of simple functions such that

$$\lim_n \int_0^T [|a(t)||f(t) - f_n(t)| + b^2(t)|f(t) - f_n(t)|^2]dt = 0.$$

The integrals $\int_0^t f(s)d\xi_s$ thus defined are $\mathcal{F}_t^\xi$-measurable and have the property that (P-a.s.)

$$\int_0^t f(s)d\xi_s = \int_0^t f(s)a(s)ds + \int_0^t f(s)b(s)dW_s, \quad 0 \le t \le T \tag{15.22}$$

(compare with Subsection 4.2.11).

15.1.5. Let $\nu = (\nu_t)$, $t \ge 0$, be a process with orthogonal increments, with $M(\nu_t - \nu_s) = 0$ and

$$M(\nu_t - \nu_s)^2 = \int_s^t a^2(u)du, \tag{15.23}$$

where $\int_0^T a^2(u)du < \infty$. For deterministic (measurable) functions $f(t)$ satisfying the condition

$$\int_0^T a^2(u)f^2(u)du < \infty, \tag{15.24}$$

one can also define the stochastic integral

$$\int_0^T f(s)d\nu_s \tag{15.25}$$

as the limit (in the mean square) of the corresponding integral sums

$$\sum_k f_n(s_k^{(n)})[\nu_{s_{k+1}^{(n)}} - \nu_{s_k^{(n)}}]$$

at $n \to \infty$, where the sequence of the simple functions $f_n(s)$ is such that

$$\int_0^T |f_n(s) - f(s)|^2 a^2(s)ds \to 0, \quad n \to \infty.$$

The correctness of such a definition can be established in the same way as in the case of stochastic integrals of a square integrable martingale[3] for which the corresponding predictable increasing process is absolutely continuous with probability one (see Theorem 5.10).

[3] It is useful to note that any square integrable martingale is a process with orthogonal increments.

Note two useful properties of the integral in (15.25);

$$M \int_0^T f(s) d\nu_s = 0; \tag{15.26}$$

$$M \int_0^T f_1(s) d\nu_s \int_0^T f_2(s) d\nu_s = \int_0^T f_1(s) f_s(s) a^2(s) ds \tag{15.27}$$

(it is assumed that $\int_0^T f_i^2(s) a^2(s) ds < \infty$, $i = 1, 2$).

In the case, where the process $\nu = (\nu_t)$, $t \geq 0$ is also a martingale and $a^2(u) > 0$, $0 \leq u \leq T$, the process

$$W_t = \int_0^t \frac{d\nu_s}{a(s)} \tag{15.28}$$

is a Brownian motion process (as was shown in Theorem 5.12). Discarding the assumption on the martingale property leads us to the following result.

Lemma 15.2. *Let $\nu = (\nu_t)$, $t \geq 0$, be a process with orthogonal increments, $M(\nu_t - \nu_s) = 0$,*

$$M(\nu_t - \nu_s)^2 = \int_s^t a^2(u) du.$$

If $\inf_{0 \leq u \leq T} a^2(u) > 0$ and $\int_0^T a^2(u) du < \infty$, then the process[4]

$$W_t = \int_0^t \frac{d\nu_s}{a(s)}$$

is a wide-sense Wiener process.

PROOF. It is seen that $MW_t = 0$, $MW_t^2 = t$. Finally, due to (15.27),

$$MW_t W_s = M \int_0^t \frac{d\nu_u}{a(u)} \int_0^s \frac{d\nu_u}{a(u)} = M \int_0^{t \vee s} \chi_{(u \leq t)} \frac{d\nu_u}{a(u)} \int_0^{t \vee s} \chi_{(v \leq s)} \frac{d\nu_v}{a(v)}$$
$$= \int_0^{t \vee s} \chi_{(u \leq t)} \chi_{(u \leq s)} du = t \wedge s.$$

□

[4] As usual,

$$\int_0^t \frac{d\nu_s}{a(s)} = \int_0^T \chi_{(s \leq t)} \frac{d\nu_s}{a(s)}.$$

15.1.6. Let the deterministic (measurable) functions $a_0(t), a_1(t)$ and $b(t)$ be such that

$$\int_0^T |a_i(t)|dt < \infty, \quad \int_0^T b^2(t)dt < \infty, \quad i = 0, 1. \tag{15.29}$$

Consider the linear equation

$$x_t = x_0 + \int_0^t [a_0(s) + a_1(s)x_s]ds + \int_0^t b(s)dW_s, \tag{15.30}$$

where $W = (W_s)$, $s \geq 0$, is a wide-sense Wiener process, and x_0 is a random variable uncorrelated with W with $Mx_0^2 < \infty$ (as in the case of the Wiener process, Equation (15.30) will be written symbolically as $dx_t = [a_0(t) + a_1(t)x_t]dt + b(t)dW_t$).

If $W = (W_s)$, $s \geq 0$, is a Wiener process, then, according to Theorem 4.10, Equation (15.30) has a unique continuous (P-a.s.) solution given by the formula

$$\begin{aligned} x_t &= \exp\left\{\int_0^t a_1(u)du\right\}\left\{x_0 + \int_0^t \exp\left[-\int_0^s a_1(u)du\right] a_0(s)ds \right. \\ &\quad \left. + \int_0^t \exp\left[-\int_0^s a_1(u)du\right] b(s)dW_s\right\}. \end{aligned} \tag{15.31}$$

The stochastic integral on the right-hand side of (15.31) is defined for a wide-sense Wiener process as well. (15.31) in the case of the Wiener process W_s holds true also in the mean square sense. Hence, it also holds true in the mean square sense when W_s is a wide-sense Wiener process which proves the existence of a solution of equation (15.30) with a wide-sense Wiener process given by (15.31). It is not difficult to convince oneself, using (15.17), that the process x_t, $0 \leq t \leq T$, is continuous in the mean square. Let y_t, $0 \leq t \leq T$, be another similar solution of Equation (15.30). Then $\Delta_t = x_t - y_t$, $0 \leq t \leq T$, satisfies the equation

$$\Delta_t = \int_0^t a_1(s)\Delta_s ds$$

and, therefore, is a continuous (P-a.s.) process, whence

$$|\Delta_t| \leq \int_0^t |a_1(s)||\Delta_s|ds.$$

By Lemma 4.13, $\Delta_s = 0$ (P-a.s.), $0 \leq t \leq T$. Hence

$$P\left\{\sup_{0 \leq t \leq T} |x_t - y_t| > 0\right\} = 0.$$

Now let $W = (W_1, \dots, W_n)$ be an n-dimensional wide-sense Wiener process (each of the processes $W_i = (W_i(t))$, $t \ge 0$, $i = 1, \dots, n$, is a wide-sense Wiener process, and the components W_i, W_j at $i \neq j$ are uncorrelated).

Let there be given the random vector $x_0 = (x_1(0), \dots, x_n(0))$, uncorrelated with W, $\sum_{i=1}^n Mx_i^2(0) < \infty$, the vector function $a_0(t) = (a_{01}(t), \dots, a_{0n}(t))$, and the matrices $a_1(t) = \|a_{ij}^1(t)\|$ and $b(t) = \|b_{ij}(t)\|$ of dimension $(n \times n)$. We shall also assume that for the elements $a_{0i}(t)$, $a_{ij}^1(t)$ and $b_{ij}(t)$ the associated conditions given by (15.29) are satisfied. Then, as in the case $n = 1$, the equation

$$x_t = x_0 + \int_0^t [a_0(s) + a_1(s)x_s] + \int_0^t b(s)dW_s \tag{15.32}$$

has the unique continuous (in the mean square) solution $x_t = (x_1(t), \dots, x_n(t))$ given by the formula

$$x_t = \Phi_0^t \left\{ x_0 + \int_0^t (\Phi_0^s)^{-1} a_0(s)ds + \int_0^t (\Phi_0^s)^{-1} b(s)dW_s \right\}, \tag{15.33}$$

where Φ_0^t is the fundamental matrix

$$\frac{d\Phi_0^t}{dt} = a_1(t)\Phi_0^t, \quad \Phi_0^0 = E_{(n\times n)}. \tag{15.34}$$

For the process x_t so obtained, let $n_t = Mx_t$, $\Gamma(t,s) = M(x_t - n_t)(x_s - n_s)^*$, $\Gamma_t = \Gamma(t,t)$.

Theorem 15.1. *The vector n_t and the matrix Γ_t are solutions of the differential equations*

$$\frac{dn_t}{dt} = a_0(t) + a_1(t)n_t, \tag{15.35}$$

$$\frac{d\Gamma_t}{dt} = a_1(t)\Gamma_t + \Gamma_t a_1^*(t) + b(t)b^*(t). \tag{15.36}$$

The matrix $\Gamma(t,s)$ is given by the formula

$$\Gamma(t,s) = \begin{cases} \Phi_s^t \Gamma_s, & t \ge s, \\ \Gamma_t (\Phi_t^s)^*, & t \le s, \end{cases} \tag{15.37}$$

where $\Phi_s^t = \Phi_0^t(\Phi_0^s)^{-1}$, $t \ge s$.

PROOF. Equation (15.35) can be obtained by averaging both sides in (15.32). It follows from (15.33) that the solution of Equation (15.35) is defined by the formula

$$n_t = \Phi_0^t \left\{ n_0 + \int_0^t (\Phi_0^s)^{-1} a_0(s)ds \right\}. \tag{15.38}$$

Next, let $V_t = x_t - n_t$. Then it follows from (15.33) and (15.38) that

$$V_t = \Phi_0^t \left\{ V_0 + \int_0^t (\Phi_0^s)^{-1} b(s) dW_s \right\}, \tag{15.39}$$

from which, due to the lack of correlation of x_0 and W, we obtain

$$\begin{aligned} \Gamma_t &= MV_tV_t^* \\ &= \Phi_0^t \left\{ MV_0V_0^* + M \int_0^t (\Phi_0^s)^{-1} b(s) dW_s \left(\int_0^t (\Phi_0^s)^{-1} b(s) dW_s \right)^* \right\} (\Phi_0^t)^*. \end{aligned}$$

Since the components of the process W are uncorrelated, from (15.15) and (15.16) it follows that

$$\begin{aligned} & M \int_0^t (\Phi_0^s)^{-1} b(s) dW_s \left(\int_0^t (\Phi_0^s)^{-1} b(s) dW_s \right)^* \\ &= \int_0^t (\Phi_0^s)^{-1} b(s) b^*(s) [(\Phi_0^s)^{-1}]^* ds. \end{aligned}$$

Therefore,

$$\Gamma_t = \Phi_0^t \left\{ \Gamma_0 + \int_0^t (\Phi_0^s)^{-1} b(s) b^*(s) [(\Phi_0^s)^{-1}]^* ds \right\} (\Phi_0^t)^*.$$

By differentiating the right-hand side of this relation and taking into account (15.34) we arrive at the required equation, (15.36).

Let us now establish (15.37). Let $t \geq s$. Then

$$\begin{aligned} \Gamma(t,s) &= MV_tV_s^* \\ &= \Phi_0^t \Bigg\{ MV_0V_0^* \\ &\quad + M \left[\int_0^t (\Phi_0^u)^{-1} b(u) dW_u \right] \left[\int_0^t \chi_{(s \geq u)} (\Phi_0^u)^{-1} b(u) dW_u \right] (\Phi_0^s)^* \Bigg\} \\ &= \Phi_s^t \Phi_0^s \left\{ \Gamma_0 + \int_0^s (\Phi_0^u)^{-1} b(u) b^*(u) [(\Phi_0^u)^{-1}]^* du \right\} (\Phi_0^s)^* = \Phi_s^t \Gamma_s. \end{aligned}$$

The other side of (15.37) can be verified for $t \leq s$ in the same fashion. □

15.1.7. For the process x_t, $0 \leq t \leq T$, satisfying Equation (15.30), for $t > s$ let

$$R(t,s) = \Gamma(t,s)\Gamma_s^+.$$

For $s < u < t$, let us show that

$$R(t,s) = R(t,u)R(u,s). \tag{15.40}$$

In order to prove this relation it suffices to consider the case where $x_0 = 0$, $a_0(s) \equiv 0$, and W_s is a Wiener process. Then it follows from the theorem on normal correlation (Theorem 13.1) that

$$M(x_t|x_u) = R(t,u)x_u.$$

It follows from (15.33) that the process x_t is a Markov process, and, in particular,

$$M(x_t|x_s, x_u) = M(x_t|x_u) \quad (P\text{-a.s.}).$$

Consequently,

$$M(x_t - R(t,u)x_u|x_s, x_u) = 0,$$

and, therefore,

$$M(x_t x_s^* \Gamma_s^+ - R(t,u)x_u x_s^* \Gamma_s^+) = 0,$$

which proves (15.40).

Thus, for the process x_t, $0 \le t \le T$, satisfying Equation (15.32), the function $R(t,s)$ satisfies (15.40). The converse holds true, in a certain sense, as well.

Theorem 15.2. *Let $x = (x_1(t), \dots, x_n(t))$, $0 \le t \le T$ be a random process with the first two moments $n_t = Mx_t$ and $\Gamma(t,s) = M[(x_t - n_t)(x_s - n_s)^*]$ given. Assume that the matrix $R(t,s) = \Gamma(t,s)\Gamma_s^+$ satisfies (15.40) and that the following assumptions are satisfied*

(1) there exist a vector $a_0(t)$ and matrices $a_1(t)$ and $B(t)$ such that their elements belong to $L_1[0,t]$;

(2) the elements of the matrices $R(t,s)$ are continuous over t $(t > s)$, and

$$R(t,s) = R(s,s) + \int_s^t a_1(u)R(u,s)du;$$

(3) the elements of the matrices $\gamma_t = \Gamma(t,t)$ are continuous and

$$\Gamma_t = \Gamma_0 + \int_0^t [a_1(u)\Gamma_u + \Gamma_u a_1^*(u)]du + \int_0^t B(u)du;$$

(4) the elements of the vectors n_t are continuous over t, and

$$n_t = n_0 + \int_0^t [a_0(u) + a_1(u)n_u]du.$$

Then there exists a wide-sense Wiener process $W_t = (W_1(t), \dots, W_n(t))$, such that (P-a.s.) for all t, $0 \le t \le T$,

$$x_t = x_0 + \int_0^t [a_0(s) + a_1(s)x_s]ds + \int_0^t B^{1/2}(s)dW_s. \tag{15.41}$$

PROOF. Let $\tilde{W}_t$, $0 \le t \le T$, be some n-dimensional wide-sense Wiener process, and let $\tilde{x}_0$ be an n-dimensional vector with the same first two moments as x_0 and independent of $\tilde{W}_t$, $0 \le t \le T$. Assume that for almost all s, $0 \le s \le T$, the matrices $B(s)$ are nonnegative definite. Let the process $\tilde{x}_t$, $0 \le t \le T$, be a solution of the equation (Subsection 15.1.6)

$$\tilde{x}_t = \tilde{x}_0 + \int_0^t [a_0(s) + a_1(s)\tilde{x}_s]ds + \int_0^t B^{1/2}(e)d\tilde{W}_s.$$

Then, due to Theorem 15.1 and assumptions (1)–(4), the first two moments in the processes x_t and $\tilde{x}_t$ coincide. Therefore, the first two moments in the processes

$$\begin{aligned} \nu_t &= x_t - x_0 - \int_0^t [a_0(s) + a_1(s)x_s]ds, \\ \tilde{\nu}_t &= \tilde{x}_t - \tilde{x}_0 - \int_0^t [a_0(s) + a_1(s)\tilde{x}_s]ds, \end{aligned} \tag{15.42}$$

also coincide.

But $\tilde{\nu}_t = \int_0^t B^{1/2}(s)d\tilde{W}_s$ is a process with orthogonal increments and, hence, so also is the process ν_t, $0 \le t \le T$.

If the matrices $B(t)$ are positive definite for almost all t, $0 \le t \le T$, the process

$$W_t = \int_0^t B^{-1/2}(s)d\nu_s,$$

by the multidimensional version of Lemma 15.2, is a wide-sense Wiener process. Hence $\nu_t = \int_0^t B^{1/2}(s)dW_s$, which, together with (15.42), proves (15.41) in this case.

If the matrices $B(t)$ for almost all t, $0 \le t \le T$, are nonnegative definite, then

$$W_t = \int_0^t [B^{1/2}(s)]^+ d\nu_s + \int_0^t [E - (B^{1/2}(s))^+(B^{1/2}(s))]dz_s,$$

where z_t, $0 \le t \le T$, is an n-dimensional wide-sense Wiener process uncorrelated with the initial process x_t, $0 \le t \le T$. (Such a process exists, if the initial probability space is sufficiently 'rich'). Then, as in Lemma 10.4, we can show that the process W_t, $0 \le t \le T$, thus defined is a wide-sense Wiener process.

Let us show now that the assumption made on the nonnegative definiteness of the matrices $B(t)$ (for almost all t, $0 \le t \le T$) is a consequence of conditions (2) and (3) of the theorem.

The properties of the matrices $B(t)$ depend only on the properties of the first two moments of the process x_t, $0 \le t \le T$; hence, without loss of generality, this process can be considered Gaussian. Then, by the theorem on normal correlation, the matrix

$$\Gamma(t+\Delta,t+\Delta)-\Gamma(t+\Delta,t)\Gamma^+(t,t)\Gamma^*(t+\Delta,t), \quad 0\le t\le t+\Delta\le T,$$

is symmetric and nonnegative definite. By the properties of pseudo-inverse matrices (see Section 13.1),

$$\Gamma^+(t,t)=\Gamma^+(t,t)\Gamma(t,t)\Gamma^+(t,t), \quad (\Gamma^+(t,t))^*=(\Gamma^*(t,t))^+=\Gamma^+(t,t).$$

Hence, the matrix

$$\begin{aligned}&\Gamma(t+\Delta,t+\Delta)-\Gamma(t+\Delta,t)\Gamma^+(t,t)\Gamma(t,t)(\Gamma^+(t,t))^*\Gamma^*(t+\Delta,t)\\ &=\Gamma(t+\Delta,t+\Delta)-R(t+\Delta,t)\Gamma(t,t)R^*(t+\Delta,t)\end{aligned} \tag{15.43}$$

is also symmetric and nonnegative definite. After simple transformations we find from (12.43), (2), (3) and the formula $\Gamma(u,t)\Gamma^+(t,t)\Gamma(t,t)=\Gamma(u,t)$, $u\ge t$ (see the proof of Theorem 13.1), that

$$B(t)=\lim_{\Delta\downarrow 0}\frac{1}{\Delta}\{\Gamma(t+\Delta,t+\Delta)-R(t+\Delta,t)\Gamma(t,t)R^*(t+\Delta,t)\}$$

(for almost all t, $0\le t\le T$). Consequently, the matrices $B(t)$ for almost all t are nonnegative definite. □

EXAMPLE. Let $W=(W_t)$, $0\le t\le 1$, be a wide-sense Wiener process and let

$$\xi_t=W_1\cdot t+W_t$$

(i.e., $d\xi_t=W_1dt+dW_t$, $\xi_0=0$).

Using the theorem above, we shall show that there exists a wide-sense Wiener process $\overline{W}_t$, $0\le t\le 1$, such that (P-a.s.)

$$\xi_t=\int_0^t\frac{3\xi_s}{1+3s}ds+\overline{W}_t$$

(compare with Theorem 7.12).

Indeed, in our case $M\xi_t\equiv 0$ and $\Gamma(t,s)=M\xi_t\xi_s=3ts+t\wedge s$. We obtain from this, for $t\ge s>0$,

$$R(t,s)=\frac{3ts+s}{3s^2+s}=\frac{3t+1}{3s+1}.$$

This function satisfies the condition of (15.40) and it is easy to see that

$$a_0(t)\equiv 0,\quad a_1(t)=\frac{3}{+3t},\quad B(t)\equiv 1.$$

Note that in our case the values of $\overline{W}_t$ are $\mathcal{F}_t^\xi$-measurable for all t, $0\le t\le 1$.

15.2 Optimal Linear Filtering for some Classes of Nonstationary Processes

15.2.1. Let $W_1 = (W_{11}, \dots, W_{1k})$ and $W_2 = (W_{21}, \dots, W_{2l})$ be mutually uncorrelated wide-sense Wiener processes. We shall discuss the random process $(\theta, \xi) = [\theta_t, \xi_t]$, $t \geq 0$, whose components $\theta_t = [\theta_1(t), \dots, \theta_k(t)]$ and $\xi_t = [\xi_1(t), \dots, \xi_l(t)]$, $t \geq 0$, satisfy the system of stochastic equations

$$\begin{aligned} d\theta_t &= [a_0(t) + a_1(t)\theta_t + a_2(t)\xi_t]dt + b_1(t)dW_1(t) + b_2(t)dW_2(t), \\ d\xi_t &= [A_0(t) + A_1(t)\theta_t + A_2(t)\xi_t]dt + B_1(t)dW_1(t) + B_2(t)dW_2(t) \end{aligned} \tag{15.44}$$

where the coefficients satisfy the conditions of Subsection 10.3.1. Assume as well that the vector of the initial values of (θ_0, ξ_0) is uncorrelated with the processes W_1 and W_2, with $M(\theta_0^*\theta_0 + \xi_0^*\xi_0) < \infty$.

Taking advantage of the results of Chapter 10, let us construct optimal (in the mean square sense) linear estimates of the unobservable component θ_t from the observations $\xi_0^t = \{\xi_s, s \leq t\}$.

Definition. We shall say that the vector $\lambda_t = [\lambda_1(t, \xi), \dots, \lambda_k(t, \xi)]$ is a *linear estimate* of the vector θ_t from ξ_0^t, if the values of $\lambda_j(t, \xi)$ belong[5] to a closed linear subspace generated by the variables ξ_s, $s \leq t$; $j = 1, \dots, k$.

The linear estimate $\lambda_t = [\lambda_1(t, \xi), \dots, \lambda_k(t, \xi)]$ will be called *optimal* if for any other linear estimate $\overline{\lambda}_t = [\overline{\lambda}_1(t, \xi), \dots, \overline{\lambda}_k(t, \xi)]$ the following holds:

$$M[\theta_j(t) - \lambda_j(t, \xi)]^2 \leq M[\theta_j(t) - \overline{\lambda}_j(t, \xi)]^2, \quad j = 1, \dots, k.$$

Note that the value of $\lambda_j(t, \xi)$ is frequently written $\hat{M}(\theta_j(t)|\mathcal{F}_t^\xi)$ and called the wide-sense conditional mathematical expectation of the random variable $\theta_j(t)$ with respect to the σ-algebra $\mathcal{F}_t^\xi$.

15.2.2.

Theorem 15.3. *The optimal linear estimate λ_t of the vector θ_t from the observations ξ_0^t can be defined from the system of equations*

$$\begin{aligned} d\lambda_t &= [a_0(t) + a_1(t)\lambda_t + a_2(t)\xi_t]dt + [(b \circ B)(t) + \gamma_t A_1^*(t)] \\ &\quad \times (B \circ B)^{-1}(t)[d\xi_t - (A_0(t) + A_1(t)\lambda_t + A_2(t)\xi_t)dt], \end{aligned} \tag{15.45}$$

$$\begin{aligned} \dot{\gamma}_t &= a_1(t)\gamma_t + \gamma_t a_1^*(t) + (b \circ b)(t) \\ &\quad -[(b \circ B)(t) + \gamma_t A_1^*(t)](B \circ B)^{-1}(t)[(b \circ B)(t) + \gamma_t A_1^*(t)]^*, \end{aligned} \tag{15.46}$$

with

$$\lambda_0 = M\theta_0 + \mathrm{cov}\,(\theta_0, \xi_0)\mathrm{cov}^+(\xi_0, \xi_0)(\xi_0 - M\xi_0), \tag{15.47}$$

[5] In the sense of convergence in the mean square.

$$\gamma_0 = \text{cov}\,(\theta_0, \theta_0) - \text{cov}\,(\theta_0, \xi_0)\text{cov}\,^+(\xi_0, \xi_0)\text{cov}\,^*(\theta_0, \xi_0). \tag{15.48}$$

In this case $\gamma_t = M[(\theta_t - \lambda_t)(\theta_t - \lambda_t)^*]$.

PROOF. Let $(\tilde{\theta}_t, \tilde{\xi}_t)$, $t \geq 0$ be the Gaussian process satisfying (15.44) where, instead of the processes (W_1, W_2), the mutually independent Wiener processes $(\tilde{W}_1, \tilde{W}_2)$ are considered. Assume that the first two moments in $(\tilde{\theta}_0, \tilde{\xi}_0)$ are the same as those in the vector (θ_0, ξ_0) and that $(\tilde{\theta}_0, \tilde{\xi}_0)$ does not depend on the processes (W_1, W_2). Let

$$\tilde{\lambda}_t = M(\tilde{\theta}_t | \mathcal{F}_t^{\tilde{\xi}}), \quad \tilde{\gamma}_t = M[(\tilde{\theta}_t - \tilde{\lambda}_t)(\tilde{\theta}_t - \tilde{\lambda}_t)^*].$$

Then, according to Theorem 10.3, $\tilde{\lambda}_t$ and $\tilde{\gamma}_t$ satisfy the system of equations given by (15.45) and (15.46) with the substitution of ξ_t for $\tilde{\xi}_t$ and λ_t for $\tilde{\lambda}_t$, and with $\gamma_t \equiv \tilde{\gamma}_t$. It follows from (15.45) that the estimate λ_t is linear (compare with (15.33)).

Let us show now that the estimate λ_t is optimal. Let $q_j(t, \xi)$ be some linear estimate of $\theta_j(t)$ from ξ_0^t, and let $q_j^{(n)}(t, \xi)$ be a sequence of linear estimates from $\xi_{t_0^{(n)}}, \ldots, \xi_{t_n^{(n)}}$, where

$$T^{(n)} = \{t_0^{(n)}, \ldots, t_n^{(n)}\} \subseteq T^{(n+1)} = \{t_0^{(n+1)}, \ldots, t_{n+1}^{(n+1)}\}, t_0^{(n)} \equiv 0, t_n^{(n)} \equiv t,$$

such that

$$q_j(t, \xi) = \text{l.i.m.}_n q_j^{(n)}(t, \xi).$$

Set $\tilde{\lambda}_j^{(n)}(t, \tilde{\xi}) = M(\tilde{\theta}_j(t) | \mathcal{F}_{t,n}^{\tilde{\xi}})$ where $\mathcal{F}_{t,n}^{\tilde{\xi}} = \sigma\{\omega : \tilde{\xi}_{t_0^{(n)}}, \ldots \tilde{\xi}_{t_n^{(n)}}\}$, and denote by $\lambda_j^{(n)}(t, \xi)$ the estimate obtained from $\tilde{\lambda}_j^{(n)}(t, \xi)$ by means of the substitution of the values of $\tilde{\xi}_{t_0^{(n)}}, \ldots, \tilde{\xi}_{t_n^{(n)}}$ for $\xi_{t_0^{(n)}}, \ldots, \xi_{t_n^{(n)}}$. By Lemma 14.1, the linear estimate $\lambda_j^{(n)}(t, \xi)$ is an optimal linear estimate of θ_t from the values of $\xi_{t_0^{(n)}}, \ldots, \xi_{t_n^{(n)}}$, i.e.,

$$M[\theta_j(t) - \lambda_j^{(n)}(t, \xi)]^2 \leq M[\theta_j(t) - q_j^{(n)}(t, \xi)]^2.$$

But

$$M[\lambda_j(t, \xi) - \lambda_j^{(n)}(t, \xi)]^2 = M[\tilde{\lambda}_j(t, \tilde{\xi}) - \tilde{\lambda}_j^{(n)}(t, \tilde{\xi})]^2.$$

It can be established in the same way as in the proof of Lemma 10.1 that

$$\lim_n M[\tilde{\lambda}_j(t, \tilde{\xi}) - \tilde{\lambda}_j^{(n)}(t, \tilde{\xi})]^2 = 0.$$

Hence,

$$\begin{aligned} M[\theta_j(t) - \lambda_j(t, \xi)]^2 &= \lim_n M[\theta_j(t) - \lambda_j^{(n)}(t, \xi)]^2 \\ &\leq \lim_n M[\theta_j(t) - q_j^{(n)}(t, \xi)]^2 = M[\theta_j(t) - q_j(t, \xi)]^2, \end{aligned}$$

which proves the optimality of the estimate $\lambda_j(t,\xi)$, $j=1,\dots,k$. □

Note. It can be verified in similar fashion that the optimal (in the mean square sense) linear estimates of interpolation and extrapolation for the process (θ_t,ξ_t) satisfying the system of equations given by (15.44) can be obtained from the corresponding estimates for the case of the Gaussian processes $(\tilde{\theta}_t,\tilde{\xi}_t)$.

15.2.3. We present now two examples illustrating the possibilities of the application of Theorem 15.3.

These examples are particularly useful in the sense that the processes considered are given in the form of a system of equations different from (15.44), the system considered above.

EXAMPLE 1. Let y_t and z_t be mutually independent Wiener processes. Consider the process (θ_t,ξ_t), $t\ge 0$, satisfying the system of stochastic equations

$$\begin{aligned} d\theta_t &= -\theta_t dt + (1+\theta_t)dy_t \\ d\xi_t &= \theta_t dt + dz_t, \end{aligned} \tag{15.49}$$

where $\xi_0=0$ and θ_0 is a random variable independent of the Wiener processes y_t, z_t, $t\ge 0$, with $M\theta_0=m$ and $M(\theta_0-m)^2=\gamma>0$.

Set

$$W_1(t)=\int_0^t \frac{1+\theta_s}{\sqrt{M(1+\theta_s)^2}}dy_s,\quad W_2(t)=z_t.$$

These two processes are mutually uncorrelated wide-sense Wiener processes, and

$$\begin{aligned} d\theta_t &= -\theta_t dt + \sqrt{M(1+\theta_t)^2}dW_1(t), \\ d\xi_t &= \theta_t dt + dW_2(t). \end{aligned} \tag{15.50}$$

Unlike (15.49), this system is a particular case of the system of equations given by (15.44). Hence, by Theorem 15.3, the optimal linear estimate λ_t of values of θ_t from $\xi_0^t=(\xi_s, s\le t)$ and the filtering error $\gamma_t=M[\theta_t-\lambda_t]^2$ can be defined from the equations

$$\begin{aligned} d\lambda_t &= -\lambda_t dt + \gamma_t(d\xi_t-\lambda_t dt), \quad \lambda_0=m, \\ \dot{\gamma}_t &= -2\gamma_t + M(1+\theta_t)^2 - \gamma_t^2, \quad \gamma_0=\gamma. \end{aligned}$$

For the complete solution of the problem it is necessary to compute

$$M(1+\theta_t)^2 = 1+2n_t+\Delta_t+n_t^2,$$

where

$$n_t=M\theta_t,\quad \Delta_t=M(\theta_t-n_t)^2,$$

$$n_t = n_0 - \int_0^t n_s ds.$$

We find from (15.50) that

$$n_t = n_0 - \int_0^t n_s ds$$

and, due to the Itô formula

$$\begin{aligned}\Delta_t &= M(\theta_t - n_t)^2 \\ &= M\left\{(\theta_0 - n_0)^2 - 2\int_0^t (\theta_s - n_s)^2 ds \right. \\ &\quad \left. + \int_0^t (1+\theta_s)^2 ds + 2\int_0^t (\theta_s - n_s)(1+\theta_s)dy\right\} \\ &= \Delta_0 - 2\int_0^t \Delta_s ds + \int_0^t (1 + \Delta_s + 2n_s + n_s^2)ds.\end{aligned}$$

Therefore, the optimal linear estimate λ_t and the error γ_t can be defined from the system of equations

$$\begin{aligned}d\lambda_t &= -\lambda_t dt + \gamma_t(d\xi_t - \lambda_t dt), \\ \dot{\gamma}_t &= -2\gamma_t - \gamma_t^2 + 1 + \Delta_t + 2n_t + n_t^2, \\ \dot{n}_t &= -n_t, \\ \dot{\Delta}_t &= -\Delta_t + 1 + 2n_t + n_t^2,\end{aligned} \tag{15.51}$$

where $\lambda_0 = n_0 = m$ and $\gamma_0 = \Delta_0 = \gamma$.

EXAMPLE 2. Again let y and z_t be mutually independent Wiener processes, and let the process (θ_t, ξ_t), $t \geq 0$, be defined from the equations

$$\begin{aligned}d\theta_t &= -\theta_t dt + dy_t, \\ d\xi_t &= -\theta_t^3 dt + dz_t,\end{aligned} \tag{15.52}$$

where $\xi_0 = 0$ and θ_0 is Gaussian, $M\theta_0 = 0$, $M\theta_0^2 = \frac{1}{2}$, independent of the processes y_t and z_t. Consider the problem of linear estimation of the variables θ_t and θ_t^3 from $\xi_0^t = \{\xi_s, s \leq t\}$.

Let $\theta_1(t) = \theta_t$ and $\theta_2(t) = \theta_t^3$. With the aid of the Itô formula we can easily see that

$$d\theta_2(t) = -3\theta_2(t)dt + 3\theta_1(t)dt + 3\theta_1^2(t)dy_t.$$

Thus, $\theta_1(t)$ and $\theta_2(t)$ satisfy the system of stochastic equations

$$\begin{aligned}d\theta_1(t) &= -\theta_1(t)dt + dy_t, \\ d\theta_2(t) &= [-3\theta_2(t) + 3\theta_1(t)]dt + 3\theta_1^2(t)dy_t.\end{aligned} \tag{15.53}$$

Let

$$W_1(t) = y_t, \quad W_2(t) = \sqrt{2}\int_0^t \theta_1^2(s)dy_s - \frac{y_t}{\sqrt{2}}, \quad W_3(t) = z_t.$$

It is not difficult to verify that $W_1(t), W_2(t)$ and $W_3(t)$ are mutually uncorrelated wide-sense Wiener processes. Therefore, the processes $\theta_1(t)$, $\theta_2(t)$ and ξ_t satisfy the system of equations

$$\begin{aligned} d\theta_1(t) &= -\theta_1(t)dt + dW_1(t), \\ d\theta_2(t) &= [-3\theta_2(t) + 3\theta_1(t)]dt + \frac{3}{2}dW_1(t) + \frac{3}{\sqrt{2}}dW_2(t), \\ d\xi_t &= \theta_2(t)dt + dW_3(t), \end{aligned} \tag{15.54}$$

where $\xi_0 = 0$ and the vector $(\theta_1(0), \theta_2(0))$ has the following moments:

$$M\theta_1(0) = M\theta_2(0) = 0, \quad M\theta_1^2(0) = \frac{1}{2}, \quad M\theta_1(0)\theta_2(0) = M\theta_0^4 = \frac{3}{4},$$

$$M\theta_2^2(0) = M\theta_0^6 = \frac{15}{8}.$$

(15.54) is of the type (15.44), and, therefore, optimal linear estimates for $\theta_1(t) = \theta_t$ and $\theta_2(t) = \theta_t^3$ can be found from the system of equations given by (15.45) and (15.46).

15.3 Linear Estimation of Wide-Sense Stationary Random Processes with Rational Spectra

15.3.1. The object of this section is to show how Theorem 15.3 can be applied to the construction of optimal linear estimates for the processes listed in the title of the section. The pertinent results for random sequences were discussed in Section 14.1. Let $\eta = (\eta_t)$, $-\infty < t < \infty$, be a real stationary (wide sense) process permitting the spectral representation

$$\eta_t = \int_{-\infty}^{\infty} e^{i\lambda t}\frac{P_{n-1}(i\lambda)}{Q_n(i\lambda)}\Phi(d\lambda), \tag{15.55}$$

where $\Phi(d\lambda)$ is the orthogonal spectral measure, $M\Phi(d\lambda) = 0$,

$$M|\Phi(d\lambda)|^2 = \frac{d\lambda}{2\pi}, \quad P_{n-1}(z) = \sum_{k=0}^{n-1} b_k z^k, \quad Q_n(z) = z^n + \sum_{k=0}^{n-1} a_k z^k,$$

and the real parts of the roots of the equation $Q_n(z) = 0$ are negative. Consider the processes

$$\eta_j(t) = \int_{-\infty}^{\infty} e^{i\lambda t} W_j(i\lambda)\Phi(d\lambda), \quad j = 1, \ldots, n, \tag{15.56}$$

where the frequency characteristics of $W_j(z)$, $j = 1, \ldots, n$ are selected in the following special way:

$$W_j(z) = z^{-(n-j)} W_n(z) + \sum_{k=j}^{n-1} \beta_k z^{-(k-j+1)}, \quad j = 1, \ldots, n-1, \tag{15.57}$$

and

$$W_n(z) = -z^{-1} \sum_{k=0}^{n-1} a_k W_{k+1}(z) + z^{-1}\beta_n \tag{15.58}$$

with

$$\beta_1 = b_{n-1}, \quad \beta_j = b_{n-j} - \sum_{i=1}^{j-1} \beta_i a_{n-j+i}, \quad j = 2, \ldots, n. \tag{15.59}$$

It follows from (15.57) and (15.58) that

$$\begin{aligned} W_j(z) &= z^{-1}[W_{j+1}(z) + \beta_j], \quad j = 1, \ldots, n-1, \\ W_n(z) &= z^{-1} \left[-\sum_{k=0}^{n-1} a_k W_{k+1}(z) + \beta_n \right]. \end{aligned} \tag{15.60}$$

We obtain from this

$$W_n(z) = z^{-1} \left[-\sum_{k=0}^{n-1} a_k \left(z^{-(n-k-1)} W_n(z) + \sum_{j=k+1}^{n-1} \beta_j z^{-(j-k)} \right) + \beta_n \right],$$

and, therefore,

$$W_n(z) = P_{n-1}^{(n)}(z)/Q_n(z), \tag{15.61}$$

where $P_{n-1}^{(n)}(z)$ is a polynomial of degree less than n.

Then, we obtain from (15.60) and (15.61)

$$W_j(z) = \frac{P_{n-1}^{(j)}(z)}{Q_n(z)}, \quad j = 1, \ldots, n-1, \tag{15.62}$$

where the polynomials $P_{n-1}^{(j)}(z)$ have degree less than $n-1$, and, due to (15.59),

$$W_1(z) = \frac{P_{n-1}(z)}{Q_n(z)}. \tag{15.63}$$

Therefore, the process $\eta_1(t) = \eta_t$, $t \geq 0$.

Theorem 15.4. *The stationary wide-sense process $\eta_1(t) = \eta_t$, permitting the spectral representation given by (15.55), is a component of the n-dimensional stationary (wide-sense) process $\tilde{\eta}_t = (\eta_1(t), \dots, \eta_n(t))$ satisfying the linear stochastic equations*

$$\begin{aligned} d\eta_j(t) &= \eta_{j+1}(t)dt + \beta_j dW_t, \quad j = 1, \dots, n-1, \\ d\eta_n(t) &= -\sum_{j=0}^{n-1} a_j \eta_{j+1}(t)dt + \beta_n dW_t \end{aligned} \tag{15.64}$$

with the wide-sense Wiener process

$$W_t = \int_{-\infty}^{\infty} \frac{e^{i\lambda t} - 1}{i\lambda} \Phi(d\lambda) \tag{15.65}$$

and the coefficients $\beta_1, \dots, \beta_n$ given by (15.59). In this case $M\eta_j(0)W_t = 0$, $t \geq 0$, $j = 1, \dots, n$.

In order to prove this theorem we shall need the following lemma.

Lemma 15.3. *Let $W(z)$ be some frequency characteristic with $\int_{-\infty}^{\infty} |W(i\lambda)|^2 d\lambda < \infty$, and let*

$$\zeta_t = \int_{-\infty}^{\infty} e^{i\lambda t} W(i\lambda) \Phi(d\lambda), \tag{15.66}$$

where $\Phi(d\lambda)$ is the orthogonal spectral measure with $M\Phi(d\lambda) = 0$ and $M|\Phi(d\lambda)|^2 = d\lambda/2\pi$. Then with probability one,

$$\int_0^t |\zeta_s| ds < \infty, \quad t < \infty, \tag{15.67}$$

$$\int_0^t \zeta_s ds = \int_{-\infty}^{\infty} \frac{e^{i\lambda t} - 1}{i\lambda} W(i\lambda) \Phi(d\lambda). \tag{15.68}$$

PROOF. The integrability of $|\zeta_s|$ follows from the Fubini theorem and the estimate

$$\begin{aligned} \int_0^t M|\zeta_s| ds &\leq \int_0^t (M\zeta_s^2)^{1/2} ds \leq \left(t \int_0^t M\zeta_s^2 ds \right)^{1/2} \\ &= t \left(\frac{1}{2\pi} \int_{-\infty}^{\infty} |W(i\lambda)|^2 d\lambda \right)^{1/2} < \infty. \end{aligned}$$

Therefore, the integral $\int_0^t \zeta_s ds$ exists and, due to (15.66),

$$\int_0^t \zeta_s ds = \int_0^t \int_{-\infty}^{\infty} e^{i\lambda s} W(i\lambda) \Phi(d\lambda) ds. \tag{15.69}$$

Let us show that in the right-hand side of (15.69) a change of integration order is possible:

$$\int_0^t \int_{-\infty}^{\infty} e^{i\lambda s} W(i\lambda)\Phi(d\lambda)ds = \int_{-\infty}^{\infty} \left(\int_0^t e^{i\lambda s} ds \right) W(i\lambda)\Phi(d\lambda). \tag{15.70}$$

Let the function $\varphi(\lambda)$ be such that $\int_{-\infty}^{\infty} |\varphi(\lambda)|^2 d\lambda < \infty$. Then, due to (15.5) and the Fubini theorem,

$$\begin{aligned} & M \int_0^t \int_{-\infty}^{\infty} e^{i\lambda s} W(i\lambda)\Phi(d\lambda)ds \overline{\int_{-\infty}^{\infty} \varphi(\lambda)\Phi(d\lambda)} \\ &= \frac{1}{2\pi} \int_0^t \int_{-\infty}^{\infty} e^{i\lambda s} W(i\lambda)\overline{\varphi}(\lambda)d\lambda ds \\ &= \frac{1}{2\pi} \int_{-\infty}^{\infty} \left(\int_0^t e^{i\lambda s} ds \right) W(i\lambda)\overline{\varphi}(\lambda)d\lambda \\ &= M \int_{-\infty}^{\infty} \left(\int_0^t e^{i\lambda s} ds \right) W(i\lambda)\Phi(d\lambda) \overline{\int_{-\infty}^{\infty} \varphi(\lambda)\Phi(d\lambda)}, \end{aligned}$$

which by virtue of the arbitrariness of $\varphi(\lambda)$ proves (15.70).

To complete the proof it remains only to note that

$$\frac{e^{i\lambda t} - 1}{i\lambda} = \int_0^t e^{i\lambda s} ds.$$

□

15.3.2.

PROOF OF THEOREM 15.4. It is clear that

$$\eta_j(t) - \eta_j(0) = \int_{-\infty}^{\infty} [e^{i\lambda t} - 1] W_j(\lambda)\Phi(d\lambda), \quad j = 1, \dots, n-1,$$

and, according to (15.60),

$$\eta_j(t) - \eta_j(0) = \int_{-\infty}^{\infty} \frac{e^{i\lambda t} - 1}{i\lambda} W_{j+1}(i\lambda)\Phi(d\lambda) + \beta_j \int_{-\infty}^{\infty} \frac{e^{i\lambda t} - 1}{i\lambda} \Phi(d\lambda). \tag{15.71}$$

By Lemma 15.3,

$$\begin{aligned} \int_{-\infty}^{\infty} \frac{e^{i\lambda t} - 1}{i\lambda} W_{j+1}(i\lambda)\Phi(d\lambda) &= \int_0^t \int_{-\infty}^{\infty} e^{i\lambda s} W_{j+1}(i\lambda)\Phi(d\lambda)ds \\ &= \int_0^t \eta_{j+1}(s)ds, \end{aligned} \tag{15.72}$$

and, by Lemma 15.1, the process

$$W_t = \int_{-\infty}^{\infty} \frac{e^{i\lambda t} - 1}{i\lambda} \Phi(d\lambda) \tag{15.73}$$

is a wide-sense Wiener process. Hence, we obtain from (15.71)–(15.73), for $t > s$,

$$\eta_j(t) - \eta_j(s) = \int_s^t \eta_{j+1}(u)dy + \beta_j[W_t - W_s], \quad j = 1, \dots, n-1,$$

which in differential form is: $d\eta_j(t) = \eta_{j+1}dt + \beta_j dW_t$.

The last equation in this system of equations given by (15.64) can be established in similar fashion.

We shall now verify the lack of correlation between the variables $\eta_j(0)$ and W_t for $t \geq 0$ and $j = 1, \dots, n$. For this purpose we shall write the system of equations given by (15.64) in matrix form

$$d\tilde{\eta}_t = A\tilde{\eta}_t dt + BdW_t, \tag{15.74}$$

with the matrices

$$A = \begin{pmatrix} 0 & 1 & 0 & \dots & 0 \\ 0 & 0 & 1 & \dots & 0 \\ \dots & \dots & \dots & \dots & \dots \\ -a_0 & -a_1 & \dots & \dots & -a_{n-1} \end{pmatrix}, \quad B = \begin{pmatrix} \beta_1 \\ \dots \\ \beta_n \end{pmatrix}.$$

Note that (15.74) remains valid for $t \geq T$ $(T < 0)$ as well if, instead of W_t, we consider the wide-sense Wiener process

$$W_t(T) = \int_{-\infty}^{\infty} \frac{e^{i\lambda t} - e^{i\lambda T}}{i\lambda} \Phi(d\lambda), \tag{15.75}$$

i.e.,

$$\tilde{\eta}_0 = \tilde{\eta}_T + \int_T^0 A\tilde{\eta}_u du + BW_0(T).$$

But $MW_tW_0(T) = 0$ (see the Parseval equality in Lemma 15.1). Therefore $M\tilde{\eta}_0 W_t = M\tilde{\eta}_T W_t + \int_T^0 AM\tilde{\eta}_u W_t du$. By solving this equation for $M\tilde{\eta}_T W_t$, $T \leq 0$, we find that

$$M\tilde{\eta}_0 W_t = e^{-AT} M\tilde{\eta}_T W_t. \tag{15.76}$$

The eigenvalues of the matrix A lie within the left-half plane, and the elements of the vector $M\tilde{\eta}_T W_t$ are bounded and independent of T. Hence,

$$\lim_{T \to -\infty} M\tilde{\eta}_0 W_t = 0.$$

To complete the proof it remains only to show that the process $\tilde{\eta}_t$ is a wide-sense stationary process (for $t \geq 0$).

It follows from (15.56) that $M\tilde{\eta}_t \equiv 0$. Next, according to Theorem 15.1, the matrix $\Gamma_t = M\tilde{\eta}_t\tilde{\eta}_t^*$ is a solution of the differential equation

$$\dot{\Gamma}_t = A\Gamma_t + \Gamma_t A^* + BB^*. \tag{15.77}$$

It is seen from (15.56) that the matrices Γ_t do not depend on t. Let $\Gamma \equiv \Gamma_t$. Then the matrix Γ satisfies the system of algebraic equations

$$A\Gamma + \Gamma A^* + BB^* = 0. \tag{15.78}$$

Taking advantage of Equation (15.77) and of the fact that the eigenvalues of the matrix A lie within the left-half plane, it is not difficult to show that the solution of the system of equations given by (15.78) is unique and given by the formula

$$\Gamma = \int_{-\infty}^{0} e^{-Au} BB^* e^{-A^*u} du. \tag{15.79}$$

Finally, it follows from (15.74) that the matrix $\Gamma(t,s) = M\tilde{\eta}\tilde{\eta}_s^*$ is given by the formula

$$\Gamma(t,s) = \begin{cases} e^{A(t-s)}\Gamma, & t \geq s, \\ \Gamma e^{A^*(t-s)}, & s \geq t. \end{cases} \tag{15.80}$$

This proves that the process $\tilde{\eta}_t$, $t \geq 0$ is a wide-sense stationary process. □

15.3.3. Consider the partially observable wide-sense stationary process $v_t = (\theta_t, \xi_t) = [(\theta_1(t), \dots, \theta_k(t)), (\xi_1(t), \dots, \xi_l(t))]$, $-\infty < t < \infty$, permitting the spectral representation

$$v_t = \int_{-\infty}^{\infty} e^{i\lambda t} W(i\lambda) \Phi(d\lambda), \tag{15.81}$$

where $W(z)$ is the matrix of dimension $(k+l) \times n$ with elements

$$W_{rq}(z) = P_{n_{rq}-1}^{(rq)}(z) / Q_{n_{rq}}^{(rq)}(z), \tag{15.82}$$

where $P_{n_{rq}-1}^{(rq)}(z)$ and $Q_{n_{rq}}^{(rq)}(z)$ are polynomials of degree $n_{rq} - 1$ and n_{rq} (respectively) with the coefficient of $z^{n_{rq}}$ in $Q_{n_{rq}}^{(rq)}(z)$ being equal to one, and where the roots of the equation $Q_{n_{rq}}^{(rq)}(z) = 0$ lie within the left-half plane. The measure $\Phi(d\lambda) = (\Phi_1(d\lambda), \dots, \Phi_n(d\lambda))$ is a vectorial measure with uncorrelated components $M\Phi_j(d\lambda) = 0$ and $M|\Phi_j(d\lambda)|^2 = d\lambda/2\pi$.

It is assumed that θ_t is an unobservable component to be estimated from the observations ξ_s, $0 \leq s \leq T$. In the case $t = T$ we have a filtering problem; in the case $T \geq t$ we have an interpolation problem; in the case $t \geq T$, we have an extrapolation problem.

For the sake of brevity we shall consider only the problem of optimal (in the mean square sense) linear filtering. In order to apply Theorem 15.3 it suffices to show that the process $v_t = (\theta_t, \xi_t)$, $t \geq 0$, can be represented as a component of the process satisfying a system of equations of the type given by (15.44).

Using Theorem 15.4, we find that the vector v_t is a component of the vector $(\hat{\theta}_t, \xi_t)$ having the dimension

$$N = \sum_{q=1}^{n} \sum_{r=1}^{n_q} n_{rq}, \tag{15.83}$$

where n_{rq} is the degree of the fraction denominator $W_{rq}(z)$ and n_q is the number of noncoincident elements of W_{rq} in the column of index q in the matrix $W(z)$.

It is obvious that the vector $\hat{\theta}_t$ contains all the components of the vector θ_t. Hence, by estimating the vector $\hat{\theta}_t$ the problem of estimating the vector θ_t is also solved. By Theorem 15.4, $(\hat{\theta}_t, \xi_t)$, $t \geq 0$, satisfies the system of stochastic equations

$$\begin{aligned} d\hat{\theta}_t &= [a_1\hat{\theta}_t + a_2\xi_t]dt + bdW_t \\ d\xi_t &= [A_1\hat{\theta}_t + A_2\xi_t]dt + BdW_t, \end{aligned} \tag{15.84}$$

with matrix coefficients of the appropriate dimensions and the vector wide-sense Wiener process $W_t = (W_1(t), \dots, W_n(t))$.

If the matrix BB^* is positive definite, then we can apply Theorem 15.3.

Indeed, for this purpose it suffices to establish that there exist mutually uncorrelated wide-sense Wiener processes

$$W_1(t) = (W_{11}(t), \dots, W_{1,n-l}(t)), \quad W_2(t) = (W_{21}(t), \dots, W_{2l}(t))$$

such that

$$bW_t = b_1W_1(t) + b_2W_2(t), \quad BW_t = B_1W_1(t) + B_2W_2(t). \tag{15.85}$$

The feasibility of such a representation can be proved in the same way as in Lemma 10.4. In this case, the matrices b_1, b_2, B_1 and B_2 can be defined by the equalities

$$b_1b_1^* + b_2b_2^* = bb^*, \quad b_1B_1^* + b_2B_2^* = bB^*, \quad B_1B_1^* + B_2B_2^* = BB^*, \tag{15.86}$$

Note. If the matrix BB^* is singular, then according to the result of Section 10.4, there is a possibility of obtaining linear (nonoptimal) estimates for $\hat{\theta}_t$, close (in the mean square sense) to optimal linear estimates.

15.3.4. Let us indicate here an example illustrating the techniques of finding optimal linear estimates. Let

$$W(z) = \begin{pmatrix} \frac{\sqrt{c_1}}{z+\alpha} & 0 \\ \frac{\sqrt{c_1}}{z+\alpha} & \frac{\sqrt{c_2}}{z+\beta} \end{pmatrix}, \quad \alpha > 0, \beta > 0, c_i > 0, i = 1, 2.$$

Then,

$$\theta_t = \sqrt{c_1} \int_{-\infty}^{\infty} \frac{e^{i\lambda t}}{i\lambda + \alpha} \Phi_1(d\lambda),$$

$$\xi_t = \sqrt{c_1}\int_{-\infty}^{\infty} \frac{e^{i\lambda t}}{i\lambda+\alpha}\Phi_1(d\lambda) + \sqrt{c_2}\int_{-\infty}^{\infty} \frac{e^{i\lambda t}}{i\lambda+\beta}\Phi_2(d\lambda).$$

If we let

$$\eta_t = \sqrt{c_2}\int_{-\infty}^{\infty} \frac{e^{i\lambda t}}{i\lambda+\beta}\Phi_2(d\lambda),$$

then $\xi_t = \theta_t + \eta_t$ and the problem of estimating θ_t from $\xi_0^t = (\xi_s, s \le t)$ is a conventional problem of estimating the 'signal' θ_t in additive 'noise' η_t. According to Theorem 15.4, there exist mutually uncorrelated wide-sense Wiener processes $W_1(t)$ and $W_2(t)$ such that

$$d\theta_t = -\alpha\theta_t dt + \sqrt{c_1}dW_1(t), \quad d\eta_t = -\beta\eta_t dt + \sqrt{c_2}dW_2(t).$$

Therefore, the partially observable process (θ_t, ξ_t), $t \ge 0$ satisfies:

$$\begin{aligned} d\theta_t &= -\alpha\theta_t dt + \sqrt{c_1}dW_1(t); \\ d\xi_t &= [-(\alpha-\beta)\theta_t - \beta\xi_t]dt + \sqrt{c_1}dW_1(t) + \sqrt{c_2}dW_2(t). \end{aligned}$$

Applying Theorem 15.3 to this system, we find that the optimal linear estimate λ_t and its error $\gamma_t = M(\theta_t - \lambda_t)^2$ can be found from the system of equations

$$\begin{aligned} d\lambda_t &= -\alpha\lambda_t dt + \frac{c_1 + \gamma_t(\beta-\alpha)}{c_1+c_2}[d\xi_t - ((\beta-\alpha)\lambda_t - \beta\xi_t)dt], \\ \dot{\gamma}_t &= -2\alpha\gamma_t + c_1 - \frac{[c_1+\gamma_t(\beta-\alpha)]^2}{c_1+c_2}. \end{aligned} \tag{15.87}$$

Let us find the initial conditions λ_0 and $\gamma_0 = M(\theta_0 - \lambda_0)^2$. By Theorem 15.3,

$$\lambda_0 = \frac{M\theta_0\xi_0}{M\xi_0^2}\xi_0, \quad \gamma_0 = M\theta_0^2 - \frac{(M\theta_0\xi_0)^2}{M\xi_0^2}.$$

Let

$$d_{11} = M\theta_0^2, \quad d_{12} = M\theta_0\xi_0, \quad d_{22} = M\xi_0^2, \quad D = \begin{pmatrix} d_{11} & d_{12} \\ d_{12} & d_{22} \end{pmatrix}.$$

By (15.78),

$$AD + DA^* + BB^* = 0$$

where

$$A = \begin{pmatrix} -\alpha & 0 \\ \beta-\alpha & -\beta \end{pmatrix}, \quad B = \begin{pmatrix} \sqrt{c_1} & 0 \\ \sqrt{c_1} & \sqrt{c_2} \end{pmatrix}.$$

Hence

$$\begin{aligned} -2\alpha d_{11} + c_1 &= 0, \\ (\beta-\alpha)d_{11} - (\beta+\alpha)d_{12} + c_1 &= 0, \\ 2(\beta-\alpha)d_{12} - 2\beta d_{22} + c_1 + c_2 &= 0, \end{aligned}$$

and

$$d_{11} = \frac{c_1}{2\alpha}, \quad d_{12} = \frac{c_1}{2\alpha}, \quad d_{22} = \frac{\alpha c_2 + \beta c_1}{2\alpha\beta}.$$

Thus, the optimal linear estimate of θ_t from $\xi_0^t = \{\xi_s, s \le t\}$ can be found from (15.87), solvable under the conditions

$$\lambda_0 = \frac{c_1\beta}{\alpha c_2 + \beta c_1}\xi_0, \quad \gamma_0 = \frac{c_1 c_2 \alpha}{2\alpha(\alpha c_2 + \beta c_1)}. \tag{15.88}$$

If we wish to estimate θ_t from $\xi_{-T}^t = \{\xi_s, -T \le s \le t\}$ where $T > 0$, then λ_t and γ_t can also be obtained from (15.87) with

$$\lambda_{-T} = \frac{c_1\beta}{\alpha c_2 + \beta c_1}\xi_{-T}, \quad \gamma_{-T} = \frac{c_1 c_2 \alpha}{2\alpha(\alpha c_2 + \beta c_1)}. \tag{15.89}$$

Letting $T \to \infty$, it is easy to show from (15.87) and (15.89) that the optimal linear estimate $\tilde{\lambda}_t$ and the estimation error $\tilde{\gamma} \equiv M[\tilde{\lambda}_t - \theta_t]^2$ of the value of θ_t from $\xi_{-\infty}^t = \{\xi_s, -\infty < s \le t\}$ can be defined by the equalities

$$\tilde{\lambda}_t = \delta_1\xi_t + \int_{-\infty}^t e^{-\delta_2(t-s)}[\delta_0 - \delta_1\delta_2]\xi_s ds,$$

where

$$\delta_0 = \delta_1\beta, \quad \delta_1 = \frac{\tilde{\gamma}(\beta-\alpha)+c_1}{c_1+c_2}, \quad \delta_2 = \delta_1(\beta-\alpha)+\alpha,$$

and

$$\tilde{\gamma} = \begin{cases} \frac{\sqrt{(\alpha^2 c_2+\beta^2 c_1)(c_1+c_2)}-\alpha c_2-\beta c_1}{\beta-\alpha}, & \alpha \ne \beta, \\ \frac{c_1 c_2}{2\alpha(c_1+c_2)}, & \alpha = \beta\,. \end{cases}$$

In particular, for $\alpha = \beta$, i.e., when the spectra of the signal and the noise are 'similar', we have:

$$\tilde{\lambda}_t = \frac{c_1}{c_1+c_2}\xi_t.$$

15.4 Comparison of Optimal Linear and Nonlinear Estimates

15.4.1. Let θ_t, $t \geq 0$, be a Markov process with two states 0 and 1, $P(\theta_0 = 1) = \pi_0$, whose transient probability $P_{1\alpha}(t,s) = P(\theta_t = 1|\theta_s = \alpha)$, $\alpha = 0, 1$, satisfies the Kolmogorov equation

$$\frac{dP_{1\alpha}(t,s)}{dt} = \lambda(1 - 2P_{1\alpha}(t,s)), \quad \lambda > 0, t > s. \tag{15.90}$$

We shall assume that the process θ_t (called a 'telegraph signal') is unobservable, and that what is observed is the process

$$\xi_t = \int_0^t \theta_s ds + W_t, \tag{15.91}$$

where W_t, $t \geq 0$ is a Wiener process independent of θ_t, $t \geq 0$.

Using the problem of filtering θ_t from $\xi_0^t = \{\xi_s, s \leq t\}$ as an example, we shall compare optimal linear and nonlinear estimates.

The optimal (in the mean square sense) nonlinear estimate π_t of the value of θ_t from $\{\xi_s, s \leq t\}$, is the conditional mathematical expectation $\pi_t = M(\theta_t|\mathcal{F}_t^\xi) = P(\theta_t = 1|\mathcal{F}_t^\xi)$.

According to (9.86), π_t, $t \geq 0$ is a solution of the stochastic equation

$$d\pi_t = \lambda(1 - 2\pi_t)dt + \pi_t(1 - \pi_t)(d\xi_t - \pi_t dt). \tag{15.92}$$

In particular, it is seen from this equation that the optimal estimate π_t is actually nonlinear.

In order to construct the optimal linear estimate λ_t it suffices to consider the filtering problem for the process $\tilde{\theta}_t$ from the values of $\{\tilde{\xi}_s, s \leq t\}$, where $\tilde{\xi}_t = \int_0^t \tilde{\theta}_s ds + \tilde{W}_t$, $\tilde{W}_t$ is some Wiener process and $\tilde{\theta}_s$ is a Gaussian process independent of $\tilde{W}_t$, $t \geq 0$, and having the same first two moments as the process θ_t, $t \geq 0$.

Making use of Equation (15.90), in standard fashion we find that $n_t = M\theta_t$ satisfies the equation

$$\frac{dn_t}{dt} = \lambda(1 - 2n_t), \quad n_0 = \pi_0, \tag{15.93}$$

and the correlation function $K(t,s)$ can be defined by the equality $K(t,s) = K(s,s)e^{-2\lambda|t-s|}$ where $K(s,s) = M[\theta_s - n_s]^2 = n_s - n_s^2$. In solving Equation (15.93) we find $n_t = \frac{1}{2}[1 - (1 - 2n_0)e^{-2\lambda t}]$.

Consequently,

$$M(\theta_t - n_t)^2 \equiv K(t,t) = \frac{1}{4}[1 - (1 - 2\pi_0)^2 e^{-4\lambda t}]$$

and $\lim_{t\to\infty} M(\theta_t - n_t)^2 = \frac{1}{4}$.

It is not difficult to see now that the required Gaussian process $\tilde{\theta}_t$, $t \geq 0$, having $M\tilde{\theta}_t = n_t$ and $M(\tilde{\theta}_t - n_t)(\tilde{\theta}_s - n_s) = K(t,s)$, can be constructed in terms of the solution of the stochastic differential equation

$$d\tilde{\theta}_t = \lambda(1 - 2\tilde{\theta}_t)dt + \sqrt{\lambda}dW_1(t), \tag{15.94}$$

where $W_1(t)$ is a Wiener process independent of $\tilde{W}_t$, $t > 0$ (see also Theorem 15.2). Then, setting $W_2(t) = \tilde{W}_t$, we obtain

$$d\tilde{\xi}_t = \tilde{\theta}_t dt + dW_2(t). \tag{15.95}$$

Applying Theorem 15.3 to the system of equations given by (15.94) and (15.95), we find that $\lambda_t = \hat{M}(\theta_t|\mathcal{F}_t^{\xi})$ and $\gamma_t = M(\theta_t - \lambda_t)^2$ satisfy the system of equations

$$d\lambda_t = \lambda(1 - 2\lambda_t)dt + \gamma_t(d\xi_t - \lambda_t dt), \quad \lambda_0 = n_0, \tag{15.96}$$

$$\dot{\gamma}_t = -4\lambda\gamma_t + \lambda - \gamma_t^2, \quad \gamma_0 = n_0 - n_0^2. \tag{15.97}$$

We can show (see also Theorem 16.2) that $\lim_{t\to\infty} \gamma_t = \gamma(\lambda)$ exists, where $\gamma(\lambda)$ is the unique positive solution of the equation

$$\gamma^2(\lambda) + 4\lambda\gamma(\lambda) - \lambda = 0. \tag{15.98}$$

Hence,

$$\gamma(\lambda) = \sqrt{\lambda + 4\lambda^2} - 2\lambda, \tag{15.99}$$

and, therefore,

$$\gamma(\lambda) = \begin{cases} \sqrt{\lambda} + O(\lambda), & \lambda \downarrow 0, \\ \frac{1}{4} + O(1/\lambda), & \lambda \uparrow \infty. \end{cases} \tag{15.100}$$

15.4.2. Let us find now the value of $\delta(\lambda) = \lim_{t\to\infty} M(\theta_t - \pi_t)^2$ for the optimal nonlinear estimates π_t, $t \geq 0$.

According to Theorem 7.12, the process $\overline{W} = (\overline{W}_t, \mathcal{F}_t^{\xi})$, $t \geq 0$, defined by

$$\overline{W}_t = \xi_t - \int_0^t \pi_s ds, \tag{15.101}$$

is a Wiener process. Hence, Equation (15.92) can be rewritten as

$$d\pi_t = \lambda(1 - 2\pi_t)dt + \pi_t(1 - \pi_t)d\overline{W}_t, \quad \pi_0 = n_0. \tag{15.102}$$

Next, since $M(\theta_t - \pi_t)^2 = M\pi_t(1 - \pi_t)$, to find $\delta(\lambda)$ one has to know how to find $\lim_{t\to\infty} M\pi_t(1 - \pi_t)$ for the process π_t, $t \geq 0$, with the differential given by (15.102).

According to Theorem 4.6, Equation (15.102) has a unique strong ($\mathcal{F}_t^{\overline{W}}$-measurable at each $t \geq 0$) solution. We can show that this solution is a Markov process whose one-dimensional distribution density $q(t,x) = dP(\pi_t \leq x)/dx$ satisfies the forward Kolmogorov equation

$$\frac{\partial q(t,x)}{\partial t} = -\frac{\partial}{\partial x}[\lambda(1-2x)q(t,x)]+\frac{1}{2}\frac{\partial^2}{\partial x^2}[x^2(1-x)^2 q(t,x)], \quad t \geq 0. \quad (15.103)$$

Due to the fact that the process π_t, $t \geq 0$, is (in the terminology of Markov chain theory) positive recurrent[6]

$$\delta(\lambda) = \lim_{t\to\infty} M\pi_t(1-\pi_t) = \lim_{t\to\infty} \int_0^1 x(1-x)q(t,x)dx$$

exists and

$$\delta(\lambda) = \int_0^1 x(1-x)q(x)dx, \quad (15.104)$$

where $q(x)$ is the unique probability ($q(x) \geq 0$, $\int_0^1 q(x)dx = 1$) solution of the equation

$$\frac{d}{dx}[\lambda(1-2x)q(x)] = \frac{1}{2}\frac{d^2}{dx^2}[x^2(1-x)^2 q(x)]. \quad (15.105)$$

It is easy to find that this solution is given by the formula

$$q(x) = \frac{\exp\left\{-\frac{2\lambda}{x(1-x)}\right\}\frac{1}{x^2(1-x)^2}}{\int_0^1 \exp\left\{-\frac{2\lambda}{y(1-y)}\right\}\frac{dy}{y^2(1-y)^2}}. \quad (15.106)$$

Hence,

$$\delta(\lambda) = \frac{\int_0^1 \exp\left(-\frac{2\lambda}{x(1-x)}\right)\frac{dx}{x(1-x)}}{\int_0^1 \exp\left(-\frac{2\lambda}{x(1-x)}\right)\frac{x}{x^2(1-x)^2}},$$

or, by virtue of the symmetry of the integrands with respect to the point $x = \frac{1}{2}$,

$$\delta(\lambda) = \frac{\int_0^{1/2} \exp\left(-\frac{2\lambda}{x(1-x)}\right)\frac{dx}{x(1-x)}}{\int_0^{1/2} \exp\left(-\frac{2\lambda}{x(1-x)}\right)\frac{dx}{x^2(1-x)^2}}. \quad (15.107)$$

Let us investigate $\lim_{\lambda\uparrow 0}\delta(\lambda)$. Substituting in (15.107) the variables

$$y = \frac{2\lambda}{x(1-x)} - 8\lambda,$$

we find that

$$\delta(\lambda) = \frac{2\lambda\int_0^\infty e^{-y}\sqrt{\frac{y+8\lambda}{y}}\frac{dy}{y+8\lambda}}{\int_0^\infty e^{-y}\sqrt{\frac{y+8\lambda}{y}}dy}. \quad (15.108)$$

Since, for $0 < c < \infty$,

[6] See Lemmas 9.3 and 9.4 in Chapter 4, [147].

$$\int_0^\infty e^{-y}\sqrt{\frac{y+8c}{y}}dy < \infty,$$

by the Lebesgue theorem on dominated convergence (Theorem 1.4)

$$\lim_{\lambda\downarrow 0}\int_0^\infty e^{-y}\sqrt{\frac{y+8\lambda}{y}}dy = \int_0^\infty e^{-y}dy = 1.$$

Next,

$$2\lambda\int_0^\infty e^{-y}\frac{dy}{\sqrt{y(y+8\lambda)}} = 2\lambda\left[\int_0^1 e^{-y}\frac{dy}{\sqrt{y(y+8y)}} + d(\lambda)\right],$$

where

$$d(\lambda) = \int_1^\infty e^{-y}\frac{dy}{\sqrt{y(y+8\lambda)}}, \quad d(0) = \lim_{\lambda\downarrow 0} d(\lambda) = \int_1^\infty \frac{e^{-y}}{y}dy < 1.$$

Hence, by the theorem on the mean ($e^{-1} \le c(\lambda) \le 1$),

$$2\lambda\int_0^\infty e^{-y}\frac{dy}{\sqrt{y(y+8\lambda)}} - 2\lambda\left[c(\lambda)\int_0^1 \frac{dy}{\sqrt{y(y+8\lambda)}} + d(\lambda)\right].$$

But

$$\int_0^1 \frac{dy}{\sqrt{y(y+8\lambda)}} = -\ln\lambda\left[1+\frac{\ln 8}{\ln\lambda} - \frac{[2\sqrt{1+8\lambda}+2+8\lambda]}{\ln\lambda}\right];$$

therefore,

$$\delta(\lambda) = -2\lambda\ln\lambda\,[c(\lambda)+\mathrm{O}(1/\ln\lambda)], \quad \lambda\downarrow 0. \tag{15.109}$$

Just as we showed the existence of

$$\lim_{t\to\infty} M\pi_t(1-\pi_t) = \int_0^1 x(1-x)q(x)dx,$$

we can also show that the limits

$$\lim_{t\to\infty} M(1-2\pi_t)^2, \quad \lim_{t\to\infty} M\pi_t^2(1-\pi_t)^2,$$

exist, and that

$$\lim_{t\to\infty} M(1-2\pi_t)^2 = \frac{1}{\lambda}\lim_{t\to\infty} M\pi_t^2(1-\pi_t)^2. \tag{15.110}$$

Note that one can arrive at (15.110) in the following way. By the Itô formula, from (15.102) it follows that

$$\begin{aligned}\pi_t(1-\pi_t) &= n_0(1-n_0)+\lambda\int_0^t(1-2\pi_s)^2ds-\int_0^t\pi_s^2(1-\pi_s)^2ds\\&\quad+\int_0^t(1-2\pi_s)\pi_s(1-\pi_s)d\overline{W}_s.\end{aligned}$$

It follows from this that

$$M\pi_t(1-\pi_t)=n_0(1-n_0)+\lambda\int_0^t M(1-2\pi_s)^2ds-\int_0^t M\pi_s^2(1-\pi_s)^2ds,$$

or,

$$\frac{d[M\pi_t(1-\pi_t)]}{dt}=\lambda M(1-2\pi_t)^2-M\pi_t^2(1-\pi_t)^2. \qquad (15.111)$$

But it is natural to expect that $\lim_{t\to\infty}d[M\pi_t(1-\pi_t)]/dt=0$. Together with (15.111), this leads to (15.110). Noting now that $(1-2x)^2=1-4x(1-x)$, we obtain from (15.110)

$$\lim_{t\to\infty}M\pi_t(1-\pi_t)=\frac{1}{4}-\lim_{t\to\infty}\frac{M\pi_t^2(1-\pi_t)^2}{4\lambda}=\frac{1}{4}+\mathrm{O}(1/\lambda). \qquad (15.112)$$

Thus, by combining estimates (15.109) and (15.112), we obtain

$$\delta(\lambda)=\begin{cases}-2\lambda\ln\lambda(c(\lambda+\mathrm{O}(1/\ln\lambda)), & \lambda\downarrow 0\\ \frac{1}{4}+\mathrm{O}(1/\lambda), & \lambda\uparrow\infty.\end{cases} \qquad (15.113)$$

Along with (15.100) for the effectiveness value $\varepsilon(\lambda)=\gamma(\lambda)/\delta(\lambda)$ of the optimal nonlinear estimate with respect to the optimal linear estimate we find the following expression

$$\varepsilon(\lambda_)=\begin{cases}-\frac{1}{2\sqrt{\lambda}\ln\lambda}[c(\lambda)+\mathrm{o}(1)], & \lambda\downarrow 0\\ 1+\mathrm{o}(1), & \lambda\uparrow\infty.\end{cases} \qquad (15.114)$$

It is seen from this that for small λ, (i.e., when the average occupation time of the 'telegraph signal' in the 0 and 1 states is long) the linear filter is inferior to the nonlinear filter with respect to the mean square error. In the case $\lambda\uparrow\infty$, the two filters are equivalent and function equally 'poorly':

$$\delta(\lambda)\sim\lim_{t\to\infty}M(\theta_t-n_t)^2=\frac{1}{4};\quad \gamma(\lambda)\sim\lim_{t\to\infty}M(\theta_t-n_t)^2=\frac{1}{4};\quad \lambda\to\infty;$$

i.e., for large λ they yield the same error as an a priori filter for which the average value of n_t is taken as an estimate of the value of θ_t.

Since $\lim_{t\to\infty}M(\theta_t-n_t)^2=\frac{1}{4}$ at all $\lambda>0$, it is seen from (15.100) that for small λ the optimal linear filter functions 'well' (from the point of view of asymptotic 'tracking' of the process θ_t in comparison with the a priori filter), i.e.,

$$\frac{\lim_{t\to\infty}M(\theta_t-\lambda_t)^2}{\lim_{t\to\infty}M(\theta_t-n_t)^2}=4\sqrt{\lambda}+\mathrm{O}(\lambda),\quad \lambda\downarrow 0.$$

Under these conditions (i.e., for small λ) the nonlinear filter provides, however, a higher accuracy of 'tracking':

$$\frac{\lim_{t\to\infty} M(\theta_t - \pi_t)^2}{\lim_{t\to\infty} M(\theta - n_t)^2} = 8\lambda \ln \frac{1}{\lambda}[c(\lambda) + \mathrm{O}\,(1/\ln\lambda)], \quad \lambda \downarrow 0.$$

This remark points to the fact observed in filtering problems that the 'gain' obtained with the aid of an optimal nonlinear filter increases as the 'tracking' accuracy of an optimal filter improves.

Notes and References. 1

15.1–15.3. In these sections the general equations of optimal filtering for linear estimation of random processes have been used.

15.4 Optimal linear estimates and nonlinear estimates have been compared by Stratonovich [296] and Liptser [193].

Notes and References. 2

15.4. Related results can be found in Khasminskii and Lazareva [149, 150], Khasminskii, Lazareva and Stapleton [151], and Khasminskii and Zeitouni [152].

16. Application of Optimal Nonlinear Filtering Equations to some Problems in Control Theory and Estimation Theory

16.1 An Optimal Control Problem Using Incomplete Data

16.1.1. In this section the results obtained in Section 14.3 for linear control problems (using incomplete data) with quadratic performance index are extended to the case of continuous time.

We shall assume that the partially observable controlled process $(\theta, \xi) = [(\theta_1(t), \ldots, \theta_k(t)); (\xi_1(t), \ldots, \xi_l(t))]$, $0 \le t \le T$, is given by the stochastic equations

$$\begin{aligned} d\theta_t &= [c(t)u_t + a(t)\theta_t]dt + b(t)dW_1(t), \\ d\xi_t &= A(t)\theta_t dt + B(t)dW_2(t). \end{aligned} \tag{16.1}$$

The matrices $c(t), a(t), b(t), A(t), B(t)$ have the dimensions $(k \times r)$, $(k \times k)$, $(k \times k)$, $(l \times k)$, $(l \times l)$, respectively; their elements $c_{ij}(t)$, $a_{ij}(t)$, $b_{ij}(t)$, $A_{ij}(t)$, $B_{ij}(t)$ are deterministic functions of time, with

$$|c_{ij}(t)| \le c, \quad |a_{ij}(t)| \le c, \quad |b_{ij}(t)| \le c,$$

$$\int_0^T A_{ij}^2(t)dt < \infty, \quad \int_0^T B_{ij}^2(t)dt < \infty$$

for all admissible values i, j. We shall also assume that the elements of the matrices $(B(t)B^*(t))^{-1}$ are uniformly bounded. The independent Wiener processes $W_1 = (W_{11}(t), \ldots, W_{1k}(t))$, $W_2 = (W_{21}(t), \ldots, W_{2l}(t))$, $0 \le t \le T$ in (16.1) do not depend on the Gaussian vector θ_0 ($M\theta_0 = m_0$, $\operatorname{cov}(\theta_0, \theta_0) = \gamma_0$), and $\xi_0 = 0$.

The vector $u_t = [u_1(t,\xi), \ldots, u_r(t,\xi)]$ in (16.1) is called a *control action* at time t. The measurable processes $u_j(t,\xi)$, $j = 1, \ldots, r$, are assumed to be such that

$$M \int_0^T \sum_{j=1}^r (u_j(t,\xi))^4 dt < \infty, \tag{16.2}$$

and the values of $u_j(t,\xi)$ are $\mathcal{F}_t^\xi$-measurable.

The controls $u = (u_t)$, $0 \le t \le T$, for which the system of equations given by (16.1) has a unique strong solution and for which the condition given by (16.2) is satisfied, will be called henceforth *admissible controls*.

16.1.2. To formulate the optimality criterion, let us introduce a performance index into our consideration.

Let h and $H(t)$ be symmetric nonnegative definite matrices of the order $(k \times k)$. Denote by $R(t)$ symmetric uniformly[1] positive definite matrices (of dimension $(r \times r)$). Assume that the elements of the matrices $H(t)$ are $R(t)$ are measurable bounded functions of t.

Consider the performance functional

$$V(u;T) = M\left\{\theta_T^* h\theta_T + \int_0^T [\theta_t^* H(t)\theta_t + u_t^* R(t)u_t]dt\right\} \tag{16.3}$$

for each admissible control $u = (u_t)$, $0 \le t \le T$.

The admissible control $\tilde{u}$ is called *optimal* if

$$V(\tilde{u};T) = \inf_u V(u;T), \tag{16.4}$$

where 'inf' is taken over the class of all admissible controls.

For admissible controls u, set

$$m_t^u = M(\theta_t|\mathcal{F}_t^\xi), \quad \gamma_t^u = M[(\theta_t - m_t^u)(\theta_t - m_t^u)^*],$$

where θ_t and ξ_t are the processes corresponding to this control, and which are defined by the system of equations given by (16.1).

Theorem 16.1. *In the class of admissible controls the optimal control $\tilde{u} = (\tilde{u}_t)$, $0 \le t \le T$, exists and is defined by the formulae*

$$\tilde{u}_t = -R^{-1}(t)c^*(t)P(t)\tilde{m}_t, \quad 0 \le t \le T, \tag{16.5}$$

where the nonnegative definite symmetric[2] *matrix $P(t) = \|P_{ij}(t)\|$ of order $(k \times k)$, $0 \le t \le T$, is the solution of the Ricatti equation*

$$\begin{aligned} -\frac{dP(t)}{dt} &= a^*(t)P(t) + P(t)a^*(t) + H(t) \\ &\quad -P(t)c(t)R^{-1}(t)c^*(t)P(t), \quad P(T) = h, \end{aligned} \tag{16.6}$$

and the vector $\tilde{m}_t$ is defined by the system of equations

$$\begin{aligned} d\tilde{m}_t &= [c(t)\tilde{u}_t + a(t)\tilde{m}_t]dt \\ &\quad +\gamma_t A^*(t)(B(t)B^*(t))^{-1}[d\xi_t - A(t)\tilde{m}_t dt], \quad \tilde{m}_0 = m_0 = M\theta_0. \end{aligned} \tag{16.7}$$

---

[1] The elements of the matrices $R^{-1}(t)$ are uniformly bounded.

[2] The nonnegative definiteness and symmetry of the matrix $P(t)$ satisfying Equation (16.6) can be proved in the same way as in the case of discrete time (see Section 14.3).

$$\begin{aligned}\dot{\gamma}_t &= a(t)\gamma_t + \gamma_t a^*(t) + b(t)b^*(t) - \gamma_t A^*(t)(B(t)B^*(t))^{-1}A(t)\gamma_t,\\ \gamma_0(t) &= \operatorname{cov}(\theta_0, \theta_0).\end{aligned}\tag{16.8}$$

In this case,

$$V(\tilde{u};T) = p(0) + m_0^* P(0) m_0 + \operatorname{Tr}\left[\int_0^T H^{1/2}(t)\gamma_t H^{1/2}(t)dt + h^{1/2}\gamma_T h^{1/2}\right],\tag{16.9}$$

where

$$p(t) = \int_t^T \sum_{i,j=1}^k D_{ij}(s)P_{ij}(s)ds,\tag{16.10}$$

and $D_{ij}(t)$ are elements of the matrix

$$D(t) = \gamma_t A^*(t)[B(t)B^*(t)]^{-1}A(t)\gamma_t.\tag{16.11}$$

PROOF. First of all note that, under the assumptions made above,

$$M\left[\sup_{0\le t\le T}\sum_{j=1}^k \theta_j^4(t)\right] < \infty,$$

which is proved as in Lemma 12.1. Next, in the same way as in the proof of Theorem 14.2, it can be established that

$$\begin{aligned}V(u;T) &= M\left\{\theta_T^* h\theta_T + \int_0^T [\theta_t^* H(t)\theta_t + u_t^* R(t)u_t]dt\right\}\\ &= M\left\{M(\theta_T^* h\theta_T|\mathcal{F}_T^\xi) + \int_0^T [M(\theta_t^* H(t)\theta_t|\mathcal{F}_t^\xi) + u_t^* R(t)u_t]dt\right\}\\ &= M\left\{(m_T^u)^* h m_T^u + \int_0^t [(m_t^u)^* H(t)m_t^u + u_t^* R(t)u_t]dt\right.\\ &\qquad \left.+ \operatorname{Tr}\left[h^{1/2}\gamma_T^u h^{1/2} + \int_0^T H^{1/2}(t)\gamma_t^u H^{1/2}(t)dt\right]\right\}.\end{aligned}\tag{16.12}$$

It should be noted that the function γ_t^u does not depend on the control u and coincides with the function γ_t satisfying Equation (16.8) (see Theorem 12.1). Hence,

$$\begin{aligned}V(u;T) &= \operatorname{Tr}\left[h^{1/2}\gamma_T h^{1/2} + \int_0^T H^{1/2}(t)\gamma_t H^{1/2}(t)dt\right]\\ &\quad + M\left\{(m_t^u)^* h m_t^u + \int_0^T [(m_t^u)^* H(t)m_t^u + u_t^* R(t)u_t]dt\right\},\end{aligned}\tag{16.13}$$

where, according to the same Theorem 12.1, m_t^u, $0 \le t \le T$, is obtained from the equation

$$dm_t^u = [c(t)u_t + a(t)m_t^u]dt + \gamma_t (B(t)B^*(t))^{-1}[d\xi_t^u - A(t)m_t^u dt], \quad m_t^u = m_0, \tag{16.14}$$

with the process ξ_t^u, $0 \le t \le T$, defined by (16.1).

According to the vector version of Lemma 11.3, the process $\overline{W}^u = (\overline{W}_t^u, \mathcal{F}_t^{\xi^u})$, $0 \le t \le T$,

$$\overline{W}_t^u = \int_0^t B^{-1}(s)[d\xi_s^u - A(s)m_s^u du], \tag{16.15}$$

is a Wiener process. Hence, from (16.14) and (16.15),

$$dm_t^u = [c(t)u_t + a(t)m_t^u]dt + \gamma_t A^*(t)(B^*(t))^{-1} d\overline{W}_t^u. \tag{16.16}$$

16.1.3. To solve the primary problem we shall consider the following auxiliary problem.

Let $(\Omega, \mathcal{F}, P)$ be some probability space, with $(\mathcal{F}_t)$, $0 \le t \le T$, a nondecreasing family of sub-σ-algebras of $\mathcal{F}$, $z = (z_t, \mathcal{F}_t)$ an r-dimensional Wiener process, and $u = (u_t, \mathcal{F}_t)$, an r-dimensional process satisfying the condition

$$M \int_0^T \sum_{j=1}^r u_j^4(t, \omega)dt < \infty, \tag{16.17}$$

where $(u_1(t,\omega), \dots, u_r(t,\omega)) = u_t$. Let us associate the control $u = (u_t, \mathcal{F}_t)$, $0 \le t \le T$, with the governed process

$$d\mu_t^u = [c(t)u_t + a(t)\mu_t^u]dt + \gamma_t A^*(t)(B^*(t))^{-1}dz_t, \tag{16.18}$$

where $c(t), a(t), A(t)$ and $B(t)$ are the matrices introduced above, and $\mu_0^u = m_0$. As before, we shall call the control $u = (u_t, \mathcal{F}_t)$, $0 \le t \le T$, *admissible* if for this control (16.17) is satisfied and Equation (16.18) has a unique strong solution.

Let the functional

$$\overline{V}(u;T) = M\left\{ (\mu_T^u)^* h(\mu_T^u) + \int_0^T [(\mu_t^u)^* H(t)\mu_t^u + u_t^* R(t) u_t]dt \right\} \tag{16.19}$$

be the performance index. We shall show that in this problem the optimal control $\tilde{u} = (\tilde{u}_t, \mathcal{F}_t)$ is defined by the formulae

$$\tilde{u}_t = -R^{-1}(t)c^*(t)P(t)\tilde{\mu}_t, \tag{16.20}$$

where $\tilde{\mu}_t$, $0 \le t \le T$ is found from the equation

$$d\tilde{\mu}_t = [a(t) - c(t)R^{-1}(t)c^*(t)P(t)]\tilde{\mu}_t dt + \gamma_t A^*(t)(B^*(t))^{-1}dz_t, \quad \tilde{\mu}_0 = m_0. \tag{16.21}$$

For this purpose introduce the function

$$Q(t,x) = x^*P(t)x + p(t), \quad x \in \mathbb{R}^k, \quad 0 \le t \le T, \tag{16.22}$$

where $P(t)$ is defined by (16.6) and $p(t)$ by (16.10).

Lemma 16.1. *The function $Q(t,x) = x^*P(t)x + p(t)$ is a solution of the differential equation*

$$\Phi(t,x,Q(t,x)) = 0, \tag{16.23}$$

where

$$\begin{aligned}\Phi(t,x,Q(t,x)) &= x^*H(t)x + x^*a^*(t)\operatorname{grad}_x Q(t,x)\\ &+\frac{1}{2}\sum_{i,j=1}^{k} D_{ij}(t)\frac{\partial^2 Q(t,x)}{\partial x_i \partial x_j} + \frac{\partial Q(t,x)}{\partial t}\\ &+\min_u[u^*R(t)u + u^*c^*(t)\operatorname{grad}_x Q(t,x)]\end{aligned}$$

with $u = (u_1,\ldots,u_r)$, $Q(T,x) = x^*hx$.

PROOF. Because of the positive definiteness of the matrices $R(t)$, $0 \le t \le T$, the quadratic form

$$J(u;t) = u^*R(t)u + u^*c^*(t)\operatorname{grad}_x Q(t,x)$$

is positive definite and attains its minimum on the vector

$$\tilde{u}_t(x) = (\tilde{u}_1(t,x),\ldots,\tilde{u}_r(t,x))$$

satisfying the system of linear algebraic equations

$$\operatorname{grad}_u J(u;t) = 0.$$

Since $\operatorname{grad}_u J(u;t) = 2R(t)u + c^*(t)\operatorname{grad}_x Q(t,x)$,

$$u_t(x) = -\frac{1}{2}R^{-1}(t)c^*(t)\operatorname{grad}_x Q(t,x).$$

But

$$\operatorname{grad}_x Q(t,x) = 2P(t)x. \tag{16.24}$$

Hence,

$$\tilde{u}_t(x) = -R^{-1}(t)c^*(t)P(t)x. \tag{16.25}$$

Due to (16.6) and (16.22),

$$\begin{aligned}\frac{\partial}{\partial t}Q(t,x) &= x^*\frac{dP(t)}{dt}x^* + \frac{dp(t)}{dt}\\ &= x^*[-a^*(t)P(t) - P(t)a(t) - H(t) + P(t)c(t)R^{-1}(t)c^*(t)P(t)]x\\ &\quad -\sum_{i,j=1}^{k} D_{ij}(t)P_{ij}(t)\end{aligned} \tag{16.26}$$

and

$$\frac{\partial^2 Q(t,x)}{\partial x_i \partial x_j} = 2P_{ij}(t). \tag{16.27}$$

(16.24)–(16.27) together with the equality $J(\tilde{u};t) = \min_u J(u;t)$, indicate that the function $Q(t,x) = x^* P(t)x + p(t)$ satisfies Equation (16.23). □

Let us show now that for the auxiliary problem the control defined by (16.20) is optimal.

It is seen from (16.23) that

$$\Phi(t, \tilde{\mu}_t, Q(t, \tilde{\mu}_t)) = 0. \tag{16.28}$$

Let now $u_t = (u_1(t), \dots, u_r(t))$, $0 \le t \le T$, be any admissible control and $\mu_t = (\mu_1(t), \dots, \mu_k(t))$ be defined by

$$d\mu_t = [c(t)u_t + a(t)\mu_t]dt + \gamma_t A_t^* (B^*(t))^{-1} dz_t. \tag{16.29}$$

Then it follows from (16.23) and the inequality $J(\tilde{u};t) \le J(u;t)$ that

$$\Phi(t, \mu_t, Q(t, \mu_t)) \ge 0. \tag{16.30}$$

By applying the Itô formula to $Q(t, \tilde{\mu}_t)$ we obtain

$$\begin{aligned} Q(T,\tilde{\mu}_T) - Q(0,\tilde{\mu}_0) &= \int_0^T \left[\frac{\partial Q(s,\tilde{\mu}_s)}{\partial s} + (c(s)\tilde{u}_s + a(s)\tilde{\mu}_s)^* \operatorname{grad}_{\tilde{\mu}} Q(s,\tilde{\mu}_s) \right. \\ &\quad \left. + \frac{1}{2} \sum_{i,j=1}^k D_{ij}(s) \frac{\partial^2 Q(s,\tilde{\mu}_s)}{\partial \tilde{\mu}_i \partial \tilde{\mu}_j} \right] ds \\ &\quad + \int_0^t [\operatorname{grad}_{\tilde{\mu}} Q(s,\tilde{\mu}_s)]^* \gamma_s A^*(s)(B^*(s))^{-1} dz_s. \end{aligned} \tag{16.31}$$

Taking into account (16.28) we find that

$$\begin{aligned} Q(T,\tilde{\mu}_T) - Q(0,\tilde{\mu}_0) &= -\int_0^T [(\tilde{\mu}_s)^* H(s)\tilde{\mu}_s + (\tilde{u}_s)^* R(s)\tilde{u}_s] ds \\ &\quad + \int_0^T [\operatorname{grad}_{\tilde{\mu}} Q(s,\tilde{\mu}_s)]^* \gamma_s A^*(s)(B^*(s))^{-1} dz_s. \end{aligned} \tag{16.32}$$

Taking now the mathematical expectation on both sides of this equality and taking into account the equality $\tilde{\mu}_0 = m_0$, we obtain

$$Q(0,m_0) = M\left\{ (\tilde{\mu}_T)^* h \tilde{\mu}_T + \int_0^T [(\tilde{\mu}_t)^* H(t)\tilde{\mu}_t + (\tilde{\mu}_t)^* R(t)\tilde{u}_t] dt \right\}. \tag{16.33}$$

Similarly, applying the same technique to $Q(t,\mu_t)$, we find that

$$Q(0,m_0) \leq M\left\{(\tilde{\mu}_T)^* h\tilde{\mu}_T + \int_0^T [(\mu_t)^* H(t)\mu_t + (u_t)^* R(t)u_t]dt\right\}. \quad (16.34)$$

Comparing (16.33) with (16.34) we obtain

$$\overline{V}(\tilde{u};T) = Q(0,m_0) \leq \overline{V}(u;T). \quad (16.35)$$

The control $\tilde{u}$ defined by (16.20) is admissible since the linear equation given by (16.21) has a solution, which is unique and strong (Theorem 4.10). (16.17) is satisfied by the vector version of Theorem 4.6. Together with (16.35) this proves that the control $\tilde{u}$ is optimal in the class of admissible controls.

16.1.4.
COMPLETION OF THE PROOF OF THEOREM 16.1. Let us consider the processes

$$\overline{W}^u = (\overline{W}_t^u, \mathcal{F}_t^{\xi^u}), \quad 0 \leq t \leq T,$$

in more detail.

It follows from (16.14) and (16.1) that with probability one the values of $\theta_t^u - m_t^u$ and $\theta_t^0 - m_t^0$ coincide (the index 0 corresponds to the 'zero' control $u_t \equiv 0$, $0 \leq t \leq T$). Hence, it is seen from (16.15) that with probability one all the processes $\overline{W}_t^u$ coincide ($\overline{W}_t^u = \overline{W}_t^0$) and, therefore, Equation (16.16) can be written as follows:

$$dm_t^u = [c(t)u_t + a(t)m_t^u]dt + \gamma_t A^*(t)(B^*(t))^{-1} d\overline{W}_t^0.$$

Let now $\overline{u}$ be any admissible control, and let $\xi^{\overline{u}} = (\xi_t^{\overline{u}})$, $0 \leq t \leq T$, be an associated process where

$$\mathcal{F}_t^{\xi^{\overline{u}}} = \sigma\{\omega; \xi_s^{\overline{u}}, s \leq t\}$$

Let us take advantage of the results of Subsection 16.1.3, setting $\mathcal{F}_t = \mathcal{F}_t^{\xi^{\overline{u}}}$ and $z_t = \overline{W}_t^0$. Let $\overline{U}$ be the class of all admissible controls $u = (u_t)$, $0 \leq t \leq T$, which are $\mathcal{F}_t^{\xi^{\overline{u}}}$-measurable at any time t. Since for any $\overline{u}$

$$\mathcal{F}_t^{\xi^{\overline{u}}} \supseteq \mathcal{F}_t^{\overline{W}^{\overline{u}}} \equiv \mathcal{F}_t^{\overline{W}^0}, \quad 0 \leq t \leq T,$$

the control $\tilde{u}$ given by (16.20) belongs to $\overline{U}$ for any $\overline{u}$ (the admissibility of the control $\tilde{u}$ follows from Theorem 4.10 and the vector version of Theorem 4.6). Hence (see (16.35)), $\overline{V}(\tilde{u};T) \leq \overline{V}(u;T)$ for all $u \in \overline{U}$ and, in particular, $\overline{V}(\tilde{u};T) \leq \overline{V}(\overline{u};T)$. By virtue of the arbitrariness of the control $\overline{u}$ it follows that the control $\tilde{u}$ is optimal.

Finally, note that (16.9) follows from (16.13) and the equalities

$$\overline{V}(\tilde{u};T) = Q(0,m_0) = m_0^* P(0)m_0 + p(0). \qquad \square$$

Note. As in the case of discrete time (Section 14.3), the theorem proved above exhibits the so-called 'principle of separation' (which holds true in a more general situation, for which see [313]), according to which the optimal control problem with incomplete data decomposes into two problems: a filtering problem and a control problem with complete data for a certain system.

16.1.5. Consider a particular case of the system of equations given by (16.1). Let $b(t) \equiv 0$, $A(t) \equiv E(k \times k)$, and $B(t) \equiv 0$. Then in the control problem of the process $\theta = (\theta_t)$, $0 < t \leq T$, with

$$\frac{d\theta_t}{dt} = c(t)u_t + a(t)\theta_t, \tag{16.36}$$

where θ_0 is a deterministic vector with performance functional

$$V(u;T) = \theta_T^* h \theta_T + \int_0^T [\theta_t^* H(t)\theta_t + u_t^* R(t) u_t] dt,$$

the optimal control $\tilde{u} = (\tilde{u}_t)$, $0 \leq t \leq T$, exists and is given by the formula

$$\tilde{u}_t = -R^{-1}(t)c^*(t)P(t)\theta_t, \tag{16.37}$$

where $P(t)$ is a solution of Equation (16.6). In this case

$$V(\tilde{u};T) = \inf_u V(u;T) = \theta_0^* P(0)\theta_0. \tag{16.38}$$

This result can be obtained by the same techniques as in the proof of Theorem 16.1. It can also be obtained from this theorem by a formal passage to the limit if we set $B(t) \equiv \varepsilon E$, $\varepsilon \downarrow 0$.

16.2 Asymptotic Properties of Kalman–Bucy Filters

16.2.1. Consider the Gaussian partially observable random process $(\theta, \xi) = [(\theta_1(t), \ldots, \theta_k(t)), (\xi_1(t), \ldots, \xi_l(t))]$, $t \geq 0$, satisfying the system of stochastic equations

$$\begin{aligned} d\theta_t &= [a_1\theta_t + a_2\xi_t]dt + b_1 dW_1(t) + b_2 dW_2(t) \\ d\xi_t &= [A_1\theta_t + A_2\xi_t]dt + B_1 dW_1(t) + B_2 dW_2(t), \end{aligned} \tag{16.39}$$

with the constant matrices $a_1, a_2, A_1, A_2, b_1, b_2, B_1$ and B_2 of the orders $(k \times k)$, $(k \times l)$, $(l \times k)$, $(l \times l)$, $(k \times k)$, $(k \times l)$, $(l \times k)$ and $(l \times l)$, respectively. The mutually independent Wiener processes $W_1 = (W_{11}(t), \ldots, W_{1k}(t))$ and $W_2(t) = (W_{21}(t), \ldots, W_{2l}(t))$, $t \geq 0$ are supposed (as usual) to be independent of a Gaussian vector with initial values (θ_0, ξ_0).

If the matrix $(B \circ B) = B_1 B_1^* + B_2 B_2^*$ is positive definite, then, according to Theorem 10.3, the mean vector $m_t = M(\theta_t | \mathcal{F}_t^\xi)$ and the covariance matrix

$$\gamma_t = M[(\theta_t - m_t)(\theta_t - m_t)^*] \tag{16.40}$$

satisfy the system of equations

$$dm_t = [a_1 m_t + a_2\xi_t]dt + [(b \circ B) + \gamma_t A_1^*](B \circ B)^{-1}[d\xi_t - (A_1 m_t + A_2\xi_t)dt], \tag{16.41}$$

$$\dot{\gamma}_t = a_1\gamma_t + \gamma_t a_1^* - [(b \circ B) + \gamma_t A_1^*](B \circ B)^{-1}[(b \circ B) + \gamma_t A_1^*] + (b \circ b), \tag{16.42}$$

where $(b \circ b) = b_1 b_1^* + b_2 b_2^*$ and $(b \circ B) = b_1 B_1^* + b_2 B_2^*$.

The components of the vector $m_t = M(\theta_t|\mathcal{F}_t^\xi)$ are the best (in the mean square sense) estimates of the corresponding components of the vector θ_t from the observations ξ_0^t. The elements of the matrix γ_t and its trace $\operatorname{Tr}\gamma_t$ exhibit the accuracy of 'tracking' the unobservable states θ_t by the estimate m_t. In this case, as in the analogous problem for the case of discrete time, the critical question (with respect to applications) is: when does the matrix γ_t converge as $t \uparrow \infty$? The present section deals with the investigation of the existence of $\lim_{t\to\infty} \gamma_t$ and the techniques for its computation.

16.2.2. Before giving a precise formulation, let us note first that by setting

$$\begin{aligned} a &= a_1 - (b \circ B)(B \circ B)^{-1}A_1, \\ b &= [(b \circ b) - (b \circ B)(B \circ B)^{-1}(b \circ B)^*]^{1/2}, \\ B &= [B \circ B]^{1/2}, \quad A = A_1, \end{aligned} \tag{16.43}$$

Equation (16.42) can be rewritten in a more convenient form:

$$\dot{\gamma}_t = a\gamma_t + \gamma_t a^* + bb^* - \gamma_t A^*(BB^*)^{-1}A\gamma_t. \tag{16.44}$$

This equation coincides with the equation for the covariance of the pair of Gaussian processes (θ, ξ) satisfying the system

$$\begin{aligned} d\theta_t &= a\theta_t dt + bdW_1(t), \\ d\xi_t &= A\theta_t dt + BdW_2(t). \end{aligned} \tag{16.45}$$

So, in terms of the behavior of the matrices γ_t for $t \to \infty$ it is enough to consider the simpler system of equations given by (16.45) instead of (16.39).

Theorem 16.2. *Let the system of equations given by (16.45) satisfy the following conditions:*

(1) the rank of the block matrix

$$G_1 = \begin{pmatrix} A \\ Aa \\ \cdots \\ Aa^{k-1} \end{pmatrix} \tag{16.46}$$

of dimension $(kl \times k)$ is equal to k;

(2) the rank of the block matrix

$$G_2 = (b\,ab \dots a^{k-1}b) \tag{16.47}$$

of dimension $(k \times lk)$ is equal to k;

(3) the matrix BB^* is nonsingular.

Then, for $\gamma_t = M(\theta_t - m_t)(\theta_t - m_t)^*$, $\lim_{t\to\infty} \gamma_t = \gamma$ exists. This limit γ does not depend on the initial value γ_0 and is the unique (in the class of positive definite matrices) solution of the equation

$$a\gamma + \gamma a^* + bb^* - \gamma A^*(B^*B)^{-1}A\gamma = 0. \tag{16.48}$$

Before proving this theorem we shall prove some auxiliary lemmas.

16.2.3.

Lemma 16.2. *Let D and Δ be matrices of dimensions $(l \times k)$ and $(k \times k)$, respectively. We shall form the block matrix (of order $(nl \times k)$)*

$$D_n = \begin{pmatrix} D \\ D\Delta \\ \dots \\ D\Delta^{n-1} \end{pmatrix}.$$

Then the matrices $D_n^* D_n$ and $\int_0^T e^{-\Delta^* t} D^* D e^{-\Delta t} dt$, $0 < T < \infty$, are either both singular or both nonsingular.

PROOF. According to Lemma 14.4, the matrices $D_n^* D_n$ and $D_k^* D_k$, $n \geq k$, are either both singular or both nonsingular. If the matrix $D_k^* D_k$ is singular, then, by that lemma, there exists a nonzero vector $x = (x_1, \dots, x_n)$ such that $D\Delta^j x = 0$, $j = 0, 1, \dots, k, k+1, \dots$.

But, then,

$$De^{-\Delta t}x = \sum_{j=0}^{\infty} \frac{(-t)^j}{j!}(D\Delta^j x) = 0,$$

and, therefore,

$$x^* \int_0^T e^{-\Delta^* t} D^* D e^{-\Delta t} dt x = 0, \tag{16.49}$$

which proves the singularity of the matrix $\int_0^T e^{-\Delta^* t} D^* D e^{-\Delta t} dt$.

Otherwise, let (16.49) be satisfied. Then, obviously, $x^* e^{-\Delta^* t} D^* D e^{-\Delta t} x \equiv 0$, $0 \leq t \leq T$. Hence,

$$De^{-\Delta t}x \equiv 0 \tag{16.50}$$

and (after differentiation over t)

$$D\Delta e^{-\Delta t}x \equiv 0,$$

$$\dots$$

$$D\Delta^{k-1}e^{-\Delta t} \equiv 0. \tag{16.51}$$

It follows from (16.50) and (16.51) for $t=0$ that $D\Delta^j x = 0$, $j=0,\dots,k-1$, which is equivalent to the equality $x^* D_k^* D_k x = 0$. □

Corollary. *Let $\tilde{D}_k = (D\,\Delta D\dots\Delta^{k-1}D)$ be a block matrix of order $(k \times kl)$ where D and Δ are matrices of dimensions $(k \times l)$ and $(k \times k)$, respectively. Then the matrices $\tilde{D}_k\tilde{D}_k^*$ and $\int_0^T e^{-\Delta t}DD^*e^{-\Delta^* t}dt$ are either both singular or both nonsingular.*

Lemma 16.3. *If the matrix G_2 has rank k then, for $t>0$, the matrices γ_t defined by Equation (16.44) are positive definite.*

PROOF. The matrix γ_t is the covariance matrix of the conditionally Gaussian distribution $P(\theta_t \le a|\mathcal{F}_t^\xi)$. If this distribution has a (P-a.s.) density then obviously the matrix γ_t will be positive definite. Considering the system of equations given by (16.45) and taking into account Corollary 1 of Theorem 7.23 (see Subsection 7.9.5), we see that the distribution $P(\theta_t \le a|\mathcal{F}_t^\xi)$, $t>0$, has a density (P-a.s.) if the distribution $P(\theta_t \le a)$ also has a density, which is equivalent to the condition of positive definiteness of the matrix

$$\Gamma_t = \operatorname{cov}(\theta_t, \theta_t).$$

According to Theorem 15.1, the matrices Γ_t are solutions of the differential equation

$$\dot{\Gamma}_t = a\Gamma_t + \Gamma_t a^* + bb^*. \tag{16.52}$$

From this we find

$$\Gamma_t = e^{at}\Gamma_0 e^{a^*t} + e^{at}\left[\int_0^t e^{-as}bb^*e^{-a^*s}ds\right]e^{a^*t}.$$

But, by virtue of the corollary to Lemma 16.2, the matrices Γ_t, $t>0$, are positive definite, since so also is the matrix $G_2G_2^*$ (rank $G_2 = k$). □

Lemma 16.4. *If the rank of the matrix G_1 is equal to k, then the elements of all the matrices γ_t are uniformly bounded.*

PROOF. Consider the auxiliary problem of control of the deterministic process $x_t = (x_1(t),\dots,x_k(t))$, $0 \le t \le T$, satisfying the equation

$$\frac{dx_t}{dt} = a^*x_t + A^*u_t, \quad x_0 = x, \tag{16.53}$$

with the performance functional

$$V(u;T) = x_T^* \gamma_0 x_T + \int_0^T [x_t^* bb^* x_t + u_t^* BB^* u_t] dt.$$

The controls u_t, $0 \le t \le T$, are chosen from the class of admissible controls (see the previous section).

According to (16.37), the optimal control $\tilde{u}_t$ exists and is given by the formula

$$\tilde{u}_t = -(BB^*)^{-1} A \gamma_{T-t} \tilde{x}_t, \tag{16.54}$$

where $\tilde{x}_t$ is the solution of Equation (16.53) with $u_t = \tilde{u}_t$, $0 \le t \le T$. In this case $V(\tilde{u};T) = x^* \gamma_T x$. Since the elements of the matrices γ_t are continuous functions, to prove the lemma it suffices to show that all the elements of the matrices γ_T for $T > 1$ are uniformly bounded.

Since rank $G_1 = k$, the matrix $G_1^* G_1$ is nonsingular and, by Lemma 16.2, so is the matrix

$$\int_0^1 e^{-a^* t} A^* A e^{-at} dt.$$

Take now a special control

$$\hat{u}_t = \begin{cases} -Ae^{-at} (\int_0^1 e^{-a^* s} A^* A e^{-as})^{-1} x, & 0 \le t \le 1, \\ 0, & t > 1, \end{cases}$$

and let $\hat{x}_t$ be the solution of Equation (16.53) with $u_t = \hat{u}_t$. By solving this equation we find that $\hat{x}_t \equiv 0$, $t \ge 1$. But then, because of optimality of the control $\tilde{u}_t$, $0 \le t \le T$, $T > 1$,

$$x^* \gamma_T x \le \int_0^1 [\hat{x}_t bb^* \hat{x}_t + \hat{u}_t^* BB^* \hat{u}_t] dt < \infty. \qquad \square$$

Lemma 16.5. *Let γ_t^0 be the solution of (16.44) with $\gamma_0^0 = \gamma_0 = 0$ and* rank $G_1 = k$. *Then $\gamma^0 = \lim_{t\to\infty} \gamma_t^0$ exists and is the nonnegative definite symmetric matrix satisfying the equation*

$$a\gamma^0 + \gamma^0 a^* + bb^* - \gamma^0 A^* (BB^*)^{-1} A \gamma^0 = 0. \tag{16.55}$$

If, in addition, rank $G_2 = k$, *then γ^0 is a positive definite matrix.*

PROOF. By virtue of the assumption that rank $G_1 = k$ it follows from the previous lemma that the elements of all the matrices γ_t^0, $t \ge 0$, are uniformly bounded.

We shall show that for $\gamma_0 = 0$ the function $x^* \gamma_T^0 x$ is monotone nondecreasing in T. Let $T_2 > T_1$. Then, denoting by $u_t(T_i)$ and $x_t(T_i)$ the optimal controls and their associated processes in the auxiliary control problems, $i = 1, 2, \ldots$, we find that

$$
\begin{aligned}
x^*\gamma_{T_2}x &= \int_0^{T_2} [(x_t(T_2))^* bb^* x_t(T_2) + (u_t(T_2))^* BB^* u_t(T_2)]dt \\
&\geq \int_0^{T_1} [(x_t(T_2))^* bb^* x_t(T_2) + (u_t(T_2))^* BB^* u_t(T_2)]dt \\
&\geq \int_0^{T_1} [(x_t(T_1))^* bb^* x_t(T_1) + (u_t(T_1))^* BB^* u_t(T_1)]dt = x^*\gamma_{T_1}x.
\end{aligned}
$$

From boundedness and monotonicity of the functions $x^*\gamma_T^0 x$ follows the existence of the matrix $\gamma^0 = \lim_{T\to\infty}\gamma_T^0$ with the properties stated.

If, in addition, rank $G_2 = k$, then, by Lemma 16.3, the matrices γ_t^0 are nonsingular and consequently the matrix $\gamma^0 = \lim_{t\to\infty}\gamma_t^0$ is also nonsingular. □

16.2.4.

PROOF OF THEOREM 16.2. Set $\gamma^0 = \lim_{t\to\infty}\gamma_t^0$ for $\gamma_0 = 0$, and set

$$
\overline{u}_t = -(BB^*)^{-1}A\gamma^0\overline{x}_t, \tag{16.56}
$$

where $\overline{x}_t$ is the solution of Equation (16.53) with $u_t = \overline{u}_t$ and $\overline{x}_0 = x$. We shall show that $\overline{x}_t \to 0$, $t\to\infty$. For this purpose it is enough, for example, to show that

$$
\lim_{t\to\infty} \overline{x}_t^*\gamma^0\overline{x}_t = 0, \tag{16.57}
$$

since the matrix γ^0 is symmetric and positive definite.

Due to (16.53), (16.55) and (16.56),

$$
\begin{aligned}
\frac{d}{dt}(\overline{x}_t^*\gamma^0\overline{x}_t) &= \overline{x}_t^*\gamma_0[a^* - A^*(BB^*)^{-1}A]\gamma^0\overline{x}_t \\
&\quad +\overline{x}_t^*[a - \gamma^0A^*(BB^*)^{-1}A\gamma^0]\overline{x}_t - \overline{x}_t^*\gamma^0A^*(BB^*)^{-1}A\gamma^0\overline{x}_t \\
&= -\overline{x}_t^*bb^*\overline{x}_t - \overline{x}_t^*\gamma^0A^*(BB^*)^{-1}(BB^*)(BB^*)^{-1}A\gamma^0\overline{x}_t \\
&= -[\overline{x}_t^*bb^*\overline{x}_t + \overline{u}_t^*BB^*\overline{u}_t].
\end{aligned}
$$

Therefore, by Lemma 16.5,

$$
\begin{aligned}
0 \leq \overline{x}_T^*\gamma^0\overline{x}_T &= x^*\gamma^0x - \int_0^T [\overline{x}_t^*bb^*\overline{x}_t + \overline{u}_t^*BB^*\overline{u}_t]dt \\
&\leq x\gamma^0x - \int_0^T [\tilde{x}_t^*bb^*\tilde{x}_t + \tilde{u}_t^*BB^*\tilde{u}_t]dt \\
&= x^*[\gamma^0 - \gamma_T^0]x \to 0, \quad T\to\infty,
\end{aligned} \tag{16.58}
$$

where $\tilde{u}_t$ is the optimal control defined in (16.54).

It also follows from (16.58) that

$$
\lim_{T\to\infty}\int_0^T [\overline{x}_t^*bb^*\overline{x}_t + \overline{u}_t^*BB^*\overline{u}_t]dt = x^*\gamma^0x. \tag{16.59}
$$

Next, let γ_0 be an arbitrary nonnegative definite matrix. Then

$$\begin{aligned}
&\overline{x}_T^*\gamma_0\overline{x}_T + \int_0^T [\overline{x}_t^* bb^*\overline{x}_t + \overline{u}_u^* BB^*\overline{u}_t]dt \\
&\geq x^*\gamma_T x = \tilde{x}_T^*\gamma_0\tilde{x}_T + \int_0^T [\tilde{x}_t^* bb^*\tilde{x}_t + \tilde{u}_t^* BB^*\tilde{u}_t]dt \\
&\geq \int_0^T [\tilde{x}_t^* bb^*\tilde{x}_t + \tilde{u}_t^* BB^*\tilde{u}_t]dt \\
&\geq \int_0^T [\breve{x}_t^* bb^*\breve{x}_t + \breve{u}_t^* BB^*\breve{u}_t]dt = x^*\gamma_T^0 x,
\end{aligned} \tag{16.60}$$

where $\breve{u}_t = -(BB^*)^{-1}A\gamma_{T-1}^0\breve{x}_t$, and $\breve{x}_t$ is the solution of Equation (16.53) with $u_t = \breve{u}_t$. It follows from these inequalities and (16.59) that

$$\lim_{T\to\infty} \overline{x}_T^*\gamma_0\overline{x}_T + x^*\gamma^0 x \geq \overline{\lim}_{T\to\infty} x^*\gamma_T x \geq \underline{\lim}_{T\to\infty} x^*\gamma_T x \geq \lim_{T\to\infty} x^*\gamma_T^0 x.$$

But, according to (16.57), $\lim_{T\to\infty} \overline{x}_T^*\gamma_0\overline{x}_T = 0$ and $\lim_{T\to\infty} x^*\gamma_T^0 x = x^*\gamma^0 x$ (see Lemma 16.5). Hence $\lim_{T\to\infty} x^*\gamma_T x$ $(= x^*\gamma x)$ does exist,

$$\lim_{T\to\infty} x^*\gamma_T x = x^*\gamma^0 x,$$

and

$$\gamma = \lim_{T\to\infty} \gamma_T = \gamma^0.$$

The limit matrix $\gamma = \lim_{T\to\infty} \gamma_T$ does not depend on the value of γ_0 and satisfies Equation (16.48).

The uniqueness of the solution of this equation (in the class of positive definite matrices) can be proved as in Theorem 14.3. □

Note. If the eigenvalues of the matrix a lie in the left-hand plane, then one can remove the first assumption, (I), of Theorem 16.2, since the $\operatorname{Tr}\gamma_t \leq \operatorname{Tr} M\theta_t\theta_t^* < \infty$, $t \geq 0$.

16.3 Computation of Mutual Information and Channel Capacity of a Gaussian Channel with Feedback

16.3.1. Let $(\Omega, \mathcal{F}, P)$ be some probability space, with $(\mathcal{F}_t)$, $0 \leq t \leq T$, a system of nondecreasing sub-σ-algebras of $\mathcal{F}$. Let $\theta = (\theta_t, \mathcal{F}_t)$, $0 \leq t \leq T$, be some *transmitted information* to be transmitted over a channel with Gaussian *white noise.* To make this description precise we suppose a Wiener process $W = (W_t, \mathcal{F}_t)$, independent of the process $\theta = (\theta_t, \mathcal{F}_t)$, $0 \leq t \leq T$, to be given. If the *received signal* $\xi = (\xi_t, \mathcal{F}_t)$ has the form

$$d\xi_t = a_t(\theta)dt + dW_t, \quad \xi_0 = 0, \tag{16.61}$$

i.e.,

$$\xi_t = \int_0^t a_s(\theta)ds + W_t, \tag{16.62}$$

then the *message* θ is said to be transmitted over the *white Gaussian channel without feedback*[3]. The functionals $a_s(\theta)$, $0 \le s \le T$, with $P(\int_0^T |a_s(\theta)|ds < \infty) = 1$ determine the coding and are assumed to be nonanticipative.

In the case where the *received signal* $\xi = (\xi_t, \mathcal{F}_t)$, $0 \le t \le T$, permits the representation

$$d\xi_t = a_t(\theta, \xi)dt + dW_t, \quad \xi_0 = 0, \tag{16.63}$$

with the nonanticipative functional $a_t(\theta, \xi)$, $0 \le t \le T$,

$$P\left(\int_0^T |a_t(\theta,\xi)|dt < \infty\right) = 1,$$

then the transmission is said to occur over the *white Gaussian channel with noiseless feedback*.

Therefore, in the case of *noiseless feedback*, the received signal ξ is sent back and can be taken into account in the future in transmitting the *information* θ.

Let $(\theta, \mathcal{B}_\theta)$ be a measure space to which the values of the signal $\theta = (\theta_t)$ $0 \le t \le T$, belong.

We shall denote by $(C_T, \mathcal{B}_T)$ the measure space of continuous functions on $[0, T]$, $x = (x_t)$, $0 \le t \le T$, with $x_0 = 0$. Let μ_W, μ_ξ and $\mu_{\theta,\xi}$ be measures corresponding to the processes W, ξ and (θ, ξ).

If a certain coding $a_t(\theta, \xi)$, $0 \le t \le T$, is chosen, then it is natural to ask how much information $I_T(\theta, \xi)$ is contained in the *received signal* $\xi = \{\xi_s, s \le t\}$ about the *transmitted signal* $\theta = \{\theta_s, s \le t\}$. By definition

$$I_T(\theta, \xi) = M \ln \frac{d\mu_{\theta,\xi}}{d[\mu_\theta \times \mu_\xi]}(\theta, \xi), \tag{16.64}$$

setting $I_T(\theta, \xi) = \infty$ if the measure $\mu_{\theta,\xi}$ is not absolutely continuous with respect to the measure $\mu_\theta \times \mu_\xi$.

Theorem 16.3. *Let the following conditions be satisfied:*

(1) Equation (16.63) has a strong (i.e., $\mathcal{F}_t^{\theta,W}$-measurable for each t, $0 \le t \le T$) solution;

---

[3] In the engineering literature, instead of writing (16.62) its formal analog, $\dot{\xi}(t) = a_t(\theta) + \dot{W}_t$, is used; $\dot{W}_t$ is called *white Gaussian noise*.

(2)

$$\int_0^t M a_t^2(\theta,\xi)dt < \infty.$$

Then

$$I_T(\theta,\xi) = \frac{1}{2} M \int_0^T [a_t^2(\theta,\xi) - \overline{a}_t^2(\xi)]dt, \tag{16.65}$$

where

$$\overline{a}_t(\xi) = M[a_t(\theta,\xi)|\mathcal{F}_t^\xi]. \tag{16.66}$$

PROOF. According to the assumptions made above and Lemmas 7.6 and 7.7, $\mu_\xi \ll \mu_W$ and $\mu_{\theta,\xi} \ll \mu_\theta \times \mu_W$. Hence, due to the note to Theorem 7.23,

$$\frac{d\mu_{\theta,\xi}}{d[\mu_\theta \times \mu_\xi]}(\theta,\xi) = \frac{d\mu_{\theta,\xi}}{d[\mu_\theta \times \mu_W]}(\theta,\xi) \Big/ \frac{d\mu_\xi}{d\mu_W}(\xi). \tag{16.67}$$

But, due to Lemmas 7.6 and 7.7,

$$\frac{d\mu_{\theta,\xi}}{d[\mu_\theta \times \mu_W]}(\theta,\xi) = \exp\left[\int_0^T a_t(\theta,\xi)d\xi_t - \frac{1}{2}\int_0^T a_t^2(\theta,\xi)dt\right], \tag{16.68}$$

$$\frac{d\mu_\xi}{d\mu_W}(\xi) = \exp\left[\int_0^T \overline{a}_t(\xi)d\xi_t - \frac{1}{2}\int_0^T \overline{a}_t^2(\xi)dt\right], \tag{16.69}$$

where

$$\overline{a}_t(x) = M[a_t(\theta,\xi)|\mathcal{F}_t^\xi]_{\xi=x}.$$

Here,

$$\int_0^T M\overline{a}_t^2(\xi)dt = \int_0^T M[M(a_t(\theta,\xi)|\mathcal{F}_t^\xi)]^2 dt \le \int_0^T M a_t^2(\theta,\xi)dt < \infty.$$

It follows from (16.67)–(16.69) that

$$\begin{aligned} \ln \frac{d\mu_{\theta,\xi}}{d[\mu_\theta \times \mu_\xi]}(\theta,\xi) &= \int_0^T [a_t(\theta,\xi) - \overline{a}_t(\xi)]d\xi_t - \frac{1}{2}\int_0^T [a_t^2(\theta,\xi) - \overline{a}_t^2(\xi)]dt \\ &= \int_0^T \Big([a_t(\theta,\xi) - \overline{a}_t(\xi)]a_t(\theta,\xi) \\ &\quad - \frac{1}{2}[a_t^2(\theta,\xi) - \overline{a}_t^2(\xi)] \Big) dt \\ &\quad + \int_0^T [a_t(\theta,\xi) - \overline{a}_t(\xi)]dW_t. \end{aligned} \tag{16.70}$$

From this, by the properties of stochastic integrals,

$$
\begin{aligned}
M \ln \frac{d\mu_{\theta,\xi}}{d[\mu_\theta \times \mu_\xi]}(\theta,\xi) &= \frac{1}{2}\int_0^T M[a_t^2(\theta,\xi) - 2a_t(\theta,\xi)\overline{a}_t(\xi) + \overline{a}_t^2(\xi)]dt \\
&= \frac{1}{2}\int_0^T M[a_t(\theta,\xi) - \overline{a}_t(\xi)]^2 dt \\
&= \frac{1}{2}\int_0^T M\left\{ M[a_t(\theta,\xi) - \overline{a}_t(\xi)]^2 \middle| \mathcal{F}_t^\xi \right\} dt \\
&= \frac{1}{2}\int_0^T M[a_t^2(\theta,\xi) - \overline{a}_t^2(\xi)]dt. \qquad (16.71)
\end{aligned}
$$

16.3.2. We use this theorem to prove the fact that (subject to 'power' limitation) feedback does not increase the channel capacity.

By definition, for a channel with feedback

$$C = \sup \frac{1}{T} I_T(\theta,\xi), \qquad (16.72)$$

where 'sup' is taken over all the information θ and the nonanticipative functionals $\{a_t(\theta,\xi),\ 0 \le t \le T\}$ for which Equation (16.63) has a unique strong solution and

$$\frac{1}{T}\int_0^T M a_t^2(\theta,\xi)dt \le P \qquad (16.73)$$

with the constant P characterizing the power constraint of the transmitter. Due to (16.71),

$$
\begin{aligned}
0 \le I_T(\theta,\xi) &= \frac{1}{2} M \int_0^T [a_t^2(\theta,\xi) - \overline{a}_t^2(\xi)]dt \\
&\le \frac{1}{2} M \int_0^T a_t^2(\theta,\xi)dt \le \frac{PT}{2}. \qquad (16.74)
\end{aligned}
$$

Therefore,

$$C \le \frac{P}{2}. \qquad (16.75)$$

We shall show now that for a channel without feedback

$$C_0 = \sup \frac{1}{T} I_T(\theta,\xi) = \frac{P}{2}, \qquad (16.76)$$

where 'sup' is taken over all the signals θ and the nonanticipative functionals $a_t(\theta)$, $0 \le t \le T$, for which

$$\frac{1}{T}\int_0^T M a_t^2(\theta)dt \le P.$$

Since $C \ge C_0$, then it will follow from (16.75) and (16.76) that feedback does not improve the channel capacity:

$$C = C_0 = \frac{P}{2}. \qquad (16.77)$$

For this purpose we consider the following example.

EXAMPLE 1. Consider $a_t(x) = x_t$, and $\theta^\alpha = (\theta_t^\alpha)$, $0 \le t \le T$, a Gaussian stationary process with $M\theta_t^\alpha \equiv 0$ and the correlation function

$$K(t,s) = P\exp\{-\alpha|t-s|\}.$$

We shall assume that the received signal $\xi = (\xi_t)$, $0 \le t \le T$, at the channel output can be expressed as

$$d\xi_t = \theta_t^\alpha dt + dW_t, \quad \xi_0 = 0,$$

where $W = (W_t)$, $t \ge 0$, is a Wiener process independent of the process θ^α.

According to Theorem 15.2, the process θ_t^α has the differential

$$d\theta_t^\alpha = -\alpha\theta_t^\alpha dt + \sqrt{2\alpha P}dz_t,$$

where $z = (z_t)$, $t \ge 0$, is a Wiener process independent of W.

Let $m_t^\alpha = M(\theta_t^\alpha|\mathcal{F}_t^\xi)$, $\gamma_t^\alpha = M(\theta_t^\alpha - m_t^\alpha)^2$. By Theorem 10.1,

$$\begin{aligned} dm_t^\alpha &= -\alpha m_t^\alpha dt + \gamma_t^\alpha(d\xi_t - m_t^\alpha dt), \quad m_0^\alpha = 0, \\ \dot\gamma_t^\alpha &= -2\alpha\gamma_t^\alpha + 2\alpha P - (\gamma_t^\alpha)^2, \quad \gamma_0^\alpha = P. \end{aligned} \tag{16.78}$$

From (16.78) and the normality of the process θ^α, the assumptions of Theorem 16.3 are satisfied and, therefore,

$$I_T(\theta^\alpha,\xi) = \frac{1}{2}\left[\int_0^T M(\theta_t^\alpha)^2dt - \int_0^T M(m_t^\alpha)^2dt\right] = \frac{PT}{2} - \frac{1}{2}\int_0^T M(m_t^\alpha)^2dt. \tag{16.79}$$

Let us show that

$$\lim_{\alpha\uparrow\infty}\int_0^T M(m_t^\alpha)^2dt = 0. \tag{16.80}$$

By Theorem 7.12, the process $\overline{W} = (\overline{W}_t, \mathcal{F}_t^\xi)$ with $\overline{W}_t = \xi_t - \int_0^t m_s^\alpha ds$ is a Wiener process.

Therefore

$$dm_t^\alpha = -\alpha m_t^\alpha dt + \gamma_t^\alpha d\overline{W}_t,$$

and, hence,

$$m_t^\alpha = e^{-\alpha t}\int_0^t e^{\alpha s}\gamma_s^\alpha d\overline{W}_s,$$

and by the properties of stochastic integrals we obtain

$$\begin{aligned} M(m_t^\alpha)^2 &= \int_0^t e^{-2\alpha(t-s)}(\gamma_s^\alpha)^2ds \\ &\le \int_0^t e^{-2\alpha(t-s)}P^2ds = P^2\frac{1-e^{-2\alpha t}}{2\alpha}, \end{aligned} \tag{16.81}$$

since $\gamma_t^\alpha = M(\theta_t^\alpha - m_t^\alpha)^2 \le M(\theta_t^\alpha)^2 = P$. The required relation (16.80) follows from (16.81).

Thus we have proved the following theorem.

Theorem 16.4. *Let the conditions of Theorem 16.3 be satisfied. Then the capacity C of the channel with feedback coincides with the capacity C_0 of the channel without feedback and*

$$C = C_0 = \frac{P}{2}.$$

16.4 Optimal Coding and Decoding for Transmission of a Gaussian Signal Through a Channel with Noiseless Feedback

16.4.1. The theory of optimal nonlinear filtering of conditionally Gaussian processes developed in the preceding chapters enables us to find the optimal method for transmission of a Gaussian process through channels with additive white noise using instant noiseless feedback.

Assume first that the signal to be transmitted is a Gaussian random variable θ with $M\theta = m$, $D\theta = \gamma > 0$, where the parameters m and γ are known at both the transmitting and the receiving ends.

The signal $\xi = (\xi_t)$, $0 \le t \le T$ at the transmitter output is assumed to satisfy the stochastic differential equation

$$d\xi_t = A(t, \theta, \xi)dt + dW_t, \quad \xi_0 = 0, \tag{16.82}$$

where $W = (W_t)$, $0 \le t \le T$, is a Wiener process independent of θ. The nonanticipative functional $A = (A(t, \theta, \xi))$, $0 \le t \le T$, determines the coding and is assumed to be such that Equation (16.82) has a unique strong solution with

$$P\left\{\int_0^T A^2(s, \theta, \xi)ds < \infty\right\} = 1.$$

We shall also assume that the functionals $A = (A(t, \theta, \xi))$, $0 \le t \le T$, are subject to the constraints

$$\frac{1}{t}\int_0^t MA^2(s, \theta, \xi)ds \le P, \tag{16.83}$$

where P is given constant. (The coding satisfying the conditions listed above will be called *admissible*).

At each instant of time t the *output signal* $\hat{\theta}_t(\xi)$ can be constructed from the received signal $\xi_0^t = \{\xi_s, s \le t\}$.

The nonanticipative functional $\hat{\theta} = (\hat{\theta}_t(\xi))$, $0 \le t \le T$, specifying the decoding must be chosen, naturally, to *reproduce* the signal θ in the optimal manner.

Set

$$\Delta(t) = \inf M[\theta - \hat{\theta}_t(\xi)]^2, \quad 0 \le t \le T,$$

where 'inf' is taken over all the *admissible codings* $A = (A(s, \theta, \xi))$, $s \ge 0$, and the *decodings* $\hat{\theta}_t(\xi)$. The problem is to find optimal coding, decoding (if such exist, of course) and the minimal reproduction error $\Delta(t)$ as a function of time.

Since (with given coding)

$$M[\theta - \hat{\theta}_t(\xi)]^2 \ge M[\theta - m_t(\xi)]^2,$$

where $m_t = M(\theta|\mathcal{F}_t^\xi)$, then it is seen that $\Delta(t) = \inf_A M[\theta - m_t]^2$ and that the optimal decoding (of the signals ξ_0^t) is the a posteriori mean $m_t = M(\theta|\mathcal{F}_t^\xi)$.

Thus the primary problem is reduced to the problem of finding only the optimal coding.

16.4.2. Consider first the subclass of admissible coding functions $A(t, \theta, \xi)$ linearly dependent on θ:

$$A(t, \theta, \xi) = A_0(t, \xi) + A_1(t, \xi)\theta, \tag{16.84}$$

where $A_0 = (A_0(t, \xi))$ and $A_1 = (A_1(t, \xi))$, $0 \le t \le T$, are nonanticipative functionals. Let

$$\Delta^*(t) = \inf_{(A_0, A_1)} M][\theta - m_t]^2. \tag{16.85}$$

The problem is to find the optimal coding function (A_0^*, A_1^*) which attains the 'inf' in (16.85).

Let some coding (A_0, A_1) be chosen, and let $\xi = (\xi_t)$, $0 \le t \le T$, be a process satisfying the equation

$$d\xi_t = [A_0(t, \xi) + A_1(t, \xi)\theta]dt + dW_t, \quad \xi_0 = 0. \tag{16.86}$$

Then, according to Theorem 12.1, $m_t = M(\theta|\mathcal{F}_t^\xi)$ and

$$\gamma_t = M[(\theta - m_t)^2|\mathcal{F}_t^\xi]$$

satisfy the equations

$$dm_t = \gamma_t A_1(t, \xi)[d\xi_t - (A_0(t, \xi) + A_1(t, \xi)m_t)dt]. \tag{16.87}$$

$$\dot{\gamma}_t = -\gamma_t^2 A_1^2(t, \xi), \tag{16.88}$$

with $m_0 = m$, $\gamma_0 = \gamma$. Equation (16.88) has the solution (Theorem 12.2)

$$\gamma_t = \frac{\gamma}{1+\gamma\int_0^t A_1^2(s,\xi)ds},$$

and it is seen that $P(\inf_{0\le s\le T}\gamma_s > 0) = 1$. Hence, we obtain from (16.88)

$$\frac{\dot{\gamma}_t}{\gamma_t} = -\gamma_t A_1^2(t,\xi),$$

and, therefore,

$$\ln\gamma_t - \ln\gamma = -\int_0^t \gamma_s A_1^2(s,\xi)ds,$$

i.e.,

$$\gamma_t = \gamma\exp\left\{-\int_0^t \gamma_s A_1^2(s,\xi)ds\right\}. \tag{16.89}$$

Since

$$\begin{aligned} & M[A_0(t,\xi)+A_1(t,\xi)\theta]^2 \\ = \; & M\{[A_0(t,\xi)+m_tA_1(t,\xi)]+[\theta-m_t]A_1(t,\xi)\}^2 \\ = \; & M\{A_0(t,\xi)+A_1(t,\xi)m_t\}^2 + M\gamma_tA_1^2(t,\xi), \end{aligned} \tag{16.90}$$

then, due to the boundedness of (16.83),

$$\int_0^t M\gamma_sA_1^2(s,\xi)ds \le Pt. \tag{16.91}$$

Hence, by the Jensen inequality ($Me^{-\eta} > e^{-M\eta}$), (16.89) and (16.91),

$$M\gamma_t \ge \gamma e^{-Pt}, \quad 0\le t\le T. \tag{16.92}$$

Therefore, for the specified coding (A_0, A_1) we have

$$M[\theta - m_t]^2 = M\gamma_t \ge \gamma e^{-Pt} \tag{16.93}$$

and, consequently (see (16.85)),

$$\Delta^*(t) \ge \gamma e^{-Pt}. \tag{16.94}$$

For the optimal coding (A_0^*, A_1^*) the inequalities in (16.91) and (16.92) have to be equalities. This will occur if we take

$$A_1^*(t) = \sqrt{\frac{P}{\gamma}}e^{Pt/2}, \tag{16.95}$$

since then the corresponding γ_t^* (see (16.88)) will be equal to γe^{-Pt}.

Comparing (16.90) with the equality

$$\int_0^t M\gamma_s^*(A_1^*(s))^2ds = \int_0^t \gamma_s^*(A_1^*(s))^2ds = Pt,$$

we find that the equality

$$A_0^*(t,\xi^*) + A_1^*(t)m_t^*(\xi^*) = 0 \tag{16.96}$$

must also be satisfied, where, according to (16.87), the optimal decoding m_t^* can be defined by the equation

$$dm_t^* = \sqrt{P\gamma}e^{-Pt/2}d\xi_t^*, \quad m_0^* = m, \tag{16.97}$$

and the transmitted signal $\xi^* = (\xi_t^*)$, $0 \le t \le T$ (see (16.86)), satisfies the equation

$$d\xi_t^* = \sqrt{\frac{P}{\gamma}}e^{Pt/2}(\theta - m_t^*)dt + dW_t, \quad \xi_0^* = 0. \tag{16.98}$$

It is seen from (16.97) that the optimal decoding can also be expressed as follows:

$$\begin{aligned} m_t^* &= m + \sqrt{P\gamma}\int_0^t e^{-(Ps/2)}d\xi_s^* \\ &= m + \sqrt{P\gamma}\left[e^{-(Pt/2)}\xi_t^* + \frac{P}{2}\int_0^t e^{-(Ps/2)}\xi_s^* ds\right]. \end{aligned} \tag{16.99}$$

Equation (16.98) shows that the optimal coding operation involves transmitting not the message θ during all the time, but the *divergence* $\theta - m_t^*$ between the value θ and its optimal estimate m_t^* multiplied by $\sqrt{P/\gamma_t}$.

Thus we have proved the following lemma.

Lemma 16.6. *In the class of admissible linear coding functions given by (16.84) the optimal coding (A_0^*, A_1^*) exists and is given by the formulae*

$$A_1^*(t) = \sqrt{\frac{P}{\gamma}}e^{Pt/2}, \tag{16.100}$$

$$A_0^*(t,\xi^*) = -A_1^*(t)m_t^*. \tag{16.101}$$

The optimal decoding m_t^* and the transmitted signal ξ_t^* satisfy Equations (16.97) and (16.98).

The reproduction error is

$$\Delta^*(t) = \gamma e^{-Pt}. \tag{16.102}$$

Note 1. Consider the class of linear coding functions $A_0(t) + A_1(t)\theta$ which do not employ feedback. In other words, we shall assume that the functions $A_0(t)$ and $A_1(t)$ depend only on time $\int_0^T [A_0^2(t) + A_1^2(t)]dt < \infty$ and

$$\frac{1}{t}\int_0^t M[A_0(s) + A_1(s)\theta]^2 ds \le P, \quad 0 \le t \le T.$$

Since

$$M[A_0(s) + A_1(s)\theta]^2 = [A_0(s) + mA_1(s)]^2 + \gamma A_1^2(s),$$

then from the above power constraint we find that

$$\int_0^t A_1^2(s)ds \le \frac{P}{\gamma}t.$$

It follows from this that

$$\gamma_t = \frac{\gamma}{1 + \gamma \int_0^t A_1^2(s)ds} \ge \frac{\gamma}{1 + Pt},$$

and, consequently, the minimal mean square reproduction error (without the employment of feedback) is

$$\tilde{\Delta}(t) = \inf M[\theta - m_t]^2 \ge \frac{\gamma}{1 + Pt}.$$

But, for the coding functions,

$$\tilde{A}_1(t) = \sqrt{\frac{P}{\gamma}}, \quad \tilde{A}_0(t) = -A_1(t)m$$

the mean square error is equal to $\gamma/(1 + Pt)$ exactly. Hence,

$$\tilde{\Delta}(t) = \frac{\gamma}{1 + Pt}.$$

Note 2. Let us note another property of the process ξ^* which is an optimal transmitted signal. If (A_0, A_1) is some admissible coding, then, according to Theorem 7.12 and Equation (16.86),

$$d\xi_t = [A_0(t, \xi) + A_1(t, \xi)m_t]dt + d\overline{W}_t,$$

where $\overline{W} = (\overline{W}_t, \mathcal{F}_t^\xi)$ is a Wiener process.

For the optimal signal ξ^*, $A_0^*(t, \xi^*) + A_1^*(t, \xi^*)m_t^* = 0$. Hence, the process $\xi^* = (\xi_t^*)$, $0 \le t \le T$, coincides with the corresponding innovation process $\overline{W} = (\overline{W}_t, \mathcal{F}_t^{\xi^*})$. Consequently, in the optimal case the transmission is such that only the innovation process $\overline{W} = (\overline{W}_t, \mathcal{F}_t^{\xi^*})$ has to be transmitted.

16.4.3. Let us show now that the coding (A_0^*, A_1^*) found in Lemma 16.6 is also optimal in the sense that it has the greatest information $I_t(\theta, \xi)$ about θ in the received message $\xi_0^t = \{\xi_s, s \le t\}$ for each t, $0 \le t \le T$.

Let $I_t = \sup I_t(\theta, \xi)$ where 'sup' is taken over all the signals $\xi_0^t = \{\xi_s, s \le t\}$ satisfying Equation (16.82) with admissible coding functions $A = (A(t, \theta, \xi))$, $0 \le t \le T$.

Lemma 16.7. *The process $\xi^* = \{\xi_s^*, 0 \le s \le T\}$ found in Lemma 16.6 is also optimal in the sense that, for this process*

$$I_t = I_t(\theta, \xi^*) = \frac{Pt}{2}, \quad 0 \le t \le T. \tag{16.103}$$

PROOF. Let $A = (A(t, \theta, \xi))$, $0 \le t \le T$, be some admissible coding. Then it follows from Theorem 16.3 and (16.83) that

$$I_t(\theta, \xi) = \frac{1}{2} \int_0^t M[A^2(s, \theta, \xi) - \overline{A}^2(s, \xi)]ds \le \frac{1}{2} \int_0^t MA^2(s, \theta, \xi)ds \le \frac{Pt}{2}, \tag{16.104}$$

where $\overline{A}(s, \xi) = M[A(s, \theta, \xi)|\mathcal{F}_s^\xi]$.

On the other hand, let us take $A(s, \theta, \xi^*) = A_0^*(s, \xi) + A_1^*(s)\theta$ with $A_0^*(s, \xi)$ and $A_1^*(s)$ defined in Lemma 16.6. Then, due to (16.101),

$$M[A(s, \theta, \xi^*)|\mathcal{F}_s^\xi] = A_0^*(s, \xi^*) + A_1^*(s)m_s^* = 0,$$

and, therefore, according to (16.104) and (16.90),

$$I_t(\theta, \xi^*) = \frac{1}{2} \int_0^t M[A_0^*(s, \xi) + A_1^*(s)\theta]^2 ds = \frac{Pt}{2},$$

which together with (16.104), proves the required equality, (16.103). □

16.4.4. It will be shown here that the linear coding (A_0^*, A_1^*) is optimal in the class of all admissible codings.

To prove this statement we will find useful (16.105), given below: in a certain sense this inequality is analogous to the Cramer–Rao inequality.

Lemma 16.8. *Let θ be a Gaussian random variable, let $\theta \sim N(m, \gamma)$, and let $\tilde{\theta}$ be some random variable. Then*

$$M[\theta - \tilde{\theta}]^2 \ge \gamma e^{-2I(\theta, \tilde{\theta})}. \tag{16.105}$$

PROOF. Let $\varepsilon^2 = M[\theta - \tilde{\theta}]^2$. Without loss of generality, we can take $0 < \varepsilon^2 < \infty$. Consider now the ε-entropy $H_\varepsilon(\theta) = \inf\{I(\theta, \overline{\theta}) : M(\theta - \overline{\theta})^2 \le \varepsilon^2\}$. According to the known formula for the ε-entropy $H_\varepsilon(\theta)$ of the Gaussian variable θ (see formula (12) in [159])

$$H_\varepsilon(\theta) = \frac{1}{2}\ln\max\left(\frac{\gamma}{\varepsilon^2}, 1\right). \qquad (16.106)$$

Consequently,

$$I(\theta, \tilde{\theta}) \geq H_\varepsilon(\theta) \geq \frac{1}{2}\ln\frac{\gamma}{\varepsilon^2} = \frac{1}{2}\ln\frac{\gamma}{M[\theta - \tilde{\theta}]^2},$$

which proves the required inequality, (16.105). □

Theorem 16.5. *Let θ be the Gaussian random variable transmitted over the channel described by Equation (16.82). Then*

$$\Delta(t) = \Delta^*(t) - = \gamma e^{-Pt} \qquad (16.107)$$

and, therefore, in the class of all admissible codings the linear coding (A_0^*, A_1^*) found in Lemma 16.6 is optimal.

PROOF. It is clear that $\Delta(t) \leq \Delta^*(t) = \gamma e^{-Pt}$. Hence, to prove the theorem it suffices to show that

$$\Delta(t) \geq \gamma e^{-Pt}. \qquad (16.108)$$

.

Let $\xi = (\xi_t)$, $0 \leq t \leq T$, be a process corresponding to some admissible coding (see (16.83)), and let $\hat{\theta} = \hat{\theta}_t(\xi)$ be some decoding. Then, due to Lemma 16.8,

$$M[\theta - \hat{\theta}_t(\xi)]^2 \geq \gamma\exp(-2I(\theta, \hat{\theta}_t(\xi)). \qquad (16.109)$$

But, as is well known, $I(\theta, \hat{\theta}_t(\xi)) \leq I_t(\theta, \xi)$. In addition, by Lemma 16.7, $I_t(\theta, \xi) \leq I_t(\theta, \xi^*) = Pt/2$. Hence,

$$M[\theta - \hat{\theta}_t(\xi)]^2 \geq \gamma e^{-Pt},$$

which proves the required inequality, (16.108). □

16.4.5. The method used in proving Lemma 16.6 can also be used for finding optimal linear coding for the cases where: the transmitted message $\theta = (\theta_t)$, $0 \leq t \leq T$, is a Gaussian process with the differential

$$d\theta_t = a(t)\theta_t dt + b(t)d\tilde{W}_t, \qquad (16.110)$$

the Wiener process $\tilde{W} = (\tilde{W}_t)$, $0 \leq t \leq T$, does not depend on the Gaussian random variable θ_0 with the prescribed values $M\theta_0 = m$ and $D\theta_0 = \gamma > 0$; and $|a(t)| \leq K$, $|b(t)| \leq K$.

We shall assume (compare with (16.86)) that the process $\xi = (\xi_t)$, $0 \leq t \leq T$, obtained at the channel output is the unique strong solution of the equation

$$d\xi_t = [A_0(t, \xi) + A_1(t, \xi)\theta_t]dt + dW_t, \quad \xi_0 = 0, \qquad (16.111)$$

where the Wiener process $W = (W_t)$, $0 \le t \le T$, does not depend on $\tilde{W}$; θ_0 and the (nonanticipative) coding functions $A_0(t, \xi)$ and $A_1(t, \xi)$ satisfy the conditions

$$P\left\{\int_0^T A_0^2(t,\xi)dt < \infty\right\} = 1, \qquad \sup_{x \in C, t \le T} |A_1(t,x)| < \infty;$$

and the power constraint

$$M[A_0(t,\xi) + A_1(t,\xi)\theta_t]^2 \le P,$$

for the prescribed constant P.

Let

$$\Delta^*(t) = \inf M[\theta_t - \hat{\theta}_t(\xi)]^2,$$

where 'inf' is taken over all the described admissible coding functions and decodings $\hat{\theta}_t(\xi)$. It is clear that

$$\Delta^*(t) = \inf_{(A_0, A_1)} M[\theta_t - m_t]^2,$$

where $m_t = M(\theta_t | \mathcal{F}_t^\xi)$.

Write

$$\gamma_t = M[(\theta_t - m_t)^2 | \mathcal{F}_t^\xi]. \tag{16.112}$$

Then

$$\Delta^*(t) = \inf_{(A_0, A_1)} M\gamma_t. \tag{16.113}$$

If the coding (A_0, A_1) is given, then, by Theorem 12.1,

$$dm_t = a(t)m_t dt + \gamma_t A_1(t,\xi)[d\xi_t - (A_0(t,\xi) + A_1(t,\xi)m_t)dt], \tag{16.114}$$

$$\dot{\gamma}_t = 2a(t)\gamma_t - \gamma_t^2 A_1^2(t,\xi) + b^2(t) \tag{16.115}$$

with $m_0 = m$, $\gamma_0 = \gamma$.

As in (16.90), we find that

$$M[A_0(t,\xi) + A_1(t,\xi)m_t]^2 + M[\gamma_t A_1^2(t,\xi)] \le P. \tag{16.116}$$

Note that Equation (16.115) is equivalent to the integral equation

$$\begin{aligned}\gamma_t &= \gamma \exp\left\{2\int_0^t a(s)ds - \int_0^t \gamma_s A_1^2(s,\xi)ds\right\} \\ &\quad + \int_0^t b^2(s) \exp\left\{2\int_0^t a(u)du - \int_s^t \gamma_u A_1^2(u,\xi)du\right\} ds.\end{aligned}$$

Due to the Jensen inequality ($Me^{-\eta} \ge e^{-M\eta}$) we obtain

$$M[\theta_t - m_t]^2 \geq \gamma \exp\left\{2\int_0^t a(s)ds - \int_0^t M\gamma_s A_1^2(s,\xi)ds\right\}$$
$$+ \int_0^t b^2(s)\exp\left\{2\int_s^t a(u)du - \int_s^t M\gamma_u A_1^2(u,\xi)du\right\}ds, \tag{16.117}$$

which, together with the inequality $M\gamma_t A_1^2(t,\xi) \leq P$ (following from (16.116)) yields for $M\gamma_t$ the estimate from below:

$$M\gamma_t \geq \gamma \exp\left\{2\int_0^t \left[a(s) - \frac{P}{2}\right] ds\right\}$$
$$+ \int_0^t b^2(s)\exp\left\{2\int_s^t \left[a(u) - \frac{P}{2}\right] du\right\} ds. \tag{16.118}$$

We shall indicate now the coding (A_0^*, A_1^*) for which in (16.118) equality is attained. Since, by assumption, $\gamma_0 = \gamma > 0$, it follows that $P\{\inf_{t\leq T}\gamma_t > 0\} = 1$ (Theorem 12.7), and consequently for all t, $0 \leq t \leq T$, we can define the functions

$$A_1^*(t,\xi^*) = \sqrt{\frac{P}{\gamma_t^*}}, \tag{16.119}$$

$$A_0^*(t,\xi^*) = -A_1^*(t,\xi)^* m_t^*, \tag{16.120}$$

where

$$m_t^* = M(\theta_t|\mathcal{F}_t^{\xi^*}), \quad \gamma_t^* = M[(\theta - m_t^*)^2|\mathcal{F}_t^{\xi^*}], \quad \xi^* = (\xi_t^*), \quad 0 \leq t \leq T,$$

is the solution of the equation

$$d\xi_t^* = [A_0^*(t,\xi^*) + A_1^*(t,\xi^*)\theta_t]dt + dW_t, \quad \xi_0^* = 0. \tag{16.121}$$

It should be emphasized that, due to (16.119), $(A_1^*(t,\xi^*))^2\gamma_t^* = P$ and, therefore (see (16.115)),

$$\dot{\gamma}_t^* = [2a(t) - P]\gamma_t^* + b^2(t), \quad \gamma_0^* = \gamma. \tag{16.122}$$

This linear equation has the unique solution

$$\gamma_t^* = \gamma \exp\left\{2\int_0^t \left[a(s) - \frac{P}{2}\right] ds\right\} + \int_0^t b^2(s)\exp\left\{2\int_0^t \left[a(u) - \frac{P}{2}\right] du\right\} ds, \tag{16.123}$$

which does not depend on the signals ξ.

Comparing (16.113), (16.118) and (16.123) we see that

$$\Delta^*(t) = \gamma_t^*, \quad 0 \leq t \leq T. \tag{16.124}$$

Thus we have the following theorem.

Theorem 16.6. *In transmitting, according to the scheme (16.111), the Gaussian process θ_t subject to Equation (16.110), the optimal transmission is described by the equation*

$$d\xi_t^* = \sqrt{\frac{P}{\gamma_t^*}}[\theta_t - m_t^*]dt + dW_t, \quad \xi_0^* = 0, \tag{16.125}$$

where the optimal decoding $m_t^* = M(\theta_t|\mathcal{F}_t^{\xi^*})$ is defined by the equation

$$dm_t^* = a(t)m_t^* dt + \sqrt{P\gamma_t^*}d\xi_t^*, \quad m_0^* = m, \tag{16.126}$$

and

$$\dot{\gamma}_t = [2a(t) - P]\gamma_t^* + b^2(t), \quad \gamma_0^* = \gamma. \tag{16.127}$$

The minimal reproduction error is

$$\begin{aligned} \Delta^*(t) &= \gamma \exp\left\{2\int_0^t \left[a(s) - \frac{P}{2}\right] ds \right. \\ &\left. + \int_0^t b^2(s) \exp\left\{2\int_s^t \left[a(u) - \frac{P}{2}\right] du\right\} ds\right\}. \end{aligned} \tag{16.128}$$

Corollary. *If $a(t) \equiv b(t) \equiv 0$, then (compare with (16.102))*

$$\Delta^*(t) = \gamma e^{-Pt}.$$

Note 1. If, in transmitting according to the scheme given in (16.111) feedback is not used, then the optimal coding functions $\tilde{A}_0(t)$ and $\tilde{A}_1(t)$ are given by the formulae

$$\tilde{A}_1(t) = \sqrt{\frac{P}{D\theta_t}}, \quad \tilde{A}_0(t) = -\tilde{A}_1(t)M\theta_t.$$

In this case the mean square reproduction error $\tilde{\Delta}(t)$ is found from the equation

$$\dot{\tilde{\Delta}}(t) = 2a(t)\tilde{\Delta}(t) + b^2(t) - \frac{P}{D\theta_t}\tilde{\Delta}^2(t), \quad \tilde{\Delta}(0) = \gamma.$$

In order to compare the values of $\tilde{\Delta}^*(t)$ and $\tilde{\Delta}(t)$ let us consider the following example.

EXAMPLE 2. Let $a(t) \equiv -1$, $\gamma = \frac{1}{2}$, $m = 0$, i.e., let the process θ_t, $t \geq 0$, be a stationary Gaussian Markov process with $d\theta_t = -\theta_t dt + d\tilde{W}_t$ and $\theta_0 \sim N(0, \frac{1}{2})$. Then $M\theta_t \equiv 0$, $D\theta_t \equiv \frac{1}{2}$ and $\dot{\tilde{\Delta}}(t) = -2\tilde{\Delta}(t) + 1 - 2P\tilde{\Delta}^2(t)$, $\tilde{\Delta}(0) = \frac{1}{2}$. It is easy to show from this that

$$\tilde{\Delta}_P = \lim_{t\to\infty} \tilde{\Delta}(t) = \frac{\sqrt{1+2P}-1}{2P}.$$

At the same time, according to (16.128),

$$\Delta^*(t) = \frac{1}{2+P} + e^{-(2+P)t}\left[\frac{1}{2} - \frac{1}{2+P}\right],$$

and, therefore, $\Delta_P^* = \lim_{t\to\infty} \Delta^*(t) = 1/(2+P)$. Hence,

$$\frac{\Delta_P^*}{\tilde{\Delta}} = \frac{2P}{(2+P)(\sqrt{1+2P}-1)},$$

and, therefore,

$$\frac{\Delta_P^*}{\tilde{\Delta}_P} \sim \begin{cases} \sqrt{2/P}, & P\to\infty, \\ 1, & P\to 0. \end{cases}$$

In other words, feedback yields a much smaller reproduction error for large P than is the case without feedback. For small P the reproduction errors are asymptotically (for $t\to\infty$) equivalent in the two cases.

Note 2. The coding (A_0^*, A_1^*) found in Theorem 16.6 is also optimal in the sense that

$$I_t(\theta,\xi^*) = \sup I_t(\theta,\xi), \tag{16.129}$$

where 'sup' is taken over all admissible linear codings, and $I_t(\theta,\xi)$ is defined in (16.64). (16.29) can be proved in the same way as Lemma 16.7.

16.4.6. Consider now the coding functions $A_t(\theta_t,\xi)$ which are not linear in θ_t. The constraints on $A_t(a,x)$ guaranteeing the existence of a unique strong solution to the equation

$$d\xi_t = A_t(\theta_t,\xi)dt + dW_t \tag{16.130}$$

will now be made more stringent.

Thus we assume that $A_t(a,x)$, $t\le T$, $a\in\mathbb{R}^1$, $x\in C$ satisfies

$$A_t^2(a,x) \le L_1(1+a^2+x_t^2) + L_2\int_0^t (1+x_s^2)dK(s), \tag{16.131}$$

and, for arbitrary $t\le T$, $a',a''\in[-N,N]$, $N<\infty$, $x',x''\in C$,

$$\begin{aligned}[A_t(a',x') - A_t(a'',x'')]^2 \le\ & L_1(a'-a'')^2 + L_3(N)(x_t'-x_t'')^2 \\ & + L_4(N)\int_0^t (x_s'-x_s'')^2 dK(s), \end{aligned} \tag{16.132}$$

where $L_1, L_2, L_3(N), L_4(N)$ are certain constants ($L_3(N)$ and $L_4(N)$ depend on N) and $K(s)$ is a monotone nondecreasing right continuous function such that $0\le K(s)\le 1$.

(16.131) and (16,132) ensure the uniqueness and existence of a strong solution to (16.130); this is proved in the same way as in Theorems 4.6 and 4.9, bearing in mind that

$$\sup_{t\le T} M\xi_t^{2k} < \infty, \quad k = 1, 2, \ldots .$$

Theorem 16.7. *Suppose that a Gaussian process θ_t governed by equation (16.110) is being transmitted according to the scheme given by (16.130), where the functionals $A_t(a, x)$ satisfy the requirements of (16.131) and (16.132) and the constraint*

$$MA_t^2(\theta_t, \xi) \le P. \tag{16.133}$$

Then the optimal transmission of the process θ_t is described by (16.125)–(16.128).

16.4.7. The proof of Theorem 16.7 (to be given in Subsection 16.4.8) will be based on the fact that for each t the mean square error of the estimate is bounded from below by $\Delta^*(t)$, given by (16.128) (see Theorem 16.5).

In order to obtain such a lower bound let us formulate first some auxiliary results. Introduce the following notation:

(1) $\rho_t(\beta) = dP(\theta_t \le \beta)/d\beta$;
(2) $\pi_t(\beta) = dP(\theta_t \le \beta | \mathcal{F}_t^\xi)/d\beta$;
(3) $I(\theta_0^t, \xi_0^t)$ will be the *mutual information between* θ_0^t *and* ξ_0^t;
(4) $I(\theta_t, \xi_0^t)$ will be the *mutual information* between θ_t and ξ_0^t;
(5)

$$\mathsf{F}(\theta_t) = \int_{-\infty}^{\infty} \left(\frac{\partial}{\partial\beta}\rho_t(\beta)\right)^2 \rho_t^{-1}(\beta) d\beta$$

will be the *Fisher information*;

(6)

$$\mathsf{F}(\theta_t, \xi_0^t) = \int_{-\infty}^{\infty} \left(\frac{\partial}{\partial\beta}\pi_t(\beta)\right)^2 \pi_t^{-1}(\beta) d\beta$$

will be the *Fisher conditional information.*

Lemma 16.9. *Assume that the functional $A_t(a, x)$ is uniformly bounded together with its partial derivatives $\partial^i A_t(a_1 x)/\partial a^i$, $i = 1, 2, 3$, $x \in C$. Then*

$$I(\theta_t, \xi_0^t) = I(\theta_0^t, \xi_0^t) - \frac{1}{2}\int_0^t b^2(s)[M\mathsf{F}(\theta_s, \xi_0^2) - \mathsf{F}(\theta_s)]ds. \tag{16.134}$$

PROOF. Note first of all, that due to (16.110), the variance Γ_t of the random variable θ_t is given by the equation (Theorem 15.1)

$$\frac{d\Gamma_t}{dt} = 2a(t)\Gamma_t + b^2(t), \tag{16.135}$$

with the initial condition $\Gamma_0 = \gamma > 0$. Hence, for all $t \le T$, the variables Γ_t are positive and the Gaussian distributions $P(\theta_t \le \beta)$, $t \le T$, have a density $\rho_t(\beta)$ which satisfies the forward equation of Kolmogorov

$$\frac{\partial \rho_t(\beta)}{\partial t} = -a(t)\frac{\partial}{\partial \beta}(\beta \rho_t(\beta)) + \frac{1}{2}b^2(t)\frac{\partial^2 \rho_t(\beta)}{\partial \beta^2}. \tag{16.136}$$

By virtue of Corollary 1 to Theorem 7.23, and because of the existence of the density $\rho_t(\beta)$, the conditional density $\pi_t(\beta)$ exists and is given by the formula

$$\begin{aligned}\pi_t(\beta) &= \rho_t(\beta)\tilde{M}\left\{\exp\left[\int_0^t (A_s(\tilde{\theta}_s,\xi) - \overline{A}_s(\xi))d\overline{W}_s \right.\right.\\ &\qquad \left.\left. -\frac{1}{2}\int_0^t (A_s(\tilde{\theta}_s,\xi) - \overline{A}_s(\xi))^2 ds\right]\middle|\tilde{\theta}_t = \beta\right\},\end{aligned} \tag{16.137}$$

where $\overline{A}_s(\xi) = M(A_s(\theta_s,\xi)|\mathcal{F}_s^\xi)$, $(\overline{W}_t, \mathcal{F}_t^\xi)$ is a Wiener process, and the process $\tilde{\theta}_t$ is given on the probability space $(\tilde{\Omega}, \tilde{\mathcal{F}}, \tilde{P})$, which is identical to the primary probability space $(\Omega, \mathcal{F}, P)$ and has the same distribution as θ_t.

According to the theorem on normal correlation (Theorem 13.1), the process $\tilde{\theta}_s$, $s \le t$, permits the representation

$$\tilde{\theta}_s = M\theta_s + \Gamma_t^{-1}\operatorname{cov}(\theta_s,\theta_t)(\tilde{\theta}_t - M\theta_t) + \tilde{\eta}_s,$$

where $\tilde{\eta}_s$, $s \le t$, is independent of $\tilde{\theta}_t$. Let

$$\tilde{A}_s(a,b,x) = A_s(M\theta_s + \Gamma_t^{-1}\operatorname{cov}(\theta_s,\theta_t)(b - M\theta_t) + a, x).$$

Then (16.137) for $\pi_t(\beta)$ can be rewritten as follows:

$$\begin{aligned}\pi_t(\beta) &= \rho_t(\beta)\tilde{M}\exp\left[\int_0^t (\tilde{A}_s(\tilde{\eta}_s,\beta,\xi) - \overline{A}_s(\xi))d\overline{W}_s \right.\\ &\qquad \left. -\frac{1}{2}\int_0^t (\tilde{A}_s(\tilde{\eta}_s,\beta,\xi) - A_s(\xi))^2 ds\right].\end{aligned} \tag{16.138}$$

From (16.138) and the assumptions on $A_t(a,x)$, it follows that:

(1°) the density $\pi_t(\beta)$ is twice continuously differentiable (P-a.s.) with respect to β, $-\infty < \beta < \infty$;

(2°)

$$M\int_0^T b^2(t)\mathsf{F}(\theta_t,\xi_0^t)dt < \infty;$$

(3°) the density $\pi_t(\beta)$ satisfies the equation (see Theorem 8.6))

$$\begin{aligned} d_t\pi_t(\beta) &= \left[-a(t)\frac{\partial}{\partial\beta}(\beta\pi_t(\beta)) + \frac{b^2(t)}{2}\frac{\partial^2}{\partial\beta^2}(\pi_t(\beta))\right]dt \\ &\quad +\pi_t(\beta)[A_t(\beta,\xi) - \overline{A}_t(\xi)]d\overline{W}_t, \quad \pi_0(\beta) = \rho_0(\beta). \end{aligned} \tag{16.139}$$

Let us now estimate the information $I(\theta_t, \xi_0^t)$. By definition

$$\begin{aligned} I(\theta_t, \xi_0^t) &= M \ln \frac{\pi_t(\theta_t)}{\rho_t(\theta_t)} = MM\left(\ln \frac{\pi_t(\theta_t)}{\rho_t(\theta_t)}\bigg| \mathcal{F}_t^\xi\right) \\ &= M\int_{-\infty}^{\infty} \pi_t(\beta) \ln \frac{\pi_t(\beta)}{\rho_t(\beta)} d\beta. \end{aligned} \tag{16.140}$$

Let $\varphi_t(\beta) = \pi_t(\beta)\ln(\pi_t(\beta)/\rho_t(\beta))$. Using (16.140), (16.136) and the identity $\pi_0(\beta) = \rho_0(\beta)$, the Itô formula gives us

$$\begin{aligned} \varphi_t(\beta) &= \int_0^t \left[\ln\frac{\pi_s(\beta)}{\rho_s(\beta)} + 1\right]\left[-a(s)\frac{\partial}{\partial\beta}(\beta\pi_s(\beta)) + \frac{b^2(s)}{2}\frac{\partial^2}{\partial\beta^2}(\pi_s(\beta))\right]ds \\ &\quad + \int_0^t \left[\ln\frac{\pi_s(\beta)}{\rho_s(\beta)} + 1\right]\pi_s(\beta)[A_s(\beta,\xi) - \overline{A}_s(\xi)]d\overline{W}_s \\ &\quad + \frac{1}{2}\int_0^t \pi_s(\beta)(A_s(\beta,\xi) - \overline{A}_s(\xi))^2 ds \\ &\quad - \int_0^t \frac{\pi_s(\beta)}{\rho_s(\beta)}\left[-a(s)\frac{\partial}{\partial\beta}(\beta\rho_s(\beta)) + \frac{b^2(s)}{2}\frac{\partial^2}{\partial\beta^2}(\rho_s(\beta))\right]ds. \end{aligned} \tag{16.141}$$

According to (16.140),

$$I(\theta_t, \xi_0^t) = M\int_{-\infty}^{\infty} \varphi_t(\beta)d\beta.$$

With this in mind, let us integrate the right-hand side of Equation (16.141) with respect to the measure $d\beta\, dP$. We obtain

$$\begin{aligned} & M\int_{-\infty}^{\infty} \frac{\pi_s(\beta)}{\rho_s(\beta)}\left[-a(s)\frac{\partial}{\partial\beta}(\beta\rho_s(\beta)) + \frac{b^2(s)}{2}\frac{\partial^2}{\partial\beta^2}(\rho_s(\beta))\right]d\beta \\ &= \int_{-\infty}^{\infty}\left[-a(s)\frac{\partial}{\partial\beta}(\beta\rho_s(\beta)) + \frac{b^2(s)}{2}\frac{\partial^2}{\partial\beta^2}(\rho_s(\beta))\right]d\beta = 0. \end{aligned}$$

Thus the integral (with respect to the measure $d\beta\, dP$) of the last member on the right in (16.141) is zero for all $t \le T$.

Next we find, using the Fubini theorem, that

$$
\begin{aligned}
&\frac{1}{2}M\int_{-\infty}^{\infty}\int_0^t \pi_s(\beta)(A_s(\beta,\xi)-\overline{A}_s(\xi))^2 ds\,d\beta\\
&=\frac{1}{2}\int_0^t M\int_{-\infty}^{\infty}\pi_s(\beta)(A_s(\beta,\xi)-\overline{A}_s(\xi))^2 d\beta\,ds\\
&=\frac{1}{2}\int_0^t MM[(A_s(\theta_s,\xi)-\overline{A}_s(\xi))^2|\mathcal{F}_s^\xi]ds\\
&=\frac{1}{2}\int_0^t M(A_s(\theta_s,\xi)-\overline{A}_s(\xi))^2 ds = I(\theta_0^t,\xi_0^t),
\end{aligned}
$$

where the last equation follows from Theorem 16.3.

Note that under the assumptions made,

$$
M\int_{-\infty}^{\infty}\int_0^t \left\{\left[\ln\frac{\pi_s(\beta)}{\rho_s(\beta)}+1\right][A_s(\beta,\xi)-\overline{A}_s(\xi)]\pi_s(\beta)\right\}^2 ds\,d\beta<\infty.
$$

Therefore, it is easy to deduce that the integral (with respect to the measure $d\beta\,dP$) of the third member from the right in the right-hand side of (16.141) is zero. Finally, it is easy to verify (by integrating by parts) that the $(d\beta\,dP)$ integrals of the quantities

$$
\int_0^t\left[-a(s)\frac{\partial}{\partial\beta}(\beta\pi_s(\beta))+\frac{b^2(s)}{2}\frac{\partial^2}{\partial\beta^2}(\pi_s(\beta))\right]ds,
$$

$$
\int_0^t a(s)\left(\ln\frac{\pi_s(\beta)}{\rho_s(\beta)}\right)\frac{\partial}{\partial\beta}(\beta\pi_s(\beta))ds
$$

are equal to zero.

Hence,

$$
I(\theta_t,\xi_0^t)=I(\theta_0^t,\xi_0^t)+\frac{1}{2}M\int_{-\infty}^{\infty}\int_0^t b^2(s)\frac{\partial^2}{\partial\beta^2}(\pi_s(\beta))[\ln\pi_s(\beta)-\ln\rho_s(\beta)]dsd\beta.
\tag{16.142}
$$

Using the Fubini theorem and integrating by parts, we find

$$
\begin{aligned}
&M\int_{-\infty}^{\infty}\int_0^t b^2(s)\frac{\partial^2}{\partial\beta^2}(\pi_s(\beta))\ln\pi_s(\beta)ds\,d\beta\\
&=\int_0^t b^2(s)M\int_{-\infty}^{\infty}\frac{\partial^2}{\partial\beta^2}(\pi_s(\beta))\ln\pi_s(\beta)d\beta\,ds\\
&=-\int_0^t b^2(s)M\mathsf{F}(\theta_s,\xi_0^s)ds.
\end{aligned}
\tag{16.143}
$$

Similarly,

$$
\begin{aligned}
&M\int_{-\infty}^{\infty}\int_0^t b^2(s)\frac{\partial^2}{\partial\beta^2}(\pi_s(\beta))\ln\rho_s(\beta)ds\,d\beta\\
&= M\int_0^t b^2(s)\int_{-\infty}^{\infty}\frac{\partial^2}{\partial\beta^2}(\pi_s(\beta))\ln\rho_s(\beta)d\beta\,ds\\
&= M\int_0^t b^2(s)\int_{-\infty}^{\infty}\frac{\partial^2}{\partial\beta^2}(\ln\rho_s(\beta))\pi_s(\beta)d\beta\,ds\\
&= \int_0^t b^2(s)\int_{-\infty}^{\infty}\frac{\partial^2}{\partial\beta^2}(\ln\rho_s(\beta))\rho_s(\beta)d\beta\,ds\\
&= -\int_0^t b^2(s)\int_{-\infty}^{\infty}\frac{\partial}{\partial\beta}(\ln\rho_s(\beta))\frac{\partial}{\partial\beta}(\rho_s(\beta))d\beta\,ds\\
&= -\int_0^t b^2(s)\mathsf{F}(\theta_s)ds.
\end{aligned}
\tag{16.144}
$$

The statement of the lemma now follows from (16.142)–(16.144). □

Let

$$
m_t = M(\theta_t|\mathcal{F}_t^\xi),\quad \gamma_t = M[(\theta_t-m_t)^2|\mathcal{F}_t^\xi],\quad \Delta(t)=M(\theta_t-m_t)^2.
$$

Lemma 16.10. *If θ_t is a Gaussian variable such that $\Gamma_t = D\theta_t > 0$, then*

$$
\mathsf{F}(\theta_t) = \Gamma_t^{-1}. \tag{16.145}
$$

If, in addition, we assume the hypotheses of Lemma 16.9, then

$$
M\mathsf{F}(\theta_t,\xi_0^t) \geq \Delta^{-1}(t). \tag{16.146}
$$

PROOF. (16.145) follows from immediate calculations. The inequality given by (16.146) follows from the two explicit identities

$$
\int_{-\infty}^{\infty} m_t\frac{\partial}{\partial\beta}(\pi_t(\beta))d\beta = 0,\qquad -\int_{-\infty}^{\infty}\beta\frac{\partial}{\partial\beta}(\pi_t(\beta))d\beta = 1,
$$

according to which

$$
1 = \int_{-\infty}^{\infty}(m_t-\beta)\left[\frac{\partial}{\partial\beta}(\pi_t(\beta))\pi_t^{-1}(\beta)\right]\pi_t(\beta)d\beta, \tag{16.147}
$$

and from the Cauchy–Schwarz inequality applied to (16.147). Actually, it follows from (16.147) that

$$
\begin{aligned}
1 &\leq \left(\int_{-\infty}^{\infty}(m_t-\beta^2)\pi_t(\beta)d\beta\cdot\int_{-\infty}^{\infty}\left(\frac{\partial}{\partial\beta}(\pi_t(\beta))\right)^2\pi_t^{-1}(\beta)d\beta\right)^{1/2}\\
&= (\gamma_t\mathsf{F}(\theta_t,\xi_0^t))^{1/2}.
\end{aligned}
\tag{16.148}
$$

Thus, taking expectations on both sides of (16.148) and using the Cauchy–Schwarz inequality, we obtain

$$1 \le [M(\gamma_t \mathsf{F}(\theta_t, \xi_0^t))^{1/2}]^2 \le M\gamma_t M\mathsf{F}(\theta_t, \xi_0^t) = \Delta(t) M\mathsf{F}(\theta_t, \xi_0^t).$$

The required relation, (16.146), follows from this if we can now show that $\Delta(t) > 0$.

But, since $I(\theta_t, m_t) \le I(\theta_t, \xi_0^t) \le I(\theta_0^t, \xi_0^t)$ ($I(\theta_t, m_t)$ is the mutual information between θ_t and m_t), Lemma 16.8, Theorem 16.3 and (16.133) imply

$$\begin{aligned}\Delta(t) &\ge \Gamma_t \exp\{-2I(\theta_t, m_t)\} \ge \Gamma_t \exp\{-2I(\theta_0^t, \xi_0^t)\} \\ &\ge \Gamma_t e^{-Pt} > 0, \quad t \le T.\end{aligned}$$

□

Corollary. *Under the assumptions of Lemma 16.9,*

$$I(\theta_t, \xi_0^t) \le I(\theta_0^t, \xi_0^t) - \frac{1}{2}\int_0^t b^2(s)(\Delta^{-1}(s) - \Gamma_s^{-1})ds. \tag{16.149}$$

16.4.8.

PROOF OF THEOREM 16.7. It is enough to show that $\Delta(t) \ge \Delta^*(t)$, where $\Delta^*(t)$ is given by (16.128). Assume first that the assumptions of Lemma 16.9 are satisfied. Then, as a consequence of Lemma 16.8, Theorem 16.3, the relation $I(\theta_t, m_t) \le I(\theta_t, \xi_0^t)$, and (16.149), we find that

$$\Delta(t) \ge \Gamma_t \exp\left\{-Pt + \int_0^t b^2(s)(\Delta^{-1}(s) - \Gamma_s^{-1})ds\right\}. \tag{16.150}$$

On the other hand, since $\Delta(t) = M(\theta_t^2 - m_t^2)$, the quantities $\Delta(t)$, $t \ge 0$, can be estimated by taking the expectation of $(\theta_t^2 - m_t^2)$. Note that m_t, $t \ge 0$, permits the Itô differential (Theorem 8.1)

$$dm_t = a(t)m_t dt + \psi_t(\xi)d\overline{W}_t, \quad m_0 = M\theta_0, \tag{16.151}$$

where

$$\psi_t(\xi) = M[\theta_t(A_t(\theta_t, \xi) - \overline{A}_t(\xi))|\mathcal{F}_t^\xi].$$

According to the Itô formula and using (16.110) and (16.151), we find that

$$\begin{aligned}(\theta_t^2 - m_t^2) &= (\theta_0^2 - m_0^2) + \int_0^t [2a(s)(\theta_s^2 - m_s^2) + b^2(s) - \psi_s^2(\xi)]ds \\ &\quad + 2\int_0^t b(s)\theta_s d\tilde{W}_s - 2\int_0^t \psi_s(\xi)m_s d\overline{W}_s.\end{aligned}$$

Taking the expectations on both sides we obtain

$$\Delta(t) = \gamma + \int_0^t [2a(s)\Delta(s) + b^2(s) - M\psi_s^2(\xi)]ds. \tag{16.152}$$

Let

$$u_t = \ln \frac{\Delta(t)}{\Gamma_t} + Pt - \int_0^t b^2(s)(\Delta^{-1}(s) - \Gamma_s^{-1})ds. \tag{16.153}$$

According to (16.150), the fact that the variables u_t are nonnegative and $u_0 = 0$, (16.135) and (16.153) imply

$$\frac{du_t}{dt} = P - = \Delta^{-1}(t)M\psi_t^2(\xi). \tag{16.154}$$

From this it follows that

$$Pt \geq \int_0^t \Delta^{-1}(s)M\psi_s^2(\xi)ds. \tag{16.155}$$

Equation (16.1532) for $\Delta(t)$ is equivalent to the following integral equation:

$$\begin{aligned}\Delta(t) &= \gamma \exp\left\{\int_0^t [2a(u) - \Delta^{-1}(u)M\psi_u^2(\xi)]du\right\} \\ &+ \int_0^t \exp\left\{\int_s^t [2a(u) - \Delta^{-1}(u)M\psi_u^2(\xi)]du\right\} b^2(s)ds.\end{aligned}$$

From this and from (16.155) we have

$$\begin{aligned}\Delta(t) &\geq \gamma \exp\left\{2\int_0^t \left[a(u) - \frac{P}{2}\right] du\right\} \\ &+ \int_0^t \exp\left\{2\int_s^t \left[a(u) - \frac{P}{2}\right] du\right\} b^2(s)ds = \Delta^*(t)\end{aligned}$$

(see (16.128)). Thus, if $A_t(a,x)$ satisfies the conditions of Lemma 16.9, one has $\Delta(t) \geq \Delta^*(t)$.

We shall show that this inequality holds true also in the case where $A_t(a,x)$ only satisfies the requirements of Theorem 16.7. For this purpose we approximate $A_t(a,x)$ by a sequence of functionals $(A_t^{(n)}(a,x),\ n = 1,2,\ldots)$ which for any n satisfy the assumptions of Lemma 16.9 and, in addition, $A_t(\theta_t, \xi) = \text{l.i.m.}_n A_t^{(n)}(\theta_t, \xi)$. Let $\xi^{(n)} = (\xi_t^{(n)})$, $t \leq T$, be the process defined by

$$d\xi_t^{(n)} = A_t^{(n)}(\theta_t, \xi_t^{(n)})dt + dW_t, \quad \xi_0^{(n)} = 0.$$

It is possible to show that for any $t \leq T$,

$$\lim_n M(\xi_t - \xi_t^{(n)})^2 = 0, \quad \lim_n M[A_t^{(n)}(\theta_t, \xi^{(n)})]^2 = MA_t^2(\theta_t, \xi).$$

Set

$$P_n(t) = \max[P, M[A_t^{(n)}(\theta_t, \xi^{(n)})]^2].$$

Then it is seen that $P_n(t) \to P$ as $n \to \infty$ for each $t \le T$. Let $m_t^{(n)} = M(\theta_t|\mathcal{F}_t^{\xi^{(n)}})$ and $\Delta_n(t) = M(\theta_t - m_t^{(n)})^2$. Since the functional $A_t^{(n)}(a,x)$ satisfies the hypothesis of Lemma 16.9 and $M[A_t^{(n)}(\theta_t, \xi^{(n)})]^2 \le P_n(t)$, we have that $\Delta_n(t) \ge \Delta_n^*(t)$, where

$$\begin{aligned}\Delta_n^*(t) &= \gamma \exp\left\{2\int_0^t \left[a(u) - \frac{P_n(u)}{2}\right] du\right\} \\ &\quad + \int_0^t \exp\left\{2\int_s^t \left[a(u) - \frac{P_n(u)}{2}\right] du\right\} b^2(s)ds.\end{aligned}$$

Clearly, $\lim_n \Delta_n^*(t) = \Delta^*(t)$ (see (16.128)).

Let us construct a sequence of decoding functionals $\{\lambda_t^{(k,N)}(x), k, N = 1, 2, \dots\}$ for which

$$\lim_k \lim_N \lim_n M\left[(\theta_t - \lambda_t^{(k,N)}(\xi^{(n)}))\right]^2 = M(\theta_t - m_t)^2. \tag{16.156}$$

Then, by the optimality of decoding of $m_t^{(n)}$, we have

$$M[\theta_t - \lambda_t^{(k,N)}(\xi^{(n)})]^2 \ge M[\theta_t - m_t^{(n)}]^2 \ge \Delta_n^*(t).$$

Taking limits in the inequality

$$M[\theta_t - \lambda_t^{(k,N)}(\xi^{(n)})]^2 \ge \Delta_n^*(t)$$

with respect to n, N and k (in that order), we obtain the required lower bound $\Delta(t) \ge \Delta^*(t)$.

Thus, in order to complete the proof of the theorem we only need to establish the existence of the functionals $\lambda_t^{(k,N)}(x)$ with the property given by (16.156).

Let $0 \equiv s_0^{(k)} < s_1^{(k)} < \cdots < s_k^{(k)} = t$ be a sequence of subdivisions such that $\max_j[s_{j+1}^{(k)} - s_j^k] \to 0$, $k \to \infty$. Define a measurable functional $\lambda_t^{(k)}(x)$ so that

$$\lambda_t^{(k)}(x) = M(\theta_t|\xi_{s_0}^{(k)}, \dots, \xi_{s_k}^{(k)})\Big|_{\xi=x}.$$

By the Lévy theorem (Theorem 1.5),

$$\lambda_t^{(k)}(\xi) \to m_t \quad (P\text{-a.s.}).$$

Also we have mean square convergence since the variables $[\lambda_t^{(k)}(\xi)]^2$, $k = 1, 2, \dots$, are uniformly integrable ($M[\lambda_t^{(k)}(\xi)]^4 \le M\theta_t^4$). The functional

$$\lambda_t^{(k)}(x) = \lambda_t^{(k)}(x_{s_0}^{(k)}, \dots, x_{s_k}^{(k)})$$

can be approximated for any k by a sequence of finite, bounded functionals $\lambda_t^{(k,N)}(x)$, continuous in the variables $x_{s_0}^{(k)}, \dots, x_{s_k}^{(k)}$, in the sense that

$$\lambda_t^{(k)}(x) = \mu_W - \lim_N \lambda_r^{(k,N)}(x),$$

where μ_W is the Wiener measure on the measurable space $(C, \mathcal{B})$ of continuous functions $x = (x_t, 0 \le t \le T)$. Let μ_ξ be a measure on the same space, corresponding to the process ξ defined by (16.130). (16.133) guarantees the absolute continuity of μ_ξ with respect to the Wiener measure μ_W (Theorem 7.2). Hence

$$\lambda_t^{(k)}(x) = \mu_\xi - \lim_N \lambda_t^{(k,N)}(x).$$

Since the $\lambda_t^{(k,N)}(x)$ are bounded, it is possible to choose a sequence $(\lambda_t^{(k,N)}(x), N = 1, 2, \dots)$ so that $\lambda_t^{(k)}(\xi) = \text{l.i.m.}_N \lambda_t^{(k,N)}(\xi)$. It is not difficult to see that the $\lambda_t^{(k,N)}(x)$ so obtained have the property given by (16.156). □

16.5 Asymptotic Properties of the Linear Filter under Wrong Initial Conditions

Consider a filtering problem for a vector signal θ_t (of size k) and a vector observation ξ_t (of size ℓ) defined by the linear Itô equations with respect to independent vector Wiener processes V_t (of size k) and W_t (of size ℓ) with independent components

$$\begin{aligned} d\theta_t &= a\theta_t dt + b dV_t \\ d\xi_t &= A\theta_t dt + B dW_t, \end{aligned} \tag{16.157}$$

where a, b, A, and B are matrices of sizes $k \times k$, $k \times k$, $\ell \times k$, and $\ell \times \ell$ respectively. Assume θ_0 is a random vector with ($\|\cdot\|^2$ is the Euclidean norm) $M\|\theta_0\|^2 < \infty$. Denote by $m_0 = M\theta_0$ and $\gamma(0) = M(\theta_0 - m_0)(\theta_0 - m_0)^*$. Assume also that BB^* is a positive definite matrix. Then the Kalman filter (see Chapter 10), subject to the initial conditions m_0 and $\gamma(0)$,

$$\begin{aligned} dm_t &= am_t dt + \gamma(t)A^*(BB^*)^{-1}(d\xi_t - Am_t dt) \\ \frac{d\gamma(t)}{dt} &= a\gamma(t) + \gamma(t)a^* + bb^* - \gamma(t)A^*(BB^*)^{-1}A\gamma(t) \end{aligned} \tag{16.158}$$

creates the optimal (in the mean square sense) linear filtering estimate m_t for θ_t: $m_t = \widehat{M}(\theta_t | \xi_{[0,t]})$ and the matrix of filtering errors

$$\gamma(t) = M(\theta_t - m_t)(\theta_t - m_t)^*.$$

If θ_0 is a Gaussian vector, then m_t coincides with the conditional expectation for θ_t given the σ-algebra generated by $\xi_{[0,t]}$: $m_t = M(\theta_t|\xi_{[0,t]})$.

A crucial role in stabilizing the Kalman filter is played by the properties of the Ricatti equation for $\gamma(t)$. By Theorem 16.2, $\lim_{t\to\infty}\gamma(t) = \gamma$ exists provided that the matrices

$$G_1 = \begin{pmatrix} A \\ Aa \\ \vdots \\ Aa^{k-1} \end{pmatrix} \quad \text{and} \quad G_2 = \begin{pmatrix} b & ab & \cdots & a^{k-1}b \end{pmatrix}$$

have rank equal to k. Moreover, the matrix γ is the unique solution, in the class of positive definite matrices, of the algebraic equation

$$a\gamma + \gamma a^* + bb^* - \gamma A^*(BB^*)^{-1}A\gamma = 0, \tag{16.159}$$

so that γ is independent of $\gamma(0)$.

The next lemma plays an important role in the asymptotic analysis of the Kalman filter under wrong initial conditions.

Lemma 16.11. *Assume that the matrices G_1, G_2 have rank equal to k. Then the matrix $a - \gamma A^*(BB^*)^{-1}A$ has eigenvalues with negative real parts.*

PROOF. Denote $a - \gamma A^*(BB^*)^{-1}A$ by K and rewrite (16.159) in the form

$$K\gamma + \gamma K^* + bb^* + \gamma A^*(BB^*)^{-1}A\gamma = 0. \tag{16.160}$$

Let φ be a left eigenvector of K corresponding to an eigenvalue λ (λ^*). Then, multiplying (16.160) from the left by φ and from the right by φ^*, we obtain

$$(2\mathrm{Re}\,\lambda)\varphi\gamma\varphi^* + \varphi bb^*\varphi^* + \varphi\gamma A^*(BB^*)^{-1}A\gamma\varphi^* = 0 \tag{16.161}$$

which implies $\mathrm{Re}\,\lambda \le 0$. We show that, under the assumption made,

$$\mathrm{Re}\,\lambda < 0. \tag{16.162}$$

Assume $\mathrm{Re}\,\lambda = 0$. Then $\varphi b = 0$ and $\varphi\gamma A^*(BB^*)^{-1/2} = 0$ and so $\varphi\gamma A^*(BB^*)^{-1}A = 0$. The definition of K then implies that $\varphi K = \varphi a$, that is φ (φ^*) is also a left (right) eigenvector of a (a^*). We now use the assumption that the rank of G_2 is k. By this assumption the matrix $G_2G_2^*$ is nonsingular. On the other hand, the vector φ^* is a right eigenvector of this matrix with eigenvalue zero. The contradiction obtained validates (16.162). □

16.5.1 Asymptotically Optimal Kalman Filter. Assume m_0 and $\gamma(0)$ are unknown. Let the linear filter (16.158) be supplied with wrong initial conditions $\widetilde{m}_0$ and γ, where γ is the limit value of $\gamma(t)$, $t \to \infty$. In this case, we arrive at a Kalman type filter

$$d\widetilde{m}_t = a\widetilde{m}_t dt + \gamma A^*(BB^*)^{-1}(d\xi_t - A\widetilde{m}_t dt). \tag{16.163}$$

Theorem 16.8. *Assume that the rank of G_1 and G_2 is equal to k. Then*

$$\lim_{t\to\infty} M(\theta_t - \widetilde{m}_t)(\theta_t - \widetilde{m}_t)^* = \gamma.$$

PROOF. Since

$$\begin{aligned} M(\theta_t - \widetilde{m}_t)(\theta_t - \widetilde{m}_t)^* &= M(\theta_t - m_t)(\theta_t - m_t)^* \\ &\quad + M(m_t - \widetilde{m}_t)(m_t - \widetilde{m}_t)^* \end{aligned}$$

it suffices to show that

$$\lim_{t\to\infty} M(m_t - \widetilde{m}_t)(m_t - \widetilde{m}_t)^* = 0. \tag{16.164}$$

Although the random vector θ_0 is not assumed to be Gaussian but, since in this proof only the second moments for random objects are used, one can assume, without loss of generality, that θ_0 is Gaussian with parameters m_0 and $\gamma(0)$. Then $m_t = M(\theta_t|\xi_{[0,t]})$ and therefore (see Chapter 10, Subsection 10.2)

$$\overline{W}_t = \int_0^t \frac{d\xi_t - Am_s}{B} ds \tag{16.165}$$

is an innovation Wiener process.

Putting $\Delta_t = m_t - \widetilde{m}_t$ and taking into account (16.165), we find

$$d\Delta_t = K\Delta_t dt + [\gamma(t) - \gamma]\, A^* B^{-1} d\overline{W}_t. \tag{16.166}$$

Denote $V_t = M\Delta_t\Delta_t^*$. Using the Itô formula, applied to $\Delta_t\Delta_t^*$, we arrive at the matrix differential equation

$$\frac{dV_t}{dt} = KV_t + V_tK^* + [\gamma(t) - \gamma]A^*(BB^*)^{-1}A[\gamma(t) - \gamma]^*.$$

Since $[\gamma(t) - \gamma] \to 0$, $t \to \infty$ and since by Lemma 16.11 the eigenvalues of the matrix K lie within the unit circle, we obtain $V_t \to 0$, $t \to \infty$. □

16.5.2 Kalman Model with Non-Gaussian Initial Conditions. Assume θ_0 is a non-Gaussian random vector such that $M\|\theta_0\|^2 < \infty$. In this case we also compare the estimate produced by the filter given in (16.163) with the optimal one $\pi_t = M(\theta|\xi_{[0,t]})$ defined by the Kushner–Zakai filter (see Chapter 8) under the known distribution of θ_0. If the assumptions of Theorem 16.8 are fulfilled, we apply a Kalman type filter (16.163) and obtain the filtering estimate $\widetilde{m}_t$. In parallel with this estimate, the optimal one $\pi_t = M(\theta_t|\xi_{[0,t]})$ is defined by the Kushner–Zakai filter (see Chapter 8).

Theorem 16.9. *Let the assumptions of Theorem 16.8 be fulfilled. Then*

1. $\lim_{t\to\infty} M(\widetilde{m}_t - \pi_t)(\widetilde{m}_t - \pi_t)^* = 0$;
2. $\theta_t - \widetilde{m}_t$, $t \to \infty$ *converges in distribution to a zero-mean Gaussian vector with covariance matrix* γ.

PROOF. 1. Evidently only $\lim_{t\to\infty} M(\theta_t-\pi_t)(\theta_t-\pi_t)^* = \gamma$ has to be checked. To this end, we use upper and lower bounds (for nonnegative definite matrices D', D'', $D' \le D''$ is taken to mean that $D'' - D'$ is a nonnegative definite matrix):

$$M(\theta_t - \pi_t)(\theta_t - \pi_t)^* \begin{cases} \le M(\theta_t - \widetilde{m}_t)(\theta_t - \widetilde{m}_t)^* \\ \ge M(\theta_t - \pi_t^\circ)(\theta_t - \pi_t^\circ)^*, \end{cases}$$

where $\pi_t^\circ = M(\theta_t|\theta_0, \xi_{[0,t]})$. Although θ_0 is a non-Gaussian vector, the conditional distribution $P(\theta_t \le x|\theta_0, \xi_{[0,t]})$ is Gaussian (P-a.s.) (see Chapter 13) and moreover π_t° is defined by the linear filter (16.158) subject to the initial conditions θ_0 and 0 (zero matrix), respectively. Denote by $\gamma^\circ(t)$ the solution of the corresponding Ricatti equation. Under the assumptions of the theorem $\lim_{t\to\infty}\gamma^\circ(t) = \gamma$. Coupled with Theorem 16.8 this yields the required conclusion.

2. $\theta_t - \pi_t^\circ$, $t \to \infty$ converges in distribution to a zero-mean Gaussian vector with covariance matrix γ. Therefore, the required statement holds provided that $\lim_{t\to\infty} M\|\pi_t^\circ - \widetilde{m}_t\|^2 = 0$ (see Theorem 4.1 in [19]). It is clear that this is implied by

$$\begin{aligned} &M(\widetilde{m}_t - \pi_t^\circ)(\widetilde{m}_t - \pi_t^\circ)^* \\ &= M(\theta_t - \widetilde{m}_t)(\theta_t - \widetilde{m}_t)^* - M(\theta_t - \pi_t^\circ)(\theta_t - \pi_t^\circ)^*. \end{aligned}$$

□

Notes and References. 1

16.1. The proof of Theorem 16.1 is essentially based on the results related to Chapter 12, concerning the equations for a posteriori means and variances in the case of conditionally Gaussian processes (see also Meditch [227], and Wonham [313]).

16.2. Theorem 16.2 was obtained by Kalman [139].

16.3. The results presented here can be found in the paper of Kadota, Zakai and Ziv [126].

16.4. The transmission of a Gaussian random variable though the channel with feedback has been discussed in Shalkwijk and Kailath [274], Zigangirov [335], Djashkov and Pinsker [54], Khasminskii (see problem 72 in the supplementary material in [147]) and Nevelson and Khasminskii [243]. The proof of Theorem 16.4 based on the employment of optimal nonlinear filtering equations is due to the authors and Katyshev (diploma paper). The proof of Lemma 16.7 and Theorem 165 is due to Ihara [95]. Theorem 16.6 has been proved by the authors, and Theorem 16.7 by Liptser [193].

Notes and References. 2

16.1. An analysis of the sensitivity of a criterion in the linear quadratic Gaussian control problem can be found in Kabanov and Di Masi [112]. Singularly perturbed two-scaled stochastic control models are investigated in Kabanov and Pergamenshchikov [121, 122] and in Kabanov and Runggaldier [123]. A control problem for a counting process is considered in Kabanov [111].

16.4. A control problem with incomplete data and information processing, closed in some sense to a coding procedure, can be found in Kuznetsov, Liptser and Serebrovski [182].

16.5. A problem of stability for nonlinear filters with correct initial conditions is studied by Kunita [168, 170] and Stettner [294]. For the case of wrong initial conditions for both linear and nonlinear filters see Ocone and Pardoux [249], Delyon and Zeitouni [52], Atar and Zeitouni [9, 10], see also Budhiraja and Ocone [33], Makowski and Sowers [224].

17. Parameter Estimation and Testing of Statistical Hypotheses for Diffusion-Type Processes

17.1 Maximum Likelihood Method for Coefficients of Linear Regression

17.1.1. Let $\xi = (\xi_t)$, $0 \le t \le T$, be a random process with

$$\xi_t = \sum_{i=1}^{N} \alpha_i(t)\theta_i + \eta_t, \tag{17.1}$$

where $\theta = (\theta_1, \dots, \theta_N)$ is a vector column of the unknown parameters, $-\infty < \theta_i < \infty$, $i = 1, \dots, N$, and $\alpha_t = (\alpha_1(t), \dots, \alpha_N(t))$ is a known vector function with the measurable deterministic components $\alpha_i(t)$, $i = 1, \dots, N$. The random process $\eta = (\eta_t)$, $-\infty < t < \infty$, is assumed stationary, $M\eta_0 = 0$, Gaussian, with the rational spectral density

$$f(\lambda) = \left| \frac{P_{n-1}(i\lambda)}{Q_n(i\lambda)} \right|^2, \tag{17.2}$$

where

$$P_{n-1}(z) = \sum_{j=0}^{n-1} b_j z^j, \quad b_{n-1} \neq 0, \quad Q_n(z) = \sum_{j=0}^{n} a_j z^j, \quad a_n = 1$$

and the roots of the equation $Q_n(z) = 0$ lie within the left half-plane.

Starting from the optimal filtering equations deduced earlier, we shall find maximum likelihood estimates of the vector θ from the observations $\xi_0^T = \{\xi_s, 0 \le s \le T\}$.

17.1.2. We shall assume that the functions $\alpha_j(t)$ have derivatives[1] $g_j(t)$, $j = 1, \dots, N$, and

$$\int_0^T g_j^2(t)dt < \infty. \tag{17.3}$$

According to Theorem 15.4, the process $\eta = (\eta_t)$, $0 \le t \le T$, is a component of the n-dimensional process $(\eta_1(t), \dots, \eta_n(t))$, where $\eta_t = \eta_1(t)$, satisfying the equations

---

[1] More precisely, $\alpha_j(t) = \int_0^t g_j(s)ds$.

$$d\eta_j(t) = \eta_{j+1}(t)dt + \beta_j dW_t, \quad j = 1, \ldots, n-1, \tag{17.4}$$

$$d\eta_n(t) = -\sum_{j=0}^{n-1} a_j \eta_{j+1}(t)dt + \beta_n dW_t, \tag{17.5}$$

where $W = (W_t)$, $0 \leq t \leq T$, is a Wiener process independent of $\eta_j(0)$, $j = 1, \ldots, n$, and the numbers β_j, $j = 1, \ldots, N$, are given by the formulae

$$\beta_1 = b_{n-1}, \quad \beta_j = b_{n-1} - \sum_{i=1}^{j-1} \beta_i a_{n-j+i}, \quad j = 2, \ldots, n.$$

According to the assumption, $\beta_1 = b_{n-1} \neq 0$ and

$$d\xi_t = \left[\sum_{i=1}^{N} g_i(t)\theta_i + \eta_2(t)\right] dt + \beta_1 dW_t. \tag{17.6}$$

Hence, if $g_t = (g_1(t), \ldots, g_N(t))$ is a vector row function, and $\theta = (\theta_1, \ldots, \theta_N)$ is a vector column, then

$$d\xi_t = [g_t\theta + \eta_2(t)]dt + \beta_1 dW_t \tag{17.7}$$

and

$$\begin{aligned} d\eta_j(t) &= \eta_{j+1}(t)dt + \beta_j dW_t, \quad j = 2, \ldots, n-1, \\ d\eta_n(t) &= \left[-a_0(\xi_t - \alpha_t\theta) - \sum_{j=0}^{n-1} a_j \eta_{j+1}(t)\right] dt + \beta_n dW_t. \end{aligned} \tag{17.8}$$

In the system of equations given by (17.7) and (17.8), the components $\eta_2(t), \ldots, \eta_n(t)$ are unobservable. The process ξ_t is observable.

We shall fix some $\theta \in \mathbb{R}^N$ and denote by

$$\begin{aligned} m_j^\theta(t, \xi) &= M[\eta_j(t) | \xi_s, 0 \leq s \leq t], \quad j = 2, \ldots, n, \\ \gamma_{ij}^\theta(t) &= M[(\eta_j(t) - m_i^\theta(t, \xi))(\eta_j(t) - m_j^\theta(t, \xi))], \quad i, j = 2, \ldots, n, \end{aligned}$$

the associated processes ξ_t and $\eta_j(t)$.

According to the equations of Theorem 10.3, the covariances $\gamma_{ij}^\theta(t)$ do not depend on θ. Here $\gamma_{ij}(t) \equiv \gamma_{ij}^\theta(t)$ satisfy (10.82) and

$$\begin{aligned} dm_j^\theta(t, \xi) &= m_{j+1}^\theta(t, \xi)dt + \frac{\beta_1\beta_j + \gamma_{2j}(t)}{\beta_1^2}[d\xi_t - (g_t\theta + m_2^\theta(t, \xi))dt], \\ & \quad j = 2, \ldots, n-1. \end{aligned} \tag{17.9}$$

$$dm_n^\theta(t,\xi) = \left[-a_0(\xi_t - \alpha_t\theta) - \sum_{j=1}^{n-1} a_j m_{j+1}^\theta(t,\xi)\right] dt$$
$$+\frac{\beta_1\beta_n + \gamma_{2n}(t)}{\beta_1^2}[d\xi_t - (g_t\theta + m_2^\theta(t,\xi))dt]. \qquad (17.10)$$

Next, by Theorem 7.17, the process $\xi = (\xi_t)$, $0 \le t \le T$, permits the differential

$$d\xi_t = [g_t\theta + m_2^\theta(t,\xi)]dt + \beta_1 d\overline{W}_t, \qquad (17.11)$$

where $\overline{W} = (\overline{W}_t, \mathcal{F}_t^\xi)$ is a Wiener process and

$$P\left\{\int_0^T (m_2^\theta(t,\xi))^2 dt < \infty\right\} = 1.$$

Along with the process $\xi = (\xi_t)$, $0 \le t \le T$, we shall consider the process

$$\tilde{\xi}_t = \tilde{\xi}_0 + \beta_1\overline{W}_t, \quad \tilde{\xi}_0 = \eta_1(0), \qquad (17.12)$$

and the processes $m_j^\theta(t,\tilde{\xi})$, $j = 2,\dots,n-1$, satisfying the system of equations given by (17.9) and (17.10) where, instead of ξ, the process $\tilde{\xi}$ is used.

Let μ^θ and $\tilde{\mu}$ be measures on $(C_T, \mathcal{B}_T)$ corresponding[2] to the processes $\xi = (\xi_t)$ and $\tilde{\xi} = (\tilde{\xi}_t)$, $0 \le t \le T$, defined by (17.11) and (17.12). Due to Theorem 7.19, Lemma 4.10 and the fact that ξ_0 and $\tilde{\xi}_0 = \eta_1(0)$ are Gaussian random variables ($D\xi_0 = D\tilde{\xi}_0 > 0$), the measures μ^θ and $\tilde{\mu}$ are equivalent and

$$\frac{d\mu^\theta}{d\tilde{\mu}}(\xi) = \exp\left\{\frac{\xi_0\alpha_0\theta}{\delta^2} - \frac{1}{2}\frac{(\alpha_0\theta)^2}{\delta^2} + \int_0^T \frac{g_t\theta + m_2^\theta(t,\xi)}{\beta_2^2} d\xi_t\right.$$
$$\left.-\frac{1}{2}\int_0^T \frac{[g_t\theta + m_2^\theta(t,\xi)]^2}{\beta_1^2} dt\right\}, \qquad (17.13)$$

where $\delta^2 = M\tilde{\xi}_0^2$ $(= M\eta_1^2(0))$.

Let us examine the structure of the functions $m_2^\theta(t,\xi)$ occurring in (17.13). It is easy to deduce from Equations (17.9) and (17.10)[3] that

$$m_2^\theta(t,\xi) = \nu_0(t,\xi) + \nu_1(t)\theta, \qquad (17.14)$$

where the $\nu_0(t,\xi)$ are $\mathcal{F}_t^\xi$-measurable for each t, and

$$\nu_1(t) = (\nu_{11}(t),\dots,\nu_{1N}(t))$$

is a deterministic vector (row) function.

---

[2] $\mathcal{B}_T$ is the Borel σ-algebra in the space C_T of continuous functions $x = (x_s)$, $0 \le s \le T$.

[3] For the pertinent considerations for the case of discrete time, see Section 14.2.

We obtain from (17.13) and (17.14)

$$\frac{d\mu^\theta}{d\tilde{\mu}}(\xi) = \exp\left\{\frac{\xi_0\alpha_0\theta}{\delta^2} - \frac{1}{2}\frac{(\alpha_0\theta)^2}{\delta^2} + \int_0^T \frac{(g_t+\nu_1(t))\theta+\nu_0(t,\xi)}{\beta_1^2}d\xi_t \right.$$
$$\left. -\frac{1}{2}\int_0^T \frac{[(g_t+\nu_1(t))\theta+\nu_0(t,\xi)]^2}{\beta_1^2}dt\right\}. \tag{17.15}$$

Suppose that the matrix

$$D_T = \frac{\alpha_0^*\alpha_0}{\delta^2} + \int_0^T \frac{[g_t+\nu_1(t)]^*[g_t+\nu_1(t)]}{\beta_1^2}dt \tag{17.16}$$

is positive definite. Then by differentiating we find from (17.15) that the vector

$$\hat{\theta}_T(\xi) = D_T^{-1}\left\{\frac{\alpha_0^*\xi_0}{\delta^2} + \int_0^T \frac{[g_t+\nu_1(t)]^*}{\beta_1^2}(d\xi_t - \nu_0(t,\xi)dt)\right\} \tag{17.17}$$

maximizes (17.15) and, consequently, is the maximum likelihood estimate of the vector θ.

17.1.3. We examine now some properties of the estimates $\hat{\theta}_T(\xi)$. It follows from (17.16), (17.17) and (17.11) that

$$\hat{\theta}_T(\xi) = D_T^{-1}\left\{\frac{\alpha_0^*\alpha_0\theta}{\delta^2} + \int_0^T \frac{[g_t+\nu_1(t)]^*}{\beta_1^2}[g_t+\nu_1(t)]\theta dt\right\}$$
$$+D_T^{-1}\left\{\frac{\alpha_0^*\alpha_0\eta_1(0)}{\delta^2} + \int_0^t \frac{[g_t+\nu_1(t)]^*}{\beta_1^2}d\overline{W}_t\right\}$$
$$= \theta + D_T^{-1}\left\{\frac{\alpha_0^*\alpha_0\eta_1(0)}{\delta^2} + \int_0^t \frac{[g_t+\nu_1(t)]^*}{\beta_1^2}d\overline{W}_t\right\}, \tag{17.18}$$

and, therefore,

$$M\hat{\theta}_T(\xi) = \theta, \tag{17.19}$$
$$M[(\hat{\theta}_T(\xi)-\theta)(\hat{\theta}_T(\xi)-\theta)^*] = D_T^{-1}. \tag{17.20}$$

After simple transformations we find that

$$\frac{d\mu^\theta}{d\tilde{\mu}}(\xi) = \exp\left\{\theta^* D_T\hat{\theta}_T(\xi) - \frac{1}{2}\theta^* D_T\theta\right\}. \tag{17.21}$$

It follows from this that the estimate $\hat{\theta}_T(\xi)$ is a sufficient statistic (see Section 1.5). Finally, as in the case of discrete time (see Section 14.2), it can be shown that the estimate $\hat{\theta}_T(\xi)$ is efficient.

Thus we have the following theorem.

Theorem 17.1. *Let the matrix D_T defined by (17.16) be positive definite. Then (17.17) gives the maximum likelihood estimate $\hat{\theta}_T(\xi)$ of the vector θ in the scheme given by (17.1). This estimate is unbiased and efficient.*

17.1.4.
EXAMPLE. Let us estimate the mean θ of the stationary Gaussian process ξ_t, $-\infty < t < \infty$, with spectral density

$$f(\lambda) = \left| \frac{i\lambda + 1}{(i\lambda)^2 + i\lambda + 1} \right|^2,$$

from the observations $\xi_0^T = \{\xi_s, 0 \le s \le T\}$.

Let $\eta_t = \xi_t - \theta$. Then η_t is stationary Gaussian with $M\eta_t = 0$ and spectral density $f(\lambda)$. By Theorem 15.4, the process η_t is a component of the two-dimensional process $(\eta_1(t), \eta_2(t))$, $\eta_t = \eta_1(t)$, satisfying the equations

$$\begin{aligned} d\eta_1(t) &= \eta_2(t)dt + dW_t, \\ d\eta_2(t) &= [-\eta_1(t) - \eta_2(t)]dt, \end{aligned}$$

and, therefore,

$$\begin{aligned} d\xi_t &= \eta_2(t)dt + dW_t, \\ d\eta_2 &= [\theta - \xi_t - \eta_2(t)]dt. \end{aligned}$$

For each fixed $\theta \in \mathbb{R}^1$, let

$$m^\theta(t, \xi) = M(\eta_2(t)|\mathcal{F}_t^\xi) \quad \text{and} \quad \gamma(t) = M[\eta_2(t) - m^\theta(t, \xi)]^2.$$

By Theorem 10.3 and the equations for the processes $(\xi(t), \eta_2(t))$, we obtain the following equations for $m^\theta(t, \xi)$ and $\gamma(t)$:

$$\begin{aligned} dm^\theta(t, \xi) &= [\theta - \xi_t - m^\theta(t, \xi)]dt + \gamma(t)[d\xi_t - m^\theta(t, \xi)dt], \\ \dot{\gamma}(t) &= -2\gamma(t) - \gamma^2(t). \end{aligned}$$

These equations can be solved under the initial conditions

$$\begin{aligned} m^\theta(0, \xi) &= M[\eta_2(0)|\xi_0] = M[\eta_2(0)|\eta_1(0) + \theta], \\ \gamma(0) &= M[\eta_2(0) - m^\theta(0, \xi)]^2, \end{aligned}$$

which can be derived from the theorem on normal correlation (Theorem 13.1). According to that theorem,

$$m^\theta(0, \xi) = \frac{M\eta_1(0)\eta_2(0)}{M\eta_1^2(0)}(\xi_0 - \theta),$$

$$\gamma(0) = M\eta_2^2(0) = \frac{(M\eta_1(0)\eta_2(0))^2}{M\eta_1^2(0)}.$$

In order to find the moments $M\eta_1^2(0)$, $M\eta_2^2(0)$, $M\eta_1(0)\eta_2(0)$ we shall take advantage of the stationarity of the process $(\eta_1(t), \eta_2(t))$, $-\infty < t < \infty$, and of the fact that the matrix

$$\Gamma \equiv M \begin{pmatrix} \eta_1^2(t) & \eta_1(t)\eta_2(t) \\ \eta_1(t)\eta_2(t) & \eta_2^2(t) \end{pmatrix}$$

is the unique solution of the system of equations (Theorem 15.4)

$$A\Gamma + \Gamma A^* + BB^* = 0$$

with

$$A = \begin{pmatrix} 0 & 1 \\ -1 & 1 \end{pmatrix}, \quad B = \begin{pmatrix} 1 \\ 0 \end{pmatrix}.$$

We find from this that

$$M\eta_1^2(0) - 1, \quad M\eta_2^2(0) = \frac{1}{2}.$$

$$M\eta_1(0)\eta_2(0) = -\frac{1}{2}, \quad m^\theta(0,\xi) = \frac{1}{2}(\theta - \xi_0), \quad \gamma(0) = \frac{1}{4}.$$

Thus, it is easy to verify that

$$\begin{aligned} m^\theta(t,\xi) &= \exp\left\{-\int_0^t (1+\gamma(s))ds\right\} \\ &\times \left\{\frac{1}{2}(\theta - \xi_0) + \int_0^t \exp\left[\int_0^s (1+\gamma(u))du\right](\theta - \xi_s)ds\right. \\ &\left. + \int_0^t \exp\left[\int_0^s (1+\gamma(u))du\right]\gamma_s d\xi_s\right\}. \end{aligned}$$

It follows from this formula (see (17.14)) that

$$m(t,\xi) = \nu_0(t,\xi) + \nu_1(t)\theta,$$

and it is easy to compute that

$$\begin{aligned} \nu_0(t,\xi) &= \exp\left\{-\int_0^t (1+\gamma(s))ds\right\}\left\{-\frac{\xi_0}{2} - \int_0^t \exp\left[\int_0^s (1+\gamma(u))du\right]\xi_u du\right. \\ &\left. + \int_0^t \exp\left[\int_0^s (1+\gamma(u))du\right]\gamma_u d\xi_u\right\}, \\ \nu_1(t) &= \exp\left\{-\int_0^t (1+\gamma(s))ds\right\}\left\{\frac{1}{2} + \int_0^t \exp\left[\int_0^s (1+\gamma(u))du\right]ds\right\}. \end{aligned}$$

Since $D\xi_0 = M\eta_1^2(0) = 1$, from (17.16) we see that $D_T = 1 + \int_0^T \nu_1^2(t)dt > 0$ (in our case $g_t \equiv 0$) and that the maximum likelihood estimate $\hat{\theta}_T(\xi)$ for $M\theta_t$ of the process ξ_t is given by the formula

$$\hat{\theta}_T(\xi) = \frac{\xi_0 + \int_0^T \nu_1(t)(d\xi_t - \nu_0(t,\xi)dt)}{1 + \int_0^T \nu_1^2(t)dt}.$$

17.2 Parameter Estimation of the Drift Coefficient for Diffusion-Type Processes

17.2.1. Let θ be an unknown parameter, $-\infty < \theta < \infty$, and let $\xi = (\xi_t, \mathcal{F}_t)$, $0 \le t \le T$, be the diffusion-type process with the differential

$$d\xi_t = \theta a_t(\xi)dt + dW_t, \quad \xi_0 = 0, \tag{17.22}$$

where $W = (W_t, \mathcal{F}_t)$ is a Wiener process and $a_t(x)$ is a nonanticipative functional, $0 \le t \le T$, $x \in C_T$.

Consider the problem of estimating the parameter θ in the drift coefficient $\theta a_t(\xi)$ from the observations $\xi_0^T = \{\xi_s, s \le T\}$.

We shall assume that the functionals $a_t(x)$ satisfy the conditions

$$P_\theta\left(\int_0^T a_t^2(\xi)dt < \infty\right) = P_\theta\left(\int_0^T a_t^2(W)dt < \infty\right) = 1, \tag{17.23}$$

where the index θ in P_θ emphasizes the fact that the distribution of the process ξ is being considered for the prescribed value θ.

According to Theorem 7.7, the measures μ_ξ^θ and μ_W ($\mu_\xi^\theta(B) = P_\theta\{\omega{:}\xi \in\}$, $B \in \mathcal{B}_T$), defined on $(C_T, \mathcal{B}_T)$ are equivalent and

$$\frac{d\mu_\xi^\theta}{d\mu_W}(\xi) = \exp\left\{\theta\int_0^T a_t(\xi)d\xi_t - \frac{\theta^2}{2}\int_0^T a_t^2(\xi)dt\right\}. \tag{17.24}$$

It follows from this that, under the condition $P_\theta\{\int_0^T a_t^2(\xi)dt > 0\} = 1$, $\theta \in \mathbb{R}^1$, the maximum likelihood estimate $\hat{\theta}_T(\xi)$ is given by the formula

$$\hat{\theta}_T(\xi) = \frac{\int_0^T a_t(\xi)d\xi_t}{\int_0^T a_t^2(\xi)dt}. \tag{17.25}$$

Let us investigate some properties of this estimate.

Theorem 17.2. *Suppose the following conditions are satisfied:*

$$\sup_{\theta_1 \le \theta \le \theta_2} \int_0^T M_\theta a_t^{16}(\xi)dt < \infty, \tag{17.26}$$

$$\sup_{\theta_1 \le \theta \le \theta_2} M_\theta\left(\int_0^T a_t^2(\xi)dt\right)^{-16} < \infty, \tag{17.27}$$

for any θ_1, θ_2 ($-\infty < \theta_1 < \theta_2 < \infty$).

Then the bias $b_T(\theta) = M_\theta[\hat{\theta}_T(\xi) - \theta]$ and the mean square error $B_T(\theta) = M_\theta[\hat{\theta}_T(\xi) - \theta]^2$ are defined by the formulae

$$b_T(\theta) = \frac{d}{d\theta} M_\theta \left(\int_0^T a_t^2(\xi) dt \right)^{-1}, \tag{17.28}$$

$$B_T(\theta) = M_\theta \left(\int_0^T a_t^2(\xi) dt \right)^{-1} + \frac{d^2}{d\theta^2} M_\theta \left(\int_0^T a_t^2(\xi) dt \right)^{-2}. \tag{17.29}$$

17.2.2. As a preliminary we prove the following two lemmas.

Lemma 17.1. *Let $\delta = \delta(x)$ be a $\mathcal{B}_T$-measurable function with*

$$\sup_{\theta_1 \le \theta \le \theta_2} M_\theta \delta^4(\xi) < \infty$$

for any θ_1, θ_2 $(-\infty < \theta_1 < \theta_2 < \infty)$. If

$$\sup_{\theta_1 \le \theta \le \theta_2} M_\theta \int_0^T a_t^8(\xi) dt < \infty, \quad -\infty < \theta_1 < \theta_2 < \infty, \tag{17.30}$$

then the function $M_\theta \delta(\xi)$ is differentiable over θ and

$$\frac{d}{d\theta} M_\theta \delta(\xi) = M_\theta \left[\delta(\xi) \int_0^T a_t(\xi) dW_t \right]. \tag{17.31}$$

PROOF. Let

$$\varphi(\theta, W) = \frac{d\mu_\xi^\theta}{d\mu_W}(W) = \exp\left\{ \theta \int_0^T a_t(W) dW_t - \frac{\theta^2}{2} \int_0^T a_t^2(W) dt \right\}.$$

The function $\varphi(\theta, W)$ is differentiable over θ and (P-a.s.)

$$\frac{\partial \varphi(\theta, W)}{\partial \theta} = \left[\int_0^T a_t(W) dW_t - \theta \int_0^T a_t^2(W) dt \right] \varphi(\theta, W). \tag{17.32}$$

Let $-\infty < \theta_1 < \theta_2 < \infty$. Then, due to (17.32),

$$\begin{aligned} M_{\theta_2}\delta(\xi) - M_{\theta_1}\delta(\xi) &= M\delta(W)[\varphi(\theta_2, W) - \varphi(\theta_1, W)] \\ &= M\delta(W) \int_{\theta_1}^{\theta_2} \left[\int_0^T a_t(W) dW_t - \theta \int_0^T a_t^2(W) dt \right] \varphi(\theta, W) d\theta. \end{aligned}$$

Note that, according to the assumptions of the lemma,

$$\int_{\theta_1}^{\theta_2} M\left|\delta(W)\left[\int_0^T a_t(W)dW_t - \theta\int_0^T a_t^2(W)dt\right]\right|\varphi(\theta,W)d\theta$$
$$= \int_{\theta_1}^{\theta_2} M_\theta\left|\delta(\xi)\left[\int_0^T a_t(\xi)d\xi_t - \theta\int_0^T a_t^2(\xi)dt\right]\right|d\theta$$
$$= \int_{\theta_1}^{\theta_2} M_\theta\left|\delta(\xi)\int_0^T a_t(\xi)dW_t\right|d\theta$$
$$\leq \int_{\theta_1}^{\theta_2}\left[M_\theta\delta^2(\xi)M_\theta\int_0^T a_t^2(\xi)dt\right]^{1/2} d\theta < \infty.$$

Hence, by the Fubini theorem,

$$M\delta(W)\int_{\theta_1}^{\theta_2}\left[\int_0^T a_t(W)dW_t - \theta\int_0^T a_t^2(W)dt\right]\varphi(\theta,W)d\theta$$
$$= \int_{\theta_1}^{\theta_2} M_\theta\delta(\xi)\left[\int_0^T a_t(\xi)d\xi_t - \theta\int_0^T a_t^2(\xi)dt\right]d\theta$$
$$= \int_{\theta_1}^{\theta_2} M_\theta\left[\delta(\xi)\int_0^T a_t(\xi)dW_t\right]d\theta,$$

and, therefore,

$$M_{\theta_2}\delta(\xi) - M_{\theta_1}\delta(\xi) = \int_{\theta_1}^{\theta_2}\left[M_\theta\left(\delta(\xi)\int_0^T a_t(\xi)dW_t\right)\right]d\theta. \tag{17.33}$$

It follows from this that $M_\theta\delta(\xi)$ is an absolutely continuous function. We shall show now that, in (17.33),

$$M_\theta\delta(\xi)\left[\int_0^T a_t(\xi)dW_t\right] = M_\theta\delta(\xi)\left[\int_0^T a_t(\xi)d\xi_t - \theta\int_0^T a_t^2(\xi)dt\right]$$

is continuous in θ.

Let

$$\delta_1(\xi) = \delta(\xi)\int_0^T a_t(\xi)d\xi_t, \quad \delta_2(\xi) = \delta(\xi)\int_0^T a_t^2(\xi)dt.$$

Then

$$M_\theta\delta(\xi)\left[\int_0^T a_t(\xi)dW_t\right] = M_\theta\delta_1(\xi) - \theta M_\theta\delta_2(\xi),$$

and to prove continuity it suffices to establish that

$$\sup_{\theta_1 \le \theta \le \theta_2} M_\theta \delta_i^2(\xi) < \infty, \quad i = 1, 2, \tag{17.34}$$

for any $\theta_1 < \theta_2$. Indeed, when these conditions are satisfied the functions $M_\theta \delta_i(\xi)$, $i = 1, 2$, as has been shown, will be absolutely continuous and, consequently, continuous.

We have

$$M_\theta \delta_1^2(\xi) \le \left\{ M_\theta \delta^4(\xi) M_\theta \left[\int_0^T a_t(\xi) d\xi_t \right]^4 \right\},$$

where, due to the Hölder inequality ($p = 4$, $q = 4/3$),

$$\begin{aligned} & M_\theta \left[\int_0^T a_t(\xi) d\xi_t \right]^4 \\ &= M_\theta \left[\int_0^T a_t(\xi) dW_t + \theta \int_0^T a_t^2(\xi) dt \right]^4 \\ &\le 2^3 \left[M_\theta \left(\int_0^T a_t(\xi) dW_t \right)^4 + \theta^4 M_\theta \left(\int_0^T a_t^2(\xi) dt \right)^4 \right] \\ &\le 2^3 \left[36T \int_0^T M_\theta a_t^4(\xi) dt + \theta^4 T^3 \int_0^T M a_t^8(\xi) dt \right]. \end{aligned} \tag{17.35}$$

(Here the estimate

$$M_\theta \left(\int_0^T a_t(\xi) dW_t \right)^4 \le 36T \int_0^T M_\theta a_t^4(\xi) dt,$$

proved in Lemma 4.12 is used). The required estimate of (17.34) with $i = 1$ follows from (17.35) and (17.30). The estimate of (17.34) with $i = 2$ can be established analogously.

Lemma 17.2. *Let $\delta(x)$ be a $\mathcal{B}_T$-measurable function and let*

$$\sup_{\theta_1 \le \theta \le \theta_2} M_\theta \delta^8(\xi) < \infty \tag{17.36}$$

for any $\theta_1 < \theta_2$. If

$$\sup_{\theta_1 \le \theta \le \theta_2} M_\theta \int_0^T a_t^{16}(\xi) dt < \infty, \tag{17.37}$$

then the function $M_\theta \delta(\xi)$ is twice differentiable over θ and

$$\frac{d^2 M_\theta \delta(\xi)}{d\theta^2} = M_\theta \delta(\xi) \left[\left(\int_0^T a_t(\xi) dW_t \right)^2 - \int_0^T a_t^2(\xi) dt \right]. \tag{17.38}$$

PROOF. Due to (17.31) and the definition of the functions $\delta_1(\xi)$ and $\delta_2(\xi)$ (see the proof of Lemma 17.1),

$$\frac{d}{d\theta} M_\theta \delta(\xi) = M_\theta \delta_1(\xi) - \theta M_\theta \delta_2(\xi). \tag{17.39}$$

Hence to prove the existence of the second derivative $d^2 M_\theta \delta(\xi)/d\theta^2$ it is enough to verify, due to Lemma 17.1, that

$$\sup_{\theta_1 \le \theta \le \theta_2} M_\theta \delta_i^4(\xi) < \infty, \quad i = 1, 2, \tag{17.40}$$

for any $\theta_1 < \theta_2$.

Due to the Cauchy–Schwarz inequality,

$$M_\theta \delta_1^4(\xi) = \left[M_\theta \delta^8(\xi) M_\theta \left(\int_0^T a_t(\xi) d\xi_t \right)^8 \right]^{1/2}.$$

Using the Hölder inequality ($p = 8$, $q = 8/7$) and Lemma 4.12, we find that

$$\begin{aligned} & M_\theta \left(\int_0^T a_t(\xi) d\xi_t \right)^8 \\ &= M_\theta \left[\int_0^T a_t(\xi) dW_t + \theta \int_0^T a_t^2(\xi) dt \right]^8 \\ &\le 2^7 \left[M_\theta \left(\int_0^T a_t(\xi) dW_t \right)^8 + \theta^8 M_\theta \left(\int_0^T a_t^2(\xi) dt \right)^8 \right] \\ &\le 2^7 \left[28^4 T^3 \int_0^T M a_t^8(\xi) dt + \theta^8 T^7 \int_0^T M_\theta a_t^{16}(\xi) dt \right]. \end{aligned} \tag{17.41}$$

Analogously, it can be shown that

$$\begin{aligned} M_\theta \delta_2^4(\xi) &\le \left[M_\theta \delta^8(\xi) M_\theta \left(\int_0^T a_t^2(\xi) dt \right)^8 \right]^{1/2} \\ &\le \left[M_\theta \delta^8(\xi) T^7 \int_0^T M_\theta a_t^{16}(\xi) dt \right]^{1/2} \end{aligned}$$

We obtain the required inequalities given by (17.40) from the above inequalities and the assumptions of the lemma. To complete the proof it remains only to note that (17.38) follows from (17.39) and (17.31). □

17.2.3.
PROOF OF THEOREM 17.2. Due to (17.22) and (17.25)

$$\hat{\theta}_T(\xi) = \theta + \frac{\int_0^T a_t(\xi)dW_t}{\int_0^T a_t^2(\xi)dt}. \tag{17.42}$$

Hence the bias

$$b_T(\theta) = M_\theta[\hat{\theta}_T(\xi) - \theta] = M_\theta \frac{\int_0^T a_t(\xi)dW_t}{\int_0^T a_t^2(\xi)dt}. \tag{17.43}$$

By the assumptions of the theorem and (17.31),

$$M_\theta \frac{\int_0^T a_t(\xi)dW_t}{\int_0^T a_t^2(\xi)dt} = \frac{d}{d\theta} M_\theta \left[\int_0^T a_t^2(\xi)dt\right]^{-1},$$

which, together with (17.43), proves (17.28).

Next we obtain from (17.42)

$$B_T(\theta) = M_\theta[\hat{\theta}_T(\xi) - \theta]^2 = M_\theta \left[\int_0^T a_t(\xi)dW_t\right]^2 \left[\int_0^T a_t^2(\xi)dt\right]^{-2}.$$

But, by Lemma 17.2,

$$\begin{aligned}
&\frac{d^2 M_\theta[\int_0^T a_t^2(\xi)dt]^{-2}}{d\theta^2} \\
&= M_\theta \left[\int_0^T a_t^2(\xi)dt\right]^{-2} \left\{\left(\int_0^T a_t(\xi)dW_t\right)^2 - \int_0^T a_t^2(\xi)dt\right\} \\
&= B_T(\theta) - M\left[\int_0^T a_t^2(\xi)dt\right]^{-1},
\end{aligned}$$

which is equivalent to (17.29). □

Note. A more detailed investigation of the values of $b_T(\theta)$ and $B_T(\theta)$ for the case where $a_t(x) = x_t$ is carried out in the next section.

17.3 Parameter Estimation of the Drift Coefficient for a One-Dimensional Gaussian Process

17.3.1. We shall assume that the observable process $\xi = (\xi_t, \mathcal{F}_t)$, $0 \le t \le T$, has the differential

$$d\xi_t = \theta\xi_t dt + dW_t, \quad \xi_0 = 0 \tag{17.44}$$

(compare with (17.22)), where θ is the unknown parameter, $-\infty < \theta < \infty$.

According to (17.25), the maximum likelihood estimate

$$\hat{\theta}_T(\xi) = \frac{\int_0^T \xi_t d\xi_t}{\int_0^T \xi_t^2 dt} = \frac{\xi_T^2 - T}{2\int_0^T \xi_t^2 dt}, \tag{17.45}$$

since, due to the Itô formula, $\int_0^T \xi_t d\xi_t = \frac{1}{2}[\xi_T^2 - T]$.

Let us calculate the bias $b_T(\theta) = M_\theta(\hat{\theta}_T(\xi) - \theta)$ and the mean square error $B_T(\theta) = M_\theta[\hat{\theta}_T(\xi) - \theta]^2$.

We introduce the auxiliary function

$$\rho_T(\theta, a) = \left[\frac{2\sqrt{\theta^2 + 2a}}{(\sqrt{\theta^2+2a}+\theta)e^{-\sqrt{\theta^2+2a}T} + (\sqrt{\theta^2+2a}-\theta)e^{\sqrt{\theta^2+2a}T}}\right]^{1/2}. \tag{17.46}$$

Theorem 17.3. *The bias $b_T(\theta)$ and the mean square error $B_T(\theta)$ are given by the formulae*

$$b_T(\theta) = \int_0^\infty \frac{\partial}{\partial\theta}\left\{\exp\left(-\frac{\theta T}{2}\right)\rho_T(\theta, a)\right\} da \tag{17.47}$$

$$\begin{aligned} B_T(\theta) &= \exp\left(-\frac{\theta T}{2}\right)\int_0^T \rho_T(\theta, a)da \\ &+ \int_0^T a\frac{\partial^2}{\partial\theta^2}\left\{\exp\left(-\frac{\theta T}{2}\right)\rho_T(\theta, a)\right\} da. \end{aligned} \tag{17.48}$$

PROOF. In order to find the values of $b_T(\theta)$ and $B_T(\theta)$ we shall take advantage of (17.28) and (17.29), obtained in Theorem 17.2. As a preliminary we shall verify that the assumptions of this theorem are satisfied.

The process $\xi = (\xi_t, \mathcal{F}_t)$, $0 \le t \le T$, with differential given by (17.44) is Gaussian with $M_\theta\xi_t = 0$ and variance $\Gamma_t(\theta) = M_\theta\xi_t^2$, satisfying the equation (see Theorem 15.1)

$$\frac{d\Gamma_t(\theta)}{dt} = 2\theta\Gamma_t(\theta) + 1, \quad \Gamma_0(\theta) = 0.$$

We find from this

$$\Gamma_t(\theta) = \frac{1}{2\theta}(e^{2\theta_t} - 1),$$

which implies (17.26) of Theorem 17.2.

To verify (17.27) and to compute the mathematical expectations $M_\theta[\int_0^T \xi_t^2 dt]^{-1}$ and $M_\theta[\int_0^T \xi_t^2 dt]^{-2}$ used in finding $b_T(\theta)$ and $B_T(\theta)$, we shall proceed as follows.

Let $a > 0$ and

$$\psi_T(\theta, a) = M_\theta \exp\left\{-a\int_0^T \xi_t^2 dt\right\}. \tag{17.49}$$

If we assume

$$\int_0^\infty a^{k-1}\psi_T(\theta, a)da < \infty, \quad -\infty < \theta < \infty, \quad k = 1, 2, \ldots, \tag{17.50}$$

then the moments $M_\theta[\int_0^T \xi_t^2 dt]^{-k}$, $k-1, 2, \ldots$, can be found using the function $\psi_T(\theta, a)$:

$$M_\theta\left[\int_0^T \xi_t^2 dt\right]^{-k} = \frac{1}{(k-1)!}\int_0^\infty a^{k-1}\psi_T(\theta, a)da. \tag{17.51}$$

Actually, if for any integer $k = 1, 2, \ldots$, (17.50) is satisfied, then, by the Fubini theorem,

$$\begin{aligned}
\int_0^\infty a^{k-1}\psi_T(\theta, a)da &= \int_0^\infty a^{k-1} M_\theta \exp\left(-a\int_0^T \xi_t^2 dt\right) da \\
&= M_\theta \int_0^\infty a^{k-1} \exp\left(-a\int_0^T \xi_t^2 dt\right) da \\
&= (k-1)! M_\theta \left(\int_0^T \xi_t^2 dt\right)^{-k}, \quad k = 1, 2, \ldots.
\end{aligned}$$

Therefore, let us find the functions $\psi_T(\theta, a)$ and verify the validity of the inequalities (17.50).

17.3.2.

Lemma 17.3. *The function*

$$\psi_T(\theta, a) = \exp\left(-\frac{\theta T}{2}\right)\rho_T(\theta, a), \tag{17.52}$$

where $\rho_T(\theta, a)$ is defined in (17.46).

PROOF. Let $\lambda = \sqrt{\theta^2 + 2a}$, $\theta \le a < \infty$. Denote by μ_{ξ^θ} and μ_{ξ^λ} the measures on $(C_T, \mathcal{B}_T)$ corresponding to the processes ξ^θ and ξ^λ having the differentials

$$\begin{aligned}
d\xi_t^\theta &= \theta\xi_t^\theta dt + dW_t, \quad \xi_0^\theta = 0, \\
d\xi_t^\lambda &= \lambda\xi_t^\lambda dt + dW_t, \quad \xi_0^\lambda = 0.
\end{aligned}$$

According to Theorem 7.19, the measures μ_{ξ^θ} and μ_{ξ^λ} are equivalent and

$$\frac{d\mu_{\xi^\theta}}{d\mu_{\xi^\lambda}}(\xi^\lambda) = \exp\left\{(\theta-\lambda)\int_0^T \xi_t^\lambda d\xi_t^\lambda - \frac{\theta^2-\lambda^2}{2}\int_0^T (\xi_t^\lambda)^2 dt\right\}.$$

Hence,

$$\begin{aligned}\psi_T(\theta,a) &= M_\theta \exp\left\{-a\int_0^T \xi_t^2 dt\right\} = M\exp\left\{-a\int_0^T (\xi_t^\theta)^2 dt\right\} \\ &= M\exp\left\{-a\int_0^T (\xi_t^\lambda)^2 dt + (\theta-\lambda)\int_0^T \xi_t^\lambda d\xi_t^\lambda \right. \\ &\quad \left. -\frac{\theta^2-\lambda^2}{2}\int_0^T (\xi_t^\lambda)^2 dt\right\}. \end{aligned} \tag{17.53}$$

Using

$$a + \frac{\theta^2-\lambda^2}{2} = 0, \tag{17.54}$$

we obtain, finally,

$$\begin{aligned}\psi_T(\theta,a) &= M\exp\left\{[\theta-\lambda]\int_0^T \xi_t^\lambda d\xi_t^\lambda\right\} \\ &= M\exp\left\{\frac{\theta-\lambda}{2}[(\xi_T^\lambda)^2 - T]\right\} \\ &= \exp\left(\frac{\lambda-\theta}{2}T\right) M\exp\left\{\frac{\theta-\lambda}{2}(\xi_T^\lambda)^2\right\}.\end{aligned}$$

The random variable ξ_T^λ is Gaussian, $N(0, 1/2\lambda(e^{2\lambda T}-1))$ and, therefore (Lemma 11.6),

$$M\exp\left\{\frac{\theta-\lambda}{2}(\xi_T^\lambda)^2\right\} = \left[\frac{2\lambda}{((\lambda-\theta)(e^{2\lambda T}-1)+2\lambda}\right]^{1/2}.$$

This, together with (17.53), leads to the following representation:

$$\psi_T(\theta,a) = e^{(\lambda-\theta/2)T}\left[\frac{2\lambda}{(\lambda-\theta)(e^{2\lambda T}-1)+2\lambda}\right]^{1/2}, \tag{17.55}$$

where, according to (17.54), $\lambda = \sqrt{2a+\theta^2}$. After simple transformations we obtain the desired representation, (17.52), from (17.55). □

Note. If $\theta = 0$, $a = \frac{1}{2}$, then

$$\begin{aligned}\psi_T\left(0,\frac{1}{2}\right) &= M\exp\left\{-\frac{1}{2}\int_0^T W_t^2 dt\right\} = \rho_T\left(0,\frac{1}{2}\right) \\ &= \sqrt{\frac{2}{e^T+e^{-T}}} = \frac{1}{\sqrt{\cosh T}}\end{aligned}$$

(compare with the example from Section 7.7).

COMPLETION OF THE PROOF OF THEOREM 17.3. By analyzing (17.52), we find that the inequalities given by (17.50) are satisfied for any $k = 1, 2, \dots$. Hence, (17.47) and (17.48) follow from (17.28), (17.29), (17.51) and (17.52). □

17.3.3.

Theorem 17.4. *The maximum likelihood estimate $\hat{\theta}_T(\xi)$ is strongly consistent, i.e., for each θ, $-\infty < \theta < \infty$,*

$$P_\theta\left\{\lim_{T\to\infty} \hat{\theta}_T(\xi) = \theta\right\} = 1. \tag{17.56}$$

PROOF. We obtain from (17.49)

$$M_\theta \exp\left\{-\int_0^T \xi_t^2 dt\right\} = \psi_T(\theta, 1),$$

where

$$\begin{aligned}\psi_T(\theta,1) &= \exp\left\{\left(-\frac{\theta}{2} - \frac{\sqrt{2+\theta^2}}{2}\right)T\right\} \\ &\quad \times \left\{\frac{2\sqrt{2+\theta^2}}{(\sqrt{\theta^2+2}-\theta) + (\sqrt{\theta^2+2}+\theta)\exp(-2T\sqrt{2+\theta^2})}\right\}^{1/2}.\end{aligned}$$

Since $\lim_{T\to\infty} \psi_T(\theta, 1) = 0$, $-\infty < \theta < \infty$, then

$$P_\theta\left(\int_0^\infty \xi_t^2 dt = \infty\right) = 1. \tag{17.57}$$

It is seen that

$$\hat{\theta}_T(\xi) = \theta + \frac{\int_0^T \xi_t dW_t}{\int_0^T \xi_t^2 dt}. \tag{17.58}$$

Hence, in order to prove (17.56) it suffices to show that

$$P_\theta\left(\lim_{T\to\infty} \frac{\int_0^T \xi_t dW_t}{\int_0^T \xi_t^2 dt} = 0\right) = 1, \quad -\infty < \theta < \infty.$$

This follows from the following general statement.

Lemma 17.4. *Let the Wiener process $W = (W_t, \mathcal{F}_t)$, $t \geq 0$, be given on a probability space and let there also be given the random process $f = (f_t, \mathcal{F}_t)$, $t \geq 0$, such that:*

(1)

$$P\left(\int_0^T f_t^2 dt < \infty\right) = 1, \quad 0 < T < \infty;$$

(2)

$$P\left(\int_0^\infty f_t^2 dt = \infty\right) = 1.$$

Then the random process $z = (z_s, \mathcal{G}_s)$, $s \geq 0$, *with* $z_s = \int_0^{\tau_s} f_t dW_t$, $\mathcal{G}_s = \mathcal{F}_{\tau_s}$, *where* $\tau_s = \inf(t : \int_0^t f_u^2 du > s)$, *is a Wiener process and with probability one*[4]

$$\lim_{t\to\infty} \frac{\int_0^t f_u dW_u}{\int_0^t f_u^2 du} = 0. \tag{17.59}$$

PROOF. Let x_t denote $\int_0^t f_u dW_u$. By the Itô formula we obtain the following representation for $z_t = x_{\tau_t}$:

$$e^{i\lambda z_t} = e^{i\lambda z_s} + i\lambda \int_{\tau_s}^{\tau_t} e^{i\lambda x_u} f_u dW_u - \frac{\lambda^2}{2}\int_{\tau_s}^{\tau_t} e^{i\lambda x_u} f_u^2 du.$$

Then by a change of variables in the Lebesgue integral

$$\int_{\tau_s}^{\tau_t} e^{i\lambda x_u} f_u^2 du = \int_s^t e^{i\lambda x_{\tau_u}} du = \int_s^t e^{i\lambda z_u} du. \tag{17.60}$$

From the above and the equation

$$M\left(\int_{\tau_s}^{\tau_t} e^{i\lambda(x_u - x_s)} f_u dW_u \middle| \mathcal{G}_s\right) = 0 \quad (P\text{-a.s.})$$

we obtain the equation for $V_t = M(e^{i\lambda(z_t - z_s)}|\mathcal{G}_s)$, $t > s$:

$$V_t = 1 - \frac{\lambda^2}{2}\int_s^t V_u du,$$

i.e., $M(e^{i\lambda(z_t - z_s)}|\mathcal{G}_s) = e^{-\lambda^2(t-s)/2}$ (P-a.s.).

Thus $(z_t, \mathcal{G}_t)$, $t \geq 0$, is a Gaussian martingale with $M[(z_t - z_s)^2|\mathcal{G}_s] = t - s$ which has right continuous trajectories having limits to the left.

By virtue of Theorem 1.10 and the equation $M(z_t - z_s)^4 = 3(t-s)^2$ which follows from the normality of the variable $z_t - z_s$ with $M(z_t - z_s) = 0$ and $M(z_t - z_s)^2 = t - s$, the process $(z_t, \mathcal{G}_t)$, $t \leq 0$ has continuous trajectories (more precisely continuous modification).

Consequently, $(z_t, \mathcal{G}_t)$, $t \geq 0$, is a Wiener process (see Theorem 4.1).

Next let us prove (17.59). Let

$$\eta_t = \frac{\int_0^t f_u dW_u}{\int_0^t f_u^2 du}$$

---

[4] Under $\int_0^t f_u dW_u$ a continuous modification of the stochastic integral is understood which exists for any $f \in P_t$ according to (4.47).

and define $\tau_s = \inf\{t : \int_0^t f_u^2 du > s\}$. Since τ_s, $s \geq 0$, is a right continuous nondecreasing function of s (Lemma 5.6), to prove (17.59) it suffices to establish that with probability one $\eta_{\tau_s} \to 0$, $s \to \infty$. But for $s > 0$,

$$\eta_{\tau_s} = \frac{\int_0^{\tau_s} f_u dW_u}{\int_0^{\tau_s} f_u^2 du} = \frac{z_s}{s},$$

and the law of the iterated logarithm (1.35) implies that with probability one $\lim_{s\to\infty} z_s/s = 0$.

Lemma 17.4 and, therefore, Theorem 17.4, also, have been proved. □

17.4 Two-Dimensional Gaussian Markov Processes: Parameter Estimation

17.4.1. Suppose that on the interval $0 \leq t \leq T$ we observe the two-dimensional Gaussian Markov stationary process $\xi_t = (\xi_1(t), \xi_2(t))$ with zero mean $M\xi_1(t) = M\xi_2(t) = 0$, $-\infty < t < \infty$, and differential

$$d\xi_t = A\xi_t dt + dW_t. \tag{17.61}$$

Here $W_t = (W_1(t), W_2(t))$ is a Wiener process with independent components independent of ξ_0, and

$$A = \begin{pmatrix} -\theta_1 & -\theta_2 \\ \theta_2 & -\theta_2 \end{pmatrix} \tag{17.62}$$

is a matrix composed of the coordinates of the vector $\theta = (\theta_1, \theta_2)$ with $\theta_1 > 0$ and $-\infty < \theta_2 < \infty$ where θ_2 is to be estimated from the observations $\xi_0^T = \{\xi_s, 0 \leq s \leq T\}$.

We shall construct the maximum likelihood estimates $\hat{\theta}_1(T,\xi)$ and $\hat{\theta}_2(T,\xi)$ of the unknown parameters θ_1 and θ_2

Theorem 17.5.

(1) The maximum likelihood estimate $\hat{\theta}_1(T,\xi)$ is the solution of the equation

$$\frac{1}{\hat{\theta}_1(T,\xi)} - 2\hat{\theta}_1(T,\xi)\left[\xi_1^2(0) + \xi_2^2(0) + \frac{1}{2}\int_0^T [\xi_1^2(t) + \xi_2^2(t)]dt\right]$$
$$= \int_0^T [\xi_1(t)d\xi_1(t) + \xi_2(t)d\xi_2(t)]. \tag{17.63}$$

(2) The estimate

$$\hat{\theta}_2(T,\xi) = \frac{\int_0^T [\xi_1(t)d\xi_2(t) - \xi_2(t)d\xi_1(t)]}{\int_0^T [\xi_1^2(t) + \xi_2^2(t)]dt}. \tag{17.64}$$

(3) The conditional distribution[5]

$$P_\theta(\hat{\theta}_2(T,\xi) \le a | \xi_1^2(t) + \xi_2^2(t), t \le T)$$

is (P_θ-a.s.) Gaussian with the parameters

$$M_\theta[\hat{\theta}_2(T,\xi) | \xi_1^2(t) + \xi_2^2(t), t \le T] = \theta_2, \tag{17.65}$$

$$M_\theta[(\hat{\theta}_2(t,\xi) - \theta)^2 | \xi_1^2(t) + \xi_2^2(t), t \le T] = \left[\int_0^T (\xi_1^2(t) + \xi_2^2(t))dt\right]^{-1}. \tag{17.66}$$

In particular, the random variable distribution

$$[\hat{\theta}_2(T,\xi) - \theta]\sqrt{\int_0^T [\xi_1^2(t) + \xi_2^2(t)]dt}$$

does not depend upon $\theta = (\theta_1, \theta_2)$ and is Gaussian, $N(0,1)$.

17.4.2. Before proving this theorem we shall make two auxiliary statements.

Lemma 17.5. *For each t, $0 \le t \le T$, the Gaussian vector $(\xi_1(t), \xi_2(t))$ has independent components with $D\xi_i(y) \equiv 1/2\theta_1$, $i = 1,2$.*

PROOF. We shall note first of all that the assumption of the stationarity of the process ξ_t, $-\infty < t < \infty$, implies $\theta_1 > 0$, since the eigenvalues of the matrix A must lie within the left-half plane.

Let $\Gamma \equiv M\xi_t\xi_t^*$. Then by Theorem 15.1, the matrix

$$\Gamma = \begin{pmatrix} \Gamma_{11} & \Gamma_{12} \\ \Gamma_{12} & \Gamma_{22} \end{pmatrix}$$

is the unique solution of the equation $A\Gamma + \Gamma A^* + E = 0$, i.e.,

$$-2\theta_1\Gamma_{11} - 2\theta_2\Gamma_{12} + 1 = 0,$$

$$-2\theta_1\Gamma_{12} + \theta_2(\Gamma_{11} - \Gamma_{22}) = 0,$$

$$2\theta_1\Gamma_{12} - 2\theta_1\Gamma_{22} + 1 = 0.$$

From this we find $\Gamma_{11} = \Gamma_{22} = 1/2\theta_1$, $\Gamma_{12} = 0$. □

Corollary. *The distribution function*

$$F_\theta(x_1, x_2) = P_\theta(\xi_1(0) \le x_1, \xi_2(0) \le x_2)$$

[5] P_θ denotes the probability distribution corresponding to a fixed $\theta = (\theta_1, \theta_2)$.

has the density

$$f_\theta(x_1,x_2) = \frac{\partial^2 F_\theta(x_1,x_2)}{\partial x_1 \partial x_2} = \frac{\theta_1}{\pi}\exp\{-\theta_1^2(x_1^2+x_2^2)\}. \tag{17.67}$$

To formulate the following statement we shall introduce some notation.

Let $(C_T^2, \mathcal{B}_T^2)$ $(C_T \times C_T, \mathcal{B}_T \times \mathcal{B}_t)$ be the measurable space of the functions $c = \{(c_1(t), c_2(t)), 0 \le t \le T\}$ where each function $c_i(t)$, $i = 1, 2$, is continuous. We shall denote the functions in C_T^2 with $c_1(0) = x_1$, $c_2(0) = x_2$ by c^x where $x = (x_1, x_2)$. Let μ_ξ^θ be the measure on $(C_T^2, \mathcal{B}_T^2)$ corresponding to the process $\xi = (\xi_t)$, $0 \le r \le T$ with the prescribed $\theta = (\theta_1, \theta_2)$, and let μ_{W^x} and $\mu_{\xi^x}^\theta$ be the measures on $(C_T^2, \mathcal{B}_T^2)$ corresponding to the process $W_t^x = x + W_t$ (i.e., $W_i^x(t) = \xi + W_i(t)$, $i = 1, 2$) and the process ξ^x with the differential

$$d\xi_t^x = A\xi_t^x dt + dW_t, \quad \xi_0^x = x. \tag{17.68}$$

If the set $B \in \mathcal{B}_T^2$, then

$$\mu_\xi^\theta(\Gamma) = \int_{\{x\in\mathbb{R}^2 : c^x \in B\}} \mu_{\xi^x}^\theta(B) f_\theta(x_1,x_2)dx_1dx_2. \tag{17.69}$$

Indeed, the solutions of Equations (17.61) and (17.68) are given by the formulae

$$\begin{aligned}\xi_t &= e^{At}\left[\xi_0 + \int_0^t e^{-As}dW_s\right],\\ \xi_t^x &= e^{At}\left[x + \int_0^t e^{-As}dW_s\right],\end{aligned}$$

respectively.

Hence, from the independence of the random variables ξ_0 and $\int_0^T e^{-As}dW_s$, it follows that

$$P_\theta\{\xi \in B | \xi_0 = x\} = P_\theta\{\xi^x \in B\}\mu_{\xi^x}^\theta(B),$$

which obviously proves (17.69).

Introduce a new measure[6] ν on $(C_T^2, \mathcal{B}_T^2)$ by defining for $B \in \mathcal{B}_T^2$

$$\nu(\Gamma) = \int_{\{x\in\mathbb{R}^2 : c^x \in B\}} \mu_{W^x}(B)dx_1dx_2. \tag{17.70}$$

For brevity, instead of (17.70) we shall write $d\nu(x, y^x) = d\mu_{W^x}(y^x)dx_1dx_2$, $y^x \in C_T^2$.

By Theorem 7.19, the measures $\mu_{\xi^x}^\theta$ and μ_{W^x} are equivalent and

[6] Note that the measure ν introduced is nonnegative and σ-finite.

$$\frac{d\mu^\theta_{\xi^x}}{d\mu_{W^x}}(W^x) = \exp\left[\int_0^T (W^x_t)^* A^* dW^x_t - \frac{1}{2}\int_0^T (W^x_t)^* A^* A W^x_t dt\right]. \quad (17.71)$$

Hence, by the Fubini theorem we obtain from (17.69) and (17.70)

$$\mu^\theta_\xi(\Gamma) = \int_\Gamma \frac{d\mu^\theta_{\xi^x}}{d\mu_{W^x}}(y^x) f_\theta(x_1,x_2) d\nu(x,y^x),$$

where $f_\theta(x_1,x_2)$ is defined by (17.67). From this follows the absolute continuity of the measure μ^θ_ξ with respect to ν and the density:

$$\frac{d\mu^\theta_\xi}{d\nu}(\xi) = \frac{\theta_1}{\pi}\exp\left\{-\theta_1^2(\xi_1^2(0)+\xi_2^2(0)) + \int_0^T \xi_t^* A^* d\xi_t - \frac{1}{2}\int_0^T \xi_t^* A^* A \xi_t dt\right\}. \quad (17.72)$$

Thus we have proved the following lemma.

Lemma 17.6. *The measure μ^θ_ξ is absolutely continuous with respect to the measure ν and its density $d\mu^\theta_\xi(\xi)/d\nu$ is defined by (17.72).*

17.4.3.

PROOF OF THEOREM 17.5. Formulae (17.63) and (17.64) for the maximum likelihood estimates $\hat\theta_1(T,\xi)$ and $\hat\theta_2(T,\xi)$ follow from (17.72), since they provide the minimum of $\ln(d\mu^\theta_\xi(\xi)/d\nu)$, as can be verified by direct calculation.

Let us go on now to prove the concluding point of the theorem.

Let $\eta_t = \xi_1^2(t)+\xi_2^2(t)$. With the aid of the Itô formula it can be calculated that

$$\begin{aligned} d\eta_t &= 2\xi_1(t)d\xi_1(t) + 2\xi_2(t)d\xi_2(t) + 2dt \\ &= 2\xi_1(t)[-\theta_1\xi_1(t) - \theta_2\xi_2(t)]dt + 2\xi_1(t)dW_1(t) \\ &\quad + 2\xi_2(t)[\theta_2\xi_1(t) - \theta_1\xi_2(t)]dt + 2\xi_2(t)dW_2(t) + 2dt \\ &= -2\theta_1[\xi_1^2(t)+\xi_2^2(t)]dt + 2dt + 2[\xi_1(t)dW_1(t) + \xi_2(t)dW_2(t)] \\ &= 2[1 \quad \theta_1\eta_t]dt + 2\sqrt{\eta_t}\tilde W_1(t), \end{aligned} \quad (17.73)$$

where (on the assumption that $\eta_s > 0$)

$$\tilde W_1(t) = \int_0^t \frac{\xi_1(s)}{\sqrt{\eta_s}} dW_1(s) + \int_0^t \frac{\xi_2(s)}{\sqrt{\eta_s}} dW_2(s). \quad (17.74)$$

It follows from Theorem 4.1 that $(\tilde W_1(t), \mathcal{F}_t)$, $0 \le t \le T$ is a Wiener process. Consequently, for the prescribed $\theta = (\theta_1,\theta_2)$, the aggregation $A = (\Omega, \mathcal{F}, \mathcal{F}_t P, \eta_t, \tilde W_1(t))$ provides a weak solution[7] of the stochastic differential equation

[7] See Definition 8 in Section 4.4.

$$d\eta_t = 2[1 - \theta_1\eta_t]dt + 2\sqrt{\eta_t}d\tilde{W}_1(t). \tag{17.75}$$

Let us show now that for each t, $0 \le t \le T$, η_t is $\mathcal{F}_t^{\eta_0,\tilde{W}_1}$-measurable and $P\{\inf_{t\le T}\eta_t > 0\} = 1$. In other words, the process $\eta_t = \xi_1^2(t) + \xi_2^2(t)$ is the strong solution of Equation (17.75), where the Wiener process $(\tilde{W}_1(t), \mathcal{F}_t)$, $0 \le t \le T$, was defined in (17.74).

For this purpose we shall investigate some properties of the weak solutions of the equation of the type given by (17.75). Let $A = (\Omega, \mathcal{F}, \mathcal{F}_t, P, x_t, z_t)$ be the weak solution of the equation

$$dx_t = 2[1 - ax_t]dt + 2\sqrt{x_t}dz_t, \quad a \ge 0, \tag{17.76}$$

where x_0 is such that $P(x_0 > 0) = 1$, $Mx_0 < \infty$.

We shall prove that $M\sup_{t\le T} x_t < \infty$. Let

$$\sigma_N = \begin{cases} \inf\{t \le T : \sup_{s\le t} x_s \ge N\}, & \\ T, & \text{if } \sup_{s\le T} x_s < N. \end{cases}$$

Then, due to (17.76),

$$x_{t\wedge\sigma_N} = x_0 + 2\int_0^{t\wedge\sigma_N}[1 - ax_s]ds + \int_0^{t\wedge\sigma_N}\sqrt{x_s}dz_s, \tag{17.77}$$

and, since $M\int_0^{t\wedge\sigma_N}\sqrt{x_s}dz_s = 0$, we have:

$$\begin{aligned} Mx_{t\wedge\sigma_N} &= Mx_0 + 2M\int_0^{t\wedge\sigma_N}[1 - ax_s]ds \\ &\le Mx_0 + 2M\int_0^{t\wedge\sigma_N}[1 + ax_{s\wedge\sigma_N}]ds \\ &\le Mx_0 + 2M\int_0^{t}[1 + ax_{s\wedge\sigma_N}]ds \\ &\le Mx_0 + 2T + 2a\int_0^t Mx_{s\wedge\sigma_N}ds. \end{aligned}$$

It follows from this, by Lemma 4.13, that

$$Mx_{t\wedge\sigma_N} \le (Mx_0 + 2T)e^{2aT},$$

and, therefore (Fatou lemma),

$$Mx_t \le (Mx_0 + 2T)e^{2aT}. \tag{17.78}$$

Next,

$$\sup_{t\le T} x_{t\wedge\sigma_N} \le x_0 + 2\int_0^T[1 + ax_s]ds + 2\sup_{t\le T}\left|\int_0^{t\wedge\sigma_N}\sqrt{x_s}dz_s\right|$$

and

$$M \sup_{t\le T} x_{t\wedge\sigma_N} \le Mx_0 + 2\int_0^T [1 + aMx_s]ds + 2M\sup_{t\le T}\left|\int_0^{t\wedge\sigma_N}\sqrt{x_s}dW_s\right|.$$

By the Cauchy–Schwarz inequality and (4.54)

$$\begin{aligned} M\sup_{t\le T}\left|\int_0^{t\wedge\sigma_N}\sqrt{x_s}dz_s\right| &\le \left[M\sup_{t\le T}\left|\int_0^{t\wedge\sigma_N}\sqrt{x_s}dz_s\right|^2\right]^{1/2} \\ &\le 2\left(M\int_0^{t\wedge\sigma_N} x_s ds\right)^{1/2} \le 2\left(M\int_0^T x_s ds\right)^{1/2}. \end{aligned}$$

Hence,

$$M\sup_{t\le T} x_{t\wedge\sigma_N} \le Mx_0 + 2\int_0^T [1 + aMx_s]ds + 4\left[\int_0^T Mx_s ds\right]^{1/2}.$$

Applying the Fatou lemma and using (17.78) we obtain the desired inequality, $M\sup_{t\le T} x_t < \infty$.

We shall show now that $P\{\inf_{t\le T} x_T > 0\} = 1$.

To prove this we set

$$\tau_n = \begin{cases} \inf\{t \le T : \inf_{s\le t} x_s \le x_0/(1+n)\}, & \\ \infty, & \text{if } \inf_{s\le t} x_s > x_0/(1+n). \end{cases}$$

It is easy to show from the Itô formula that

$$-\ln x_{\tau_n\wedge T} = -\ln x_0 + 2a(\tau_n\wedge T) - 2\int_0^{\tau_n\wedge T}\frac{dz_s}{\sqrt{x_s}}.$$

Hence, for $\varepsilon > 0$,

$$\begin{aligned} -\chi_{\{x_0>\varepsilon\}}\ln x_{\tau_n\wedge T} &= -\chi_{\{x_0>\varepsilon\}}\ln x_0 + \chi_{\{x_0>\varepsilon\}}2a(\tau_n\wedge T) \\ &\quad -2\int_0^{\tau_n\wedge T}\chi_{\{x_0>\varepsilon\}}\frac{dz_s}{\sqrt{x_s}} \\ &\le -\chi_{\{x_0>\varepsilon\}}\ln x_0 + 2aT - 2\int_0^{\tau_n\wedge T}\chi_{\{x_0>\varepsilon\}}\frac{dz_s}{\sqrt{x_s}}. \end{aligned} \tag{17.79}$$

Since $M\int_0^{\tau_n\wedge T}\chi_{\{x_0>\varepsilon\}}dz_s/\sqrt{x_s} = 0$, it follows that

$$-M\chi_{\{x_0>\varepsilon\}}\ln x_{\tau_n\wedge T} \le M|\chi_{\{x_0>\varepsilon\}}\ln x_0| + 2aT. \tag{17.80}$$

But

$$\begin{aligned} \chi_{\{x_0>\varepsilon\}}\ln x_{\tau_n\wedge T} &= \chi_{\{x_0>\varepsilon\}}\chi_{\{x_{\tau_n\wedge T}\le 1\}}\ln x_{\tau_n\wedge T} \\ &\quad +\chi_{\{x_0>\varepsilon\}}\chi_{\{x_{\tau_n\wedge T}>1\}}\ln x_{\tau_n\wedge T} \\ &\le \chi_{\{x_0>\varepsilon\}}\chi_{\{x_{\tau_n\wedge T}\le 1\}}\ln x_{\tau_n\wedge T} + \sup_{s\le T} x_s, \end{aligned}$$

which, together with (17.79), leads to the inequality

$$\begin{aligned} & M\chi_{\{x_0>\varepsilon\}}\chi_{\{x_{\tau_n\wedge T}\leq 1\}}|\ln x_{\tau_n^T}| \\ \leq\ & M|\chi_{\{x_0>\varepsilon\}}\ln x_0| + 2aT + M\sup_{s\leq T} x_s \quad (= c(\varepsilon) < \infty); \end{aligned}$$

from this follows the inequality

$$M\chi_{\{x_0>\varepsilon\}}\chi_{\{\tau_n\leq T\}}\chi_{\{x_{\tau_n}\leq 1\}}|\ln x_{\tau_n}| \leq c(\varepsilon) < \infty. \tag{17.81}$$

Let $\tau = \lim_{n\to\infty}\tau_n$. Then taking the limit in (17.81) as $n\to\infty$ we obtain

$$M\chi_{\{x_0>\varepsilon\}}\chi_{\{\tau\leq T\}}\chi_{\{x_\tau\leq 1\}}|\ln x_\tau| \leq c(\varepsilon) < \infty. \tag{17.82}$$

On the set $\{\tau\leq T\}$, $|\ln x_\tau| = \infty$.

Hence, due to (17.82),

$$P\{x_0 > \varepsilon, \tau\leq T, x_\tau\leq 1\} = 0.$$

But $x_\tau = 0$ on the set $\{\tau\leq T\}$; therefore,

$$P\{x_0 > \varepsilon, \tau\leq T\} = 0. \tag{17.83}$$

Finally,

$$\begin{aligned} P\{\tau\leq T\} &= P\{\tau\leq T, x_0>\varepsilon\} + P\{\tau\leq T, x_0\leq\varepsilon\} \\ &\leq P\{x_0\leq\varepsilon\}\to 0, \quad \varepsilon\downarrow 0, \end{aligned}$$

which, together with (17.83), leads to the desired relation

$$P\left\{\inf_{t\leq T} x_t = 0\right\} = P\{\tau\leq T\} = 0.$$

Therefore, the process $\eta_t = \xi_1^2(t) + \xi_2^2(t)$, $0\leq t\leq T$, is such that for any $\theta = (\theta_1,\theta_2)$, $\theta_1 > 0$, $-\infty < \theta_2 < \infty$,

$$P_\theta\left\{\inf_{t\leq T}\eta_t > 0\right\} = 1. \tag{17.84}$$

Let us use this result to prove the fact that for each t, $0\leq t\leq T$, the random variables η_t are $\mathcal{F}_t^{\eta_0,\bar{W}_1}$-measurable.

Introduce the functions

$$g_n(y) = \begin{cases} 1/(2\sqrt{y}), & 1/n\leq y<\infty, \\ 1/(2\sqrt{n}), & 0\leq y\leq 1/n, \end{cases}$$

and

$$b_n(x) = 1 + \int_1^x g_n(y)dy.$$

It is seen that $0 < g_n(y) \leq 1/2\sqrt{n}$ and $\lim_{n\to\infty} b_n(x) = \sqrt{x}$. For each $n = 1, 2, \dots$, we shall consider the equation

$$\eta_t^{(n)} = \eta_0 + 2\int_0^t [1 - \theta_1\eta_s^{(n)}]ds + 2\int_0^t b_n(\eta_s^{(n)})d\tilde{W}_1(s). \tag{17.85}$$

The coefficients of this equation satisfy the assumptions of Theorem 4.6, and, hence, this equation has the unique strong solution $\eta_t^{(n)}$, $0 \leq t \leq T$. Let

$$\sigma_n(\eta) = \begin{cases} \inf\{t \leq T : \eta_t \leq 1/n\}, & \\ T, & \text{if } \inf_{s\leq T}\eta_s > 1/n. \end{cases}$$

Then it is obvious that for each $t \leq \sigma_n(\eta)$, $\eta_t^{(n)} = \eta_t$ (P_θ-a.s.) and $\sigma_n(\eta) = \sigma_n(\eta^{(n)})$. Consequently, $\eta^{(n)}_{t\wedge\sigma_n(\eta^{(n)})} = \eta_{t\wedge\sigma_n(\eta)}$. But the variables $\eta^{(n)}_{t\wedge\sigma_n(\eta^{(n)})}$ are $\mathcal{F}_t^{\eta_0,\tilde{W}_1}$-measurable. Hence so also[8] are the variables $\eta_{t\wedge\sigma_n(\eta)}$. But, due to (17.84), $\lim_{n\to\infty}\sigma_n(\eta) = T$ (P_θ-a.s.). It follows from this that η_t are $\mathcal{F}_t^{\eta_0,\tilde{W}_1}$-measurable for each t.

By transforming expression (17.64) for $\hat{\theta}_2(T,\xi)$ we find that

$$\hat{\theta}_2(T,\xi) = \theta_2 + \frac{\int_0^T \xi_1(t)dW_2(t) - \int_0^T \xi_2(t)dW_1(t)}{\int_0^T [\xi_1^2(t) + \xi_2^2(t)]dt} = \theta_2 + \frac{\int_0^T \sqrt{\eta_t}d\tilde{W}_2(t)}{\int_0^T \eta_t dt}, \tag{17.86}$$

where

$$\tilde{W}_2(t) = -\int_0^T \frac{\xi_2(t)}{\sqrt{\eta_t}}dW_1(t) + \int_0^T \frac{\xi_1(t)}{\sqrt{\eta_t}}dW_2(t). \tag{17.87}$$

It follows from Theorem 4.2 that $[(\tilde{W}_1(T), \tilde{W}_2(t)), \mathcal{F}_t]$, $0 \leq t \leq T$, is a Wiener process. Now, $\eta_0 = \xi_1^2(0) + \xi_2^2(0) > 0$ (P-a.s.) and $M_\theta\eta_0 = 1/\theta_1 < \infty$ for all $\theta = (\theta_1, \theta_2)$ with $\theta_1 > 0$, $-\infty < \theta_2 < \infty$; hence, η_t is $\mathcal{F}_t^{\eta_0,\tilde{W}_1}$-measurable for each t. But the process $\tilde{W}_2(t)$ does not depend on η_0 and $\tilde{W}_1(t)$. Hence the processes $\eta = (\eta_t, \mathcal{F}_t)$ and $\tilde{W}_2 = (\tilde{W}_2(t), \mathcal{F}_t)$ are mutually independent. It follows from this that (P-a.s.) the conditional distribution

$$P_\theta\left\{\int_0^T \sqrt{\eta_t}d\tilde{W}_2(t) \leq y \,\middle|\, \eta_t, t \leq T\right\}$$

is Gaussian, $N(0, \int_0^T \eta_t dt)$. In particular, this proves (17.65) and (17.66), and, therefore, Theorem 17.5. □

Note. Since, for any admissible $\theta = (\theta_1, \theta_2)$,

$$P_\theta\left(\int_0^\infty [\xi_1^2(t) + \xi_2^2(t)]dt = \infty\right) = 1,$$

[8] The σ-algebras $\mathcal{F}_t^{\eta_0,\tilde{W}_1}$, $0 \leq t \leq T$ are assumed to be augmented by sets of P_θ measure zero for all admissible values $\theta = (\theta_1, \theta_2)$.

it is easy to deduce from (17.63) and (17.64) that the estimates $\hat{\theta}_i(T,\xi)$, $i = 1, 2$, are consistent, i.e., for any $\varepsilon > 0$,

$$\lim_{T\to\infty} P_\theta\{|\tilde{\theta}_i(T,\xi) - \theta_i| > \varepsilon\} = 0.$$

17.5 Sequential Maximum Likelihood Estimates

17.5.1. As in Section 17.2, let θ be the unknown parameter, $-\infty < \theta < \infty$, to be estimated from the observations of the process $\xi = (\xi_t, \mathcal{F}_t)$, $t \geq 0$, with the differential

$$d\xi_t = \theta a_t(\xi)dt + dW_t, \quad \xi_0 = 0. \tag{17.88}$$

Under the assumptions given by (17.23), the maximum likelihood estimate $\tilde{\theta}_T(\xi)$ of the parameter θ is given by (17.25). Generally speaking, this estimate is biased and its bias $b_T(\theta)$ and the mean square error $B_T(\theta)$ are defined (under the assumptions given by (17.26) and (17.27)) by (17.28) and (17.29), respectively. According to the Cramer–Rao–Wolfowitz theorem (Theorem 7.22):

$$B_T(\theta) \geq \frac{\left\{1 + \frac{d^2}{d\theta^2} M_\theta \left[\int_0^T a_t^2(\xi)dt\right]^{-1}\right\}^2}{M_\theta \int_0^T a_t^2(\xi)dt} + \left\{\frac{d}{d\theta} M_\theta \left[\int_0^T a_t^2(\xi)dt\right]^{-1}\right\}^2, \tag{17.89}$$

where equality need not, generally speaking, be attained.

For this problem we shall study properties of sequential maximum likelihood estimates obtained with the aid of the sequential schemes $\Delta = \Delta(\tau, \delta)$ (see Section 7.8), each of which is characterized by the final time of observation $\tau = \tau(\xi)$, the $\mathcal{F}_\tau^\xi$-measurable function $\delta(\xi)$ being the estimate of the parameter θ.

Theorem 17.6. *For all θ, $-\infty < \theta < \infty$, let*

$$P_\theta\left\{\int_0^\infty a_t^2(\xi)dt = \infty\right\} = 1. \tag{17.90}$$

Then the sequential scheme $\Delta_H = \Delta(\tau_H, \delta_H)$, $0 < H < \infty$, with

$$\tau_H(\xi) = \inf\left(t : \int_0^T a_s^2(\xi)ds = H\right) \tag{17.91}$$

and

$$\delta_H(\xi) = \frac{1}{H}\int_0^{\tau_H(\xi)} a_t(\xi)d\xi_t \tag{17.92}$$

has the following properties:

$$P_\theta(\tau_H(\xi) < \infty) = 1, \quad -\infty < \theta < \infty, \tag{17.93}$$

$$M_\theta \delta_H(\xi) = \theta, \quad -\infty < \theta < \infty, \tag{17.94}$$

$$M_\theta[\delta_H(\xi) - \theta]^2 \equiv \frac{1}{H}. \tag{17.95}$$

The random variable $\delta_H(\xi)$ is Gaussian, $N(\theta, 1/H)$. In the class Δ_H of unbiased sequential schemes $\Delta(\tau, \delta)$ satisfying the condition

$$P_\theta\left\{\int_0^\tau a_t^2(\xi)dt < \infty\right\} = P_\theta\left\{\int_0^\tau a_t^2(W)dt < \infty\right\} = 1 \tag{17.96}$$

and the conditions

$$M_\theta \delta^2(\xi) < \infty, \quad M_\theta \int_0^\tau a_t^2(\xi)dt \le H, \tag{17.97}$$

where H is a given constant, $0 < H < \infty$, the scheme $\Delta_H = \Delta(\tau_H, \delta_H)$ is optimal in the mean square sense:

$$M_\theta[\delta_H(\xi) - \theta]^2 \le M_\theta[\delta(\xi) - \theta]^2. \tag{17.98}$$

PROOF. According to Theorem 7.10 and (17.96), the measures $\mu^\theta_{\tau,\xi}$ and $\mu^\theta_{\tau,W}$ corresponding to the processes ξ (for fixed θ) and W are equivalent and

$$\frac{d\mu^\theta_{\tau,\xi}}{d\mu^\theta_{\tau,W}}(\tau(\xi), \xi) = \exp\left\{\theta \int_0^{\tau(\xi)} a_t(\xi)d\xi_t - \frac{\theta^2}{2}\int_0^{\tau(\xi)} a_t^2(\xi)dt\right\}. \tag{17.99}$$

This implies that the sequential maximum likelihood estimate is given by

$$\hat\theta_{\tau(\xi)}(\xi) = \frac{\int_0^{\tau(\xi)} a_t(\xi)d\xi_t}{\int_0^{\tau(\xi)} a_t^2(\xi)dt}. \tag{17.100}$$

Setting $\tau(\xi) = \tau_H(\xi)$ in (17.100) and writing $\delta_H(\xi) = \hat\theta_{\tau_H(\xi)}$, we obtain for the estimate $\delta_H(\xi)$ the representation given by (17.92). To verify (17.93) it is enough to note that

$$P_\theta\{\tau_H(\xi) > t\} = P\left\{\int_0^t a_2^2(\xi)ds < H\right\},$$

from which, due to (17.90), it follows that

$$P_\theta\{\tau_H(\xi) = \infty\} = P_\theta\left\{\int_0^\infty a_t^2(\xi)dt < H\right\} = 0.$$

Next,

$$\delta_H(\xi) = \theta - \frac{1}{H}\int_0^{\tau_H(\xi)} a_t(\xi)dW_t$$

and, by Lemma 17.4, the value of $[\delta_H(\xi) - \theta]\sqrt{H}$ is Gaussian distributed, $N(0,1)$ for each θ.

Finally, according to Theorem 7.22, for any unbiased scheme $\Delta = \Delta(\tau, \delta)$ satisfying (17.96) and (17.97) it follows that

$$M_\theta[\delta(\xi) - \theta]^2 \geq \left[M_\theta \int_0^{\tau(\xi)} a_t^2(\xi)dt\right]^{-1} \geq \frac{1}{H}, \quad -\infty < \theta < \infty.$$

The comparison of this inequality with (17.95) indicates that the scheme $\Delta_H = \Delta(\tau_H, \delta_H)$ is optimal in the mean square sense. □

17.5.2. (17.95) reveals the meaning of the constant $H > 0$ in the definition of the schemes $\Delta_H = \Delta(\tau_H, \delta_H)$: if it is required to construct the sequential scheme for which the error variance (for all θ, $-\infty < \theta < \infty$) is equal to a given value $\varepsilon > 0$, then the scheme $\Delta_H = \Delta(\tau_H, \delta_H)$ with $H = 1/\varepsilon$ can be taken as the desired scheme.

According to the statements of Theorem 17.6, this scheme has some definite advantages: it is unbiased, and the fact that the distribution of the value $(\delta_H(\xi) - 0)\sqrt{H}$ is Gaussian, $N(0,1)$, makes it possible to construct confidence intervals for θ.

An essential question, however, arises: are these advantages simply due to the fact that the average observation time $M_\theta \tau_H$ is too long? In the theorem given below for the case[9] $a_t(x) = x_t$, the estimates of this average time are given as functions of the prescribed error variance

Theorem 17.7. *Let the observable process ξ_t, $t \geq 0$, have the differential*

$$d\xi_t = \theta\xi_t dt + dW_t. \tag{17.101}$$

Then for the sequential scheme $\Delta_H = \Delta(\tau_H, \delta_H)$, $H > 0$, with all $n = 1, 2, \ldots$,

$$M_\theta \tau_H^n(\xi) < \infty, \quad -\infty < \theta < \infty, \tag{17.102}$$

and

$$M_\theta \tau_H(\xi) \leq 2[|\theta|H + 2\sqrt{H}] + \sqrt{8(\theta^2 H^2 + 4H) + 2H}, \quad -\infty > \theta < \infty. \tag{17.103}$$

In the case $\theta < 0$ the following lower estimate holds for $M_\theta \tau_H(\xi)$:

$$M_\theta \tau_H(\xi) \geq -2\theta H. \tag{17.104}$$

[9] It follows from Theorem 17.4 that $P_\theta\{\int_0^\infty \xi_t dt = \infty\} = 1$, $|\theta| < \infty$.

PROOF. First of all we note that in our case the estimate

$$\delta_H(\xi) = \frac{1}{H}\int_0^{\tau_H(\xi)} \xi_t d\xi_t$$

can be written as follows:

$$\delta_H(\xi) = \frac{\xi^2_{\tau_H(\xi)} - \tau_H(\xi)}{2H},$$

since $\int_0^t \xi_s d\xi_s = \frac{1}{2}(\xi_t^2 - t)$.

In order to prove (17.102), we note that, by the Itô formula,

$$\xi_t^2 = 2\theta\int_0^t \xi_s^2 ds + 2\int_0^t \xi_s dW_s + t. \tag{17.105}$$

We obtain from this

$$H = \int_0^{\tau_H(\xi)} \xi_t^2 dt = 2\theta\int_0^{\tau_H(\xi)}\left(\int_0^t \xi_s^2 ds\right)dt + 2\int_0^{\tau_H(\xi)}\left(\int_0^t \xi_s dW_s\right)dt + \frac{\tau_h^2(\xi)}{2},$$

and, consequently,

$$\begin{aligned}\tau_H^2 &\le 2H - 4\theta\int_0^{\tau_H(\xi)}\left(\int_0^T \xi_s^2 ds\right)dt - 4\int_0^{\tau_H(\xi)}\left(\int_0^T \xi_s dW_s\right)dt \\ &\le 2H + 4|\theta|\tau_H(\xi) + 4\tau_H(\xi)\sup_{0\le t\le \tau_H(\xi)}\left|\int_0^t \xi_s dW_s\right|. \end{aligned}\tag{17.106}$$

Let

$$\beta = \sup_{0\le t\le \tau_H(\xi)}\left|\int_0^t \xi_s dW_s\right|.$$

Then we obtain from (17.106)

$$\tau_H^2(\xi) - 4\tau_H(\xi)[|\theta|H + \beta] - 2H \le 0,$$

and, therefore, for each θ,

$$\tau_H(\xi) \le 2[|\theta|H + \beta] + \sqrt{4[|\theta|H+\beta]^2 + 2H}. \tag{17.107}$$

By Theorem 3.2, for $p > 1$,

$$M_\theta\beta^p = M_\theta\left(\sup_{0\le t\le\tau_H(\xi)}\left|\int_0^t \xi_s dW_s\right|^p\right) \le \left(\frac{p}{p-1}\right)^p M_\theta\left|\int_0^{\tau_H(\xi)} \xi_s dW_s\right|^p.$$

Hence, for $p = 2m$,

$$\begin{aligned} M_\theta \beta^{2m} &\le \left(\frac{2m}{2m-1}\right)^{2m} M_\theta \left| \int_0^{\tau_H(\xi)} \xi_s dW_s \right|^{2m} \\ &= \left(\frac{2m}{2m-1}\right)^{2m} (2m-1)!!H^m < \infty, \end{aligned} \tag{17.108}$$

since the random variable $\int_0^{\tau_H(\xi)} \xi_s dW_s \sim N(0, H)$.

From (17.107) and (17.108) we obtain the inequality $M_\theta[\tau_H(\xi)]^n < \infty$, $-\infty < \theta < \infty$, $n = 1, 2, \ldots$. In particular, for the case $n = 1$,

$$\begin{aligned} M_\theta \tau_H(\xi) &\le 2[|\theta|H + (M_\theta \beta^2)^{1/2}] + \sqrt{8(\theta^2 H^2 + M_\theta \beta^2) + 2H} \\ &\le 2[|\theta|H + 2\sqrt{H}] + \sqrt{8(\theta^2 H^2 + 4H) + 2H}. \end{aligned}$$

To deduce (17.104) it suffices to note that in the case $\theta < 0$, the inequality

$$\tau_H(\xi) \ge -2\theta H - \int_0^{\tau_H(\xi)} \xi_s dW_s$$

follows from (17.105). Averaging both sides of this inequality, we obtain (17.104). □

17.6 Sequential Testing of Two Simple Hypotheses for Itô Processes

17.6.1. On the probability space $(\Omega, \mathcal{F}, P)$, let there be given a nondecreasing family of σ-algebras $\mathcal{F}_t$, $t \ge 0$, $\mathcal{F}_t \subseteq \mathcal{F}$, the Wiener process $W = (W_t, \mathcal{F}_T)$ and the unobservable process $\theta = (\theta_t, \mathcal{F}_t)$, $t \ge 0$, independent of W. Assume further that one of the hypotheses holds on the observable process $\xi = (\xi_t, \mathcal{F}_t)$, $t \ge 0$:

$$H_0 : d\xi_t = dW_t, \quad \xi_0 = 0; \tag{17.109}$$

$$H_1 : d\xi_t = \theta_t dt + dW_t, \quad \xi_0 = 0. \tag{17.110}$$

In other words, if the process θ is interpreted as a signal and the Wiener process as noise, then the problem being considered involves testing two hypotheses on the presence (hypothesis H_1) or the absence (hypothesis H_0) of the signal θ in the observations of the process ξ.

We shall discuss the sequential scheme $\Delta = \Delta(\tau, \delta)$ of hypothesis testing characterized by the time τ at the end of the observation and a function of the final decision δ. It is supposed that $\tau = \tau(x)$ is a Markov time (with respect to the system $\mathcal{B}_t = \sigma\{x : x_s, s \le t\}$, where the $x = (x_t)$, $t \ge 0$ are continuous functions with $x_0 = 0$) and the function $\delta = \delta(x)$ is $\mathcal{B}_\tau$-measurable and takes only two values: 0 and 1. The decision $\delta(x) = 0$ will be identified with a decision in favor of hypothesis H_0. If $\delta(x) = 1$, then hypothesis H_1 will be accepted.

For each scheme $\Delta = \Delta(\tau, \Delta)$, denote[10]

$$\alpha(\Delta) = P_1(\delta(\xi) = 0), \quad \beta(\Delta) = P_0\{\delta(\xi) = 1\},$$

called error probabilities of the first and second kind.

It is well known[11] that for the case $\theta_t \equiv c \neq 0$ in the class $\Delta_{\alpha,\beta}$ of sequential schemes $\Delta = \Delta(\tau, \delta)$, with $\alpha(\Delta) \leq \alpha$, $\beta(\Delta) \leq \beta$ (α and β are given constants, $\alpha + \beta < 1$) where $M_0\tau(\xi) < \infty$ and $M_1\tau(\xi) < \infty$, there exists a scheme $\tilde{\Delta} = \Delta(\tilde{\tau}, \tilde{\delta})$, optimal in the sense that

$$M_0\tilde{\tau} \leq M_0\tau, \quad M_1\tilde{\tau} \leq M_1\tau, \tag{17.111}$$

for any other scheme $\Delta = \Delta(\tau, \delta) \in \Delta_{\alpha,\beta}$.

It appears that in a certain sense this result can be extended to a more general class of random processes $\theta = (\theta_t, \mathcal{F}_t)$, $t \geq 0$.

We shall assume that the process $\theta = (\theta_t, \mathcal{F}_t)$, $t \geq 0$, satisfies the condition

$$M|\theta_t| < \infty, \quad t < \infty \tag{17.112}$$

and that

$$P_1\left\{\int_0^\infty m_t^2(\xi)dt = \infty\right\} = P_0\left\{\int_0^\infty m_t^2(\xi)dt = \infty\right\} = 1, \tag{17.113}$$

where the functional $m_t(x)$, $t \geq 0$, is such that, for almost all $t \geq 0$,

$$m_t(\xi) = M_1(\theta_t|\mathcal{F}_t^\xi) \quad (P\text{-a.s.}).$$

We shall denote by $\Delta_{\alpha,\beta}$, the class of sequential schemes $\Delta = \Delta(\tau, \delta)$ with $\alpha(\Delta) \leq \alpha$ and $\beta(\Delta) \leq \beta$, where $\alpha + \beta < 1$ and

$$M_0 \int_0^{\tau(\xi)} m_t^2(\xi)dt < \infty, \quad M_1 \int_0^{\tau(\xi)} m_t^2 dt < \infty. \tag{17.114}$$

Theorem 17.8. *Let (17.112) and (17.113) be satisfied. Then in the class $\Delta_{\alpha,\beta}$ there exists a scheme $\tilde{\Delta} = \Delta(\tilde{\tau}, \tilde{\delta})$, optimal in the sense that for any other scheme $\Delta = \Delta(\tau, \delta) \in \Delta_{\alpha,\beta}$,*

$$M_0 \int_0^{\tilde{\tau}(\xi)} m_t^2(\xi)dt \leq M_0 \int_0^{\tau(\xi)} m_t^2(\xi)dt,$$
$$M_1 \int_0^{\tilde{\tau}(\xi)} m_t^2(\xi)dt \leq M_1 \int_0^{\tau(\xi)} m_t^2(\xi)dt. \tag{17.115}$$

The scheme $\tilde{\Delta} = \delta(\tilde{\tau}, \tilde{\delta})$ can be defined by the relations

[10] P_i denotes the probability distribution for the case where the process ξ being considered satisfies hypothesis H_i, $i = 0, 1$. M_i will denote the corresponding average.

[11] See, for example, Chapter 4, Section 2 in [282].

$$\tilde{\tau}(\xi) = \inf\{t : \lambda_t(\xi) \notin (A, B)\}, \tag{17.116}$$

$$\tilde{\delta}(\xi) = \begin{cases} 1, & \lambda_{\tilde{\tau}(\xi)} \geq B, \\ 0, & \lambda_{\tilde{\tau}(\xi)} \leq A, \end{cases} \tag{17.117}$$

where

$$\lambda_t(\xi) = \int_0^t m_s(\xi)d\xi_s - \frac{1}{2}\int_0^t m_s^2(\xi)ds, \quad A = \ln\frac{\alpha}{1-\beta}, \quad B = \ln\frac{1-\alpha}{\beta}.$$

In this case

$$M_0 \int_0^{\tilde{\tau}(\xi)} m_t^2(\xi)dt = 2\omega(\beta, \alpha),$$

$$M_1 \int_0^{\tilde{\tau}(\xi)} m_t^2(\xi)dt = 2\omega(\alpha, \beta), \tag{17.118}$$

where

$$\omega(x, y) = (1-x)\ln\frac{1-x}{y} + x\ln\frac{x}{1-y}. \tag{17.119}$$

Before proving this we shall make a few auxiliary observations.

17.6.2.
Lemma 17.7. *For the scheme* $\tilde{\Delta} = \Delta(\tilde{\tau}, \tilde{\delta})$,

$$P_0(\tilde{\tau}(\xi) < \infty) = P_1(\tilde{\tau}(\xi) < \infty) = 1.$$

PROOF. In the case of hypothesis H_0, $\xi_t = W_t$ and

$$P_0(\tilde{\tau}(\xi) < \infty) = P(\tilde{\tau}(W) < \infty).$$

Let

$$\sigma_n(W) = \inf\left\{t : \int_0^t m_s^2(W)ds \geq n\right\}.$$

Then

$$\lambda_{\tilde{\tau}(W)\wedge\sigma_n(W)}(W) = \int_0^{\tilde{\tau}(W)\wedge\sigma_n(W)} m_t(W)dW_t - \frac{1}{2}\int_0^{\tilde{\tau}(W)\wedge\sigma_n(W)} m_t^2(w)dt$$

and $A \leq \lambda_{\tilde{\tau}(W)\wedge\sigma_n(W)}(W) \leq B$. Consequently,

$$A \leq \int_0^{\tilde{\tau}(W)\wedge\sigma_n(W)} m_t(W)dW_t - \frac{1}{2}M\int_0^{\tilde{\tau}(W)\wedge\sigma_n(W)} m_s^2(W)ds \leq B.$$

Hence,

$$M\int_0^{\tilde{\tau}(W)\wedge\sigma_n(W)} m_s^2(W)ds \leq 2(B - A) < \infty, \tag{17.120}$$

since $0 < \alpha + \beta < 1$ and, therefore.

$$B - A = \ln\left[\frac{1-\alpha}{\alpha}, \frac{1-\beta}{\beta}\right] < \infty.$$

We obtain from (17.120) and (17.113)

$$M \int_0^{\tilde{\tau}(W)} m_s^2(W)ds \leq 2(B - A) < \infty.$$

Since

$$M \int_0^{\tilde{\tau}(W)} m_s^2(W)ds \geq M\chi_{\{\tilde{\tau}(W)=\infty\}} \int_0^{\infty} m_s^2(W)ds,$$

by (17.113) it follows that $P(\tilde{\tau}(W) < \infty) = 1$.

The equality $P_1(\tilde{\tau}(\xi) < \infty) = 1$ can be proved in a similar way. It should be noted that, according to Theorem 7.12, the process ξ_t, $t \geq 0$, with differential given by (17.110), permits also the differential

$$d\xi_t = m_t(\xi)dt + d\overline{W}_t \tag{17.121}$$

for some Wiener process $\overline{W} = (\overline{W}_t, \mathcal{F}_t^{\xi})$, $t \geq 0$. Therefore, in the case of hypothesis H_1,

$$\lambda_t(\xi) = \int_0^t m_s(\xi)d\overline{W}_s + \frac{1}{2}\int_0^t m_s^2(\xi)ds. \tag{17.122}$$

Corollary. *The random variable $\lambda_{\tilde{\tau}(\xi)}(\xi)$ takes (P_0- and P_1-a.s.) only two values: A or B.*

Lemma 17.8. *For the scheme $\tilde{\Delta} = \Delta(\tilde{\tau}, \tilde{\delta})$ defined by (17.116) and (17.117), $\alpha(\tilde{\Delta}) = \alpha$, $\beta(\tilde{\Delta}) = \beta$.*

PROOF. Since

$$\alpha(\tilde{\Delta}) = P_1\{\tilde{\delta}(\xi) = 0\} = P_1\{\lambda_{\tilde{\tau}(\xi)}(\xi) = A\}$$

and

$$\beta(\tilde{\Delta}) = P_0(\tilde{\delta}(\xi) = 1\} = P_0\{\lambda_{\tilde{\tau}(\xi)}(\xi) = B\},$$

then, in order to prove the lemma, it is necessary to establish that

$$P_1\{\lambda_{\tilde{\tau}(\xi)}(\xi) = A\} = \alpha, \quad P_0 = \{\lambda_{\tilde{\tau}(\xi)}(\xi) = B\} = \beta, \tag{17.123}$$

where

$$A = \ln\frac{\alpha}{1-\beta}, \quad B = \ln\frac{1-\alpha}{\beta}. \tag{17.124}$$

For this purpose consider two solutions $a(x), b(x)$, $A \leq x \leq B$ of the differential equations

$$a''(x) + a'(x) = 0, \quad a(A) = 1, \quad a(B) = 0, \tag{17.125}$$

$$b''(x) + b'(x) = 0, \quad b(A) = 0, \quad b(B) = 1. \tag{17.126}$$

It is seen that

$$a(x) = \frac{e^A(e^{B-x} - 1)}{e^B - e^A}, \quad b(x) = \frac{e^x - e^A}{e^B - e^A}, \tag{17.127}$$

and, due to (17.124),

$$a(0) = \alpha, \quad b(0) = \beta. \tag{17.128}$$

We shall show that $P_1\{\lambda_{\tilde{\tau}(\xi)}(\xi) = A\} = \alpha$. For this purpose let $\sigma_n(\xi) = \inf\{t : \int_0^t m_s^2(\xi)ds \geq n\}$. Then, taking into account (17.122) and (17.125) and applying the Itô formula to $a(\lambda_t(\xi))$, we find

$$\begin{aligned} a(\lambda_{\tilde{\tau}(\xi)\wedge\sigma_n(\xi)}(\xi)) &= a(0) + \int_0^{\tilde{\tau}(\xi)\wedge\sigma_n(\xi)} a'(\lambda_t(\xi))m_t(\xi)d\overline{W}_t \\ &\quad + \frac{1}{2}\int_0^{\tilde{\tau}(\xi)\wedge\sigma_n(\xi)} [a'(\lambda_t(\xi)) + a''(\lambda_t(\xi))]m_t^2(\xi)dt \\ &= \alpha + \int_0^{\tilde{\tau}(\xi)\wedge\sigma_n(\xi)} a'(\lambda_t(\xi))m_t(\xi)d\overline{W}_t. \end{aligned} \tag{17.129}$$

But

$$\begin{aligned} & M_1\int_0^{\tilde{\tau}(\xi)\wedge\sigma_n(\xi)} [a'(\lambda_t(\xi))m_t(\xi)]^2dt \\ \leq & \sup_{A\leq x\leq B} [a'(x)]^2 M_1 \int_0^{\tilde{\tau}(\xi)\wedge\sigma_n(\xi)} m_t^2(\xi)dt \\ \leq & \; n \sup_{A\leq x\leq B} [a'(x)]^2 < \infty. \end{aligned}$$

Hence.

$$M_1\int_0^{\tilde{\tau}(\xi)\wedge\sigma_n(\xi)} a'(\lambda_t(\xi))m_t(\xi)d\overline{W}_t = 0$$

and, therefore, taking the mathematical expectation $M_1(\cdot)$ in (17.129) we obtain

$$M_1 a(\lambda_{\tilde{\tau}(\xi)\wedge\sigma_n(\xi)}(\xi)) = \alpha.$$

The function $a(x)$, $A \leq x \leq B$, is bounded and $\lim_{n\to\infty}\sigma_n(\xi) = \infty$ (P-a.s.). Hence, by the dominated convergence theorem (Theorem 1.4), $Ma(\lambda_{\tilde{\tau}(\xi)}(\xi)) = \alpha$. Using Lemma 17.7 and its corollary, we find that

$$\begin{aligned} \alpha &= M_1 a(\lambda_{\tilde{\tau}(\xi)}(\xi)) \\ &= 1\cdot P_1\{\lambda_{\tilde{\tau}(\xi)}(\xi) = A\} + 0\cdot P\{\lambda_{\tilde{\tau}(\xi)}(\xi) = B\} = P_1\{\lambda_{\tilde{\tau}(\xi)}(\xi) = A\}. \end{aligned}$$

The formula $P_0\{\lambda_{\tilde{\tau}(\xi)}(\xi) = B\} = \beta$ is proved in a similar manner. □

Lemma 17.9. *For the scheme $\tilde{\Delta} = \Delta(\tilde{\tau}, \tilde{\delta})$, the formulae given by (17.118) hold true.*

PROOF. Denote by $g_0(x)$ and $g_1(x)$, $A \le x \le B$, the solutions of the differential equations

$$g_i''(x) + (-1)^{1+i} \cdot g_i'(x) = -2, \quad g_i(A) = g_i(B) = 0, \quad i = 0, 1. \tag{17.130}$$

An easy calculation yields:

$$g_0(x) = 2\left\{\frac{(e^B - e^{A+B-x})(B-A)}{e^B - e^A} + A - x\right\} \tag{17.131}$$

and

$$g_1(x) = 2\left\{\frac{(e^B - e^x)(B-A)}{e^B - e^A} - B + x\right\}. \tag{17.132}$$

Taking into account (17.124) and (17.119), we find

$$-g_0(0) = 2\omega(\beta, \alpha), \tag{17.133}$$

$$g_1(0) = 2\omega(\alpha, \beta). \tag{17.134}$$

Suppose the hypothesis H_0 is valid and $\sigma_n(W) = \inf\{t : \int_0^t m_s^2(W)ds \ge n\}$, $n = 1, 2, \ldots$. Then, applying the Itô formula to $g_0(\lambda_t(W))$, we obtain

$$\begin{aligned} g_0(\lambda_{\tilde{\tau}(w)\wedge\sigma_N(W)}(W)) &= g_0(0) + \int_0^{\tilde{\tau}(W)\wedge\sigma_n(W)} g'(\lambda_t(W))m_t(W)dW_t \\ &\quad -\frac{1}{2}\int_0^{\tilde{\tau}(W)\wedge\sigma_n(W)} [g'(\lambda_t(W)) - g''(\lambda_t(W))]m_t^2(W)dt \\ &= g_0(0) + \int_0^{\tilde{\tau}(W)\wedge\sigma_n(W)} g'(\lambda_t(W))m_t(W)dW_t \\ &\quad + \int_0^{\tilde{\tau}(W)\wedge\sigma_n(W)} m_t^2(W)dt. \end{aligned} \tag{17.135}$$

Since

$$M\int_0^{\tilde{\tau}(W)\wedge\sigma_n(W)} g'(\lambda_t(W))m_t(W)dW_t = 0,$$

then, by averaging both sides of (17.135), we arrive at the equality

$$M\int_0^{\tilde{\tau}(W)\wedge\sigma_n(W)} m_t^2(W)dt = -g_0(0) + Mg_0(\lambda_{\tilde{\tau}(W)\wedge\sigma_n(W)}(W)). \tag{17.136}$$

Passing in (17.136) to the limit we obtain the desired equality,

$$M\int_0^{\tilde{\tau}(W)} m_t^2(W)dt = -g_0(0) = 2\omega(\beta,\alpha).$$

The equality

$$M_1\int_0^{\tilde{\tau}(\xi)} m_t^2(\xi)dt = g_1(0) = 2\omega(\alpha,\beta)$$

can be proved in similar fashion. □

17.6.3.

PROOF OF THEOREM 17.8. Let $\Delta = \Delta(\tau,\delta)$ be any scheme belonging to the class $\Delta_{\alpha,\beta}$. Denote by $\mu_{\tau,\xi}$ and $\mu_{\tau,W}$ the restriction of the measures μ_ξ and μ_W, corresponding to the process ξ with differential given by (17.110) and the Wiener process W, to the σ-algebra $\mathcal{B}_t$. Then, due to the conditions of (17.112)–(17.114) and (17.121), we find from Theorem 7.10 that $\mu_{\tau,\xi}\sim\mu_{\tau,W}$,

$$\ln\frac{d\mu_{\tau,\xi}}{d\mu_{\tau.W}}(\tau,W) = \int_0^{\tau(W)} m_s(W)dW_s - \frac{1}{2}\int_0^{\tau(W)} m_s^2(W)ds, \tag{17.137}$$

and

$$\ln\frac{d\mu_{\tau,W}}{d\mu_{\tau,\xi}}(\tau,\xi) = -\int_0^{\tau\xi)} m_s(\xi)d\xi_s + \frac{1}{2}\int_0^{\tau(\xi)} m_s^2(\xi)ds. \tag{17.138}$$

There follows from this that

$$M_0\ln\frac{d\mu_{\tau,W}}{d\mu_{\tau,\xi}}(\tau,\xi) = \frac{1}{2}M_0\int_0^{\tau(\xi)} m_s^2(\xi)ds = \frac{1}{2}M\int_0^{\tau(W)} m_s^2(W)ds; \tag{17.139}$$

$$M_1\ln\frac{d\mu_{\tau,\xi}}{d\mu_{\tau,W}}(\tau,\xi) = \frac{1}{2}M_1\int_0^{\tau(\xi)} m_s^2(\xi)ds. \tag{17.140}$$

Making use of the Jensen inequality, we obtain

$$\begin{aligned}
\frac{1}{2}M_1\int_0^{\tau(\xi)} m_t^2(\xi)dt &= M_1\ln\frac{d\mu_{\tau,\xi}}{d\mu_{\tau,W}}(\tau,\xi) = -M_1\ln\frac{d\mu_{\tau,W}}{d\mu_{\tau,\xi}}(\tau,\xi)\\
&= -M_1\left\{M_1\left[\ln\frac{d\mu_{\tau,W}}{d\mu_{\tau,\xi}}(\tau,\xi)\middle|\,\delta(\xi)\right]\right\}\\
&\ge -M_1\left\{\ln M_1\left[\frac{d\mu_{\tau,W}}{d\mu_{\tau,\xi}}(\tau,\xi)\middle|\,\delta(\xi)\right]\right\}\\
&= -P_1\{\delta(\xi)=1\}\ln M_1\left[\frac{d\mu_{\tau,W}}{d\mu_{\tau,\xi}}(\tau,\xi)\middle|\,\delta(\xi)=1\right]\\
&\quad -P_1\{\delta(\xi)=0\}\ln M_1\left[\frac{d\mu_{\tau,W}}{d\mu_{\tau,\xi}}(\tau,\xi)\middle|\,\delta(\xi)=0\right]\\
&= -P_1\{\delta(\xi)=1\}\ln\frac{P_1\{\delta(\xi)=1\}M_1\left[\frac{d\mu_{\tau,W}}{d\mu_{\tau,\xi}}(\tau,\xi)|\delta(\xi)=1\right]}{P_1\{\delta(\xi)=1\}}\\
&\quad -P_1\{\delta(\xi)=0\}\ln\frac{P_1\{\delta(\xi)=0\}M_1\left[\frac{d\mu_{\tau,w}}{d\mu_{\tau,\xi}}(\tau,\xi)|\delta(\xi)=0\right]}{P_1\{\delta(\xi)=0\}}.
\end{aligned} \tag{17.141}$$

Note now that, because of the equivalence $\mu_{\tau,\xi} \sim \mu_{\tau,W}$, for $i = 0, 1$,

$$\begin{aligned} P_0\{\delta(\xi) = i\} &= P\{\delta(W) = i\} \\ &= M_1 \chi_{\{\delta(\xi)=i\}} \frac{d\mu_{\tau,W}}{d\mu_{\tau,\xi}}(\tau,\xi) \\ &= M_1 \left\{ \chi_{\{\delta(\xi)=i\}} M_1 \left[\frac{d\mu_{\tau,W}}{d\mu_{\tau,\xi}}(\tau,\xi) \middle| \delta(\xi) = i \right] \right\} \\ &= P_1\{\delta(\xi) = i\} M_1 \left[\frac{d\mu_{\tau,W}}{d\mu_{\tau,\xi}}(\tau,\xi) \middle| \delta(\xi) = i \right]. \end{aligned}$$

This implies that (17.141) can be transformed as follows:

$$\begin{aligned} \frac{1}{2} M_1 \int_0^{\tau(\xi)} m_t^2(\xi) dt &\geq -P_1\{\delta(\xi) = 1\} \ln \frac{P_0\{\delta(\xi) = 1\}}{P_1\{\delta(\xi) = 1\}} \\ &\quad - P_1\{\delta(\xi) = 0\} \ln \frac{P_0\{\delta(\xi) = 0\}}{P_1\{\delta(\xi) = 0\}} \\ &= P_1\{\delta(\xi) = 1\} \ln \frac{P_1\{\delta(\xi) = 1\}}{P_0\{\delta(\xi) = 1\}} \\ &\quad + P_1\{\delta(\xi) = 0\} \ln \frac{P_1\{\delta(\xi) = 0\}}{P_0\{\delta(\xi) = 0\}} \\ &\geq (1-\alpha) \ln \frac{1-\alpha}{\beta} + \alpha \ln \frac{\alpha}{1-\beta} \\ &= \frac{1}{2} M_1 \int_0^{\tilde{\tau}(\xi)} m_t^2(\xi) dt, \end{aligned}$$

where the last equality follows from Lemma 17.9.

The inequality

$$M_0 \int_0^{\tilde{\tau}(\xi)} m_t^2(\xi) dt \geq M_0 \int_0^{\tilde{\tau}(\xi)} m_t^2(\xi) dt$$

can be proved in similar fashion. □

Corollary. *Let $\theta_t \equiv s(t)$, where $s(t)$, $t \geq 0$, is a deterministic differentiable function, such that $\int_0^\infty s^2(t)dt = \infty$ and $s(t)s'(t) \geq 0$. (It follows from the assumptions that the function $\Phi(t) = \int_0^t s^2(u)du$ is convex downwards, $\Phi(0) = 0$, $\Phi(\infty) = \infty$).*

Let α, β be given numbers, $0 < \alpha + \beta < 1$, and let $\Delta_{\alpha,\beta}$ be the class of sequential schemes considered above. Denote by $\Delta_T = (T, \delta_T)$ the scheme belonging to the class $\Delta_{\alpha,\beta}$ and having fixed duration of observation equal to T, $0 < T < \infty$ (the Neyman–Pearson test is an example of such a scheme). Then the optimal scheme $\tilde{\Delta} = (\tilde{\tau}, \tilde{\delta}) \in \Delta_{\alpha,\beta}$ has $M_0\tilde{\tau} \leq T$, $M_1\tilde{\tau} \leq T$.

Indeed, by the theorem proved above $M_i \int_0^{\tilde{\tau}(\xi)} s^2(t)dt \leq \Phi(t)$, $i = 0, 1$, from which, by the Jensen inequality, $\Phi(T) \geq M_i\Phi(\tilde{\tau}(\xi)) \geq \Phi(M_i\tilde{\tau}(\xi))$, and, therefore, $T \geq M_i\tau(\xi)$, $i = 0, 1$.

17.7 Some Applications to Stochastic Approximation

17.7.1. Let θ be the unknown parameter, $-\infty < \theta < \infty$, to be estimated from the observations of the process $\xi = (\xi_t)$, $t \geq 0$, with the differential

$$d\xi_t = [A_0(t,\xi) + A_1(t,\xi)\theta]dt + B(t,\xi)dW_t, \quad \xi_0 = 0. \tag{17.142}$$

The nonanticipative functionals $A_0(t,x)$, $A_1(t,x)$, $B(t.x)$ prescribed on $[0,\infty) \times C$, where C is the space of continuous functions $x = (x_t)$, $t > 0$, assumed to be such that:

(1) $\int_0^T [A_0^2(t,x) + A_1^2(t,x) + B^2(t,x)]dt < \infty$, $T < \infty$, $x \in C$.
(2) $B^2(t,x) \geq d > 0$, $t < \infty$, $x \in C$;
(3) $\int_0^\infty (A_1^2(t,x)/B^2(t,x))dt = \infty$, $x \in C$;
(4) for $B(t,x)$ (4.110) and (4.111) are satisfied.

If the parameter θ were a Gaussian random variable $N(0,\alpha^2)$, independent of the Wiener process W_t, $t \geq 0$, then, according to (12.34) and (12.35), the conditional mathematical expectation $m_t = M(\theta_t|\mathcal{F}_t^\xi)$ and the conditional variance $\gamma_t = M[(\theta_t - m_t)^2|\mathcal{F}_t^\xi]$ would be given by the formulae

$$m_t = \gamma_t \int_0^t \frac{A_1(s,\xi)}{B^2(s,\xi)}[d\xi_s - A_0(s,\xi)ds], \quad \gamma_t = \left[\frac{1}{\alpha^2} + \int_0^t \frac{A_1^2(s,\xi)}{B^2(s,\xi)}ds\right]^{-1}, \tag{17.143}$$

which follow from the equations

$$dm_t = \frac{\gamma_t A_1(t,\xi)}{B^2(t,\xi)}[d\xi_t - (A_0(t,\xi) + A_1(t,\xi)m_t)dt], \quad m_0 = 0, \tag{17.144}$$

$$\dot{\gamma}_t = -\frac{\gamma_1^2 A_1^2(t,\xi)}{B^2(t,\xi)}, \quad \gamma_0 = \alpha^2. \tag{17.145}$$

(Note that with $\alpha^2 = \infty$ and $\int_0^t (A_1^2(s,x)/B^2(s,x))ds > 0$, $x \in C$, the estimate m_t defined by (17.143) is a maximum likelihood estimate for the parameter θ).

In the case where nothing is known about the probabilistic nature of the parameter θ, it is natural to pose the question as to whether the estimate m_t^2, $t \geq 0$, defined by the equation

$$dm_t^\alpha = A_1(t,\xi)\gamma_t B^{-2}(t,\xi)\{d\xi_t - (A_0(t,\xi) + A_1(t,\xi)m_t^\alpha)dt\}, \tag{17.146}$$

where $0 < \alpha^2 \leq \infty$, converges in a suitable sense to the true value of the parameter θ.

It follows from (17.143) that

$$m_t^\alpha - \theta = \gamma_t \left[-\frac{1}{\alpha^2} + \int_0^t \frac{A_1(s,\xi)}{B(s,\xi)} dW_s \right].$$

Hence, due to (3), above,

$$\overline{\lim}_{t\to\infty} |m_t^\alpha - \theta| \le \overline{\lim}_{t\to\infty} \left| \int_0^t \frac{A_1(s,\xi)}{B(s,\xi)} dW_s \right| \Big/ \int_0^t \frac{A_1^2(s,\xi)}{B^2(s,\xi)} ds. \qquad (17.147)$$

But it follows from Lemma 17.4 that the upper limit in the right-hand side of (17.147) is zero (P_θ-a.s.) for any θ. Consequently, if the true value of the unknown parameter is θ, then (P_θ-a.s.) $m_t^\alpha \to \theta$, $t \to \infty$, where the process m_t^α, $t \ge 0$, can be defined by Equation (17.146); this is a typical example of a stochastic approximation algorithm.

It is interesting to know how 'fast' the process m_t^α, $t \ge 0$, converges to the estimated value of θ. Since $m_t^\alpha \to \theta$ with P_θ-probability one, then for P_θ-almost all ω and for $\varepsilon > 0$ there will be a least time $\tau_\varepsilon(\omega;\alpha)$, such that $|m_t^\alpha - \theta| \le \varepsilon$ for all $t \ge \tau_\varepsilon(\omega,\alpha)$. (Note that the time $\tau = \tau_\varepsilon(\omega;\alpha)$ is not Markov).

We shall investigate the mathematical expectation $M_\theta \tau_\varepsilon(\omega;\alpha)$ of the time $\tau_\varepsilon(\omega;\alpha)$ needed for the estimation of the unknown parameter to within ε, restricting ourselves to the case $A_0 \equiv 0$, $A_1 \equiv 1$ $B \equiv 1$, $\alpha = \infty$.

Therefore, let the observable process ξ_t, $t \ge 0$, have the differential

$$d\xi_t = \theta dt + dW_t. \qquad (17.148)$$

For the sake of simplicity of writing we shall let $m_t = m_t^\infty$, $\tau_\varepsilon(\omega) = \tau_\varepsilon(\omega;\infty)$. In the present case the stochastic approximation algorithm, (17.146), takes the following form

$$dm_t = \frac{1}{t}\{d\xi_t - m_t dt\}. \qquad (17.149)$$

Since this equation has the solution

$$m_t = \frac{\xi_t}{t} = \theta + \frac{W_t}{t},$$

we have:

$$\tau_\varepsilon(\omega) = \inf\left\{ t : \left|\frac{W_s}{s}\right| \le \varepsilon, s \ge t \right\}.$$

Theorem 17.9. *For any* θ, $-\infty < \theta < \infty$,

$$P_\theta\left\{\tau_\varepsilon(\omega) \le \frac{x}{\varepsilon^2}\right\} = P\left\{\sup_{0\le t\le 1} |W_t| < \sqrt{x}\right\}$$

and

$$M_\theta \tau_\varepsilon(\omega) = \frac{C}{\varepsilon^2},$$

where C *is some constant,* $-0 < C < \infty$.

PROOF. Let us take advantage of the fact that each of the processes

$$W_t^* = \begin{cases} tW_{1/t}, & t > 0, \\ 0, & t = 0 \end{cases}$$

$$W^{**}(t) = \sqrt{d} W_{t/d},$$

$(d > 0)$ is a Brownian motion process (see Subsection 1.4.4). Then[12]

$$\begin{aligned}
P_\theta\left\{\tau_\varepsilon(\omega) \le \frac{x}{\varepsilon^2}\right\} &= P\left\{|W_t| \le t\varepsilon, t > \frac{x}{\varepsilon^2}\right\} \\
&= P\left\{|W_t^*| \le t\varepsilon, t > \frac{x}{\varepsilon^2}\right\} \\
&= P\left\{t|W_{1/t}| \le t\varepsilon, t > \frac{x}{\varepsilon^2}\right\} \\
&= P\left\{|W_{1/t}| \le \varepsilon, t > \frac{x}{\varepsilon^2}\right\} \\
&= P\left\{|W_s| \le \varepsilon, 0 < s < \frac{\varepsilon^2}{x}\right\} \\
&= P\{|W_{t-\varepsilon/x^2}| \le \varepsilon, 0 < t < 1\} \\
&= P\left\{\frac{\sqrt{x}}{\varepsilon}|W_{t-\varepsilon^2/x}| \le \frac{\varepsilon\sqrt{x}}{\varepsilon}, 0 < t < 1\right\} \\
&= P\{|W_t| \le \sqrt{x}, 0 \le t \le 1\} \\
&= P\left\{\sup_{0\le t\le 1} |W_t| < \sqrt{x}\right\}.
\end{aligned}$$

It is well known[13] that

$$P\left\{\sup_{0\le t\le 1} |W_t| < \sqrt{x}\right\} = \sum_{k=-\infty}^{\infty} (-1)^k \frac{1}{\sqrt{2\pi}} \int_{-\sqrt{x}}^{\sqrt{x}} e^{-(1/2)(y-2k\sqrt{x})} dy. \quad (17.150)$$

Thus, the series on the right-hand side of (17.150) determines the probability distribution of the random variable $\varepsilon^2\tau_\varepsilon(\omega)$. Since

$$P_\theta\{\varepsilon^2\tau_\varepsilon(\omega) \le x\} = P\left\{\sup_{0\le t\le 1} W_t^2 \le x\right\}$$

and, from (3.8), $M \sup_{0\le t\le 1} W_t^2 \le 4$, it follows that $M_\theta \varepsilon^2 \tau_\varepsilon(\omega) < \infty$ and, consequently, $M_\theta\tau_\varepsilon(\omega) = C/\varepsilon^2$, where the constant

[12] $\{\alpha_t \le a, t > s\}$ denotes the event $\alpha_t \le a$ for all $t > s$.
[13] See, for example, [291], p. 173.

$$C = \int_0^\infty \left[1 - \frac{1}{\sqrt{2\pi}} \sum_{k=-\infty}^{\infty} (-1)^k \int_{-\sqrt{x}}^{\sqrt{x}} e^{-(1/2)(y-2k\sqrt{x})^2} dy \right] dx < \infty. \quad \square$$

Notes and References. 1

17.1. The results of Chapters 7 and 10 have been repeatedly used here.

17.2. The estimates of drift coefficient parameters for diffusion-type processes have been studied by Novikov [246] and Arato [6].

17.3. The results related to this section are due to Novikov [246].

17.4. The parameter estimation of a two-dimensional Gaussian Markov process has been discussed in Arato, Kolmogorov and Sinai [7], Arato [6], Liptser and Shiryaev [205], and Novikov [246].

The maximum likelihood sequential estimates $\delta_H(\xi)$ have been introduced by the authors. The properties of these estimates have been studied by Novikov [246] and the authors. Theorem 17.7 had been proved by Vognik.

17.6. Theorem 17.8 generalizes one of the results obtained by Laidain [183].

17.7. Theorem 17.9 was proved in [289].

Notes and References. 2

17.1–17.5 A parameter estimation for diffusion processes is considered in Kutoyants, Mourid and Bosq [180], Kutoyants [178], Kutoyants and Vostrikova [181]. For the case of a small diffusion parameter see also the book [179]. A parameter estimation and adaptive filtering are given in Yashin and Kuznetsov [324].

17.6. Theorem 17.8 has been generalized by Yashin [322].

18. Random Point Processes: Stieltjes Stochastic Integrals

18.1 Point Processes and their Compensators

18.1.1. In the previous chapters we described observable random processes $X = (\xi_t)$, $t \geq 0$, which possessed continuous trajectories and had properties analogous, to a certain extent, to those of a Wiener process. Chapters 18 and 19 will deal with the case of an observable process that is a point process whose trajectories are pure jump functions (a Poisson process with constant or variable intensity is a typical example).

18.1.2. We shall begin with some basic definitions. We assume that we are given a complete probability space $(\Omega, \mathcal{F}, P)$ with a distinguished family $F = (\mathcal{F}_t)$, $t \geq 0$, of right continuous sub-σ-algebras of $\mathcal{F}$ augmented by sets of zero probability.

Let $T = (\tau_n)$, $n \geq 1$, be a sequence of Markov times (with respect to the system $F = (\mathcal{F}_t)$, $t \geq 0$) such that[1]:

$$\begin{aligned}
&(1) \quad \tau_1 > 0 \quad (P\text{-a.s.});\\
&(2) \quad \tau_n < \tau_{n+1} \quad (\{\tau_n < \infty\} : (P\text{-a.s.}));\\
&(3) \quad \tau_n = \tau_{n+1} \quad (\{\tau_n = \infty\} : (P\text{-a.s.})).
\end{aligned} \tag{18.1}$$

We shall write $\tau_\infty = \lim_{n\to\infty} \tau_n$ for the limit point of the sequence $T = (\tau_n)$, $n \geq 1$.

The random sequence $T = (\tau_n)$, $n \geq 1$, is fully characterized by a counting process

$$N_t = \sum_{n\geq 1} I_{\{\tau_n \leq t\}}, \quad t \geq 0. \tag{18.2}$$

In this connection it is clear that the investigation of the sequence $T = (\tau_n)$, $n \geq 1$, is equivalent to that of the process $N = (N_t)$, $t \geq 0$.

Definition 1. The sequence of Markov times $T = (\tau_n)$, $n \geq 1$, satisfying (18.1) is said to be a *random point process.* The process $N=(N_t)$, $t\geq 0$, defined in (18.2) is said to be a *point process* also (*corresponding to the sequence* $T = (\tau_n)$, $n \geq 1$).

---

[1] The notation '$\xi < \eta(A; (P\text{-a.s.}))$' implies that $P(A \cap \{\xi \geq \eta\}) = 0$.

Note 1. The point processes introduced above represent a particular case of the so-called 'multivariate point processes' to be defined as random sequences $(T, \varphi) = (\tau_n, \xi_n)$, where the τ_n are Markov times satisfying (18.1) and the ξ_n are $\mathcal{F}_{\tau_n}/\mathcal{X}$-measurable random variables with values in some measurable space $(X, \mathcal{X})$.

18.1.3. We shall note some simple properties of point processes $N = (N_t)$, $t \geq 0$. It is seen from the definition that the process N is measurable (with respect to (t, ω)) and $\mathcal{F}_t$-measurable for each $t \geq 0$ (in this connection, we shall use also the notation $N = (N_t, \mathcal{F}_t)$, $t \geq 0$, for this process). Trajectories of these processes are (P-a.s.) right continuous, have limits to the left, and are piecewise constant functions with unit jumps. It is also clear that

$$\begin{aligned} N_{\tau_n} &\leq n \qquad (P\text{-a.s.}), \\ N_{\tau_\infty} &= \lim_n N_{\tau_n} \qquad (P\text{-a.s.}) \end{aligned}$$

(by definition $N_\infty(\omega) = \lim_{t\to\infty} N_t(\omega)$).

EXAMPLE 1. A simple example of a point process is a process $N_t = I\{\tau \leq t\}$, $t \geq 0$, where τ is a Markov time with $P(\tau > 0) = 1$. (In this case $\tau_1 = \tau$, $\tau_n \equiv \infty$, $n \geq 2$).

EXAMPLE 2. The Poisson process $\Pi = (\pi_t)$, $t \geq 0$, with parameter λ, that is, a process with stationary independent increments,

$$\pi_0 = 0,$$

$$P(\pi_t - \pi_s = k) = e^{-\lambda(t-s)}[\lambda(t-s)]^k/k!, \quad s \leq t, \quad k = 0, 1, \dots,$$

is a point process with respect to the family of σ-algebras $\mathcal{F}_t = \mathcal{F}_t^\pi \equiv \sigma\{\omega : \pi_s, s \leq t\}$, $t \geq 0$.

EXAMPLE 3. If $N = (N_t, \mathcal{F}_t)$, $t \geq 0$, is a point process and σ is a Markov time (with respect to $(\mathcal{F}_t)$, $t \geq 0$), the process $(N_{t\wedge\sigma}, \mathcal{F}_t)$, $t \geq 0$, is also a point process.

In this case

$$N_{t\wedge\sigma} = \sum_{n\geq 1} I_{\{\tau_n^\sigma \leq t\}},$$

where

$$\tau_n^\sigma = \begin{cases} \tau_n, & \text{if } \tau_n \leq \sigma, \\ \infty, & \text{if } \tau_n > \sigma. \end{cases}$$

18.1.4. We shall consider, together with the process $N = (N_t, \mathcal{F}_t)$, $t \geq 0$, the point processes $N^{(n)} = (N_{t\wedge\tau_n}, \mathcal{F}_t)$, $t \geq 0$, for each $n \geq 1$. Since $P(0 \leq N_{t\wedge\tau_n} \leq n) = 1$, this process (as well as any bounded and nondecreasing process) is a submartingale of class D (see Section 3.3) and, therefore, a Doob–Meyer decomposition holds for it (see the corollary to Theorem 3.8):

$$N_{t\wedge\tau_n} = m_t^{(n)} + A_t^{(n)},$$

where $(m_t^{(n)}, \mathcal{F}_t)$, $t \geq 0$, is a uniformly integrable martingale, and $(A_t^{(n)}, \mathcal{F}_t)$, $t \geq 0$, is a predictable increasing process.

By virtue of the equality

$$N_{t\wedge\tau_k-} = N_{t\wedge\tau_n\wedge\tau_k}, \quad k \leq n$$

and the uniqueness of the Doob–Meyer decomposition, it follows that

$$m_t^{(k)} = m_{t\wedge\tau_k}^{(n)}, \quad A_t^{(k)} = A_{t\wedge\tau_k}^{(n)}. \tag{18.3}$$

Since $A_t^{(n+1)} \geq A_{t\wedge\tau_n}^{(n+1)}$, $A_t^{(n+1)} \geq A_t^{(n)}$, we have that, for all $t \geq 0$, the process

$$A_t = A_t^{(1)} + \sum_{n\geq 1}[A_t^{(n+1)} - A_t^{(n)}] \tag{18.4}$$

is a right continuous, predictable increasing process and is such that $A_{t\wedge\tau_n} = A_t^{(n)}$ (compare with the proof of Theorem 3.9).

For $\tau < \tau_\infty$ we set

$$m_t = N_t - A_t. \tag{18.5}$$

Then

$$m_{t\wedge\tau_n} = N_{t\wedge\tau_n} - A_{t\wedge\tau_n} = N_t^{(n)} - A_t^{(n)} = m_t^{(n)}$$

and, therefore, for each $n \geq 1$ the family of random variables $\{m_{t\wedge\tau_n}, t < \tau_\infty\}$ forms a uniformly integrable martingale. By generalizing Definition 6 of Section 3.3, we can say that the random process $M = (m_t, \mathcal{F}_t)$ defined for $t < \sigma$ (σ is a Markov time with respect to the system $F = (\mathcal{F}_t)$, $t \geq 0$), is a *σ-local martingale* if there exists an (increasing) sequence of Markov times σ_n, $n \geq 1$, such that $P(\sigma_n < \sigma_{n+1} < \sigma) = 1$, $P(\lim_n \sigma_n = \sigma) = 1$, and, for each n, the sequence $\{m_{t\wedge\sigma_n}, t < \sigma\}$ forms a uniformly integrable martingale.

According to this definition the arguments given above prove the following:

Theorem 18.1. *A point process $N = (N_t, \mathcal{F}_t)$, $t \geq 0$, admits, for all $t < \tau_\infty$, the unique (up to stochastic equivalence) decomposition*

$$N_t = m_t + A_t, \tag{18.6}$$

where $m = (m_t, \mathcal{F}_t)$, $t < \tau_\infty$, *is a* τ_∞*-local martingale, and* $A = (A_t, \mathcal{F}_t)$, $t \geq 0$, *is a predictable increasing process.*

EXAMPLE 4. Let $N = (N_t, \mathcal{F}_t)$, $t \geq 0$, be a deterministic process with $N_t = I_{[1,\infty)}(t)$ and trivial σ-algebras $\mathcal{F}_t = \{\emptyset, \Omega\}$. Then, in the decomposition given by (18.6), $m_t = 0$, $A_t = N_t$.

EXAMPLE 5. Let $\Pi = (\pi_t, \mathcal{F}_t^\pi)$, $t \geq 0$, be a Poisson process with parameter $\lambda > 0$. Then it can be easily verified that the process $(\pi_t - \lambda_t, \mathcal{F}_t^\pi)$ is a martingale. This implies that $m_t = \pi_t - \lambda t$ and $A_t = \lambda t$ in the decomposition given by (18.6).

EXAMPLE 6. Let $\Pi = (\pi_t, \mathcal{F}_t^\pi)$, $t \geq 0$, again be a Poisson process with parameter $\lambda > 0$, Let $\tau_1 = \inf\{t \geq 0 : \pi_t = 1\}$ and let $\tilde{\Pi} = (\tilde{\pi}_t, \mathcal{F}_t^\pi)$ with $\tilde{\pi}_t = \pi_{t \wedge \tau_1}$. Then the decomposition given by (18.6) for the process $\tilde{\Pi}$ has the form

$$\tilde{\pi}_t = [\pi_t - \lambda \cdot (t \wedge \tau_1)] + \lambda \cdot (t \wedge \tau_1).$$

Definition 2. The predictable increasing process $A = (A_t, \mathcal{F}_t)$, $t \geq 1$, appearing in the decomposition (18.6) is called the *compensator* of the point process $N = (N_t, \mathcal{F}_t)$, $t \geq 0$.

It is useful to note that the increasing process $A = (A_t, \mathcal{F}_t)$, $t \geq 0$, is a predictable process if and only if it is predictable in the sense of Definition 3, Section 5.4, in other words, is a process of class Φ_3[2].

The following two definitions will play an essential role from now on.

Definition 3. A Markov time θ (with respect to the family $(\mathcal{F}_t)$, $t \geq 0$) is called *predictable* if a random point process $N_t = I_{\{\theta \leq t\}}$ is predictable.

By virtue of Theorem T52, Chapter VII in [229], the Markov time θ is predictable if and only if there exists an increasing sequence of Markov times (θ_n), $n \geq 1$, such that (P-a.s.) $\theta_n < \theta$ and $\lim_n \theta_n = \theta$.

Definition 4. A Markov time σ (with respect to the family $(\mathcal{F}_t)$, $t \geq 0$) is said to be *totally inaccessible* if $P(\theta = \sigma < \infty) = 0$ for each predictable Markov time θ.

In a specific sense the Markov times introduced above are diametrically opposite: predictable times correspond to predictable events, and totally inaccessible times fully correspond to nonpredictable events.

---

[2] See the corresponding proof in, for example, [49], Theorem T27, Chapter V.

EXAMPLE 7. Let $A = (A_t, \mathcal{F}_t)$, $t \geq 0$, be a compensator of a point process $N = (N_t, \mathcal{F}_t)$. Then, for $a > 0$, the time

$$\theta = \begin{cases} \inf(t \geq 0 : A_t \geq a) & \\ \infty & \text{if } A_\infty < a, \end{cases} \tag{18.7}$$

is a predictable Markov time (Theorem T16, chapter IV, in [49]).

EXAMPLE 8. Let $\Pi = (\pi_t, \mathcal{F}_t^\pi)$, $t \geq 0$, be a Poisson process where $\sigma = \inf(t \geq 0 : \pi_t = 1)$. The Markov time σ is totally inaccessible. Indeed, let (σ_n), $n \geq 1$, be a sequence of Markov times such that $\sigma_n < \sigma$ and $\lim_n \sigma_n = \sigma$ on the set of positive probability. The process $\pi_t^{(n)} = \pi_{t+\sigma_n} - \pi_{\sigma_n}$ is also a Poisson process (by virtue of the strong Markovian property of the Poisson process Π). Therefore, the time $\sigma^{(n)} = \inf(t \geq 0 : \pi_t^{(n)} = 1)$ of the first jump of such a process has an exponential distribution. But, since $\pi_{\sigma_n} = 0$, $\sigma^{(n)} = \sigma - \sigma_n$, and, therefore, $P(\sigma = \lim \sigma_n) = 0$. The contradiction thus obtained demonstrates that the time σ is totally inaccessible.

18.1.5. Let

$$A_{t-} = \lim_{s \uparrow t} A_s \quad \text{and} \quad \Delta A_t = A_t - A_{t-}.$$

Since the trajectories of compensator $A = (A_t)$, $t \geq 0$, are (P-a.s.) nondecreasing right continuous functions, the number of jumps of A_t, $t \geq 0$, is at most countable. The lemma which follows shows that the magnitude of these jumps does not exceed unity,

Lemma 18.1. *With probability one*

$$\sup_{t \leq \tau_\infty} \Delta A_t = \sup_{t < \tau_\infty} \Delta A_t \leq 1. \tag{18.8}$$

PROOF. We shall establish first that $\Delta A_{\tau_\infty} = A_{\tau_\infty} - A_{(\tau_\infty)-} = 0$. Indeed, since $A_{\tau_\infty} \geq A_{\tau_n}$, $A_{\tau_\infty} \geq \lim_n A_{\tau_n}$. On the other hand, by virtue of (18.4) and the Fatou lemma (Theorem 1.2)

$$\begin{aligned} A_{\tau_\infty} &= A_{\tau_\infty}^{(1)} + \sum_{n \geq 1} [A_{\tau_\infty}^{(n+1)} - A_{\tau_\infty}^{(n)}] \\ &= \lim_k A_{\tau_k}^{(1)} + \sum_{n \geq 1} \lim_k [A_k^{(n+1)} - A_{\tau_k}^{(n)}] \\ &\leq \lim_k \left\{ A_{\tau_k}^{(1)} + \sum_{n \geq 1} [A_{\tau_k}^{(n+1)} - A_{\tau_k}^{(n)}] \right\} \\ &= \lim_k A_{\tau_k}. \end{aligned}$$

Hence, $A_{\tau_\infty} = \lim_k A_{\tau_k}$ and, therefore, $A_{\tau_\infty} = A_{(\tau_\infty)-}$, i.e., $\Delta A_{\tau_\infty} = 0$. Let

$$\theta = \inf\left(t \le \tau_\infty : \sup_{s\le t} \Delta A_s > 1\right),$$

assuming $\theta = \tau_\infty$ if $\sup_{s\le\tau_\infty} \Delta A_s \le 1$. Then, in order to prove the lemma it suffices to establish that (P-a.s.)

$$A_{\theta\wedge\tau_k} - A_{(\theta\wedge\tau_k)-} \le 1, \quad k = 1, 2, \dots. \tag{18.9}$$

Since the time θ is predictable (Example 7), there exists an increasing sequence of Markov times (θ_n), $n \ge 1$, such that $\theta_n < \theta$ and $\lim_n \theta_n = \theta$ (P-a.s.).

Thus we have from the decomposition

$$N_{t\wedge\tau_k} = m_{t\wedge\tau_k} + A_{t\wedge\tau_k}$$

(for a uniformly integrable martingale $(m_{t\wedge\tau_k}, \mathcal{F}_t)$ and an integrable process $A_{t\wedge\tau_k}$, $t \ge 0$) that, for each $j < n$,

$$M(A_{\theta\wedge\tau_k} - A_{\theta_n\wedge\tau_k}|\mathcal{F}_{\theta_j\wedge\tau_k}) = M(N_{\theta\wedge\tau_k} - N_{\theta_n\wedge\tau_k}|\mathcal{F}_{\theta_j}). \tag{18.10}$$

From this, letting $n \to \infty$, by the Lebesgue theorem on dominated convergence (Theorem 1.4) we find

$$M(A_{\theta\wedge\tau_k} - A_{(\theta\wedge\tau_k)-}|\mathcal{F}_{\theta_j}) = M(N_{\theta\wedge\tau_k} - N_{(\theta\wedge\tau_k)-}|\mathcal{F}_{\theta_j}) \le 1. \tag{18.11}$$

By virtue of Theorem T35, Chapter 3, in [49][3]

$$\mathcal{F}_{\theta-} = \sigma\left(\bigcup_j \mathcal{F}_{\theta_j}\right). \tag{18.12}$$

Hence, from (18.11), by Lévy's theorem (Theorem 1.5) we obtain

$$M(A_{\theta\wedge\tau_k} - A_{(\theta\wedge\tau_k)-}|\mathcal{F}_{\theta-}) \le 1. \tag{18.13}$$

But the values of $A_{\theta\wedge\tau_k}$ are $\mathcal{F}_{\theta-}$-measurable (see Section 3.4, and also Theorem T34, Chapter X, in [229]). Consequently, the value $\Delta A_{\theta\wedge\tau_k} = A_{\theta\wedge\tau_k} - A_{(\theta\wedge\tau_k)-}$ is also $\mathcal{F}_{\theta-}$-measurable and, by virtue of (18.13), $\Delta A_{\theta\wedge\tau_k} \le 1$ (P-a.s.), which was to be proved. □

Lemma 18.2. *Let σ be a Markov time (with respect to the family $(\mathcal{F}_t)$, $t \ge 0$). Then (Δ denotes the symmetric difference of sets)*

$$MN_{\sigma\wedge\tau_\infty} = MA_{\sigma\wedge\tau_\infty}, \tag{18.14}$$

[3] For the definition of the σ-algebras $\mathcal{F}_t$ see the note to Theorem 3.10.

$$P[(\{N_\sigma < \infty\} \Delta \{A_\sigma < \infty\}) \cap (\{\sigma < \tau_\infty\})] = 0. \tag{18.15}$$

PROOF. Since the martingale $m_{t\wedge\tau_n} = N_{t\wedge\tau_n} - A_{t\wedge\tau_n}$, $t \geq 0$, is uniformly integrable, we have (Theorem 3.6) $MN_{\sigma\wedge\tau_n} = MA_{\sigma\wedge\tau_n}$. But $\lim_n N_{\sigma\wedge\tau_n} = N_{\sigma\wedge\tau_\infty}$ and $\lim_n A_{\sigma\wedge\tau_n} = A_{\sigma\wedge\tau_\infty}$; hence, by virtue of the monotone convergence theorem (Theorem 1.1), $MN_{\sigma\wedge\tau_\infty} = MA_{\sigma\wedge\tau_\infty}$.

Next, since

$$\begin{aligned}
&\{N_\sigma < \infty\} \Delta \{A_\sigma < \infty\} \\
&= (\{N_\sigma < \infty\} \cap \{A_\sigma = \infty\}) \cup (\{N_\sigma = \infty\} \cap \{A_\sigma < \infty\}),
\end{aligned}$$

in order to prove (18.15) we need to show that

$$P\{N_\sigma < \infty, A_\sigma = \infty, \sigma < \tau_\infty\} = P\{N_\sigma = \infty, A_\sigma < \infty, \sigma < \tau_\infty\} = 0. \tag{18.16}$$

We have $MA_{\sigma\wedge\tau_n} = MN_{\sigma\wedge\tau_n} \leq n < \infty$. Hence $P\{A_{\sigma\wedge\tau_n} = \infty\} = 0$ and

$$\begin{aligned}
&P\{N_\sigma < \infty, A_\sigma = \infty, \sigma < \tau_\infty\} \\
&= P\{N_{\sigma\wedge\tau_n} < \infty, A_{\sigma\wedge\tau_n} = \infty, \sigma \leq \tau_n, \sigma < \tau_\infty\} \\
&\quad + P\{N_\sigma < \infty, A_\sigma = \infty, \tau_n < \sigma < \tau_\infty\} \\
&= P\{N_\sigma < \infty, A_\sigma = \infty, \tau_n < \sigma < \tau_\infty\} \\
&\leq P\{\tau_n < \sigma < \tau_\infty\} \to 0, \quad n \to \infty.
\end{aligned}$$

In order to prove that the second expression in equality (18.16) is equal to zero probability, we shall consider the Markov times

$$\theta_n = \begin{cases} \inf(t \leq \tau_\infty : A_t \geq n), & \\ \tau_\infty & \text{if } A_{\tau_\infty} < n. \end{cases}$$

By virtue of Lemma 18.1, $A_{t\wedge\theta_n} \leq n + 1$. Hence,

$$\begin{aligned}
&P\{N_\sigma = \infty < \infty, A_\sigma < \tau_\infty\} \\
&= P\{N_{\sigma\wedge\theta_n} = \infty, A_{\sigma\wedge\theta_n} < \infty, \sigma < \tau_\infty, \sigma \leq \theta_n\} \\
&\quad + P\{N_\sigma = \infty, A_\sigma < \infty, \sigma < \tau_\infty, \sigma > \theta_n\} \\
&\leq P\{N_{\sigma\wedge\theta_n} = \infty\} + P\{\theta_n < \sigma < \tau_\infty\}.
\end{aligned}$$

But $MN_{\sigma\wedge\theta_n} = MA_{\sigma\wedge\theta_n} \leq n + 1$ and, therefore, $P\{N_{\sigma\wedge\theta_n} = \infty\} = 0$. Finally, since $\theta_n \uparrow \tau_\infty$ and $\sigma < \tau_\infty$, then $P\{\theta_n < \sigma < \tau_\infty\} \to 0$, $n \to \infty$. □

18.1.6. Processes with continuous compensators constitute an important class of point processes. The structure of such processes will be described in the lemma which follows.

Lemma 18.3. *A necessary and sufficient condition for the compensator A_t, $t \geq 0$, of a point process $N = (N_t, \mathcal{F}_t)$, $t \geq 0$, to be (P-a.s.) continuous on*

$[0, \tau_\infty]$ *is that the process be left quasicontinuous on* $[0, \tau_\infty)$, *i.e., that for any nondecreasing sequence of Markov times* (σ_n), $n \geq 1$,

$$\lim_n N_{\sigma_n \wedge \tau_\infty}, = N_{\lim_n \sigma_n \wedge \tau_\infty} \quad (P\text{-a.s.}). \tag{18.17}$$

PROOF.

Necessity. Let $\sigma = \lim_n \sigma_n$. Then, from the following equality,

$$M[N_{\tau_k \wedge \sigma} - N_{\tau_k \wedge \sigma_n}] = M[A_{\tau_k \wedge \sigma} - A_{\tau_k \wedge \sigma_n}],$$

and the continuity of A_t, $t \geq 0$, we have that $N_{\tau_k \wedge \sigma} = \lim_n N_{\tau_k \wedge \sigma_n}$. From this we have

$$N_{\tau_\infty \wedge \sigma} = \lim_k \lim_n N_{\tau_k \wedge \sigma_n} = \lim_n \lim_k N_{\tau_k \wedge \sigma_n} = \lim_n N_{\tau_\infty \wedge \sigma_n}.$$

Sufficiency. We shall consider the potential $\Pi^{(k)} = (\Pi_t^{(k)}, \mathcal{F}_t)$, $t \geq 0$, with

$$\Pi_t^{(k)} = M(N_{\tau_k} | \mathcal{F}_t) - N_{t \wedge \tau_k}. \tag{18.18}$$

Because of the left quasicontinuity of the process N,

$$\begin{aligned} M\Pi_{\sigma_n}^{(k)} = MN_{\tau_k} - MN_{\sigma_n \wedge \tau_k} &\to MN_{\tau_k} - MN_{\sigma \wedge \tau_k} \\ &= M[M(N_{\tau_k} | \mathcal{F}_\sigma) - N_{\sigma \wedge \tau_k}] \\ &= M\Pi_\sigma^{(k)}, \quad n \to \infty, \end{aligned}$$

i.e., the potential $\Pi^{(k)}$ is regular (in the sense of Definition 7, Section 3.4). Hence, by virtue of Theorem 3.11, the potential $\Pi^{(k)}$ permits the Doob–Meyer decomposition

$$\Pi_t^{(k)} = M_t^{(k)} + B_t^{(k)} \tag{18.19}$$

with a continuous predictable process $B^{(k)} = (B_t^{(k)}, \mathcal{F}_t)$, $t \geq 0$. It follows from the uniqueness of the Doob–Meyer decomposition (18.18), (18.19) and (18.6), that $B_t^{(k)} = -A_{t \wedge \tau_k}$. Hence, for $\tau < \tau_\infty$, the compensator A_t has (P-a.s.) continuous trajectories. This, together with the equality $A_{\tau_\infty} = A_{(\tau_\infty)-}$, proves that $P(\Delta At \neq 0, t \leq \tau_\infty) = 0$. □

Corollary 1. *The compensator* A_t, $t \geq 0$, *of the point process* $N = (N_t, \mathcal{F}_t)$, $t \geq 0$, *is continuous* ($\{t \leq \tau_\infty\}$: (P-a.s.)) *if and only if the jump times of the process* N_t, $t \geq 0$, *are totally inaccessible.*

In fact, if the compensator is continuous, the process is left quasicontinuous. Then, if δ is a jump time and with positive probability $\delta_n \uparrow \delta$ ($\leq \tau_\infty$), then because of left quasicontinuity of N_δ, $N_{\delta-} = \lim_n N_{\delta_n} = N_\delta$, which contradicts the assumption that δ is a jump time. Therefore, the time δ is totally inaccessible.

Conversely, let the Markov time δ be such that there exists a sequence of times (δ_n), $n \geq 1$, such that $\delta_n < \delta$ and $\delta_n \uparrow \delta \leq \tau_\infty$. The time δ cannot be a jump time (since by assumption jump times are totally inaccessible) and therefore, $\lim_n N_{\delta_n} = N_\delta$, i.e., the process N is left quasicontinuous; by the previous theorem, the compensator A_t, $t \geq 0$, is continuous ($\{t \leq \tau_\infty\}$: (P-a.s.)).

Corollary 2. *The point process N_t, $t < \tau_\infty$, with a continuous compensator is stochastically continuous:*

$$\lim_{s \to t} P(|N_{t\wedge\tau_\infty} N_{s\wedge\tau_\infty}| > \varepsilon) = 0, \quad \varepsilon > 0.$$

18.2 Minimal Representation of a Point Process: Processes of the Poisson Type

18.2.1. Let $N = (N_t, \mathcal{F}_t)$, $t \geq 0$, be a point process, and let

$$N_t = m_t + A_t, \quad t < \tau_\infty \tag{18.20}$$

be its Doob–Meyer decomposition.

The variables N_t are $\mathcal{F}_t$-measurable, but they may turn out to be measurable also with respect to smaller σ-algebras. Thus, for example, it can be seen that the N_t are $\mathcal{F}_t^N$-measurable ($\mathcal{F}_t^N = \sigma\{\omega : N_s, s \leq t\}$ and $\mathcal{F}_t^N \subseteq \mathcal{F}_t$). It is also obvious that the family $(\mathcal{F}_t^N)$, $t \geq 0$, is the smallest σ-algebra family with respect to which the values N_t, $t \geq 0$, are measurable; in this case the process $\overline{N} = (N_t, \mathcal{F}_t^N)$, $t \geq 0$, is also a point process. For this process we have (if the family of σ-algebras $(\mathcal{F}_t^N)$, $t \geq 0$, is right continuous) the Doob–Meyer decomposition

$$N_t = \overline{m}_t + \overline{A}_t, \quad t < \tau_\infty, \tag{18.21}$$

which is naturally called the *minimal representation* of the point process N.

The minimal representation given by (18.21) will play an essential role in the investigation of point process properties. Hence, we shall discuss in detail the question of right continuity for the family of σ-algebras $(\mathcal{F}_t^N)$, $t \geq 0$.

Lemma 18.4. *Let a space of elementary events Ω be such that for each $t \geq 0$ and $\omega \in \Omega$ there is an $\omega' \in \Omega$ such that $N_s(\omega') = N_{t\wedge s}(\omega)$ for all $s > 0$. Then the family of σ-algebras $(\mathcal{F}_t^N)$, $t \geq 0$, is right continuous: $\mathcal{F}_{t+}^N = \mathcal{F}_t^N$, $t \geq 0$.*

PROOF. It is known (see, for example, Lemma 3, Chapter I, in [285]) that under the assumptions of the lemma, the σ-algebra $\mathcal{F}_t^N$ consists of the sets

$A \in \mathcal{F}$ which possess a property implying that if $\omega \in A$ and $N_s(\omega') = N_s(\omega)$, $s \leq t$, ω' also belongs to A.

Let us take a set $A \in \mathcal{F}_{t+}^N$. Let $\omega \in A$ and ω' be such that $N_s(\omega') = N_s(\omega)$, $s \leq t$. It follows from the right continuity of the trajectories of the process N_t, $t \geq 0$, that the point ω' also belongs to the set A. Consequently, by virtue of the statement made in the preceding paragraph, the set $A \in \mathcal{F}_t^N$ and, therefore, $\mathcal{F}_{t+}^N = \mathcal{F}_t^N$, $t \geq 0$. □

Note 1. From now on we shall assume that the space of elementary outcomes Ω satisfies the conditions of Lemma 18.4. This assumption holds for the minimal representation given by (18.21). In (18.21), the structure of the compensator $\overline{A} = (\overline{A}_t, \mathcal{F}_t^N)$, $t \geq 0$, can be described as follows.

Theorem 18.2. *Let $F_1(t) = P(\tau_1 \leq t)$, and let*

$$F_i(t) = P(\tau_i \leq t | \tau_{i-1}, \dots, \tau_1), \quad i \geq 2,$$

be regular conditional distribution functions. Then the compensator $\overline{A} = (\overline{A}_t, \mathcal{F}_t^N)$, $t < \tau_\infty$, of the point process $\overline{N} = (N_t, \mathcal{F}_t^N)$, $t \geq 0$, can be defined by the formula

$$\overline{A}_t = \sum_{i \geq 1} \overline{A}_t^{(i)}, \tag{18.22}$$

where[4]

$$\overline{A}_t^{(i)} = \int_0^{t \wedge \tau_i} \frac{dF_i(u)}{1 - F_i(u-)}, \quad i \geq 1. \tag{18.23}$$

To prove this theorem we shall need two auxiliary assertions which are of interest by themselves.

18.2.2.
Lemma 18.5. *Let*

$$N_t = \sum_{n \geq 1} I_{\{\tau_n \leq t\}}, \quad t \geq 0,$$

be a point process and let $\theta = \theta(\omega)$ be a Markov time with respect to the family $(\mathcal{F}_t^N)$, $t \geq 0$, such that $P(\theta < \tau_\infty) = 1$. Then there exist Borel functions $\varphi_n = \varphi_n(t_1, \dots, t_n)$, $n \geq 1$, and a constant φ_0 such that ($\tau_0 \equiv 0$)

$$\theta(\omega) = \sum_{n \geq 1} I_{\{\tau_{n-1} \leq \theta < \tau_n\}} \cdot \varphi_{n-1}(\tau_1, \dots, \tau_{n-1}), \tag{18.24}$$

i.e., on the set $\{\theta < \tau_1\}$ the random variable $\theta(\omega)$ is a constant and, on the sets $\{\tau_{n-1} \leq \theta < \tau_n\}$ $(n \geq 1)$, $\theta(\omega) = \varphi_{n-1}(\tau_1(\omega), \dots, \tau_{n-1}(\omega))$.

---

[4] $\int_s^t f(u)dF_i(u)$ is understood as a Lebesgue–Stieltjes integral over a set $(s,t]$, i.e., $\int_s^t f(u)dF_i(u) = \int_{(s,t]} f(u)dF_i(u)$. (For more details see Section 18.4).

PROOF. We shall take advantage of the fact that the σ-algebra $\mathcal{F}_s^N$ coincides with a σ-algebra $\sigma\{\omega : N_{s\wedge t}, t \geq 0\}$, and for any Markov time θ (with respect to the system $(\mathcal{F}_s^N)$, $s \geq 0$) the σ-algebra

$$\mathcal{F}_\theta^N = \sigma\{\omega : N_{\theta\wedge t}, t \geq 0\}$$

(see, for example, Theorem 6, Chapter I, in [285]).

The random variable θ is measurable with respect to the σ-algebra $\mathcal{F}_\theta^N$ and, hence, there exist a countable set $S \subset [0,\infty)$ and a Borel function $\varphi(x_n; n \in \mathbb{N})$, such that $\theta(\omega) = \varphi(N_{t\wedge\theta(\omega)}(\omega); t \in S)$. Therefore, $\theta(\omega) = \sum_{n\geq 1} I_{\{\tau_{n-1}\leq\theta<\tau_n\}} \cdot \varphi(N_{t\wedge\theta}; t \in S)$.

Note now that, on the set $\{\tau_{n-1} \leq \theta < \tau_n\}$,

$$N_{t\wedge\theta} = N_{t\wedge\tau_{n-1}} = \sum_{k=1}^{n-1} I_{\{\tau_k\leq t\}}$$

and, consequently, on this set

$$\varphi(N_{t\wedge\theta}; t \in S) = \varphi\left(\sum_{k=1}^{n-1} I_{\{\tau_k\leq t\}}; t \in S\right).$$

The function $\varphi(\sum_{k=1}^{n-1} I_{\{\tau_k\leq t\}}; t \in S)$ can obviously be represented as $\varphi_{n-1}(\tau_1, \ldots, \tau_{n-1})$, where $\varphi_{n-1}(t_1, \ldots, t_{n-1})$ is a Borel function of $n-1$ variables. Hence,

$$\begin{aligned}\theta(\omega) &= \sum_{n\geq 1} I_{\{\tau_{n-1}\leq\theta<\tau_n\}} \cdot \varphi(N_{t\wedge\theta}; t \in S) \\ &= \sum_{n\geq 1} I_{\{\tau_{n-1}\leq\theta<\tau_n\}} \cdot \varphi_{n-1}(\tau_1, \ldots, \tau_{n-1}).\end{aligned}$$

□

Corollary. *There exist Borel functions $\theta_n(t_1, \ldots, t_n)$ and a constant θ_0 such that on the set $\{\theta < \tau_n\}$, $\theta = \theta_{n-1}(\tau_1, \ldots, \tau_{n-1})$. In particular, $\theta \wedge \tau_n = \theta_{n-1} \wedge \tau_n$ and $\theta \wedge \tau_k = \theta_{n-1} \wedge \tau_k$, $k < n$.*

Lemma 18.6. *Let $(\Omega, \mathcal{F}, P)$ be a probability space and let $(\mathcal{F}_t)$, $t \geq 0$, be a nondecreasing family of sub-σ-algebras of $\mathcal{F}$. A necessary and sufficient condition for the integrable random process $X = (x_t, \mathcal{F}_t)$, $t \geq 0$, to be a martingale is that, for any two-valued stopping time τ,*

$$Mx_\tau = Mx_0. \tag{18.25}$$

PROOF. The necessity of (18.25) follows from Theorem 3.5.

To prove sufficiency we shall assume that there exist times s and t $(s < t)$ such that the set $A = \{\omega : x_s < M(x_t|\mathcal{F}_s)\}$ has $P(A) > 0$.

Construct a time $\tau = tI_A + sI_{\overline{A}}$. Since $\{\tau = t\} = A \in \mathcal{F}_s \subseteq \mathcal{F}_t$ and $\{\tau = s\} = \overline{A} \in \mathcal{F}_s$, τ is a Markov time with respect to the system $(\mathcal{F}_t)$, $t \geq 0$. Hence,

$$\begin{aligned} Mx_\tau &= MI_A x_t + MI_{\overline{A}} x_s \\ &= M(I_A M(x_t|\mathcal{F}_s)) + MI_{\overline{A}} x_s \\ &< M(I_A x_s + I_{\overline{A}} x_s) = Mx_s, \end{aligned}$$

which fact contradicts the assumption that $Mx_\tau = Mx_0$, i.e., $P(A) = 0$.

For $A = \{\omega : x_s > M(x_t|\mathcal{F}_s)\}$, the proof of $P(A) = 0$ is given in a similar way. □

Note. The previous lemma shows that the martingale $X = (x_t, \mathcal{F}_t)$, $t \geq 0$, can be defined as a random process such that: $M|x_t| < \infty$; x_t is $\mathcal{F}_t$-measurable for each $t \geq 0$; for any two-valued stopping time τ (with respect to the family $(\mathcal{F}_t)$, $Mx_\tau = Mx_0$.

18.2.3.

PROOF OF THEOREM 18.2. To prove the theorem it suffices to show that for each n, $n = 1, 2, \ldots$, the processes $m^{(n)} = (N_{t\wedge\tau_n} - \overline{A}_{t\wedge\tau_n}, \mathcal{F}_t^N)$ (where $\overline{A}_t$ is as defined in (18.22) and (18.23)) are uniformly integrable martingales, and the process $\overline{A} = (\overline{A}_t, \mathcal{F}_t^N)$, $t \geq 0$, is a predictable process.

By virtue of Lemma 18.6, in order to prove that the process $m^{(n)}$ is a martingale, it suffices to establish that, for any stopping time θ (with respect to $(\mathcal{F}_t^N)$, $t \geq 0$),

$$MN_{\theta\wedge\tau_n} = M\overline{A}_{\theta\wedge\tau_n}. \tag{18.26}$$

We have

$$\begin{aligned} M\overline{A}_{\theta\wedge\tau_n} &= M\sum_{i\geq 1} \overline{A}^{(i)}_{\theta\wedge\tau_n} = M\sum_{i=1}^{n} \overline{A}^{(i)}_{\theta\wedge\tau_n} = M \\ &= M\sum_{i=1}^{n} \overline{A}^{(i)}_{\theta_{i-1}\wedge\tau_i} = M\sum_{i=1}^{n} M(\overline{A}_{\theta_{i-1}\wedge\tau_i}|\tau_{i-1}, \ldots, \tau_1), \end{aligned} \tag{18.27}$$

where the θ_{i-1} were defined in the corollary to Lemma 18.5.

According to (18.23),

$$\begin{aligned} M(\overline{A}_{\theta_{i-1}\wedge\tau_i}|\tau_{i-1}, \ldots, \tau_1) &= \int_0^\infty \left[\int_0^{\theta_{i-1}\wedge s} \frac{dF_i(u)}{1 - F_i(u-)}\right] dF_i(s) \\ &= \int_0^{\theta_{i-1}} \left[\int_0^s \frac{dF_i(u)}{1 - F_i(u-)}\right] dF_i(s) \\ &\quad + [1 - F_i(\theta_{i-1})] \cdot \int_0^{\theta_{i-1}} \frac{dF_i(u)}{1 - F_i(u-)}. \end{aligned} \tag{18.28}$$

Let $A(s) = F_i(s)$ and $B(s) = \int_0^s dF_i(u)/(1 - F_i(u-))$. Then, by virtue of the formula

$$A(t)B(t) = \int_0^t A(u-)dB(u) + \int_0^t B(u)dA(u)$$

(the proof of which, in a more general case, will be given in Lemma 18.7),

$$\begin{aligned}
&\int_0^{\theta_{i-1}} \left[\int_0^s \frac{dF_i(u)}{1 - F_i(u-)}\right] dF_i(s) \\
&= F_i(\theta_{i-1}) \cdot \int_0^{\theta_{i-1}} \frac{dF_i(u)}{1 - F_i(u-)} - \int_0^{\theta_{i-1}} \frac{F_i(u-)dF_i(u)}{1 - F_i(u-)} \\
&= [F_i(\theta_{i-1}) - 1] \cdot \int_0^{\theta_{i-1}} \frac{dF_i(u)}{1 - F_i(u-)} + [F_i(\theta_{i-1}) - F_i(0)]. \quad (18.29)
\end{aligned}$$

But $F_i(0) = 0$, $i = 1, 2, \ldots$, and we find from (18.28) and (18.29) that

$$M(A^{(i)}_{\theta_{i-1}\wedge\tau_i} | \tau_{i-1}, \ldots, \tau_1) = F_i(\theta_{i-1}); \tag{18.30}$$

therefore, by virtue of (18.27),

$$M\overline{A}_{\theta\wedge\tau_n} = M \sum_{i=1}^n F_i(\theta_{i-1}). \tag{18.31}$$

On the other hand, by virtue of the corollary to Lemma 18.5, $\theta \wedge \tau_i = \theta_{i-1} \wedge \tau_i$ and $\theta \wedge \tau_{i-1} = \theta_{i-1} \wedge \tau_{i-1}$. Hence,

$$\begin{aligned}
MN_{\theta\wedge\tau_n} &= MN_{\theta\wedge\tau_1} + M\sum_{i=2}^n [N_{\theta\wedge\tau_i} - N_{\theta\wedge\tau_{i-1}}] \\
&= MN_{\theta\wedge\tau_1} + M\sum_{i=2}^n [N_{\theta_{i-1}\wedge\tau_i} - N_{\theta_{i-1}\wedge\tau_{i-1}}] \\
&= MI_{\{\tau_1 \le \theta_0\}} + M\sum_{i=2}^n I_{\{\tau_i \le \theta_{i-1}\}} \\
&= F_1(\theta_0) + \sum_{i=2}^n MF_i(\theta_{i-1}). \qquad (18.32)
\end{aligned}$$

We obtain the required assertion (18.26), by comparing (18.31) with (18.32).

Further, since $|N_{t\wedge\tau_n} - \overline{A}_{t\wedge\tau_n}| \le N_{t\wedge\tau_n} + \overline{A}_{t\wedge\tau_n} \le n + \overline{A}_{\tau_n}$, and $M\overline{A}_{\tau_n} = MN_{\tau_n} \le n$, the martingale $m^{(n)}$ is uniformly integrable.

Let us establish that the process $\overline{A}_t = \sum_{i\ge1} \overline{A}_t^{(i)}$, $t \ge 0$, is predictable. To this end, it suffices to verify that each of the processes $\overline{A}_t^{(i)}$, $i = 1, 2, \ldots$, is predictable.

Let $Y = (y_t, \mathcal{F}_t^N)$, $t \geq 0$, be a nonnegative bounded martingale with trajectories having (P-a.s.) limits to the left.

Then, since $y_{s-t} = M(y_s|\mathcal{F}_{s-t}^N)$, from Lévy's theorem (Theorem 1.5) it follows that $y_{s-} = M(y_s|\mathcal{F}_{s-}^N)$. Furthermore, the variables $I(s \leq \tau_1\}$ are $\mathcal{F}_{s-}^N$-measurable. Hence,

$$\begin{aligned} M\int_0^t y_s d\overline{A}_s^{(1)} &= M\int_0^t I_{\{s\leq\tau_1\}} y_s \frac{dF_1(s)}{1-F_1(s-)} \\ &= \int_0^t M[I_{\{s\leq\tau_1\}} y_s] \frac{dF_1(s)}{1-F_1(s-)} \\ &= \int_0^t M[I_{\{s\leq\tau_1\}} y_{s-}] \frac{dF_1(s)}{1-F_1(s-)} = M\int_0^t y_{s-} d\overline{A}_s^{(1)}. \end{aligned}$$

Similarly, making use of the fact that the variables $F_i(s)$ are $\mathcal{F}_{\tau_{i-1}}^N$-measurable, we find that

$$\begin{aligned} M\int_0^t y_s d\overline{A}_s^{(i)} &= M\int_0^t I_{\{\tau_{i-1}<s\leq\tau_i\}} y_s \frac{dF_i(s)}{1-F_i(s-)} \\ &= M\int_0^t M[I_{\{\tau_{i-1}<s\leq\tau_i\}} y_s \; \mathcal{F}_{\tau_i-1}^N] \frac{dF_i(s)}{1-F_i(s-)} \\ &= M\int_0^t M[I_{\{\tau_{i-1}<s\leq\tau_i\}} y_{s-} | \mathcal{F}_{\tau_{i-1}}^N] \frac{dF_i(s)}{1-F_i(s-)} \\ &= M\int_0^t y_{s-} d\overline{A}_s^{(i)}. \end{aligned}$$

It follows, from the equalities $M\int_0^t y_s d\overline{A}_s^{(i)} = M\int_0^t y_{s-} d\overline{A}_s^{(i)}$, $i \geq 1$, thus obtained and from Lemma 3.2, that each of the processes $\overline{A}_t^{(i)}$, $t \geq 0$, is predictable; consequently, the process $\overline{A}_t = \sum_{i\geq 1} \overline{A}_t^{(i)}$, $t \geq 0$, is predictable. □

Corollary. *If the functions $F_i(t)$, $i \geq 1$, are (P-a.s.) continuous (absolutely continuous) the compensator $\overline{A}_t$, $t \geq 0$ has (P-a.s.) continuous (absolutely continuous) trajectories.*

EXAMPLE. Let $W = (W_t, \mathcal{F}_t)$, $t \geq 0$, be a Wiener process $\tau = \inf(t{\geq}0, W_t = 1)$ ($\tau = \infty$ if $\sup_{t\geq 0} W_t < 1$). Let us consider a point process $N = (N_t, \mathcal{F}_t)_2$, $t \geq 0$, with $N_t = I_{\{\tau\leq t\}}$. If we define the times

$$\tau_n = \inf\left\{t \geq 0 : W_t = 1 - \frac{1}{n}\right\},$$

we shall see that $\tau_n < \tau$ and $\lim_n \tau_n = \tau$ (P-a.s.). Hence the time τ is predictable and, therefore, the (nondecreasing) process N_t, $t \geq 0$, is also predictable. Consequently, as noted in Section 18.1, this process is predictable and, therefore, in the Doob–Meyer decomposition we have $N_t = A_t + m_t$, $A_t = N_t$, and $m_t = 0$, $t \geq 0$.

We shall consider next the minimal representation of the process $N_t = I_{\{\tau \leq t\}}$, $t \geq 0$. Let $\mathcal{F}_t^N = \sigma\{\omega : N_s, s \leq t\}$ and $\overline{N} = (N_t, \mathcal{F}_t^N)$, $t \geq 0$. We shall show that the compensator $\overline{A} = (\overline{A}_t, \mathcal{F}_t^N)$ of this process can be defined by the formula

$$\overline{A}_t = -\ln\left(1 - \sqrt{\frac{2}{\pi}}\int_{(t\wedge\tau)^{-1/2}}^{\infty} e^{-y^2/2}dy\right).$$

In fact, let $F(t) = P(\tau \leq t)$. According to (1.42),

$$F(t) = \sqrt{\frac{2}{\pi}}\int_{t^{-1/2}}^{\infty} e^{-y^2/2}dy$$

and, by virtue of (18.23), $\overline{A}_1 = -\ln(1 - F(t \wedge \tau))$.

18.2.4. In the class of point process $N = (N_t, \mathcal{F}_t)$, $t \geq 0$, there are some of relatively simple structure in which the compensator $A = (A_t, \mathcal{F}_t)$, $t \geq 0$, has the form

$$A_t = \int_0^t \lambda_s db_s, \tag{18.33}$$

where $\lambda = (\lambda_t(\omega), \mathcal{F}_t)$, $t \geq 0$, is some nonnegative predictable process, and b_t is a deterministic nonnegative right continuous and nondecreasing function.

In the case of a Poisson process (with parameter λ) $\lambda_t \equiv \lambda$, $b_t \equiv t$. Hence, the point processes whose compensators satisfy formula (18.33) are naturally called *Poisson type processes.*

Let $N = (N_t, \mathcal{F}_t)$, $t \geq 0$, be a point process with a compensator $A = (A_t, \mathcal{F}_t)$, $t \geq 0$. Consider the process $\overline{N} = (N_t, \mathcal{F}_t^N)$, $t \geq 0$, and let $N_t = \overline{A}_t + \overline{m}_t$, $t \geq 0$, be its minimal representation. It is not easy, in general, to find the compensator $\overline{A}_t$ from the compensator A_t; this is the case, however, for the Poisson type process.

Theorem 18.3. *Let the compensator of a point process $N = (N_t, \mathcal{F}_t)$, $t \geq 0$, be given by the formula $A_t = \int_0^t \lambda_s db_s$. Then*

$$\overline{A}_t = \int_0^t \overline{\lambda}_s db_s, \tag{18.34}$$

where $\overline{\lambda}_t = M(\lambda_t | \mathcal{F}_{t-}^N)$.

In this case

$$A_t = \int_0^t \lambda_s \overline{\lambda}_s^+ d\overline{A}_s \quad (\{t < \tau_\infty\} : \quad (P\text{-a.s.})), \tag{18.35}$$

where

$$\overline{\lambda}_t^+ = \begin{cases} \overline{\lambda}_t^{-1}, & \overline{\lambda}_t > 0, \\ 0, & \overline{\lambda}_t = 0. \end{cases}$$

PROOF. We shall show first that $\int_0^t \overline{\lambda}_s db_s < \infty$ ($\{t < \tau_\infty\}$: (P-a.s.)). By virtue of (18.14),

$$M \int_0^t I_{\{s \le \tau_n\}} \lambda_s db_s = MA_{t\wedge\tau_n} = MN_{t\wedge\tau_n} \le n.$$

Since the values $I_{\{s\le\tau_n\}}$ for each s are $\mathcal{F}_{s-}^N$-measurable, we have

$$M \int_0^{t\wedge\tau_n} \overline{\lambda}_s db_s = M \int_0^t I_{\{s\le\tau_n\}} \lambda_s db_s \le n,$$

and, therefore. (P-a.s.) on the sets $\{t \le \tau_n\}$, $\int_0^{t\wedge\tau_n} \overline{\lambda}_s db_s < \infty$, $n \ge 1$. Consequently, $\int_0^t \overline{\lambda}_s db_s < \infty$ ($\{t < \tau_\infty\}$: (P-a.s)).

Let $B_t = \int_0^t \overline{\lambda}_s db_s$, $0 \le t < \tau_\infty$. This process is nondecreasing and right continuous. We shall show that the process $B = (B_t, \mathcal{F}_t^N)$, $t < \tau_\infty$, is a predictable process.

In fact, if $Y = (y_t, \mathcal{F}_t^N)$ is a bounded nonnegative martingale which has limits to the left, then, by virtue of the equality $y_{s-} = M(y_s|\mathcal{F}_{s-}^N)$ and the $\mathcal{F}_{s-}^N$-measurability of the function $I_{\{s\le\tau_n\}}$, we obtain

$$\begin{aligned} M\int_0^{t\wedge\tau_n} y_s dB_s &= M\int_0^{t\wedge\tau_n} y_s \overline{\lambda}_s db_s \\ &= M\int_0^t I_{\{s\le\tau_n\}} \overline{\lambda}_s M(y_s|\mathcal{F}_{s-}^N) db_s \\ &= M\int_0^{t\wedge\tau_n} y_{s-} \overline{\lambda}_s db_s = \int_0^{t\wedge\tau_n} y_{s-} dB_s. \end{aligned}$$

Consequently, for each $n \ge 1$, the processes $(B_{\tau\wedge\tau_n} \mathcal{F}_t^N)$, $t \ge 0$, are predictable processes (Lemma 3.2) and, therefore, the process $(B_{t\wedge\tau_\infty}, \mathcal{F}_t^N)$, $t \ge 0$, is also a predictable process.

To prove that $\overline{A}_t = B_t$ it suffices to verify that the process $\overline{m} = (\overline{m}_t, \mathcal{F}_t^N)$ with $\overline{m}_t = N_t - B_t$ is a τ_∞-local martingale.

But

$$\overline{m}_{t\wedge\tau_n} = m_{t\wedge\tau_n} + \int_0^{t\wedge\tau_n} [\lambda_s - \overline{\lambda}_s] db_s$$

and

$$\begin{aligned}
M[\overline{m}_{t\wedge\tau_n} - \overline{m}_{s\wedge\tau_n}|\mathcal{F}_s^N] &= M[m_{t\wedge\tau_n} - m_{s\wedge\tau_n}|\mathcal{F}_s^N] \\
&\quad + M\left[\int_{s\wedge\tau_n}^{t\wedge\tau_n} (\lambda_u - \overline{\lambda}_u)db_u \,\middle|\, \mathcal{F}_s^N\right] \\
&= M\left[\int_s^t I_{\{u\le\tau_n\}}(\lambda_u - \overline{\lambda}_u)db_u \,\middle|\, \mathcal{F}_s^N\right] \\
&= M\left[\int_s^t I_{\{u\le\tau_n\}}M(\lambda_u - \overline{\lambda}_u|\mathcal{F}_{u-}^N)db_u \,\middle|\, \mathcal{F}_s^N\right] = 0.
\end{aligned}$$

Thus, we have proved (18.34).

To prove (18.35), we first note that

$$A_t = \int_0^t \lambda_s \overline{\lambda}_s^+ d\overline{A}_s + a_t, \tag{18.36}$$

where

$$a_t = \int_0^t \lambda_s(1 - \overline{\lambda}_s\overline{\lambda}_s^+)db_s.$$

But

$$\begin{aligned}
Ma_{t\wedge\tau_n} &= M\int_0^{t\wedge\tau_n} \lambda_s(1 - \overline{\lambda}_s\overline{\lambda}_s^+)db_s \\
&= M\int_0^t I_{\{s\le\tau_n\}}M(\lambda_s|\mathcal{F}_{s-}^N)[1 - \overline{\lambda}_s\overline{\lambda}_s^+]db_s \\
&= M\int_0^t I_{\{s\le\tau_n\}}\overline{\lambda}_s(1 - \overline{\lambda}_s\overline{\lambda}_s^+)db_s = 0.
\end{aligned}$$

Consequently, $a_t = 0$ ($\{t < \tau_\infty\}$: (P-a.s.)) which, together with (18.36), proves (18.35). □

Corollary. *If A_t is a deterministic function, then $\overline{A}_t = A_t$.*

It suffices to set $\lambda_t = 1$, $b_t = A_t$.

Note. If the assumption $A_t = \int_0^t \lambda_s db_s$ is not fulfilled, (18.35) might not hold.

Indeed, the example given at the end of Section 18.3 shows that A_t is a discontinuous function, whereas $\overline{A}_t$ is an absolutely continuous function.

18.3 Construction of Point Processes with Given Compensators: Theorems on Existence and Uniqueness

18.3.1. Consider a probability space $(\mathsf{X}, \mathcal{B}, \mu)$ where X is a space of piecewise-constant functions $x = (x_t)$, $t \ge 0$, such that $x_0 = 0$, $x_t = x_{t-} + (0 \text{ or } 1)$, $\mathcal{B}$

is a σ-algebra $\sigma\{x : x_s, s > 0\}$, and μ is probability measure on $(\mathsf{X}, \mathcal{B})$. Let $\mathcal{B}_t = \sigma\{\omega : x_s, s \le t\}$ and $\tau_i(x) = \inf\{s \ge 0 : x_s = i\}$, setting $\tau_i(x) = \infty$ if $\lim_{t\to\infty} x_t < i$, and let $\tau_\infty(x) = \lim_{i\to\infty} \tau_i(x)$.

We shall note that for each function $x = (x_t)$, $t \ge 0$, $x_t = \sum_{i\ge 1} I_{\{\tau_i(x)\le t\}}$ and the family of σ-algebras $(\mathcal{B}_t)$, $t \ge 0$, is right continuous (Lemma 18.6).

According to Section 18.1, the process $X = (x_t, \mathcal{B}_t)$, $t \ge 0$ on the probability space $(\mathsf{X}, \mathcal{B}, \mu)$, is a point process. the compensator $A = (A_t(x), \mathcal{B}_t)$ of this process can be defined (Theorem 18.2) by the formula

$$A_t(x) = \sum_{i\ge 1} A_t^{(i)}(x), \tag{18.37}$$

where

$$A_t^{(i)}(x) = \int_0^{t\wedge\tau_i(x)} \frac{dF_i(u)}{1 - F_i(u-)}, \tag{18.38}$$

and the functions

$$F_1(t) = \mu\{x : \tau_1(x) \le t\}, \quad F_i(t) = \mu\{x : \tau_i(x) \le t | \tau_{i-1}(x), \dots, \tau_1(x)\}$$

are regular conditional distribution functions, $i \ge 2$.

It follows from (18.38) that there exist Borel functions $Q_i(s_i; s_{i-1}, \dots, s_1, s_0)$, $i \ge 1$, $s_0 \equiv 0$, such that:

(A) for fixed $s_{i-1}, \dots, s_1, s_0$, the function $Q_i(s_i; s_{i-1}, \dots, s_1, s_0)$, is nondecreasing right continuous with jumps not exceeding unity and, for $s_{i-1} < \infty$,

$$\lim_{s_i \downarrow s_{i-1}} Q_i(s_i; s_{i-1}, \dots, s_1, s_0) = 0;$$

(B) $Q_i(s_i; s_{i-1}, \dots, s_1, s_0) = 0$ outside of the domain

$$\{(s_i, s_{i-1}, \dots, s_1, s_0) : s_i > s_{i-1} > \dots > s_1 > s_0\};$$

(C)

$$A_t^{(i)}(x) = Q_i(t \wedge \tau_i(x); \tau_{i-1}(x), \dots, \tau_1(x), 0).$$

It is obvious from (18.37) and (18.38) that the measure μ completely defines the compensator of a point process.

The theorem given below demonstrates that the converse also holds in a certain sense.

Theorem 18.4 (Existence Theorem). *Let $Q_i(s_i; s_{i-1}, \dots, s_1, s_0)$, $i \ge 1$, be a sequence of Borel functions satisfying (A) and (B). Then, on the measurable space $(\mathsf{X}, \mathcal{B})$, there exists a probability measure μ such that the process*

$$A = (A_t(x), \mathcal{B}_t, \mu), \quad t \geq 0, \quad with \quad A_t(x) = \sum_{i \geq 1} A_t^{(i)}(x),$$

and

$$A_t^{(i)}(x) = Q_i(t \wedge \tau_i(x) : \tau_{i-1}(x), \ldots, \tau_1(x), 0),$$

is a compensator of the point process $X = (x_t, \mathcal{B}_t, \mu)$, $t \geq 0$[5].

Before proving this theorem we shall give two lemmas.

18.3.2.

Lemma 18.7. *Let* A_t *and* B_t, $t \geq 0$, *be right continuous functions of bounded variation (on any finite interval of time). Then we have the following formulae of integration by parts for Stieltjes integrals:*

$$A_t B_t = A_0 B_0 + \int_0^t A_{s-} dB_s + \int_0^t B_s dA_s; \tag{18.39}$$

and

$$\begin{aligned} A_t B_t &= A_0 B_0 + \int_0^t A_{s-} dB_s + \int_0^t B_{s-} dA_s \\ &\quad + \sum_{s \leq t} (A_s - A_{s-})(B_s - B_{s-}). \end{aligned} \tag{18.40}$$

PROOF. We shall note first that the functions A_t and B_t have the limits to the left (at each point $t > 0$) $A_{t-} = \lim_{s \uparrow t} A_s$, $B_{t-} = \lim_{s \uparrow t} B_s$ since each of them can be represented as the difference of two nondecreasing functions. Further, all Stieltjes integrals considered, $\int_s^t f(u) dA_u$, of the Borel functions $f(u)$ can be understood as Lebesgue–Stieltjes integrals over the set $(s, t]$, i.e.,

$$\int_s^t f(u) dA_u = \int_{(s,t]} f(u) dA_u = \int f(u) I_{(s,t]}(u) dA_u.$$

Therefore, in particular, for $s \leq t$,

$$\int_0^t I_{(0,s]}(u) dA_u = A_s - A_0, \qquad \int_0^t I_{\{s\}}(u) dA_u = A_s - A_{s-}$$

and

$$\int_0^t I_{(0,s)}(u) dA_u = A_{s-} - A_0.$$

To prove (18.39) we shall note, that by virtue of the Fubini theorem.

[5] The notation $X = (x_t, \mathcal{B}_t, \mu)$, $t \geq 0$, implies that the process $(x_t, \mathcal{B}_t)$, $t \geq 0$, is being considered on a measurable space $(\mathsf{X}, \mathcal{B})$ with measure μ.

$$
\begin{aligned}
(A_t - A_0)(B_t - B_0) &= \int_{(0,t]\times(0,t]} dA_s dB_u \\
&= \int_{(0,t]\times(0,t]} I_{(s\geq u)} dA_s dB_u + \int_{(0,t]\times(0,t]} I_{(s<u)} dA_s dB_u \\
&= \int_0^t (B_s - B_0) dA_s \int_0^t (A_{u-} - A_0) dB_u \\
&= \int_0^t B_s dA_s + \int_0^t A_{u-} dB_u - B_0(A_t - A_0) - A_0(B_t - B_0).
\end{aligned}
$$

From this we immediately obtain (18.39).

Finally, to establish (18.40) we need only show that

$$\int_0^t (B_s - B_{s-}) dA_s = \sum_{s\leq t} \Delta B_s \Delta A_s,$$

where

$$\Delta B_s = B_s - B_{s-}, \quad \Delta A_s = A_s - A_{s-}.$$

Let

$$A_t^c = A_t - \sum_{s\leq t} \Delta A_s.$$

Then

$$\int_0^t (B_s - B_{s-}) dA_s - \sum_{s\leq t} \Delta A_s \Delta B_s = \int_0^t (B_s - B_{s-}) dA_s^c = 0,$$

since A_t^c is a continuous function, and B_t has no more than a countable number of discontinuity points. □

Note 1. We shall write (18.39) as

$$dA_t B_t = A_{t-} dB_t + B_t dA_t. \tag{18.41}$$

Note 2. We shall agree that $A_{0-} = B_{0-} = 0$. Then (18.39) can be written as

$$A_t B_t = \int_{[0,t]} A_{s-} dB_s + \int_{[0,t]} B_s dA_s. \tag{18.42}$$

Corollary. *Let $A_t \geq 0$. Then*

$$\int_0^t A_{s-}^{n-1} dA_s \leq \frac{A_t^n - A_0^n}{n} \leq \int_0^t A_s^{n-1} dA_s. \tag{18.43}$$

18.3.3.

Lemma 18.8. *Let a_t, $t \geq 0$, be a nondecreasing right continuous function with $A_0 = 0$ and let a_t, $t \geq 0$, be a measurable function with*

$$\int_0^t |a_s| dA_s < \infty, \quad t < \infty.$$

Then the equation

$$Z_t = Z_0 + \int_0^t Z_{s-} a_s dA_s \tag{18.44}$$

has a unique locally bounded ($\sup_{s \leq t} |Z_s| < \infty$, $t < \infty$) solution which has limits to the left and can be defined by the formula

$$Z_t = Z_0 \prod_{s \leq t} [1 + a_s \Delta A_s] \exp\left(\int_0^t a_s dA_s^c \right), \tag{18.45}$$

where $\Delta A_s = A_s - A_{s-}$, $A_s^c = A_s - \sum_{u \leq s} \Delta A_u$.

PROOF. Let

$$U_t = Z_0 \prod_{s \leq t} [1 + a_s \Delta A_s], \quad V_t = \exp\left(\int_0^t a_s dA_s^c \right).$$

Then, by virtue of (18.39),

$$\begin{aligned} Z_t = U_t V_t &= Z_0 + \int_0^t U_{s-} dV_s + \int_0^t V_s dU_s \\ &= Z_0 + \int_0^t U_{s-} V_s a_s dA_s^c + \sum_{s \leq t} V_s U_{s-} a_s \Delta A_s \\ &= Z_0 + \int_0^t Z_{s-} a_s dA_s. \end{aligned}$$

Thus, the function Z_t given by (18.45) is a solution of Equation (18.44). We shall show that this solution is unique in the class of locally bounded solutions.

Let Z_t', $t \geq 0$ be another solution. Let

$$\tilde{Z}_t = Z_t - Z_t', \quad L(t) = \sup_{s \leq t} |\tilde{Z}_s|, \quad \alpha(s) = \int_0^s |a_u| dA_u.$$

Then, for any $s \leq t$,

$$|\tilde{Z}_s| \leq \int_0^s |\tilde{Z}_{u-}| |a_u| dA_u \leq L(t) \alpha(s)$$

and, therefore, by virtue of (18.43),

$$|\tilde{Z}_s| \le \int_0^s |\tilde{Z}_{u-}||a_u|dA_u \le L(t)\int_0^s \alpha(u-)d\alpha(u) \le L(t)\frac{\alpha^2(s)}{2!}.$$

Similarly,

$$|\tilde{Z}_s| \le \frac{L(t)}{2}\int_0^t \alpha^2(u-)d\alpha(u) \le \frac{L(t)\alpha^3(s)}{3!}$$

and, in general, for $s \le t$, and for any $n \ge 1$,

$$|\tilde{Z}_s| \le \frac{L(t)}{n!}\alpha^n(s).$$

From this we have $\tilde{Z}_s \equiv 0$, $s \ge 0$, □

Corollary. *The equation*

$$Z_t = Z_0 - \int_0^t Z_{s-}dA_s \tag{18.46}$$

with $Z_0 \ge 0$ *and* $\Delta A_s \le 1$, $A_t < \infty$, $t \ge 0$, *has the unique nonnegative locally bounded solution*

$$Z_t = Z_0 \exp(-A_t^c) \cdot \prod_{s \le t}(1 - \Delta A_s). \tag{18.47}$$

18.3.4.

PROOF OF THEOREM 18.4. Consider the equations

$$\Phi_t^{(i)} = 1 - \int_0^t \Phi_{s-}^{(i)} d_s Q_i(s; s_{i-1}, \dots, s_0), \quad i \ge 1. \tag{18.48}$$

By virtue of (18.47), the $\Phi_t^{(i)} = \Phi_t^{(i)}(s_{i-1}, \dots, s_0)$ are given by the formulae

$$\Phi_t^{(i)}(s_{i-1}, \dots, s_0) = \exp(-Q_i^c(t; s_{i-1}, \dots, s_0)) \cdot \prod_{s \le t}[1 - \Delta Q_i(s; s_{i-1}, \dots, s_0)] \tag{18.49}$$

with

$$\Delta Q_i(s; s_{i-1}, \dots, s_0) = Q_i(s; s_{i-1}, \dots, s_0) - Q_i(s-; s_{i-1}, \dots, s_0),$$

$$Q_i^c(t; s_{i-1}, \dots, s_0) = Q_i(t; s_{i-1}, \dots, s_0) - \sum_{s \le t} \Delta Q_i(s; s_{i-1}, \dots, s_0).$$

Let

$$F_i(t; s_{i-1}, \dots, s_0) = 1 - \Phi_t^{(i)}(s_{i-1}, \dots, s_0). \tag{18.50}$$

It follows from (18.49) that the $F_i(t; s_{i-1}, \dots, s_0)$ are Borel functions different from zero only for $t > s_{i-1} > \cdots > s_1 > 0$ and are distribution functions (of some random variable taking on, perhaps, a value $+\infty$ as well) for fixed $s_{i-1},\dots,s_0$. Furthermore, because of (A) (Subsection 18.3.1),

$$\lim_{t\downarrow s_{i-1}<\infty} F_i(t;s_{i-1},\dots,s_0)=0. \tag{18.51}$$

Let

$$\begin{aligned} H_1(t_1) &= F_1(t_1;0),\\ H_2(t_1,t_2) &= \int_0^{t_1} F_2(t_2;s_1,0)d_{s_1}F_1(s_1;0),\\ &\vdots\\ H_i(t_i,\dots,t_1) &= \int_0^{t_1}\cdots\int_0^{t_{i-1}} F_i(t_i;s_{i-1},\dots,s_1,0)\\ &\quad \times d_{s_{i-1}}F_{i-1}(s_{i-1};s_i,\dots,s_1)\cdots\\ &\quad \cdots\times d_{s_2}F_2(s_2;s_1,0)d_{s_1}F_1(s_1;0). \end{aligned} \tag{18.52}$$

It follows from Kolmogorov's theorem on measure extension that there exist a probability space $(\Omega,\mathcal{F},P)$ and random variables $\sigma_1,\sigma_2,\dots$, given on it such that

$$P\{\sigma_1\le t_1,\dots,\sigma_i\le t_i\}=H_i(t_i,\dots,t_1).$$

In this case

$$\begin{aligned} P\{\sigma_2\le\sigma_1<\infty\} &= \int_{\{(t_1,t_2):t_2\le t_1<\infty\}} H_2(dt_2,dt_1)\\ &= \int_{\{(t_1,t_2):t_2\le t_1<\infty\}} d_{t_2}F_2(t_2;t_1,0)d_{t_1}F_1(t_1;0)\\ &= \int_{(0,\infty)} d_{t_1}F_1(t_1;0)\left[\int_{(0,t_1]} d_{t_2}F_2(t_2;t_1,0)\right]=0 \end{aligned}$$

since $F_2(t_2;t_1,0)=0$ for $0\le t_2\le t_1<\infty$.

It can be verified in a similar way that $P\{\sigma_{i+1}\le\sigma_i<\infty\}=0$. Therefore, the relation $\sigma_i=\sigma_j=\infty$ holds (P-a.s.) on the set $\{\omega:\sigma_j\le\sigma_i\}$, $j>i$.

Let us denote by Ω_0 the set of $\omega\in\Omega$ for which there are i, $i=1,2,\dots$, such that $\sigma_{i+1}(\omega)=\sigma_i(\omega)<\infty$. It is clear that

$$P(\Omega_0)\le\sum_i P\{\sigma_{i+1}=\sigma_i<\infty\}=0.$$

Let us consider now the mapping φ of a space $\Omega\setminus\Omega_0$ into a space X defined by the formula $x_t(\omega)=\sum_{i\ge1} I_{\{\sigma_i(\omega)\le t\}}$. In other words, each point $\omega\in\Omega\setminus\Omega_0$ can be associated with a step function $x\in\mathsf{X}$ such that the time $\tau_i(x)$ of its ith jump is equal to $\sigma_i(\omega)$.

We shall denote by ν a measure on $(\mathsf{X}, \mathcal{B})$ induced by the mapping φ of the space $\Omega \setminus \Omega_0$ into X: $\mu(B) = P\{\varphi^{-1}(B)\}$, $B \in \mathcal{B}$.

The process $X = (x_t, \mathcal{B}_t, \mu)$, $t \geq 0$, is a point process and, by virtue of the decomposition (18.6),

$$x_t = m_t(x) + a_t(x), \quad t < \tau_\infty(x),$$

where $a = (a_t(x), \mathcal{B}_t)$ is the compensator of the constructed process. By virtue of Theorem 18.2, $a_t(x) = \sum_{i \geq 1} a_t^{(i)}(x)$ where the $a_t^{(i)}(x)$ are defined according to (18.23). Recalling how the measure μ was constructed and the definition of $A_t^{(i)}(x)$, and noting that $a_t^{(i)}(x) = A_t^{(i)}(x)$, $i = 1, 2, \ldots$, it follows, therefore, that $a_t(x) = A_t(x)$.

18.3.5.

EXAMPLE 1. Let $A_t, t \geq 0$ be a deterministic right continuous nondecreasing function with $A_0 = 0$, $\Delta A_t \leq 1$. Then A_t, $t \geq 0$, is the compensator of a certain point process.

In fact, it suffices to set

$$Q_1(s; 0) = A_s, \quad Q_i(s_i; s_{i-1}, \ldots, s_1, 0) = A_{s_i} - A_{s_i \wedge s_{i-1}}, \quad i \geq 2.$$

EXAMPLE 2. The process $A = (A_t(x), \mathcal{B}_t)$, $t \geq 0$, with continuous (for each $x \in \mathsf{X}$) nondecreasing trajectories, $A_0(x) = 0$, is the compensator of a point process.

Indeed the processes $A_t^{(1)}(x) = A_{t \wedge \tau_1(x)}(x)$ and $A_t^{(i)}(x) = A_{t \wedge \tau_{i-1}(x)}(x) - A_{t \wedge \tau_i(x)}(x)$ are adapted to the families $(\mathcal{B}_{t \wedge \tau_i(x)})$, $t \geq 0$, $i \geq 1$, respectively, which implies the existence of the functions $Q_i(s_i; s_{i-1}, \ldots, s_0)$ satisfying (A)–(C) (Section 18.3.1).

18.3.6.

Theorem 18.5 (Theorem on Uniqueness). *Let $A = (A_t(x), \mathcal{B}_t, \mu)$ and $B = (B_t(x), \mathcal{B}_t, \nu)$, $t \geq 0$, be compensators of point processes $(x_t, \mathcal{B}_t, \mu)$ and $(x_t, \mathcal{B}_t, \nu)$, $t \geq 0$, respectively. Let $\theta = \theta(x)$ be a Markov time (with respect to the system $(\mathcal{B}_t)$, $t \geq 0$) such that μ (or ν) (P-a.s.), $A_{t \wedge \theta} = B_{t \wedge \theta}$. Then the narrowings $\mu / \mathcal{B}_{\theta \wedge \tau_\infty}$ and $\nu / \mathcal{B}_{\theta \wedge \tau_\infty}$ of the measures μ and ν by the σ-algebra $\mathcal{B}_{\theta \wedge \tau_\infty} = \sigma(\cup_n \mathcal{B}_{\theta \wedge \tau_n})$ coincide.*

We shall, prove this theorem on the basis of the two lemmas which follow.

Let $N = (x_t, \mathcal{B}_t, \mu)$, $t \geq 0$, be a point process with a compensator $A = (A_t(x), \mathcal{B}_t)$ and let $\theta = \theta(x)$ be a Markov time. Let us consider the 'stopped' process $\tilde{N} = (\tilde{x}_t, \tilde{\mathcal{B}}_t, \mu)$, $t \geq 0$, where

$$\tilde{x}_t = x_{t \wedge \theta}(x), \quad \tilde{\mathcal{B}}_t = \sigma\{x : x_{s \wedge \theta}, s \leq t\}.$$

Since the σ-algebra $\tilde{\mathcal{B}}_t = \mathcal{B}_{t \wedge \theta}$ (see, for example, Theorem 6, Chapter 1, in [285]), and the restricted measure μ on a σ-algebra $\mathcal{B}_\theta$ is $\tilde{\mu} = \mu / \mathcal{B}_\theta$, $\tilde{N} = (\tilde{x}_t, \tilde{\mathcal{B}}_t, \tilde{\mu})$, $t \geq 1$. Denote by $\tilde{A} = (\tilde{A}_t(\tilde{x}), \tilde{B}_t)$ the compensator of this point process.

Lemma 18.9. *If* $\tilde{x}_t = x_{t\wedge\theta(x)}$, $t \geq 0$, *then* $\tilde{A}_t(\tilde{x}) = A_{t\wedge\theta}(x)$.

The proof follows immediately from the fact that the process

$$(\tilde{x}_t - \tilde{A}_t, \tilde{\mathcal{B}}_t, \tilde{\mu}) = x_{t\wedge\theta} - A_{t\wedge\theta}, \mathcal{B}_{t\wedge\theta}, \mu)$$

is a local martingale. □

Lemma 18.10. *Let* $(\Omega, \mathcal{F}, P)$ *be a probability space on which a random variable* $\alpha = \alpha(\omega)$ *and a sub-*σ*-algebra* $\mathcal{G} \subseteq \mathcal{F}$ *are given. Let* $F(t,\omega) = P(\alpha \leq t|\mathcal{G})$ *be a regular conditional distribution function and let* $\tilde{F}(t,\omega)$ *be a measurable nondecreasing, right continuous (over* t *for fixed* ω*) function such that* $0 \leq \tilde{F}(t,\omega) \leq 1$ *and* $\tilde{F}(t\wedge\alpha,\omega) = F(t\wedge\alpha,\omega)$ *(*P-a.s.*) for each* $t \in \mathbb{R}$. *Then* $\tilde{F}(t,\omega) = F(t,\omega)$ *(*P-a.s.*) for each* $t \in \mathbb{R}$.

PROOF. Let us consider first the case where the σ-algebra $\mathcal{G} = \{\emptyset, \Omega\}$ is trivial, and $\tilde{F}(t,\omega)$ is independent of ω. Then the assertion of the lemma consists of the fact that the equality $\tilde{F}(t\wedge\alpha) = F(t\wedge\alpha)$ (P-a.s.), $t \in \mathbb{R}$, implies the correspondence of the functions $\tilde{F}(t)$ and $F(t)$.

Write $\beta = \inf\{t \in \mathbb{R} : F(t) = 1\}$ ($\beta = \infty$ if for all $t \in \mathbb{R}$, $F(T) < 1$). Let a point β ($\beta < \infty$) be a discontinuity point of the function $F(t)$, i.e., $1 = F(\beta) > F(\beta-)$. Then $P\{\alpha = \beta\} > 0$ and, therefore, the set $\{\omega : \alpha(\omega) = \beta\}$ is nonempty. Let us take an arbitrary point $\omega^0 \in \{\omega : \alpha(\omega) = \beta\}$. Then $\tilde{F}(t\wedge\beta) = \tilde{F}(t\wedge\alpha(\omega^0)) = F(t\wedge\alpha(\omega^0)) = F(t\wedge\beta)$ and, therefore, $\tilde{F}(t) = F(t)$ for all $t \leq \beta$. In addition, since $F(\beta) = 1$ and thus $\tilde{F}(t) = F(t)$ for all $t \in \mathbb{R}$. Let the point β ($\beta < \infty$) be a continuity point. Then $P\{\beta - \varepsilon \leq \alpha(\omega) \leq \beta\} > 0$, $\varepsilon > 0$. By taking a point

$$\omega' \in \{\omega : \beta - \varepsilon \leq \alpha(\omega) \leq \beta\},$$

as before we can show that $\tilde{F}(t) = F(t)$, $t \leq \beta - \varepsilon$. Because of the arbitrariness of $\varepsilon > 0$ we have that $\tilde{F}(t) = F(t)$ for all $t \in \mathbb{R}$.

The case with $\beta = \infty$ can be treated similarly.

Let us take now a general case. Since

$$\begin{aligned} 0 &= MI_{\{F(t\wedge\alpha,\omega)\neq\tilde{F}(t\wedge\alpha,\omega)\}}(\omega) \\ &= MM[I_{\{F(t\wedge\alpha,\omega)\neq\tilde{F}(t\wedge\alpha,\omega)\}}(\omega)|\mathcal{G}], \end{aligned}$$

for almost all (α,ω) (with respect to measure $F(d\alpha,\omega)P(d\omega)$) $F(t\wedge\alpha,\omega) = \tilde{F}(t\wedge\alpha,\omega)$. As in the case discussed above, we can deduce that, for almost all ω and any $t \in \mathbb{R}$, $F(t,\omega) = \tilde{F}(t,\omega)$. □

PROOF OF THEOREM 18.5. Using the notation given before (Lemma 18.9), we shall note that it suffices to show that the equality

$$\tilde{\mu}\{x : \tilde{A}_t(x) = \tilde{B}_t(x), t \geq 0\} = 1$$

implies that $\tilde{\mu} = \tilde{\nu}$. Hence, without loss of generality, we need only establish that $\mu\{x : A_t(x) = B_t(x), t \geq 0\} = 1$ implies $\mu = \nu$.

Let

$$\begin{aligned} F_i^A(t) &= \mu\{\tau_i(x) \leq t | \tau_{i-1}, \dots, \tau_1, \tau_0\}, \\ F_i^B(t) &= \nu\{\tau_i(x) \leq t | \tau_{i-1}, \dots, \tau_1, \tau_0\}, \quad \tau_0 \equiv 0, \quad i = 1, 2, \dots. \end{aligned}$$

Then, by virtue of (18.23), for $t \leq \tau_i(x)$ it follows that

$$dF_i^A(t) = [1 - F_i^A(t-)]dA_t^{(i)}, \quad dF_i^B(t) = [1 - F_i^B(t-)]dB_t^{(i)}.$$

But $A_t^{(i)} = B_t^{(i)}$, hence, by virtue of Lemma 18.8, (μ-a.s.)

$$F_i^A(t \wedge \tau_i) = F_i^B(t \wedge \tau_i), \quad i = 1, 2, \dots; t \geq 0,$$

and, using Lemma 18.10, this implies the equality (μ-a.s.)

$$F_i^A(t) = F_i^B(t), \quad i = 1, 2, \dots, \ t \geq 0.$$

Let us note that $F_1^A(t)$ and $F_1^B(t)$ are deterministic functions. Hence, the fact that they are equal implies that the measures μ and ν coincide on the σ-algebra $\mathcal{B}_{\tau_1}$. Furthermore, $F_2^A(t) = \mu\{\tau_2 \leq t | \tau_1\}$ coincides (μ-a.s.) with $F_2^B(t) = \nu\{\tau_2 \leq t | \tau_1\}$. Hence, by taking the relation $\mu/\mathcal{B}_{\tau_1} = \nu/\mathcal{B}_{\tau_1}$ into account, we find that $\mu\{\tau_2 \leq t, \tau_1 \leq s\} = \nu\{\tau_2 \leq t, \tau_1 \leq s\}$ and, therefore, $\mu/\mathcal{B}_{\tau_2} = \nu/\mathcal{B}_{\tau_2}$. Similarly, for any n, $\mu/\mathcal{B}_{\tau_n} = \nu/\mathcal{B}_{\tau_n}$ and, therefore, $\mu/\mathcal{B}_\infty = \nu/\mathcal{B}_\infty$ where $\mathcal{B}_\infty = \sigma(\cup_n \mathcal{B}_{\tau_n})$.

18.4 Stieltjes Stochastic Integrals

18.4.1. Let $(\Omega, \mathcal{F}, P)$ be some probability space with a distinguished family $(\mathcal{F}_t)$, $t \geq 0$, of right continuous sub-σ-algebras of $\mathcal{F}$ augmented by sets from $\mathcal{F}$ of zero probability.

Let us consider a point process $N = (N_t, \mathcal{F}_t)$ with compensator $A = (A_t, \mathcal{F}_t)$, $t \geq 0$. The trajectories $N_t(\omega)$, $t \geq 0$, are nondecreasing right continuous functions of $t < \tau_\infty$ for each $\omega \in \Omega$, and the trajectories are the same functions as those above for almost all $\omega \in \Omega$.

Let

$$C = \{(t, \omega) : t < \tau_\infty(\omega)\}, \quad C_t = \{\omega : (t, \omega) \in C\}, \quad C_\omega = \{t : (t, \omega) \in C\},$$

$$f = (f_t(\omega), \mathcal{F}_t), \quad t \geq 0,$$

be a nonnegative process of class Φ_1 (see Definition 1 in Section 5.4). Then for each $t > 0$ and for almost all $\omega \in C_t$, we can define the Stieltjes integrals

$$\int_0^t f_s(\omega) dN_s, \quad \int_0^t f_s(\omega) dA_s, \tag{18.53}$$

understood (for fixed ω) as Lebesgue–Stieltjes integrals over a set $(0,t]$.

Stieltjes integrals play just as an essential role in the theory of point process as that which Itô's stochastic integrals play in the theory of diffusion-type processes.

Before investigating the properties of such integrals we shall consider an example which may reveal the significance of predictable processes (i.e., processes of class $\Phi_3 \subseteq \Phi_1$; see Definition 3, Section 5.4) in the theory of integration in the Stieltjes sense.

EXAMPLE 3. Let $N = (N_t, \mathcal{F}_t)$, $t \geq 0$, be a Poisson process with parameter $\lambda > 0$. Then $A_t = \lambda t$ and, if $f_s = N_s$, it follows that

$$\int_0^t N_s dN_s = \sum_{s \leq t} N_s [N_s - N_{s-}] = 1 + 2 + \cdots + N_t = \frac{N_t(N_t+1)}{2}$$

and, therefore,

$$M \int_0^t N_s dN_s = \frac{1}{2}[MN_t^2 + MN_t] = \frac{\lambda^2 t^2}{2} + \lambda t.$$

Further,

$$M \int_0^t N_s dA_s = \frac{\lambda^2 t^2}{2},$$

and the integral $\int_0^t N_s dm_s$ (over the martingale $m_s = N_s - A_s$, so that $\int_0^t N_s dN_s - \int_0^t N_s dA_s$), is such that $M \int_0^t N_s dm_s = \lambda t$. Similarly,

$$M \left[\int_s^t N_u dm_u \middle| \mathcal{F}_s^N \right] = \lambda(t-s).$$

Thus, unlike the Itô stochastic integrals

$$\int_0^t f_s dW_s \quad (f \in \mathcal{M}_t),$$

the Stieltjes integrals (over the martingale $m_s = N_s - A_s$) are not, in general, martingales (for $f \in \Phi_1$).

It is not difficult to establish, however, that if we consider the integrals $\int_0^t N_s - dN_s$ and $\int_0^t N_s - dA_s$ instead of the integrals $\int_0^t N_s dN_s$ and $\int_0^t N_{s-} dA_s$, by virtue of the equalities

$$\begin{aligned} \int_0^t N_{s-} dN_s &= \sum_{s \leq t} N_{s-}[N_s - N_{s-}] = \sum_{s \leq t} N_s [N_s - N_{s-}] \\ &\quad - \sum_{s \leq t} [N_s - N_{s-}]^2 = \frac{N_t(N_t - 1)}{2} \end{aligned}$$

the expectation $M\int_0^t N_{s-}dN_s = \lambda^2 t^2/2$ and, therefore, $M\int_0^t N_{s-}dm_s = 0$. Similarly, $M[\int_s^t N_{u-}dm_u|\mathcal{F}_s] = 0$; thus, the process $(\int_0^t N_{s-}dm_s, \mathcal{F}_s)$, $t \geq 0$ is a martingale.

It follows that the stochastic integral $\int_0^t N_{s-}dm_s$ of the predictable function (N_{s-}) over the martingale (m_s) is also a martingale, This property of predictable functions explains the role they play in the investigation of Stieltjes integrals.

18.4.2. We shall note some properties of Stieltjes stochastic integrals

$$\int_0^t f_s dN_s, \quad \int_0^t f_s dA_s \quad \text{and} \quad \int_0^t f_s dm_s \quad \left(= \int_0^t f_s dN_s - \int_0^t f_s dA_s\right).$$

Theorem 18.6. *Let $f = (f_t, \mathcal{F}_t)$, $t \geq 0$, be a nonnegative process with $P(f_t < \infty) = 1$, $t \geq 0$.*

If $f \in \Phi_3$, then

$$M\int_0^{\tau_\infty} f_s dN_s = M\int_0^{\tau_\infty} f_s dA_s. \tag{18.54}$$

If $f \in \Phi_1$, there exists a nonnegative process $\tilde{f} \in \Phi_3$ such that

$$M\int_0^{\tau_\infty} f_s dN_s = M\int_0^{\tau_\infty} \tilde{f}_s dA_s. \tag{18.55}$$

If $f \in \Phi_3$, then for any C, $0 < C < \infty$, (P-a.s.)

$$\left\{\int_0^{\tau_\infty} (f_s \wedge C)dA_s < \infty\right\} = \left\{\int_0^{\tau_\infty} f_s dN_s < \infty\right\}, \tag{18.56}$$

i.e., the symmetric difference of these sets has a zero P-probability.

PROOF. We shall note first that by the integrals $\int_0^{\tau_\infty} f_s dN_s$ and $\int_0^{\tau_\infty} f_s dA_s$ we mean limits of the respective integrals $\int_0^{\tau_n} f_s dN_s$ and $\int_0^{\tau_n} f_s dA_s$ as $n \to \infty$.

To prove (18.54) we need only verify that it holds only for the functions of the form $f_s = I_{\{s \leq \tau_n\}} \cdot I_{\{a < s \leq b\}} \cdot \xi$ where ξ is a $\mathcal{F}_a$-measurable nonnegative random variable with $\xi \leq k < \infty$, since the general case follows from the given case by virtue of the monotone convergence theorem (Theorem 1.1). We have

$$\begin{aligned} M\int_0^{\tau_\infty} I_{\{s \leq \tau_n\}} \cdot I_{\{a < s \leq b\}} \cdot \xi dN_s &= M\xi[N_{b\wedge\tau_n} - N_{a\wedge\tau_n}] \\ &= M(\xi M[m_{b\wedge\tau_n} - m_{a\wedge\tau_n}|\mathcal{F}_a]) \\ &\quad + M\xi[A_{b\wedge\tau_n} - A_{a\wedge\tau_n}] \\ &= M\int_0^{\tau_\infty} I_{\{s \leq \tau_n\}} \cdot I_{\{a < s \leq b\}} dA_s, \end{aligned}$$

since the process $(m_{t\wedge\tau_n}, \mathcal{F}_t)$, $t \geq 0$, is a uniformly integrable martingale and (Theorem 3.6, Note 3)

$$M[m_{b\wedge\tau_n} - m_{a\wedge\tau_n}|\mathcal{F}_a] = 0 \quad (P\text{-a.s.}).$$

To prove (18.55) we shall introduce the following notation. Let $\mathcal{B}([0,\infty))\times\mathcal{F}$ be a σ-algebra of sets (t,ω) in $[0,\infty)\times\Omega$ and $\mathcal{P} = \mathcal{P}(F)$ be its sub-σ-algebra of predictable sets, i.e., the smallest σ-algebra on $[0,\infty)\times\Omega$, generated by nonanticipative (in other words, $F = (\mathcal{F}_t)$-adapted) processes which have left continuous trajectories on $(0,\infty)$. (It is useful to note that the σ-algebra $\mathcal{P} = \mathcal{P}(F)$ coincides with the smallest σ-algebra generated by stochastic intervals $[\tau,\infty] = \{(t,\omega) : t \geq \tau(\omega)\}$, where the τ are predictable (with respect to the system of σ-algebras $F = (\mathcal{F}_t)$) times. It also coincides (we took advantage of this fact in the proof of (18.54)) with the smallest σ-algebra generated by sets of the form $\{0\}\times B$ ($B\in\mathcal{F}_0$) and $(s,t]\times B$ where $B\in\mathcal{F}_s$; see [49], chapter IV). It can be easily shown that $\mathcal{P}(F)\supseteq\mathcal{P}(G)$, where the system of nondecreasing σ-algebras $G = (\mathcal{G}_t)$, $t\geq 0$, is such that $\mathcal{F}_t\supseteq\mathcal{G}_t$, $t\geq 0$. The system $G\equiv F^N = (\mathcal{F}_t^N)$, $t\geq 0$, where N is the point process in question, will play a particular role from now on. For simplicity of notation we shall use $\hat{\mathcal{P}}$ instead of $\mathcal{P}(F^N)$.

Let $\Phi = \{\beta\}$ be an aggregate of nonnegative predicable processes $\beta = (\beta_t(\omega),\mathcal{F}_t)$, $t\geq 0$, such that $\beta_t^2(\omega) = \beta_t(\omega)$, $\omega\in\Omega$. We shall define on the sets $B = \{(t,\omega) : \beta_t(\omega) = 1\}\in\mathcal{P}$ two σ-finite measures

$$\mathcal{N}_f(B) = M\int_0^{\tau_\infty} f_s\beta_s dN_s, \quad \mathcal{N}(B) = M\int_0^{\tau_\infty}\beta_s dN_s.$$

We shall show that the measure $\mathcal{N}_f \ll \mathcal{N}$. In fact, let $\mathcal{N}(B) = 0$. Then

$$0 = \mathcal{N}(B) = M\int_0^{\tau_\infty}\beta_s dN_s = M\sum_{n\geq 1}\beta_{\tau_n}.$$

Therefore, $P(\beta_{\tau_n} = 0) = 1$, $n\geq 1$, and

$$\mathcal{N}_f(B) = M\sum_{n\geq 1} f_{\tau_n}\beta_{\tau_n} = 0.$$

Hence, by the Radon–Nikodym theorem there exists a $\mathcal{P}$ measurable nonnegative function $\tilde{f} = (\tilde{f}_s,\mathcal{F}_s)$ such that

$$\mathcal{N}_f(B) = \int_B \tilde{f}_s(\omega)d\mathcal{N}(s,\omega),$$

i.e., for any function $\beta\in\Phi$,

$$M\int_0^{\tau_\infty} f_s\beta_s dN_s = M\int_0^{\tau_\infty}\tilde{f}_s\beta_s dN_s.$$

But, by (18.54),

$$M\int_0^{\tau_\infty}\tilde{f}_s\beta_s dN_s = M\int_0^{\tau_\infty}\tilde{f}_s\beta_s dA_s$$

and, therefore

$$M\int_0^{\mathcal{T}_\infty} f_s\beta_s dN_s = M\int_0^{\mathcal{T}_\infty} \tilde{f}_s\beta_s dA_s.$$

In particular, we obtain (18.55) from this.

Note. If we consider the restriction of measures $\mathcal{N}_f$ and $\mathcal{N}$ to the σ-algebra $\hat{\mathcal{P}} = \mathcal{P}(F^N) \subseteq \mathcal{P}(F) = \mathcal{P}$, we can similarly establish the existence of a $\hat{\mathcal{P}}$-measurable function

$$\hat{f} = (\hat{f}_s, \mathcal{F}_s^N)$$

such that

$$\mathcal{N}_f(B) = \int_B \hat{f}_s(\omega)d\mathcal{N}(s,\omega), \quad B \in \hat{\mathcal{P}},$$

and

$$M\int_0^{\mathcal{T}_\infty} f_s dN_s = M\int_0^{\mathcal{T}_\infty} \hat{f}_s dA_s.$$

It is natural to call the function $\hat{f}_s(\omega)$ the 'conditional mathematical expectation of $f_s(\omega)$ with respect to the σ-algebra $\hat{P}$ with measure $\mathcal{N}$'. In connection with this remark, we shall use also the notation $M_{\mathcal{N}}(f|\hat{\mathcal{P}})_t(\omega)$ or $M_{\mathcal{N}}(f|\hat{\mathcal{P}})_t$ for $\hat{f}_s(\omega)$.

We can make similar remarks regarding the function $\tilde{f}_s(\omega)$.

We shall introduce the notation

$$\begin{aligned} f_{\leq 1}(t) &= I_{\{f_t \Delta A_t \leq 1\}} \cdot f_t, \\ f_{>1}(t) &= I_{\{f_t \Delta A_t > 1\}} \cdot f_t. \end{aligned}$$

$$\mathsf{A} = \left\{\omega : \int_0^{\mathcal{T}_\infty} f_s dA_s < \infty\right\}, \quad \mathsf{N} = \left\{\omega : \int_0^{\mathcal{T}_\infty} f_s dN_s < \infty\right\}.$$

We shall show first that up to sets of P-measure zero $\mathsf{A} \subseteq \mathsf{N}$, i.e., $P(I_{\mathsf{A}} \leq I_{\mathsf{N}}) = 1$.

For almost all $\omega \in \mathsf{A}$,

$$\sum_{t\geq 0} f_{>1}(t)\Delta A_t = \int_0^{\mathcal{T}_\infty} f_{>1}(t)dA_t \leq \int_0^{\mathcal{T}_\infty} f_t dA_t < \infty$$

and, therefore, the sum on the left-hand side of the above inequality contains only a finite number of terms. Consequently, for almost all $\omega \in \mathsf{A}$,

$$\int_0^{\mathcal{T}_\infty} f_{>1}(t)dN_t = \sum_{t\geq 0} f_{>1}(t)\Delta A_t(\Delta A_t)^+\Delta N_t < \infty. \tag{18.57}$$

We shall next show that for almost all $\omega \in \mathsf{A}$, $\int_0^{\tau_\infty} f_{\le 1}(t)dN_t < \infty$. Let

$$\sigma_k = \inf\left\{t \geq 0 : \int_0^t f_{\le 1}(s)dA_s \geq k\right\}, \quad k = 1, 2, \ldots.$$

Then, by (18.54),

$$M\int_0^{\tau_\infty \wedge \sigma_k} f_{\le 1}(s)dN_s = M\int_0^{\tau_\infty \wedge \sigma_k} f_{\le 1}(s)dA_s \leq k+1,$$

and, therefore,

$$P\left(\int_0^{\tau_\infty \wedge \sigma_k} f_{\le 1}(s)dN_s = \infty\right) = 0.$$

Hence,

$$\begin{aligned}
&P\left(\int_0^{\tau_\infty} f_{\le 1}(s)dN_s = \infty, \int_0^{\tau_\infty} f_s dA_s < \infty\right) \\
&= P\left(\int_0^{\tau_\infty} f_{\le 1}(s)dN_s = \infty, \int_0^{\tau_\infty} f_s dA_s < \infty, \sigma_k < \infty\right) \\
&\quad + P\left(\int_0^{\tau_\infty} f_{\le 1}(s)dN_s = \infty, \int_0^{\tau_\infty} f_s dA_s < \infty, \sigma_k = \infty\right) \\
&\leq P\left(\int_0^{\tau_\infty} f_s dA_s < \infty, \sigma_k < \infty\right) + P\left(\int_0^{\tau_\infty \wedge \sigma_k} f_{\le 1}(s)dN_s = \infty\right) \\
&= P\left(\int_0^{\tau_\infty} f_{\le 1}(s)dA_s < \infty, \sigma_k < \infty\right) \to 0, \quad k \to \infty. \qquad (18.58)
\end{aligned}$$

It follows from (18.57) and (18.58) that, modulo sets of P-measure zero $\mathsf{A} \subseteq \mathsf{N}$ and, in particular, for $0 < C < \infty$,

$$\left\{\int_0^{\tau_\infty} (f_t \wedge C)dA_t < \infty\right\} \subseteq \left\{\int_0^{\tau_\infty} (f_t \wedge C)dN_t\right\}. \qquad (18.59)$$

We shall establish the inclusion

$$\left\{\int_0^{\tau_\infty} (f_t \wedge C)dA_t < \infty\right\} \subseteq \left\{\int_0^{\tau_\infty} f_t dN_t < \infty\right\}. \qquad (18.60)$$

Since (P-a.s.)

$$C\int_0^{\tau_\infty} I_{\{f_t > C\}} dA_t \leq \int_0^{\tau_\infty} (f_t \wedge C)dA_t, \qquad (18.61)$$

it follows that

$$\left\{\int_0^{\tau_\infty} (f_t \wedge C)dA_t < \infty\right\} \subseteq \left\{\int_0^{\tau_\infty} I_{\{f_t > C\}} dA_t < \infty\right\},$$

which together with the inclusion established

$$\left\{\int_0^{\tau_\infty} I_{\{f_t>C\}} dA_T < \infty\right\} \subseteq \left\{\int_0^{\tau_\infty} I_{\{f_t>C\}} dN_t < \infty\right\},$$

yields

$$\left\{\int_0^{\tau_\infty} (f_t \wedge C) dA_t < \infty\right\} \subseteq \left\{\int_0^{\tau_\infty} I_{\{f_t>C\}} dN_t < \infty\right\}. \tag{18.62}$$

But

$$\left\{\int_0^{\tau_\infty} I_{\{f_t>C\}} dN_t < \infty\right\} = \left\{\int_0^{\tau_\infty} f_t I_{\{f_t>C\}} dN_t < \infty\right\}.$$

Hence,

$$\left\{\int_0^{\tau_\infty} (f_t \wedge C) dA_t < \infty\right\} \subseteq \left\{\int_0^{\tau_\infty} f_t I_{\{f_t>C\}} dN_t < \infty\right\}. \tag{18.63}$$

From (18.59), (18.63) and the inequality $f_t \leq (f_t \wedge C) + f_t I_{\{f_t>C\}}$, we obtain the required inclusion, (18.60).

We shall establish the inverse inclusion,

$$\left\{\int_0^{\tau_\infty} f_t dN_t < \infty\right\} \subseteq \left\{\int_0^{\tau_\infty} (f_t \wedge C) dA_t < \infty\right\}. \tag{18.64}$$

To this end we shall note first that, since

$$\int_0^{\tau_\infty} (f_t \wedge C) dN_t \leq \int_0^{\tau_\infty} f_t dN_t,$$

then

$$\left\{\int_{\tau_\infty} f_t dN_t < \infty\right\} \subseteq \left\{\int_0^{\tau_\infty} (f_t \wedge C) dN_t < \infty\right\}.$$

Hence to prove (18.64) we need only show that

$$\left\{\int_0^{\tau_\infty} (f_t \wedge C) dN_t < \infty\right\} \subseteq \left\{\int_0^{\tau_\infty} (f_t \wedge C) dA_t < \infty\right\}. \tag{18.65}$$

Let us set $\theta_k = \inf\{t \geq 0 : \int_0^t (f_t \wedge C) dN_t \geq k\}$, $k = 1, 2, \ldots$. Then it is clear that

$$M \int_0^{\tau_\infty \wedge \theta_k} (f_t \wedge C) dA_t = M \int_0^{\tau_\infty \wedge \theta_k} (f_t \wedge C) dN_t \leq k + C$$

and, therefore,

$$P\left(\int_0^{\tau_\infty \wedge \theta_k} (f_t \wedge C) dA_t = \infty\right) = 0. \tag{18.66}$$

It is also obvious that, by virtue of the definition of times θ_k, $k = 1, 2, \ldots$,

$$\lim_{k\to\infty} P\left(\int_0^{\tau_\infty} (f_t \wedge C)dN_t < \infty, \theta_k < \infty\right) = 0. \tag{18.67}$$

But

$$\begin{aligned} &P\left(\int_0^{\tau_\infty} (f_t \wedge C)dA_t = \infty, \int_0^{\tau_\infty} (f_t \wedge C)dN_t < \infty\right) \\ &= P\left(\int_0^{\tau_\infty} (f_t \wedge C)dA_t = \infty, \int_0^{\tau_\infty} (f_t \wedge C)dN_t < \infty, \theta_k < \infty\right) \\ &\quad +P\left(\int_0^{\tau_\infty} (f_t \wedge C)dA_t = \infty, \int_0^{\tau_\infty} (f_t \wedge C)dN_t < \infty, \theta_k = \infty\right) \\ &\le P\left(\int_0^{\tau_\infty} (f_t \wedge C)dN_t < \infty, \theta_k < \infty\right) + P\left(\int_0^{\tau_\infty \wedge \theta_k} (f_t \wedge C)dA_t = \infty\right), \end{aligned}$$

from which (18.65) follows immediately if we make use of (18.66) and (18.67). □

Corollary 1. *Modulo sets of P-measure zero,*

$$\{A_{\tau_\infty} < \infty\} = \{N_{\tau_\infty} < \infty\}. \tag{18.68}$$

Corollary 2. *If $f \in \Phi_3$, for any C, $0 < C < \infty$,*

$$P\left(\int_0^{\tau_\infty} f_s dN_s < \infty\right) = 1 \Leftrightarrow P\left(\int_0^{\tau_\infty} (f_s \wedge C)dA_s < \infty\right) = 1. \tag{18.69}$$

Corollary 3. *Let $\sigma_\infty = \inf\{t \ge 0 : A_t = \infty\}$. Then ($P$-a.s.) $\sigma_\infty = \tau_\infty$.*

In fact, it follows from (18.15) that $\tau_\infty \le \sigma_\infty$ (P-a.s.). But if $\tau_\infty(\omega) \le \sigma_\infty(\omega)$, $N_{\tau_\infty(\omega)}(\omega) = \infty$ and $A_{\tau_\infty(\omega)}(\omega) < \infty$, which fact contradicts (18.68).

EXAMPLE. Let $f_t = N_t \cdot I_{\{t \le \tau_n\}}$. We shall show that

$$\hat{f}_t = (N_{t-} + 1) \cdot I_{\{t \le \tau_n\}}. \tag{18.70}$$

Indeed, let $\varphi \in \Phi_3$. Then

$$\begin{aligned} M\int_0^{\tau_\infty} f_t\varphi_t dN_t &= M\int_0^{\tau_n} N_t\varphi_t dN_t \\ &= M\int_0^{\tau_n} \varphi_t[N_{t-} + \Delta N_t]dN_t \\ &= M\int_0^{\tau_n} \varphi_t[N_{t-} + 1]dN_t \\ &= M\int_0^{\tau_\infty} \varphi_t I_{\{t \le \tau_n\}}[N_{t-} + 1]dA_t, \end{aligned}$$

which fact proves (18.70) because of the arbitrariness of $\varphi \in \Phi_3$.

18.4.3. Let $f \in \Phi_3$ and let

$$\int_0^t |f_s| dA_s < \infty \quad (\{t < \tau_\infty\} : (P\text{-a.s.})).$$

Then, by virtue of Theorem 18.6,

$$\int_0^t |f_s| dN_s < \infty \quad (\{t < \tau_\infty\} : (P\text{-a.s.}))$$

and, consequently the variables

$$\mathcal{M}_t = \int_0^t f_s dm_s \quad \left(= \int_0^t f_s dN_s - \int_0^t f_s dA_s \right),$$

where, as usual, $m_t = N_t - A_t$ are well defined and finite ($\{t < \tau_\infty\} : (P$-a.s.).

Theorem 18.7. *Let $f \in \Phi_3$. $P(|f_t| < \infty) = 1$, $t \geq 0$, and $\mathcal{M} = (\mathcal{M}_{t \wedge \tau_\infty}, \mathcal{F}_t)$, $t \geq 0$. Then:*

(a) $M \int_0^{\tau_\infty} |f_t| dA_t < \infty$ implies $\mathcal{M}$ is a uniformly integrable martingale;
(b) $P\{\int_0^t |f_s| dA_s = \infty, t < \tau_\infty\} = 0$ implies $\mathcal{M}$ is a τ_∞-local martingale.

PROOF. To prove (a), let ξ be a $\mathcal{F}_s$-measurable random variable $|\xi| \leq k < \infty$. As in the proof of (18.54), it can be proved that for $s < t \leq \infty$,

$$M\xi \cdot \int_{s \wedge \tau_\infty}^{t \wedge \tau_\infty} f_u dm_u = 0,$$

i.e., $\mathcal{M}$ is a martingale. In particular, $M(\mathcal{M}_{\tau_\infty} | \mathcal{F}_s) = \mathcal{M}_{\tau_\infty \wedge s}$, which fact, according to Theorem 2.7, proves that the martingale $\mathcal{M}$ is uniformly integrable.

Note. Statement (a) holds true only if $M \int_0^{\tau_\infty} |f_t|(1 - \Delta A_t) d_t < \infty$.

To prove (b), we shall need a lemma which is of interest in itself, namely:

Lemma 18.11. *Let $B = (B_t, \mathcal{F}_t)$, $t \geq 0$, be a nondecreasing right continuous predictable process and let $\beta = \inf\{t \geq 0 : B_t = \infty\}$, with $\beta = \infty$ if $\lim_{t \to \infty} B_t < \infty$.*

Then there exists a sequence of Markov times (σ_n), $n = 1, 2, \ldots$, such that $\sigma_n > \beta$, $\sigma_n \uparrow \beta$, and $B_{t \wedge \sigma_n} < n$ (P-a.s.).

PROOF. Let us write $\beta_n = \inf\{t \geq 0 : B_t \geq n\}$, setting $\beta_n = \infty$ if $\lim_{t \to \infty} B_t < n$. The times β_n are predictable (see Example 7 in Section 18.1, and (18.7)).

Therefore, for each $n = 1, 2, \ldots$, there exist times $\tilde{\beta}_n$ such that $\tilde{\beta}_n < \beta_n$ and, without loss of generality, we can assume $P(\beta_n - \tilde{\beta}_n > 2^{-n}) \le 2^{-n}$. Set $\sigma_n = \tilde{\beta}_1 \vee \cdots \vee \tilde{\beta}_n$. Then $\tilde{\beta}_n \le \sigma_n < \beta_n$ (P-a.s.) and, since $P(\beta_n - \tilde{\beta}_n > 2^{-n}) \le 2^{-n}$, by virtue of the Borel–Cantelli lemma (Section 1.1) $\lim_n \sigma_n = \lim_n \beta_n = \beta$, with $\sigma_n \uparrow \beta$. Since $\sigma_n < \beta_n$, $B_{t\wedge\sigma_n} < n$. □

We shall use this lemma to prove (b) of Theorem 18.7. Set $B_t = \int_0^t |f_s| dA_s$ and $\beta = \inf\{t \ge 0 : B_t = \infty\}$. Since $P\{\tau_\infty \le \beta\} = 1$, by virtue of the preceding lemma there exist Markov times (σ_n), $n = 1, 2, \ldots, \sigma_n < \beta, \sigma_n \uparrow \beta$, such that $\int_0^{\tau_\infty\wedge\sigma_n} |f_t| dA_t \le n$ and, therefore, $M\int^{\tau_\infty\wedge\sigma_n} |f_t| dA_t \le n$. By virtue of (a) of Theorem 18.7, the processes $(\mathcal{M}_{t\wedge\tau_\infty\wedge\sigma_n}, \mathcal{F}_t)$ are uniformly integrable martingales for each $n = 1, 2, \ldots$.

This, together with the relations $\tau_\infty \wedge \sigma_n \uparrow \tau_\infty$, $n \to \infty$, indicates that the process $\mathcal{M} = (\mathcal{M}_{t\wedge\tau_\infty}, \mathcal{F}_t)$, $t \ge 0$, is a τ_∞-local martingale. □

18.4.4. By virtue of Theorem 18.1 the process $m = (m_t, \mathcal{F}_t)$, $t < \tau_\infty$, with $m_t = N_t - A_t$, is a τ_∞-local martingale.

We can actually assert more than this, namely, that the process m is locally bounded, i.e., there exists a sequence of Markov times (σ_n), $n = 1, 2, \ldots$, such that $\sigma_n \le \sigma_{n+1}$, $\sigma_n \to \tau_\infty$ (P-a.s) and $\sup_t |m_{t\wedge\sigma_n}| \le k_n < \infty$. Indeed, letting $\sigma_n = \inf\{t \ge 0 : N_t + A_t \ge n\}$ and $\sigma_n = \tau_\infty$, if $N_{\tau_\infty} + A_{\tau_\infty} < n$, then due to the fact that $\Delta N_t \le 1$ and $\Delta A_t \le 1$ (Lemma 18.1), we have

$$|m_{t\wedge\sigma_n}| \le n + 2 \quad (= k_n).$$

The above remark implies, in particular, that the process $m = (m_t, \mathcal{F}_t)$, $t < \tau_\infty$, is a τ_∞-locally square integrable martingale, i.e., there exists a sequence of Markov times (σ_n), $1, 2, \ldots$, such that

$$\sigma_n \le \sigma_{n+1}, \quad \sigma_n \to \tau_\infty \quad (P\text{-a.s.})$$

and

$$\sup_t M m^2_{t\wedge\sigma_n} < \infty.$$

It follows from the fact that the Doob–Meyer decomposition is unique for a τ_∞-local submartingale $(m^2_{t\wedge\tau_\infty}, \mathcal{F}_t)$, $t \ge 0$ that there exists a unique (to within stochastic equivalence) predictable increasing process $\langle m \rangle_t$, $t < \tau_\infty$, such that, for any $n = 1, 2, \ldots$, the process $(m^2_{t\wedge\sigma_n} - \langle m \rangle_{t\wedge\sigma_n}, \mathcal{F}_t)$, $t \ge 0$, is a martingale.

Lemma 18.12. *The process $\langle m \rangle_t$, $t < \tau_\infty$, corresponding to the τ_∞-locally square integrable martingale $m = (m_t, \mathcal{F}_t)$, $t < \tau_\infty$, is defined by the formula*

$$\langle m \rangle_t = \int_0^t (1 - \Delta A_s) dA_s. \tag{18.71}$$

If $MN_{\tau_\infty} < \infty$, *then* $P(\tau_\infty = \infty) = 1$ *and the process* $m = (m_t, \mathcal{F}_t)$, $t \geq 0$, *is a square integrable martingale.*

PROOF. By virtue of (18.40),

$$m_t^2 = 2\int_0^t m_{s-}dm_s + \sum_{s\leq t}(\Delta m_s)^2, \tag{18.72}$$

where

$$(\Delta m_s)^2 = (\Delta N_s - \Delta A_s)^2 = \Delta N_s - 2\Delta N_s \Delta A_s + (\Delta A_s)^2.$$

Hence

$$\begin{aligned} m_t^2 &= 2\int_0^t m_{s-}dm_s + N_t - 2\int_0^t \Delta A_s dN_s + \int_0^t \Delta A_s dA_s \\ &= \int_0^t (2m_{s-} + 1 - 2\Delta A_s)dm_s + \int_0^t (1 - \Delta A_s)dA_s. \end{aligned} \tag{18.73}$$

It can easily be seen that the first integral on the right-hand side of (18.73) is a τ_∞-local martingale, and the second integral is a predictable increasing process. Because of the uniqueness of the Doob–Meyer decomposition, the above gives (18.71).

Let $MN_{\tau_\infty} < \infty$. Then $P(N_{\tau_\infty} < \infty) = 1$ and, therefore, $P(\tau_\infty = \infty) = 1$. Let $\sigma_n = \inf\{t \geq n : N_t + A_t \geq n\}$, assuming $\sigma_n = \infty$ if $N_{\tau_\infty} + A_{\tau_\infty} < n$. Then, from (18.73) and Lemma 18.1, we have

$$Mm_{t\wedge\sigma_n}^2 = M\int_0^{t\wedge\sigma_n} (1 - \Delta A_s)dA_s \leq MA_{\sigma_n} = MN_{\sigma_n} \leq MN_{\tau_\infty} < \infty.$$

From this, by virtue of the Fatou lemma, we obtain

$$Mm_t^2 \leq MN_{\tau_\infty} < \infty,$$

which fact proves that the martingale $m = (m_t, \mathcal{F}_t)$, $t \geq 0$, is square integrable. □

Corollary. *Since* $MN_{\tau_n} \leq n$, *then each of the processes* $(m_{t\wedge\tau_n}, \mathcal{F}_t)$, $t \geq 0$, $n = 1, 2, \ldots$, *is a square integrable martingale.*

18.4.5. Let us consider next the process $\mathcal{M}_t = \int_0^t f_s dm_s$. It follows from Theorem 18.7 that if $f \in \Phi_3$ and $P\{\int_0^t |f_s|dA_s < \infty, t < \tau_\infty\} > 0$, then

$$\mathcal{M} = (\mathcal{M}_{t\wedge\tau_\infty}, \mathcal{F}_t), \quad t \geq 0,$$

will be a τ_∞-local martingale. The following theorem defines the result more exactly.

Theorem 18.8. *Let* $f \in \Phi_3$, $P(|f_t| < \infty) = 1$, $t \geq 0$. *Then:*

(a)

$$M\int_0^{\tau_\infty} f_s^2(1-\Delta A_s)dA_s < \infty \Leftrightarrow P\left\{\int_0^t |f_s|dA_s = \infty, t < \tau_\infty\right\} = 0$$

and $\mathcal{M}$ is a square integrable martingale;

(b)

$$P\left\{\int_0^t f_s^2(1-\Delta A_s)dA_s = \infty, t < \tau_\infty\right\} = 0 \Leftrightarrow \mathcal{M}$$

is a τ_∞-locally square integrable martingale;

(c)

$$P\left\{\int_0^{\tau_\infty} f_s^2(1-\Delta A_s)dA_s < \infty\right\} = 1 \Rightarrow P\left\{\sup_{t\geq 0}|\mathcal{M}_{t\wedge\tau_\infty}| < \infty\right\} = 1.$$

PROOF. As to (a), let $M\int_0^{\tau_\infty} f_s^2(1-\Delta A_s)dA_s < \infty$. We shall show first that, under this condition, $P\{\int_0^t |f_s|dA_s = \infty, t < \tau_\infty\} = 0$ and, therefore, the values $\mathcal{M}_t$ are defined on the sets $\{(t,\omega) : t < \tau_\infty(\omega)\}$.

To this end, we note that, by virtue of the Cauchy–Schwarz inequality

$$\left(\int_0^t |f_s|dA_s\right)^2 \leq A_t \int_0^t f_s^2 dA_s. \tag{18.74}$$

But $N_t < \infty$ ($\{t < \tau_\infty\}$; (P-a.s.)) and, therefore, according to Lemma 18.2, $A_t < \infty$ ($\{t < \tau_\infty\}$; (P-a.s.)). It follows from this and (18.74) that the process $\mathcal{M}_t$ is defined on the set $\{(t,\omega) : t < \tau_\infty(\omega)\}$ if only we can show that the condition $P\{\int_0^{\tau_\infty} f_s^2(1-\Delta A_s)dA_s < \infty\} = 1$ implies the relation

$$P\left\{\int_0^t f_s^2 dA_s = \infty, t < \tau_\infty\right\} = 0.$$

We have

$$\begin{aligned}\int_0^t f_s^2 dA_s &= \int_0^t I_{\{\Delta A_s \leq 1/2\}} f_s^2 dA_s + \int_0^t I_{\{\Delta A_s > 1/2\}} f_s^2 dA_s \\ &= \int_0^t I_{\{\Delta A_s \leq 1/2\}} f_s^2 dA_s + \sum_{s\leq t} f_s^2 \cdot I_{\{\Delta A_s > 1/2\}} \Delta A_s.\end{aligned} \tag{18.75}$$

Since $A_t < \infty$ ($\{t < \tau_\infty\}$; (P-a.s.)), the number of jumps in A_s, $s \leq t$, of magnitude larger than $\frac{1}{2}$ can only be finite. Hence, on the set $\{t < \tau_\infty\}$,

$$\sum_{s\leq t} f_s^2 I_{\{\Delta A_s > 1/2\}} \Delta A_s < \infty.$$

Further,

$$\int_0^t I_{\{\Delta A_s \le 1/2\}} f_s^2 dA_s \le 2 \int_0^t I_{\{\Delta A_s \le 1/2\}} (1-\Delta A_s) f_s^2 dA_s$$
$$\le 2 \int_0^{\tau_\infty} f_s^2 (1-\Delta A_s) dA_s < \infty.$$

Thus, the values $\mathcal{M}_t$ are defined on the set $\{t < \tau_\infty\}$.

We shall consider next the square integrable martingales

$$\tilde{\mathcal{M}}^{(n)} = (\tilde{\mathcal{M}}_t^{(n)}, \mathcal{F}_t), \quad t \ge 0,\ n = 1, 2, \ldots,$$

with

$$\tilde{\mathcal{M}}_t^{(n)} = \int_0^t f_s dm_{s\wedge\tau_n},$$

where the integrals are to be understood as stochastic integrals over the square integrable martingales $(m_{t\wedge\tau_n}, \mathcal{F}_t)$, $t \ge 0$. (The existence of such integrals follows from (18.71), the inequality

$$M \int_0^{\tau_\infty} f_s^2 d\langle m\rangle_{s\wedge\tau_n} = M \int_0^{\tau_n} f_s^2 (1-\Delta A_s) dA_s < \infty,$$

and Theorem 5.10).

For each $t \ge 0$, the sequence $(\tilde{\mathcal{M}}_t^{(n)})$, $n = 1, 2, \ldots$, is fundamental in the mean square sense, since. due to (5.82), we have

$$M\left[\tilde{\mathcal{M}}_t^{(n)} - \tilde{\mathcal{M}}_t^{(m)}\right]^2 = M\left[\int_{t\wedge\tau_m}^{t\wedge\tau_n} f_s dm_s\right]^2$$
$$= M \int_{t\wedge\tau_m}^{t\wedge\tau_n} f_s^2 (1-\Delta A_s) dA_s \to 0 \quad (m < n, m \to \infty, n \to \infty).$$

Consequently, there exists a square integrable process $\tilde{\mathcal{M}} = (\tilde{\mathcal{M}}_t, \mathcal{F}_t)$, $t \ge 0$, such that $\tilde{\mathcal{M}}_t = \lim_n \tilde{\mathcal{M}}_t^{(n)}$. It can easily be seen that this process is a martingale and that

$$\tilde{\mathcal{M}}_{t\wedge\tau_n}^{(n)} = \tilde{\mathcal{M}}_{t\wedge\tau_n}, \quad M[\tilde{\mathcal{M}}_{t\wedge\tau_n} | \mathcal{F}_s] = \tilde{\mathcal{M}}_{s\wedge\tau_n}, \tag{18.76}$$

for $s \le t$, $n = 1, 2, \ldots$.

Since

$$\sup_n M \tilde{\mathcal{M}}_{t\wedge\tau_n}^2 \le M \int_0^{\tau_\infty} f_t^2 (1-\Delta A_t) dA_t < \infty,$$

the sequence $(\tilde{\mathcal{M}}_{t\wedge\tau_n})$, $n = 1, 2, \ldots$, is uniformly integrable (see Section 1.1), we obtain from (18.76)

$$M[\tilde{\mathcal{M}}_{t\wedge\tau_\infty}|\mathcal{F}_s] = \tilde{\mathcal{M}}_{s\wedge\tau_\infty}. \tag{18.77}$$

We shall show that $\mathcal{M}_t = \tilde{\mathcal{M}}_t$ ($\{t < \tau_\infty\}$; (P-a.s.)).

In fact, starting from the definitions of the Stieltjes stochastic integral and the stochastic integral over a square integrable martingale, we can easily show that, for simple functions $f \in \Phi_3$, the pertinent values $\mathcal{M}_{t\wedge\tau_n} = \tilde{\mathcal{M}}^{(n)}_{t\wedge\tau_n}$ and $\tilde{\mathcal{M}}_{t\wedge\tau_n}$ coincide (P-a.s.), $t \geq 0$, $n = 1, 2, \ldots$.

By passing to the limit we can find from the above that these values coincide for all the functions $f \in \Phi_3$. Thus, $\mathcal{M}_t = \tilde{\mathcal{M}}_t$ ($\{t < \tau_\infty\}$; (P-a.s.)).

It follows, from this relation, (18.77) and Theorem 3.6, that

$$\mathcal{M}_{\tau_n} = \tilde{\mathcal{M}}_{\tau_n} = M(\tilde{\mathcal{M}}_{\tau_\infty}|\mathcal{F}_{\tau_n})$$

and, therefore, by virtue of Lévy's theorem (Theorem 1.5), there exists (P-a.s.) a limit $\lim_n \mathcal{M}_{\tau_n}$ which (by definition) is to be taken as the value of the integral

$$\int_0^{\tau_\infty} f_s dm_s.$$

Therefore, for all $t \geq 0$, $\mathcal{M}_{t\wedge\tau_\infty} = \tilde{\mathcal{M}}_{t\wedge\tau_\infty}$ (P-a.s.) which fact proves that the process $\mathcal{M} = (\mathcal{M}_{t\wedge\tau_\infty}, \mathcal{F}_t)$, $t \geq 0$, is a square integrable martingale by virtue of the square integrability of the process $(\tilde{\mathcal{M}}_{t\wedge\tau_\infty})$, $t \geq 0$, and (18.77).

We shall next establish the inverse implication in (a). On the set

$$\{(t, \omega) : t < \tau_\infty(\omega)\},$$

let $\int_0^t |f_s| dA < \infty$ and let $\mathcal{M} = (\mathcal{M}_{t\wedge\tau_\infty}, \mathcal{F}_t)$ be a square integrable martingale.

By virtue of Lemma 18.7,

$$\mathcal{M}_t^2 = 2\int_0^t \mathcal{M}_{s-} d\mathcal{M}_s + \sum_{s\leq t} (\Delta\mathcal{M}_s)^2. \tag{18.78}$$

Obviously,

$$\int_0^t \mathcal{M}_{s-} dM_s = \int_0^t \mathcal{M}_{s-} f_s dm_s,$$

and

$$P\left(\sup_{t\geq 0} |\mathcal{M}_{t\wedge\tau_\infty}| < \infty\right) = 1$$

(Theorem 3.2).

On the set $\{t < \tau_\infty\}$,

$$\int_0^t |\mathcal{M}_{s-} f_s| dA_s \leq \sup_{s\geq 0} |\mathcal{M}_{s\wedge\tau_\infty}| \cdot \int_0^t |f_s| dA_s < \infty \quad (P\text{-a.s.}).$$

Consequently, by virtue of Lemma 18.7, $\mathcal{M}$ is a τ_∞-local martingale. Hence we have, from (18.78) and the Fatou lemma

$$M \sum_{s \le \tau_n} (\Delta \mathcal{M}_s)^2 \le M \mathcal{M}_{\tau_\infty}^2. \tag{18.79}$$

We find from this and (18.54) that, for any C such that $0 < C < \infty$,

$$\begin{aligned}
\infty > M\mathcal{M}_{\tau_\infty}^2 &\ge M \sum_{s \le \tau_n} (\Delta \mathcal{M}_s)^2 \\
&= M \sum_{s \le \tau_n} [f_s(\Delta N_s - \Delta A_s)]^2 \\
&\ge M \sum_{s \le \tau_n} (f_s^2 \wedge C)[\Delta N_s - 2\Delta A_s \Delta N_s + (\Delta A_s)^2] \\
&= M \int_0^{\tau_n} (f_s^2 \wedge C)[1 - 2\Delta A_s] d[N_s - A_s] \\
&\quad + M \int_0^{\tau_n} (f_s^2 \wedge C)[1 - \Delta A_s] dA_s \\
&= M \int_0^{\tau_n} (f_s^2 \wedge C)(1 - \Delta A_s) dA_s.
\end{aligned} \tag{18.80}$$

Letting $C \uparrow \infty$ and $n \to \infty$, we obtain from (18.80) the required inequality

$$M \int_0^{\tau_\infty} f_s^2 (1 - \Delta A_s) dA_s < \infty.$$

As to (b), the implication part of the theorem follows from (a) of the same theorem and from Lemma 18.11 (compare with the proof of (b) in Theorem 18.7).

The inverse implication in (b) can be established as follows. If $\mathcal{M}$ is a τ_∞-locally square integrable martingale there exist stopping times $\sigma_k \uparrow \tau_\infty$ such that for each $k = 1, 2, \ldots$, the process $(\mathcal{M}_{t \wedge \sigma_k}, \mathcal{F}_t)$, $t \ge 0$, is a square integrable martingale. Then, by virtue of (a),

$$P\left(\int_0^{\sigma_k} f_s^2 (1 - \Delta A_s) dA_s = \infty \right) = 0$$

and, hence,

$$\begin{aligned}
&P\left(\int_0^t f_s^2 (1 - \Delta A_s) dA_s = \infty, t < \tau_\infty \right) \\
&= P\left(\int_0^t f_s^2 (1 - \Delta A_s) dA_s = \infty, \bigcup_k \{t < \sigma_k\} \right) \\
&\le \sum P\left(\int_0^t f_s^2 (1 - \Delta A_s) dA_s = \infty, t < \sigma_k \right) = 0.
\end{aligned}$$

As to (c), let

$$\begin{aligned}\varphi'(t) &= I_{\{f_t^2(1-\Delta A_t)\Delta A_t>1\}} f_t, \\ \varphi''(t) &= I_{\{f_t^2(1-\Delta A_t)\Delta A_t\le 1\}} f_t.\end{aligned}$$

These processes belong to class Φ_3 and $\mathcal{M}_t = \mathcal{M}'_t + \mathcal{M}''_t$, where

$$\mathcal{M}'_t = \int_0^t \varphi'(s)d[N_s - A_s], \quad \mathcal{M}''_t = \int_0^t \varphi''(s)d[N_s - A_s].$$

We shall show that

$$P\left\{\int_0^{\tau_\infty} f_s^2(1-\Delta A_s)dA_s < \infty\right\} = 1 \text{ implies } P\left\{\sup_{t\ge 0}|\mathcal{M}'_{t\wedge\tau_\infty}| < \infty\right\} = 1. \tag{18.81}$$

We have

$$\sup_{t\ge 0}|\mathcal{M}'_{t\wedge\tau_\infty}| \le \int_0^{\tau_\infty} |\varphi'(s)|d[N_s + A_s] = \sum_{s\le\tau_\infty} |\varphi'(s)|[\Delta N_s + \Delta A_s]. \tag{18.82}$$

But (P-a.s.)

$$\begin{aligned}\sum_{s\le\tau_\infty} (\varphi'(s))^2(1-\Delta A_s)\Delta A_s &= \int_0^{\tau_\infty} I_{\{f_s^2(1-\Delta A_s)\Delta A_s>1\}} f_s^2(1-\Delta A_s)dA_s \\ &\le \int_0^{\tau_\infty} f_s^2(1-\Delta A_s)dA_s < \infty.\end{aligned}$$

Hence the number of terms in the sum $\sum_{s\le\tau_\infty}(\varphi'(s))^2(1-\Delta A_s)\Delta A_s$ is finite (P-a.s.). For $\Delta A_s = 1$ and $\Delta A_s = 0$, $\varphi'(s) = 0$. Therefore, the number of nonzero terms in the sum $\sum_{s\le\tau_\infty}|\varphi'(s)|[\Delta N_s + \Delta A_s]$ is the same as in the sum $\sum_{s\le\tau_\infty}(\varphi'(s))^2(1-\Delta A_s)\Delta A_s$, which fact, together with (18.82), proves (18.81).

We shall show next that

$$P\left\{\sup_{t\ge 0}|\mathcal{M}''_{t\wedge\tau_\infty}| < \infty\right\} = 1. \tag{18.83}$$

To this end we write

$$D_t = \int_0^t (\varphi''(s))^2(1-\Delta A_s)dA_s$$

and introduce stopping times $\theta_k = \inf\{t \ge 0 : D_t \ge k\}$, setting $\theta_k = \tau_\infty$ on the set $D_{\tau_\infty} < k$.

It is easy to see that $D_{\theta_k} \le k+1$ and

$$\lim_{k\to\infty} P\{\theta_k < \tau_\infty\} = 0. \tag{18.84}$$

By virtue of (a), the processes $(\mathcal{M}''_{t\wedge\theta_k}, \mathcal{F}_t)$, $t \geq 0$, for each $k = 1, 2, \ldots$, are square integrable martingales (with right continuous trajectories). Hence, according to Theorems 3.2 and 5.10, for $c > 0$ we have

$$P\left\{\sup_{t\geq 0} |\mathcal{M}''_{t\wedge\theta_k}| > c\right\} \leq \frac{1}{c} M(\mathcal{M}''_{\theta_k})^2 = \frac{1}{c} M D_{\theta_k} \leq \frac{k+1}{c}.$$

From this,

$$\begin{aligned} P\left\{\sup_{t\geq 0} |\mathcal{M}''_{t\wedge\tau_\infty}| > c\right\} &= P\left\{\sup_{t\geq 0} |\mathcal{M}''_{t\wedge\theta_k}| > c, \theta_k = \tau_\infty\right\} \\ &\quad + P\left\{\sup_{t\geq 0} |\mathcal{M}''_{t\wedge\tau_\infty}| > c, \theta_k < \tau_\infty\right\} \\ &\leq P\left\{\sup_{t\geq 0} |\mathcal{M}''_{t\wedge\theta_k}| > c\right\} + P\left\{\theta_k < \tau_\infty\right\} \\ &\leq \frac{k+1}{c} + P\left\{\theta_k < \tau_\infty\right\}, \end{aligned}$$

and, therefore,

$$P\left\{\sup_{t\geq 0} |\mathcal{M}''_{t\wedge\tau_\infty}| = \infty\right\} \leq P\{\theta_k < \tau_\infty\};$$

this fact, together with (18.84), proves (18.83). □

18.4.6. To conclude this section we shall formulate a result to be used in investigating the requirements for absolute continuity of the measures which correspond to point processes (Section 19.4).

Let processes $f = (f_t, \mathcal{F}_t)$, $B = (B_t, \mathcal{F}_t)$, $t \geq 0$, belong to the class Φ_3, $P(|f_t| < \infty) = 1$, $P(\int_0^t |f_s| dA_s = \infty, t < \tau_\infty)$, and the process $B = (B_t, \mathcal{F}_t)$ have nondecreasing right continuous trajectories with $B_0 = 0$. We shall form a process $Z_t = \mathcal{M}_t + B_t$, where $\mathcal{M}_t = \int_0^t f_s d[N_s - A_s]$.

Lemma 18.13. *If $P(\sup_{t\geq 0} Z_t < \infty) = 1$ and $\Delta Z_t = Z_t - Z_{t-} \leq c < \infty$, then $P(B_{\tau_\infty} < \infty) = 1$.*

PROOF. By Theorem 18.7, the process $\mathcal{M} = (\mathcal{M}_t, \mathcal{F}_t)$ is a τ_∞-local martingale. Therefore, there exists a sequence of stopping times $\sigma_n \uparrow \tau_\infty$ such that the processes $(\mathcal{M}_{t\wedge\sigma_n}, \mathcal{F}_t)$, for each $n = 1, 2, \ldots$, are uniformly integrable martingales.

We shall define the stopping times $\theta_k = \inf\{t \geq 0 : Z_t \geq k\}$, assuming $\theta_k = \infty$ on the set $\{\sup_{t\geq 0} Z_t < k\}$. By virtue of the assumptions of the lemma, $Z_{t\wedge\theta_k} \leq k + c$ and $\lim_{k\to\infty} P(\theta_k < \infty) = 0$. Hence,

$$M B_{\sigma_n\wedge\theta_k} = M\{\mathcal{M}_{\sigma_n\wedge\theta_k} + B_{\sigma_n\wedge\theta_k}\} = M Z_{\sigma_n\wedge\theta_k} \leq k + c.$$

From this, by virtue of Theorem 1.1,

$$MB_{\tau_\infty \wedge \theta_k} = M \lim B_{\sigma_n \wedge \theta_k} = \lim_n MB_{\sigma_n \wedge \theta_k} \le k + c$$

and, therefore, $P(B_{\tau_\infty \wedge \theta_k}) < \infty = 1$.

Finally, we find

$$\begin{aligned} P(B_{\tau_\infty} = \infty) &= P(B_{\tau_\infty} = \infty, \theta_k = \infty) + P(B_{\tau_\infty} = \infty, \theta_k < \infty) \\ &= P(B_{\tau_\infty \wedge \theta_k} = \infty, \theta_k = \infty) + P(B_{\tau_\infty} = \infty, \theta_k < \infty) \\ &\le P(B_{\tau_\infty \wedge \theta_k} = \infty) + P(\theta_k < \infty) \\ &= P(\theta_k < \infty) \to 0, \quad k \to \infty, \end{aligned}$$

thus proving the lemma. □

Note. The statement of the lemma holds true if $Z_t = M_t + B_t$, where M_t is a τ_∞-local martingale with right continuous trajectories.

Lemma 18.14. *Let $(\mathcal{F}_t)$, $t \ge 0$, be a nondecreasing, right continuous family of σ-algebras $\mathcal{F}_t$, let $Y = (y_t, \mathcal{F}_t)$, $t \ge 0$, be a σ-locally square integrable martingale with right continuous trajectories ($\sigma \le \infty$), and let $\langle Y \rangle = (\langle y \rangle_t, \mathcal{F}_t)$, $t \ge 0$, be a predictable increasing process such that $(y_t^2 - \langle y \rangle_t, \mathcal{F}_t)$, $t \ge 0$, is a σ-local martingale.*

Let $\langle y \rangle_\sigma = \lim_{t \to \sigma} \langle y \rangle_t$. If

$$P\{\langle y \rangle_\sigma < \infty\} = 1,$$

then $P\{sup_{t<\sigma} |y_t| < \infty\} = 1$ and there exists a $\mathcal{F}_\sigma$-measurable random variable y_σ, $P\{|y_\sigma| < \infty\} = 1$, such that

$$P\left\{\lim_{t \to \sigma} y_t = y_\sigma\right\} = 1.$$

PROOF. Since the process Y is a σ-locally square integrable martingale there exists a sequence of stopping times $\sigma_n \uparrow \sigma$ such that $Y^{(n)} = (y_{t \wedge \sigma_n}, \mathcal{F}_t)$, $t \ge 0$, for each n, $n = 1, 2, \ldots$, are square integrable martingales. Starting from the martingales $Y^{(n)}$, $n = 1, 2, \ldots$, we can define the unique process $\langle Y \rangle$ analogously to the corresponding process in Theorem 3.9 or to the compensator A_t of the point process N_t (Section 18.1). Let $f \equiv (f(t), \mathcal{F}_t)$, $t \ge 0$, be a random process belonging to class Φ_3 (Definition 3 in Subsection 5.4.1) such that $P\{\int_0^t f^2(s) d\langle y \rangle_s = \infty, t < \sigma\} = 0$. In this case we can define, by generalizing Theorem 5.10, the stochastic integral

$$\mathcal{J}_t(f) = \int_0^t f(s) dy_s$$

from the σ-locally square integrable martingale Y so that the process $\mathcal{J} = (\mathcal{J}_t(f), \mathcal{F}_t)$, $t \ge 0$, has right continuous trajectories. The process $\mathcal{J}$ will also be a σ-locally square integrable martingale with

$$\langle \mathcal{J}(f)\rangle_t = \int_0^t f^2(s)d\langle y\rangle_s.$$

The proof of this fact differs very little from the implication of assertion (b) of Theorem 18.8.

We shall prove the next assertions of the lemma. To this end we define the σ-locally square integrable martingale $Y' = (y'_t, \mathcal{F}_t), t \geq 0$, and $Y'' = (y''_t, \mathcal{F}_t)$, $t \geq 0$, by

$$y'_t = \int_0^t I_{\{\Delta\langle y\rangle_s \leq 1\}} dy_s, \quad y''_t = \int_0^t I_{\{\Delta\langle y\rangle_s > 1\}} dy_s,$$

where $\Delta\langle y\rangle_s = \langle y\rangle_s - \langle y\rangle_{s-}$. It is seen that

$$y_t - y_0 = y'_t + y''_t$$

and

$$\langle y'\rangle_t = \int_0^t I_{\{\Delta\langle y\rangle_s \leq 1\}} d\langle y\rangle_s, \quad \langle y''\rangle_t = \int_0^t I_{\{\Delta\langle y\rangle_s > 1\}} d\langle y\rangle_s.$$

We shall have proved the assertions of the lemma if we prove them for either of the processes Y' or Y''.

Let us prove first the assertions of the lemma for the process Y'.

We shall define the stopping times T_n, $n = 1, 2, \ldots$, as follows: $T_n = \inf(t : \langle y'\rangle_t \geq n)$, assuming $T_n = \sigma$ if $\langle y'\rangle_\sigma < n$. By virtue of the definition of $\langle y'\rangle_t$ and the inequality $\langle y'\rangle_t \leq \langle y\rangle_t$, we have

$$\langle y'\rangle_{t\wedge T_n} \leq n + 1, \quad \lim_{n\to\infty} P\{T_n < \sigma\} = 0.$$

It follows from this, in particular, that the processes $(y'_{t\wedge T_n}, \mathcal{F}_t)$, $t \geq 0$, for each $n = 1, 2, \ldots$, are square integrable martingales. Hence, by Theorem 3.3 we can define the random variables $y'_{\sigma\wedge T_n} = \lim_{t\to\infty} y'_{t\wedge T_n}$ (P-a.s.) with $M(y'_{\sigma\wedge T_n})^2 \leq n + 1$, and, by virtue of Theorem 3.2,

$$P\left\{\sup_{t\leq\sigma} |y'_{t\wedge T_n}| = \infty\right\} = 0.$$

We shall show that $P\{\sup_{t\leq\sigma} |y'_t| = \infty\} = 0$. We have

$$\begin{aligned} P\left\{\sup_{t\leq\sigma} |y'_t| = \infty\right\} &= P\left\{\sup_{t\leq\sigma} |y'_{t\wedge T_n}| = \infty, T_n = \sigma\right\} \\ &\quad + P\left\{\sup_{t\leq\sigma} |y'_t| = \infty, T_n < \sigma\right\} \\ &\leq P\left\{\sup_{t\leq\sigma} |y'_{t\wedge T_n}| = \infty\right\} + P\{T_n < \sigma\} \\ &= P\{T_n < \sigma\} \to 0, \quad n \to \infty. \end{aligned}$$

Set

$$\overline{y}'_\sigma = \overline{\lim}_{t\to\sigma} y'_t, \quad \underline{y}'_\sigma = \underline{\lim}_{t\to\sigma} y'_t$$

and note that $|y'_\sigma| \le \sup_{t\le\sigma} |y'_t|$, i.e., $P\{|y'_\sigma| < \infty\} = 1$.

Since $\overline{y}'_\sigma = \underline{y}'_\sigma((T_n = \sigma); (P\text{-a.s.}))$, we have

$$P\{\overline{y}'_\sigma > \underline{y}'_\sigma\} \le P\{T_n < \sigma\} \to 0, \quad n \to \infty.$$

Thus, we have proved the lemma for the process Y'.

To prove the lemma for the process Y'' we note that (P-a.s.)

$$\infty > \langle y \rangle_\sigma \ge \langle y'' \rangle_\sigma = \int_0^\sigma I_{\{\Delta\langle y\rangle_s > 1\}} d\langle y \rangle_s = \sum_{s<\sigma} I_{\{\Delta\langle y\rangle_s > 1\}} \Delta\langle y \rangle_s.$$

Therefore, the number of terms in the last sum is finite (P-a.s.). Hence,

$$y''_t = \int_0^t I_{\{\Delta\langle y\rangle_s > 1\}} dy_s = \sum_{s<\sigma} I_{\{\Delta\langle y\rangle_s > 1\}} \Delta y_s,$$

where $\Delta y_s = y_s - y_{s-}$, i.e., y''_t is a right continuous piecewise constant function of t with a finite number of discontinuity points coinciding with no σ, from which fact the required assertions follow obviously. □

18.5 The Structure of Point Processes with Deterministic and Continuous Compensators

18.5.1. Let a point process $N = (N_t, \mathcal{F}_t)$, $t \ge 0$, have a deterministic compensator A_t, $t \ge 0$. We shall write $\sigma_\infty = \inf\{t \ge 0 : A_t = \infty\}$, setting $\sigma_\infty = \infty$ if $\lim_{t\to\infty} A_t < \infty$.

Theorem 18.9. *The point process $N = (N_t, \mathcal{F}_t)$, $t < \sigma_\infty$, with a deterministic compensator is a process with independent increments and (P-a.s.) for $s \le t < \sigma_\infty$, $\lambda \in \mathbb{R}$,*

$$M(\exp[i\lambda(N_t - N_s)]|\mathcal{F}_s) = \prod_{s<u\le t} [1 + (e^{i\lambda} - 1)\Delta A_u] \times \exp[(e^{i\lambda} - 1)(A^c_t - A^c_s)], \quad (18.85)$$

where $\Delta A_t = A_t - A_{t-}$, $A^c_t = A_t - \sum_{s\le t} \Delta A_s$.

PROOF. Obviously,

$$\begin{aligned} e^{i\lambda N_t} - e^{i\lambda N_s} &= \sum_{s<u\le t} [e^{i\lambda N_u} - e^{iu\lambda N_{u-}}] \\ &= \sum_{s<u\le t} e^{i\lambda N_{u-}}(e^{i\lambda}-1)(N_u - N_{u-}) \\ &= (e^{i\lambda}-1)\int_s^t e^{i\lambda N_{u-}} dN_u \\ &= (e^{i\lambda}-1)\int_s^t e^{i\lambda N_{u-}} d[N_u - A_u] \\ &\quad +(e^{i\lambda}-1)\int_s^t e^{i\lambda N_{u-}} dA_u. \end{aligned} \tag{18.86}$$

The first integral on the right-hand side of (18.86) is a uniformly integrable martingale for $t < \sigma_\infty$. Hence, (P-a.s.)

$$M\left(\int_s^t e^{i\lambda N_{u-}} d[N_u - A_u] \middle| \mathcal{F}_s\right) = 0, \quad s \le t < \sigma_\infty. \tag{18.87}$$

We shall let $V_s(t) = M(e^{i\lambda N_t}|\mathcal{F}_s)$, $s \le t$, and note that, by virtue of the theorem on dominated convergence (Theorem 1.4), $V_s(t) = M(e^{i\lambda N_{t-}}|\mathcal{F}_s)$. Taking into account this remark, we find from (18.86) and (18.87) that

$$V_s(t) = V_s(s) + (e^{i\lambda}-1)\int_s^t V_s(u-) dA_u. \tag{18.88}$$

Since $V_s(t) \neq 0$, for

$$U_s(t) = V_s(t)/V_s(s) = M(e^{i\lambda(N_t - N_s)}|\mathcal{F}_s)$$

we obtain

$$U_s(t) = 1 + (e^{i\lambda}-1)\int_s^t U_s(u-) dA_u.$$

By virtue of Lemma 18.8, a unique solution of this equation is given by (18.85), thus proving the theorem. □

Corollary. *If $A_t \equiv t$, the point process $N = (N_t, \mathcal{F}_t)$ is a Poisson process and*

$$M(e^{i\lambda(N_t - N_s)}|\mathcal{F}_s) = \exp[(e^{i\lambda}-1)(t-s)].$$

18.5.2. We shall show now that the point process $N = (N_t, \mathcal{F}_t)$, $t \geq 0$, with continuous compensators $A = (A_t, \mathcal{F}_t)$, $t \geq 0$, can be transformed into a Poisson process by a change of time.

Let

$$\sigma(t) = \inf\{s \geq 0 : A_s > t\},$$

and let the σ-algebra $\mathcal{G}_t = \mathcal{F}_{\sigma(t)}$ and $\pi_t = N_{\sigma(t)}$.

Theorem 18.10. *Assume that the compensator A_t, $t \geq 0$, is continuous (P-a.s.) and that*

$$P\left\{\lim_{t\to\infty} A_t = \infty\right\} = 1.$$

Then the process $\Pi = (\pi_t, \mathcal{G}_t)$, $t \geq 0$, is a Poisson process (with a single parameter).

PROOF. It is clear that the process Π has piecewise constant, right continuous trajectories, where the jumps of the trajectories are integers.

Hence, to prove the theorem it suffices to show that the magnitude of jumps in the process Π is equal to unity and (P-a.s.)

$$M(e^{i\lambda[\pi_t-\pi_s]}|\mathcal{G}_s) = \exp[(e^{i\lambda} - 1)(t - s)]. \tag{18.89}$$

By virtue of (18.86),

$$e^{i\lambda\pi_t} = e^{i\lambda\pi_s} + (e^{i\lambda} - 1)\int_{\sigma(s)}^{\sigma(t)} e^{i\lambda N_{u-}} d[N_u - A_u] + (e^{i\lambda} - 1)\int_{\sigma(s)}^{\sigma(t)} e^{i\lambda N_{u-}} dA_u. \tag{18.90}$$

The process $(\int_0^t e^{i\lambda N_{u-}} d[N_u - A_u], \mathcal{F}_t)$, $t < \infty$, is a local martingale. Hence, by virtue of the equality $MA_{\sigma(t)} = t$,

$$M\left(\int_{\sigma(s)}^{\sigma(t)} e^{i\lambda N_{u-}} d[N_u - A_u]\middle| \mathcal{F}_{\sigma(s)}\right) = 0. \tag{18.91}$$

Further, making use of the continuity of the function A_t and taking into account the definition of the times $\sigma(t)$, we find (see also Section 1.1)

$$\int_{\sigma(s)}^{\sigma(t)} e^{i\lambda N_{u-}} dA_u = \int_{\sigma(s)}^{\sigma(t)} e^{i\lambda N_u} dA_u = \int_s^t e^{i\lambda N_{\sigma(u)}} du = \int_s^t e^{i\lambda\pi_u} du. \tag{18.92}$$

Let us set $V_s(t) = M(e^{i\lambda\pi_t}|\mathcal{G}_s)$. Then, by virtue of (18.90)–(18.92),

$$V_s(t) = V_s(s) + (e^{i\lambda} - 1)\int_s^t V_s(u)du,$$

from which we can obtain (18.89) as in the previous theorem.

We shall note, finally, that the magnitude of jumps in the process Π is equal to unity.

By differentiating both sides of (18.89) with respect to λ and assuming $\lambda = 0$, we find that the process $(\pi_t - t, \mathcal{G}_t)$ is a martingale. Consequently, if $\theta_1 = \inf\{t : \pi_t > 0\}$, by virtue of Theorem 3.6 we have

$$M\pi_{t\wedge\theta_1} = M(t \wedge \theta_1) \leq M\theta_1.$$

Since $P(\theta_1 > t) = P(\pi_t = 0)$, and $Me^{i\lambda\pi_t} = \exp\{(e^{i\lambda} - 1)t\}$, it is not difficult to show that $P(\pi_t = 0) = e^{-t}$ and, therefore, $M\theta_1 = 1$.

Hence $M\pi_{t\wedge\theta_1} \leq 1$ and, by the Fatou lemma, $M\pi_{\theta_1} \leq 1$. But $\pi_{\theta_1} \geq 1$ (P-a.s.). Consequently,

$$P(\pi_{\theta_1} = 1) = 1,$$

i.e., the value of the first jump in the process Π is equal to unity. Similarly, we can establish that the magnitude of the remaining jumps is also equal to unity (P-a.s.). □

Notes and References. 1

18.1. Martingale methods came into use in point-process theory after the work of Brémaud [26]. A martingale approach to point processes was discussed by Boel, Varaiya, and Wong [22], Van Schuppen [302], Jacod [103], Jacod and Memin [105], Segal [271], Davis [48], Kabanov, Liptser and Shiryaev [113], Grigelionis [88] and Segal and Kailath [273].

18.2. Theorem 18.2 is to be found in Chou and Meyer [37], and Jacod [103]. For Theorem 18.3, see also Segal and Kailath [273] and Kabanov, Liptser and Shiryaev [113].

18.3. Theorems 18.4 and 18.5 (on existence and uniqueness) are due to Orey [251], Jacod [103] and Kabanov, Liptser and Shiryaev [113]. Lemma 18.8 was proved by Doléans-Dade [56].

18.4. This section is based on Jacod [102], and Kabanov, Liptser and Shiryaev [113]. Equation (18.56), Theorem 18.8 and Lemma 18.4 are apparently presented here for the first time.

18.5. Theorem 18.9 is due to Kabanov, Liptser and Shiryaev [113], and Brémaud [27].

Notes and References. 2

18.1–18.5. The martingale approach to the investigation of point processes used in this chapter has been developed for description and analysis of multivariate point processes and integer-valued random measures, Jacod [104]. Exhaustive information on these subjects can be found in Liptser and Shiryaev [214], Jacod and Shiryaev [106].

19. The Structure of Local Martingales, Absolute Continuity of Measures for Point Processes, and Filtering

19.1 The Structure of Local Martingales

19.1.1. We established in Theorem 5.7 that any martingale (or local martingale) of a Wiener process permitted a representation as a stochastic integral (see (5.42)). We shall show in Theorem 19.1 that a similar result also holds for point processes.

Theorem 19.1. *Let $\overline{N} = (N_t, \mathcal{F}_t^N)$, $t \geq 0$, be a point process with a compensator $\overline{A} = (\overline{A}_t, \mathcal{F}_t^N)$, $t \geq 0$, and let $Y = (y_t, \mathcal{F}_t^N)$ be a τ_∞-local martingale with right continuous trajectories. Then Y permits the representation*

$$y_t = y_0 + \int_0^t f_s d[N_s - \overline{A}_s], \tag{19.1}$$

where $f = (f_t, \mathcal{F}_t^N)$ is a predictable process with

$$P\left(\int_0^t |f_s| d\overline{A}_s = \infty, t < \tau_\infty\right) = 0. \tag{19.2}$$

19.1.2. To prove the above result we shall investigate some properties of the τ_∞-local martingale Y.

Lemma 19.1. *There exists a sequence (σ_n), $n = 1, 2, \ldots$, of Markov times (with respect to the family $(\mathcal{F}_t^N)$, $t \geq 0$) such that: $\sigma_n \uparrow \tau_\infty$ as $n \to \infty$; and*

$$M \int_0^{\sigma_n} |y_t| dN_t < \infty. \tag{19.3}$$

PROOF. Let (σ_n'), $n \geq 1$, be a sequence of Markov times $\sigma_n' \uparrow \tau_\infty$, with respect to the family $(\mathcal{F}_t^N)$, $t \geq 0$, such that the processes $(y_{t\wedge\sigma_n'}, \mathcal{F}_t^N)$, $t \geq 0$, $n = 1, 2, \ldots$, are uniformly integrable martingales. Set $\sigma_n = \tau_n \wedge \sigma_n'$. Then

$$
\begin{aligned}
M\int_0^{\sigma_n}|y_t|dN_t &\le M\int_0^{\tau_n}|y_{t\wedge\sigma'_n}|dN_t\\
&= M\sum_{i=1}^n |y_{\tau_i\wedge\sigma'_n}|\\
&\le \sum_{i=1}^n M|M(y_{\sigma'_n}|\mathcal{F}^N_{\tau_i})| \le nM|y_{\sigma'_n}|<\infty,
\end{aligned}
$$

which proves (19.3). □

Note 1. Let $\sigma''_n = \inf(t : \overline{A}_t > n)$ and $\sigma''_n = \infty$ if $\overline{A}_{\tau_\infty} \le n$. If $\sigma_n = \tau_n \wedge \sigma'_n \wedge \sigma''_n$ then

$$
M\int_0^{\sigma_n}|y_{t-}|dN_t<\infty,\quad M\int_o^{\sigma_n}|y_t-y_{t-}|dN_t<\infty. \tag{19.4}
$$

It is useful to note this result of the proof (see (18.54) and the proof of Lemma 3.2)

$$
\begin{aligned}
M\int_0^{\sigma_n}|y_{t-}|dN_t &= M\int_0^{\sigma_n}|y_{t-}|d\overline{A}_t = M\int_0^{\sigma_n}|M(y_{\sigma'_n}|\mathcal{F}^N_{t-})|d\overline{A}_t\\
&\le M\int_0^{\sigma_n}M(|y_{\sigma'_n}||\mathcal{F}^N_{t-})d\overline{A}_t = M[M(|y_{\sigma'_n}||\mathcal{F}^N_{\sigma_n})\overline{A}_{\sigma_n}].
\end{aligned}
$$

Note 2. It follows from (19.3), (19.4) and Section 18.4 that we have defined the *conditional mathematical expectations* $M_{\mathcal{N}}(|y||\hat{\mathcal{P}})_t$, $M_{\mathcal{N}}(y|\hat{\mathcal{P}})_t$ $M_{\mathcal{N}}(\Delta y|\hat{\mathcal{P}})_t$ (denoted by $\widehat{|y_t|}$, $\widehat{y_t}$ and $\widehat{\Delta y_t}$, respectively; note that $\widehat{\Delta y_t} = \widehat{y_t} - y_{t-}$ since $\widehat{y_{t-}} = y_{t-}$).

Lemma 19.2. *The probability*

$$
P\left(\int_0^t \widehat{|y_s|}(1-\Delta\overline{A}_s)^+d\overline{A}_s=\infty; t<\tau_\infty\right)=0. \tag{19.5}
$$

PROOF. Let (σ_n), $n = 1, 2, \ldots$, be the times defined in Note 1. Then to prove (19.5) it suffices to show that for each $n = 1, 2, \ldots$,

$$
\int_0^{\sigma_n}\widehat{|y_s|}(1-\Delta\overline{A}_s)^+d\overline{A}_s<\infty \quad (P\text{-a.s.}). \tag{19.6}
$$

We have

$$
\begin{aligned}
\int_0^{\sigma_n}\widehat{|y_s|}(1-\Delta\overline{A}_s)^+d\overline{A}_s &= \int_0^{\sigma_n}\widehat{|y_s|}(1-\Delta\overline{A}_s)^+I_{\{\Delta\overline{A}_s\le 1/2\}}d\overline{A}_s\\
&\quad+\int_0^{\sigma_n}\widehat{|y_s|}(1-\Delta\overline{A}_s)^+I_{\{\Delta\overline{A}_s>1/2/\}}d\overline{A}_s\\
&\quad(=Q_n+R_n).
\end{aligned}
$$

But, by virtue of (18.54) and Lemma 19.1,

$$\begin{aligned} MQ_n &= M\int_0^{\sigma_n} \widehat{|y_s|}(1-\Delta\overline{A}_s)^+ I_{\{\Delta\overline{A}_s\le 1/2\}} d\overline{A}_s \\ &\le 2M\int_0^{\sigma_n} \widehat{|y_s|} d\overline{A}_s = 2M\int_0^{\sigma_n} |y_s| dN_s < \infty, \end{aligned}$$

and

$$R_n = \sum_{t\le\sigma_n} \widehat{|y_t|}(1-\Delta\overline{A}_t)^+ I_{\{\Delta\overline{A}_t>1/2\}}\Delta\overline{A}_t < \infty \quad (P\text{-a.s.}),$$

since (Lemma 18.2) the compensator $\overline{A}_t$ can have only a finite number of jumps of magnitude greater than $\frac{1}{2}$ on the interval $(0,\sigma_n)$. □

Note. For each $n = 1, 2, \ldots$,

$$P\left(\int_0^{\sigma_n} (\widehat{|y_n|} + |y_{t-}|)(1-\Delta\overline{A}_t)^+ d\overline{A}_t < \infty\right) = 1. \tag{19.7}$$

19.1.3. Let $Y = (y_t, \mathcal{F}_t^N)$, $t \ge 0$, be a uniformly integrable martingale. Then $M|y_\infty| < \infty$ ($y_\infty = \lim_{t\to\infty} y_t$) and there exists a sequence of bounded $\mathcal{F}_\infty^N$-measurable random variables $y_\infty^{(k)}$, $k = 1, 2, \ldots$, such that

$$|y_\infty^{(k)}| \le |y_\infty|, \quad M|y_\infty - y_\infty^{(k)}| \le k^{-2}.$$

Let us consider a sequence of uniformly bounded (for each k) martingales $Y^{(k)} = (y_t^{(k)}, \mathcal{F}_t^N)$, $t \ge 0$, $y_t^{(k)} = M(y_\infty^{(k)}|\mathcal{F}_t^N)$, with right continuous trajectories. (The existence of these modifications follows from the fact that the σ-algebras $\mathcal{F}_t^N$ are right continuous and from the corollary to Theorem 3.1).

According to (3.6),

$$P\left(\sup_{t\ge 0}|y_t - y_t^{(k)}| > \varepsilon\right) \le \frac{1}{\varepsilon}\cdot M|y_\infty - y_\infty^{(k)}| \le \frac{1}{\varepsilon k^2}.$$

Hence, by virtue of the Borel–Cantelli lemma, there exists a subsequence $\{k_j\}$ such that

$$P\left(\lim_{k_j\to\infty}\sup_{t\ge 0}|y_t - y_t^{(k_j)}| = 0\right) = 1. \tag{19.8}$$

(We shall number the sequence $\{k_j\}$ as $\{k\}$ from now on to avoid new notation).

Lemma 19.3. *There exists a sequence of Markov times (σ_n), $n = 1, 2, \ldots$, (with respect to the family $(\mathcal{F}_t^N)$, $t \ge 0$) such that: $\sigma_n \uparrow \gamma_\infty$ as $n \to \infty$; and, for each n,*

$$\lim_{k\to\infty} M\int_0^{\sigma_n} |\widehat{y}_t - \widehat{y_t^{(k)}}|(1-\Delta\overline{A}_t)^+ d\overline{A}_t = 0. \tag{19.9}$$

PROOF. We shall consider a random process $B = (B_t, \mathcal{F}_t^N)$ with $B_t = \int_0^t M(|y_\infty||\mathcal{F}_{t-}^N)(1-\Delta\overline{A}_s)^+ dA_s$.

By virtue of Lemma 19.2,

$$P\{\inf(t\ge 0 : B_t = \infty) \ge \tau_\infty\} = 1.$$

Hence, because of the predictability of the process B_t and Lemma 18.11, there exists a sequence of times (σ_n), $n = 1, 2, \ldots$, such that $\sigma_n \uparrow \tau_\infty$ and $B_{\sigma_n} \le n$ (P-a.s.).

From this we have

$$\begin{aligned} &M\int_0^{\sigma_n} M(|y_\infty||\mathcal{F}_{t-}^n)(1-\Delta\overline{A}_t)^+ dN_t \\ &= M\int_0^{\sigma_n} M(|y_\infty||\mathcal{F}_{t-}^N)(1-\Delta\overline{A}_t)^+ d\overline{A} = MB_{\sigma_n} \le n. \end{aligned} \tag{19.10}$$

Further, since $|y_t - y_t^{(k)}|(1-\Delta\overline{A}_t)^+ \le 2M(|y_\infty||\mathcal{F}_t^N)(1-\Delta\overline{A}_t)^+$, by virtue of the theorem on dominated convergence (Theorem 1.4), (19.8), and (19.10), we obtain

$$\lim_{k\to\infty} M\int_0^{\sigma_n} |y_t - y_t^{(k)}|(1-\Delta\overline{A}_t)^+ dN_t = 0. \tag{19.11}$$

The required relation, (19.9), follows immediately from (19.11) and the estimate

$$\begin{aligned} M\int_0^{\sigma_n} |\widehat{y}_t - \widehat{y_t^{(k)}}|(1-\Delta\overline{A}_t)^+ d\overline{A}_t &\le M\int_0^{\sigma_n} |\widehat{y - y^{(k)}}|_t \cdot (1-\Delta\overline{A}_t)^+ d\overline{A}_t \\ &= M\int_0^{\sigma_n} |y_t - y_t^{(k)}|(1-\Delta\overline{A}_t)^+ dN_t. \end{aligned}$$

□

Note. For each $n = 1, 2, \ldots$,

$$\lim_{k\to\infty} \int_0^{\sigma_n} |y_{t-1} - y_{t-}^{(k)}|(1-\Delta\overline{A}_t)^+ d\overline{A}_t = 0.$$

19.1.4. Let us consider a τ_∞-local martingale $z = (z_t, \mathcal{F}_t^N)$, $t \ge 0$, with

$$z_t = y_t - y_0 - \int_0^t (\widehat{y}_s - y_{s-})(1-\Delta\overline{A}_s)^+ d[N_s - \overline{A}_s]$$

and set $\widehat{z}_t = M_{\mathcal{N}}(z|\mathcal{P})$, $\widehat{y}_t = M_{\mathcal{N}}(y|\mathcal{P})$.

Lemma 19.4. *For each $t > 0$,*

$$(\widehat{z}_t - z_{t-1}) I_{\{\Delta\overline{A}_t < 1\}} = 0 \quad (P\text{-a.s.}). \tag{19.12}$$

PROOF. Let (σ_n), $n = 1, 2, \ldots$, be a sequence of stopping times defined in Note 1 to Lemma 19.1 and let $\varphi = (\varphi_t, \mathcal{F}_t^N)$, $t \geq 0$, be a predictable process with $|\varphi_t| \leq 1$ (P-a.s.), $t \geq 0$. Then

$$
\begin{aligned}
& M \int_0^{\sigma_n} \varphi_t(\widehat{z}_t - z_{t-}) I_{\{\Delta \overline{A}_t < 1\}} dN_t \\
&= M \int_0^{\sigma_n} \varphi_t(\widehat{y}_t - y_{t-}) I_{\{\Delta \overline{A}_t < 1\}} [1 - (1 - \Delta \overline{A}_t)^+ (\Delta N_t - \Delta \overline{A}_t)] dNt \\
&= M \int_0^{\sigma_n} \varphi_t(\widehat{y}_t - y_{t-}) I_{\{\Delta \overline{A}_t < 1\}} [1 - (1 - \Delta \overline{A}_t)^+ (1 - \Delta \overline{A}_t)] dN_t = 0.
\end{aligned}
$$

The required assertion, (19.12), follows from the above equality and from the definition of the conditional mathematical expectation $M_N(\cdot|\hat{\mathcal{P}})$ (Section 18.4). □

19.1.5. The main result related to the structure of local martingales $Y = (y_t, \mathcal{F}_t^N)$, $t \geq 0$, is the following.

Theorem 19.2.

(1) Any τ_∞-local martingale $Y = (y_t, \mathcal{F}_t^N)$, $t \geq 0$, with right continuous trajectories permits the representation ($\{t < \tau_\infty\}$; (P-a.s.))

$$
y_t = y_0 + \int_0^t (\widehat{y}_s - y_{s-})(1 - \Delta \overline{A}_s)^+ d\overline{m}_s, \tag{19.13}
$$

where $\overline{m}_s = N_s - \overline{A}_s$ and

$$
P\left(\int_0^t |\widehat{y}_s - y_{s-}| \cdot (1 - \Delta \overline{A}_s)^+ d\overline{A}_s = \infty, t < \tau_\infty \right) = 0. \tag{19.14}
$$

(2) If, in addition, Y is a τ_∞-locally square integrable martingale, then

$$
P\left(\int_0^t |\widehat{y}_s - y_{s-}|^2 (1 - \Delta \overline{A}_s)^+ d\overline{A}_s = \infty, t > \tau_\infty \right) = 0. \tag{19.15}
$$

(3) If Y is a square integrable martingale, we have

$$
M \int_0^{\tau_\infty} (\widehat{y}_s - y_{s-})^2 (1 - \Delta \overline{A}_s)^+ d\overline{A}_s < \infty, \tag{19.16}
$$

$$
y_t = y_0 + \int_0^t \frac{d\langle y, \overline{m} \rangle_s}{d\langle \overline{m} \rangle_s} d\overline{m}_s, \tag{19.17}
$$

$$
\frac{d\langle y, \overline{m} \rangle_s}{d\langle \overline{m} \rangle_s} = (\widehat{y}_s - y_{s-})(1 - \Delta \overline{A}_s)^+. \tag{19.18}
$$

PROOF. We shall note first that, due to (19.8) and Lemma 19.3, we need to prove (19.13) only for the case of uniformly bounded martingales. Thus, let $|y_t| \le d$, $t \ge 0$, $\omega \in \Omega$. We shall define the martingale $\tilde{Y} = (\tilde{y}_t, \mathcal{F}_t^N)$, $t \ge 0$, by

$$\tilde{y}_t = y_0 + \int_0^t (\widehat{y}_s - y_{s-})(1 - \Delta\overline{A}_s)^+ d[N_s - \overline{A}_s]$$

and note that

$$\begin{aligned} |\tilde{y}_t| &\le |y_0| + 2d \int_0^t (1 - \Delta\overline{A}_s)^+ d[N_s - \overline{A}_s] \\ &\le |y_0| + 4d \left\{ N_t + \int_0^t (1 - \Delta\overline{A}_s)^+ d\overline{A}_s \right\}. \end{aligned} \tag{19.19}$$

Since $N_{\tau_n} \le n$ and the process $B_t = \int_0^t (1 - \Delta\overline{A}_s^+) d\overline{A}_s$ is nondecreasing, predictable and such that $(\{t < \tau_\infty\}; (P\text{-a.s.}))$

$$B_t \le 2\overline{A}_t + \sum_{s \le t} I_{\{\Delta\overline{A}_s > 1/2\}} (1 - \Delta\overline{A}_s)^+ \Delta A_s < \infty,$$

by virtue of Lemma 18.11, there exists a sequence of times (σ_n), $n = 1, 2, \ldots$, $\sigma_n \uparrow \tau_\infty$, such that $B_{\sigma_n} \le n$; this and (19.19) imply that the martingale Y is τ_∞-locally bounded:

$$|\tilde{y}_{t\wedge\theta_n}| \le d + 4dn, \quad \theta_n = \tau_n \wedge \sigma_n, \quad n = 1, 2, \ldots.$$

We shall consider next the uniformly bounded martingale $z = (z_t, \mathcal{F}_t^N)$ with $z_t = 1 + C(y_{t\wedge\theta_n} - \tilde{y}_{t\wedge\theta_n})$ where we choose the constant $C \ne 0$ so that $P\{\inf_t Z_t > 0\} = 1$.

Then, by virtue of Lemma 19.4 for $t \ge 0$,

$$I_{\{\Delta\overline{A}_t < 1\}} \frac{\widehat{z}_t}{z_{t-}} = 1 \quad (P\text{-a.s.}). \tag{19.20}$$

Let Q be the restriction of P to the σ-algebra $\mathcal{F}_\infty^N$ and let Q' be a measure on $(\Omega, \mathcal{F}_\infty^N)$ with $dQ' = z_\infty dQ$ where $z_\infty = \lim_{t\to\infty} z_t$.

It follows from Theorem 19.2 of the next section and from (19.20) that the compensators of the point processes $\overline{N} = (N_t, \mathcal{F}_t^N, Q)$ and $\overline{N}' = (N_t, \mathcal{F}_t, Q')$ coincide. Therefore, by virtue of Theorem 18.5, $Q' = Q$, i.e., $z_\infty = 1$ (Q-a.s.) and, for any $t \ge 0$,

$$y_{t\wedge\theta_n} = \tilde{y}_{t\wedge\theta_n} \quad (P\text{-a.s.}).$$

Since $\theta_n \uparrow \tau_\infty$, we find from this that

$$y_t = \tilde{y}_t \quad (\{t < \tau_\infty\}; (P\text{-a.s.})).$$

(19.15) and (19.16) follow immediately from (19.14), Theorem 18.8 and from the fact that $(C^+)^2 C = C^+$ (see Section 13.1).

Finally, (19.17) and (19.18) hold on account of the following: $\langle \overline{m} \rangle_t = \int_0^t (1 - \Delta \overline{A}_s) d\overline{A}_s$; for the square integrable martingale $y_t = y_0 + \int_0^t f_s d\overline{m}_s$,

$$\langle y, \overline{m} \rangle_t = \int_0^t f_s (1 - \Delta \overline{A}_s) d\overline{A}_s$$

(compare with Lemma 18.12). □

19.2 Nonnegative Supermartingale: Analog of Girsanov's Theorem

19.2.1. Let $N = (N_t, \mathcal{F}_t)$, $t \geq 0$, be a point process with a compensator $A = (A_t, \mathcal{F}_t)$, $t \geq 0$, and let $f = (f_t, \mathcal{F}_t)$ be a predictable process such that $P(|f_t| < \infty) = 1$, $t \geq 0$, and

$$P\left(\int_0^t |f_s| dA_s = \infty, t < \tau_\infty \right) = 0. \tag{19.21}$$

It follows from Theorem 18.7 that the process $z = (z_t, \mathcal{F}_t)$, $t > \tau_\infty$, with

$$z_t = 1 + \int_0^t f_s dm_s, \tag{19.22}$$

is a τ_∞-local martingale

Suppose that $z_t \geq 0$. Then the process $z = (z_{t \wedge \tau_\infty}, \mathcal{F}_t)$, $t \geq 0$ is a supermartingale. In fact, let (σ_n), $n \geq 1$, be a sequence of times such that $\sigma_n \uparrow \tau_\infty$, $n \to \infty$, and the process $(z_{t \wedge \sigma_n}, \mathcal{F}_t)$ is a martingale. Then $M z_{t \wedge \sigma_n} = 1$, $M(z_{t \wedge \sigma_n} | \mathcal{F}_s) = z_{s \wedge \sigma_n}$ and, by virtue of the Fatou lemma, $M z_{t \wedge \tau_\infty} \leq 1$, $M(z_{t \wedge \tau_\infty} | \mathcal{F}_s) \leq z_{s \wedge \tau_\infty}$.

Lemma 19.5. *Let $z = (z_t, \mathcal{F}_t)$ be a nonnegative supermartingale defined as in (19.22). Then there exists a predictable process $\lambda = (\lambda_t, \mathcal{F}_t)$ such that*[1]

(1) $0 \leq \lambda_t < \infty$;
(2) $\Delta A_t = 1$ *implies* $\lambda_t = 1$; $z_{t-} = 0$ *implies* $\lambda_t = 1$;
(3) $\lambda_t \leq (\Delta A_t)^{-1}$;
(4) the process $\int_0^t \lambda_s dA_s$ *is a right continuous process* ($\{t < \tau_\infty\}$; *(P-a.s.)) in the topology of the extended real line;*

[1] Following Section 18.3, we assume $z_{0-} = 0$, $A_{0-} = 0$.

(5)

$$z_t = 1 + \int_0^t z_{s-}(\lambda_s - 1)(1 - \Delta A_s)^+ d[N_s - A_s], \quad t < \tau_\infty, \tag{19.23}$$

and

$$\begin{aligned} z_t &= \prod_{\{n\geq 0:\tau_n \leq t\}} \lambda_{\tau_n} \prod_{\substack{s\leq t \\ s\neq \tau_n}} [1 + (1 - \Delta A_s)^+(1 - \lambda_s)\Delta A_s] \\ &\quad \times \exp\left(\int_0^t (1 - \lambda_s) dA_s^c\right), \end{aligned} \tag{19.24}$$

where $\tau_0 = 0$.

PROOF. Since z is a nonnegative supermartingale, by virtue of Note 2 in Section 3.9 and (19.22), for $t < \tau_\infty$ we have

$$z_t = 1 + \int_0^t z_{s-}(z_{s-})^+ f_s dm_s. \tag{19.25}$$

Let us denote by α_t the number of jumps of magnitude 1 of the compensator A during time $[0, t]$, and denote by β_t the number of jumps in the process N which occur during time $[0, t]$ at the times of jumps of magnitude 1 of the compensator A. That is, let

$$\alpha_t = \int_0^t I_{\{\Delta A_s = 1\}} dA_s, \quad \beta_t = \int_0^t I_{\{\Delta A_s = 1\}} dN_s.$$

The definition of α_t and β_t implies, first, that $\alpha_t \geq \beta_t$, and, second, that the process $(\beta_t - \alpha_t, \mathcal{F}_t)$, $t \geq 0$, is a nonpositive local martingale with $\beta_0 - \alpha_0 = 0$; this fact, in turn, implies that $\alpha_t = \beta_t$ (P-a.s.) and, therefore,

$$\begin{aligned} m_t &= N_t - A_t = N_t - A_t - \int_0^t I_{\{\Delta A_s = 1\}} d[N_s - A_s] \\ &= \int_0^t (1 - \Delta A_s)(1 - \Delta A_s)^+ d[N_s - A_s]. \end{aligned} \tag{19.26}$$

Let

$$\Lambda_t = 1 + (z_{t-})^+ f_t[1 - \Delta A_t] \tag{19.27}$$

and

$$\lambda = \min\{|\Lambda_t|, (\Delta A_t)^{-1}\}. \tag{19.28}$$

We have from (19.25)–(19.28) that

$$z_t = 1 + \int_0^t z_{s-}(\Lambda_s - 1)(1 - \Delta A_s)^+ dm_s. \tag{19.29}$$

We shall show that in (19.29) we can replace Λ_s by the function λ_s defined in (19.28).

To this end we note that the function λ_s satisfies (1)–(3) of the theorem.

Further, since $||a|-1| \le |a-1|$ and, for $a \ge 0$ and $b \ge 1$, $|a \wedge b - 1| \le |a-1|$, it follows that

$$\begin{aligned} z_{t-}|\lambda_t - 1|(1-\Delta A_t)^+ &\le z_{t-}||\Lambda_t| - 1|(1-\Delta A_t)^+ \\ &\le z_{t-}|\Lambda_t - 1|(1-\Delta A_t)^+ \\ &= |f_t|(1-\Delta A_t)(1-\Delta A_t)^+ \le |f_t|. \end{aligned}$$

It follows from the above inequalities and (19.21) that

$$\int_0^t z_{s-}|\lambda_s - 1|(1-\Delta A_s)^+ dA_s < \infty \quad (\{t < \tau_\infty\};\ (P\text{-a.s.})),$$

and, consequently, that the process $(\tilde{z}_t, \mathcal{F}_t)$, $t < \tau_\infty$, defined by

$$\tilde{z}_t = 1 + \int_0^t z_{s-}(\lambda_s - 1)(1-\Delta A_s)^+ d[N_s - A_s] \tag{19.30}$$

is a τ_∞-local martingale (Theorem 18.7).

We shall show that, in fact, $z_t = \tilde{z}_t$ $(\{t < \tau_\infty\};\ (P\text{-a.s.}))$.

The process $(z_t - \tilde{z}_t, \mathcal{F}_t)$, $t < \tau_\infty$, is a τ_∞-local martingale and, by virtue of (19.29) and (19.30), we have

$$z_t - \tilde{z}_t = \int_0^t z_{s-}(\Lambda_s - \lambda_s)(1-\Delta A_s)^+ d[N_s - A_s]. \tag{19.31}$$

By virtue of (19.28), we have

$$\Lambda_t - \lambda_t = \Lambda_t - |\Lambda_t| \wedge (\Delta A_t)^{-1} = (\Lambda_t - |\Lambda_t|) + (|\Lambda_t| - |\Lambda_t| \wedge (\Delta A_t)^{-1}).$$

Hence $z_t - \tilde{z}_t = \Delta'_t + \Delta''_t$ where

$$\Delta'_t = \int_0^t z_{s-}(\Lambda_s - |\Lambda_s|)(1-\Delta A_s)^+ dm_s,$$

$$\Delta''_t = \int_0^t z_{s-}(|\Lambda_s| - |\Lambda_s| \wedge (\Delta A_s)^{-1})(1-\Delta A_s)^+ dm_s.$$

We shall show that $(\{t < \tau_\infty\};\ (P\text{-a.s.}))$ $\Delta'_t = \Delta''_t = 0$. To this end we shall note first that $\Lambda_{\tau_i} \ge 0$. Indeed, we have from (19.29) that

$$0 \le z_{\tau_i} = z_{\tau_i -}[1 + (\Lambda_{\tau_i} - 1)(1-\Delta A_{\tau_i})^+(1-\Delta A_{\tau_i})].$$

From this we conclude that $\Lambda_{\tau_i} \ge 0$ for $z_{\tau_i -} > 0$ and $\Delta A_{\tau_i} < 1$. If $z_{\tau_i -} = 0$ or $\Delta A_{\tau_i} = 1$, by virtue of (19.27) we have $\Lambda_{\tau_i} = 1$.

The fact that the values of Λ_{τ_i} are nonnegative and the definition of Δ'_t imply that

$$\Delta'_t = \int_0^t z_{s-}(|\Lambda_s| - \Lambda_s)(1 - \Delta A_s)^+ dA_s. \tag{19.32}$$

Hence the τ_∞-local martingale $\Delta' = (\Delta'_t, \mathcal{F}_t)$, $t < \tau_\infty$, has monotone nondecreasing trajectories with $\Delta'_0 = 0$. Therefore $\Delta'_t = 0$ ($\{t < \tau_\infty\}$; (P-a.s.)).

We shall show also that $\Delta''_t = 0$ ($\{t < \tau_\infty\}$; (P-a.s.)).

Let σ_j be a time of jump of the compensator A and let $\sigma_j \neq \tau_i$, $i = 1, 2, \ldots$. Then we have from (19.29) that

$$0 \leq z_{\sigma_j} = z_{\sigma_j -}[1 - (\Lambda_{\sigma_j} - 1)(1 - \Delta A_{\sigma_j})^+ \Delta A_{\sigma_j}].$$

This fact implies that $\Lambda_{\sigma_j} \leq (\Delta A_{\sigma_j})^{-1}$ for $\Delta A_{\sigma_j} < 1$ and $z_{\sigma_j -} > 0$. But, if $\Delta A_{\sigma_j} = 1$ or $z_{\sigma_j -} = 0$, then we have again that $\Lambda_{\sigma_i} = 1 \leq (\Delta A_{\sigma_j})^{-1}$. Hence, it follows from (19.32) and the equality $\Delta'_t = 0$ that (P-a.s.)

$$\int_0^t z_{s-}(|\Lambda_s| - |\Lambda_s| \wedge (\Delta A_s)^{-1})(1 - \Delta A_s)^+ dA_s = 0, \quad t > \tau_\infty.$$

Thus,

$$\Delta''_t = \int_0^t z_{s-}(|\Lambda_s| - |\Lambda_s| \wedge (\Delta A_s)^{-1})(1 - \Delta A_s)^+ dN_s.$$

This indicates that the τ_∞-local martingale $\Delta'' = (\Delta''_t, \mathcal{F}_t)$ has nondecreasing trajectories with $\Delta''_0 = 0$. Therefore, $\Delta''_t = 0$ ($\{t < \tau_\infty\}$; (P-a.s.)). Thus $\tilde{z}_t = z_t$, thereby proving (19.23).

We shall prove (19.24). Let $\xi = \inf\{t \geq 0 : z_t = 0\}$, setting $\xi = \infty$ if $\inf_t z_t > 0$. The values $\int_0^t |\lambda_s - 1|(1 - \Delta A_s)^+ ds$ are finite for $t < \tau_1 \wedge \xi$, since

$$\begin{aligned}\infty > \int_0^t |f_s| dA_s &\geq \int_0^t z_{s-}|\lambda_s - 1|(1 - \Delta A_s)^+ dA_s \\ &\geq \inf_{u \leq t < \tau_1 \wedge \xi} z_u \cdot \int_0^t |\lambda_s - 1|(1 - \Delta A_s)^+ dA_s.\end{aligned}$$

Hence, for $t < \tau_1 \wedge \xi$ the equation

$$z_t = 1 - \int_0^t z_{s-}(\lambda_s - 1)(1 - \Delta A_s)^+ dA_s$$

has a unique solution, namely

$$z_t = \prod_{s \leq t}[1 + (1 - \Delta A_s)^+(1 - \lambda_s)\Delta A_s] \exp\left[\int_0^t (1 - \lambda_s) dA_s^c\right], \tag{19.33}$$

by virtue of Lemma 18.8.

If $\xi > \tau_1$, from (19.23) we have

$$z_{\tau_1} = z_{\tau_1-} + z_{\tau_1-}(\lambda_{\tau_1} - 1)(1 - \Delta A_{\tau_1})^+(1 - \Delta A_{\tau_1}) = \lambda_{\tau_1} z_{\tau_1-},$$

since $\Delta A_t = 1$ implies $\lambda_t = 1$ by virtue of (2).

It follows from the above and (19.33) that

$$z_{\tau_1} = \lambda_{\tau_1} \cdot \prod_{s<\tau_1} [1 + (1 - \Delta A_s)^+(1 - \lambda_s)\Delta_s] \exp\left[\int_0^{\tau_1} (1 - \lambda_s) dA_s^c\right]. \quad (19.34)$$

Now let $\xi \le \tau_1$. Then $z_\xi = 0$. The value z_ξ can vanish at time ξ in two ways: in a continuous fashion, when $z_{\xi-} = 0$; or in a jumpwise fashion, when $z_{\xi-} > 0$ and $z_\xi = 0$.

According to (19.33), at least one of the following two relations holds for the first case:

$$\lim_{t \uparrow \sigma} \int_0^t \lambda_s dA_s^c = \infty;$$

$$\lim_{t \uparrow \sigma} \prod_{s \le t} [1 + (1 - \Delta A_s)^+(1 - \lambda_s)\Delta A_s] = 0.$$

Therefore, (19.34) is satisfied.

We shall consider the case where $z_{\xi-} > 0$, $z_\xi = 0$. Then, if $\xi < \tau_1$, we have from (19.23) that

$$1 + (1 - \lambda_\xi)(1 - \Delta A_\xi)^+ \Delta A_\xi = 0$$

and, consequently, (19.34) is satisfied. If $\xi = \tau_1$, we have from (19.23) that $\lambda_{\tau_1} = 0$ and (19.34) is again satisfied.

Thus, we have proved (19.34), which fact and (19.33) imply that (19.24) holds for $t \le \tau_1$. (Note that $\lambda_{\tau_0} \equiv \lambda_0 = 1$ in accord with the agreement $z_{0-} = 0$ and the inference that $z_{t-} = 0$ implies $\lambda_t = 1$). In the general case, (19.34) is established by induction.

Let us prove, finally, (4). It is clear that

$$\int_0^t \lambda_s dA_s = \sum_{s \le t} \lambda_s \Delta A_s + \int_0^t \lambda_s dA_s^c.$$

Since $\lambda_s \Delta A_s \le 1$, the process $\sum_{s \le t} \lambda_s \Delta A_s$ is right continuous (in the topology of the extended real line). Further, if the function $\int_0^t \lambda_s dA_s^c$ is bounded on the interval $[0, t]$, it will also be continuous on this interval. If this function is unbounded it will 'go' to infinity in a continuous way only, as was shown before. Hence, the function $\int_0^t \lambda_s dA_s$ is right continuous (P-a.s.) in the topology of the extended real line. □

Corollary. *Let $\lambda = (\lambda_t, \mathcal{F}_t)$, $t \ge 0$, be a predictable process satisfying (1)–(4) of Lemma 19.5*[2]. *Suppose the random process $z = (z_t, \mathcal{F}_t)$, $t \ge 0$, is defined*

[2] (2) implies only that the first implication is satisfied.

for $t < \tau_\infty$ *by (19.24) and for* $t \ge \tau_\infty$ *by the equality* $z_t = z_{t\wedge\tau_\infty}$ *where* $z_{\tau_\infty} = \underline{\lim}_{t\uparrow\tau_\infty} z_t$ *is a nonnegative supermartingale with* $Mz_t \le 1$ *and is a* τ_∞*-local martingale as well. In this case* ($\{t < \tau_\infty\}$; *(P-a.s.))*

$$z_t = 1 + \int_0^t z_{s-}(\lambda_s - 1)(1 - \Delta A_s)^+ d[N_s - A_s].$$

19.2.2. Let us suppose that a probability measure P' in addition to the probability measure P is given on a measurable space $(\Omega, \mathcal{F})$. We shall consider point processes $N = (N_t, \mathcal{F}_t, P)$ and $N' = (N_t, \mathcal{F}_t, P')$, and compensators $A = (A_t, \mathcal{F}_t, P)$ and $A' = (A'_t, \mathcal{F}_t, P')$, $t \ge 0$, of the processes.

It turns out that the compensators A' and A are related if we assume that the measure P' is absolutely right continuous with respect to measure P. To formulate the result we shall denote by $z_t = M(dP'/dP|\mathcal{F}_t)$, $t \ge 0$, a continuous modification of the martingale $M(dP'/dP|\mathcal{F}_t)$, $t \ge 0$, which exists by virtue of the corollary to Theorem 3.1. We shall also write $\hat{z}_t = M_N(z|\hat{\mathcal{P}})_t$ (see Section 18.4).

Theorem 19.3. *If* $P' \ll P$, *then* (P'-a.s.)

$$A'_t = \int_0^t \hat{z}_s (z_{s-})^+ dA_s, \quad t < \tau_\infty, \tag{19.35}$$

$$\Delta A_t = 1 \text{ implies } \hat{z}_t(z_{t-})^+ = 1. \tag{19.36}$$

PROOF. Let $f = (f_t, \mathcal{F}_t)$, $t \ge 0$, be a nonnegative predictable process such that $M' \int_0^{\tau_\infty} f_s dA'_s < \infty$ where M' denotes the expectation under measure P'. Then, using Theorem 18.6, we find that

$$\begin{aligned} M' = \int_0^{\tau_\infty} f_t dA'_t &= M' \int_0^{\tau_\infty} f_t dN_t = M' \sum_{n\ge1} f_{\tau_n} \\ &= Mz_{\tau_\infty} \sum_{n\ge1} f_{\tau_n} = M \sum_{n\ge1} z_{\tau_n} f_{\tau_n} \\ &= M \int_0^{\tau_\infty} z_t f_t dN_t = M \int_0^{\tau_\infty} \hat{z}_t f_t dA_t \\ &= M \int_0^{\tau_\infty} \hat{z}_t z_{t-}(z_{t-})^+ f_t dA_t \\ &\quad + M \int_0^{\tau_\infty} \hat{z}_t (1 - z_{t-}(z_{t-})^+) f_t dA_t. \end{aligned} \tag{19.37}$$

We shall show that the last term in (19.37) is equal to zero. In fact, by virtue of Theorem 18.6, we have

$$M\int_0^{\tau_\infty} \hat{z}_t(1-z_{t-}(z_{t-})^+)f_t dA_t = M\int_0^{\tau_\infty} z_t(1-z_{t-}(z_{t-})^+)f_t dN_t$$
$$= M\sum_{n\geq 1} z_{\tau_n}(1-z_{\tau_n-}(z_{\tau_n-})^+)f_{\tau_n} = 0 \tag{19.38}$$

since if $z_{\tau_n-}=0$, then $z_{\tau_n}=0$ (Note 2 to Theorem 3.5) and, therefore, also

$$z_{\tau_n}(1-z_{\tau_n-}(z_{\tau_n-})^+)=0.$$

Further, by virtue of Lemma 3.2,

$$M\int_0^{\tau_\infty} \hat{z}_t z_{t-}(z_{t-})^+ f_t dA_t = M z_{\tau_\infty}\int_0^{\tau_\infty} \hat{z}_t(z_{t-})^+ f_t dA_t$$
$$= M'\int_0^{\tau_\infty} \hat{z}_t(z_{t-})^+ f_t dA_t.$$

Hence this, together with (19.37) and (19.38), leads us to the relation

$$M'\int_0^{\tau_\infty} f_t dA'_t = M'\int_0^{\tau_\infty} \hat{z}_t(z_{t-})^+ f_t dA_t,$$

from which, in particular, (19.35) follows.

We shall prove next (19.36). Let θ be a jump time of the compensator A such that $\Delta A_\theta = 1$.

This time is predictable (with respect to measure families of σ-algebras completed by P and P' as well, since $P' \ll P$). While proving Lemma 19.5 we established, in particular, that if $\Delta A_\theta = 1$, then also $\Delta N_\theta = 1$. Therefore, (P-a.s.) and (P'-a.s.), $\Delta N_\theta = 1$.

Further, since $A_{\tau_\infty} = \lim_{t\to\tau_\infty} A_t$, the time $\theta < \tau_\infty$ (P-a.s. and P'-a.s.). Hence,

$$\Delta A'_\theta = M'(\Delta N_\theta|\mathcal{F}_{\theta-}) = 1$$

and, therefore, by virtue of (19.35), (P'-a.s.)

$$1 = \Delta A'_\theta = \hat{z}_\theta(z_{\theta-})^+\Delta A_\theta = \hat{z}_\theta(z_{\theta-})^+,$$

thus proving (19.36). □

19.2.3.

Theorem 19.4 (Analog of Girsanov's Theorem). *Let the process $\lambda=(\lambda_t,\mathcal{F}_t)$, $t\geq 0$, satisfy (1)–(4) of Lemma 19.5, and let the process $z=(z_t,\mathcal{F}_t)$, $t\geq 0$, be defined by (19.24), with $Mz_{\tau_\infty}=1$. Then the compensators $A=(A_t,\mathcal{F}_t,P)$ and $A'=(A'_t,\mathcal{F}_t,P')$, $t\geq 0$, for the point processes $N=(N_t,\mathcal{F},P)$ and $N'=(N_t,\mathcal{F}_t,P')$, $t\geq 0$, with $dP'=z_{\tau_\infty}dP$, are related (P'-a.s.) by*

$$A'_t = \int_0^t \lambda_s dA_s, \quad t\leq \tau_\infty. \tag{19.39}$$

PROOF, By virtue of Theorem 19.3, it suffices to show that

$$\lambda_t = \hat{z}_t(z_{t-})^+ \quad (P'\text{-a.s.}). \tag{19.40}$$

By virtue of (19.23),

$$z_t = z_{t-}[1 + (\lambda_t - 1)(1 - \Delta A_t)^+(\Delta N_t - \Delta A_t)]. \tag{19.41}$$

Let $g = (g_t, \mathcal{F}_t)$, $t \geq 0$, be a nonnegative predictable process such that $M \int_0^{\tau_\infty} z_t g_t dN_t < \infty$. Then, from (19.41) and Theorem 18.6, we have

$$\begin{aligned} M \int_0^{\tau_\infty} \hat{z}_t g_t dA_t &= M \int_0^{\tau_\infty} z_t g_t dN_t \\ &= M \int_0^{\tau_\infty} z_{t-}[1+(\lambda_t-1)(1-\Delta A_t)^+(\Delta N_t-\Delta A_t)] g_t dN_t \\ &= M \int_0^{\tau_\infty} z_{t-}[1 + (\lambda_t - 1)(1 - \Delta A_t)^+(1 - \Delta A_t)] g_t dN_t \\ &= M \int_0^{\tau_\infty} z_{t-}[1 + (\lambda_t - 1)(1 - \Delta A_t)^+(1 - \Delta A_t)] g_t dA_t \\ &= M \int_0^{\tau_\infty} z_{t-}\lambda_t g_t dA_t, \end{aligned} \tag{19.42}$$

where the last equality holds due to the fact that, if $\Delta A_t = 1$, then $\lambda_t = 1$ (Lemma 19.5) and, therefore, also

$$z_{t-}[1 - (\lambda_t - 1)(1 - \Delta A_t)^+(1 - \Delta A_t)] = z_{t-}\lambda_t.$$

From (19.42) we have that $\hat{z}_t = z_{t-}\lambda_t$ (P-a.s.). Consequently, $\hat{z}_t(z_{t-})^+ = z_{t-}(z_{t-})^+\lambda_t$ (P-a.s. and P'-a.s.). But, by virtue of Lemma 6.5, $P'\{\inf_{t\leq\tau_\infty} z_t > 0\} = 1$. Hence, ($P'$-a.s.) $\hat{z}_t(z_{t-})^+ = \lambda_t$, $t \leq \tau_\infty$. □

19.2.4. We shall give some simple sufficient conditions which guarantee the condition $Mz_{\tau_\infty} = 1$; this condition was given in Theorem 19.4, implying that the supermartingale $z = (z_t, \mathcal{F}_t)$, $t \leq \tau_\infty$, is a martingale (Lemma 6.4).

Lemma 19.6. *Let any of the following conditions be satisfied:*

(1) there exists a constant $C < \infty$ such that

$$\int_0^{\tau_\infty} |\lambda_t - 1|(1 - \Delta A_t)^+ dA_t \leq C \quad (P\text{-a.s.}); \tag{19.43}$$

(2) the compensator A is continuous ($\Delta A_t = 0$, $t \geq 0$) and, for some constant $C < \infty$,

$$\int_0^{\tau_\infty} (\lambda_t - 1)^2 dA_t \leq C \quad (P\text{-a.s.}); \tag{19.44}$$

(3) there exists a constant $C<\infty$ such that (P-a.s.)

$$\int_0^{\tau_\infty} \frac{(\lambda_t-1)^2}{1+|\lambda_t-1|} dA_t^c + \sum_{t<\tau_\infty} |\lambda_t-1|(1-\Delta A_t)^+ \Delta A_t \le C. \tag{19.45}$$

Then $Mz_{\tau_\infty}=1$.

PROOF. For (1), by virtue of Theorem 18.7, we need only show that

$$M\int_0^{\tau_\infty} z_{t-}|\lambda_t-1|(1-\Delta A_t)^+ dA_t<\infty. \tag{19.46}$$

To this end we note that since $z=(z_t,\mathcal{F}_t)$, $t\ge 0$, is a local martingale, there exist times $\theta_n\uparrow\tau_\infty$ such that the processes $(z_{t\wedge\theta_n},\mathcal{F}_t)$, $t\ge 0$, are uniformly integrable martingales with $Mz_{\theta_n}=1$ for each $n=1,2,\ldots$. Taking advantage of this fact, Lemma 3.2 and an earlier theorem on monotone convergence (Theorem 1.1), we obtain

$$\begin{aligned} &M\int_0^{\tau_\infty} z_{t-}|\lambda_t-1|(1-\Delta A_t)^+ dA_t\\ &=\lim_n M\int_0^{\theta_n} z_{t-}|\lambda_t-1|(1-\Delta A_t)^+ dA_t\\ &=\lim_n Mz_{\theta_n}\cdot\int_0^{\theta_n} |\lambda_t-1|(1-\Delta A_t)^+ dA_t\\ &\le C\lim_n Mz_{\theta_n}=C<\infty, \end{aligned}$$

thus proving (19.46).

As to (2), denote by $z_t(\lambda)$ the right-hand side in (19.24) and denote by $z_t(\lambda^2)$ the corresponding function when substituting λ_t^2 for λ. Since in the case in question $\Delta A_t=0$, it is seen from (19.24) that

$$z_t^2(\lambda)=z_t(\lambda^2)\exp\left[\int_0^t(1-\lambda_s)^2 dA_s\right].$$

It is easy to conclude from (19.44) and (1) that the process $(z_t(\lambda^2),\mathcal{F}_t)$ is a nonnegative supermartingale with $Mz_t(\lambda^2)\le 1$. Hence,

$$Mz_t^2(\lambda)=Mz_t(\lambda^2)\exp\left[\int_0^t(1-\lambda_s)^2 dA_s\right]\le Mz_t(\lambda^2)e^C\le e^C$$

and $Mz_{\theta_n}^2(\lambda)\le e^C$ where the times θ_n were introduced to prove (1).

The above implies that the values $(z_{\theta_n}(\lambda))$, $n\ge 1$, are uniformly integrable and, therefore,

$$Mz_{\tau_\infty}=\lim_n Mz_{\theta_n}(\lambda)=1.$$

Regarding (3), we shall introduce three processes $\lambda^{(i)} = (\lambda_t^{(i)}, \mathcal{F}_t)$, $t \geq 0$, $i = 1, 2, 3$, with

$$\lambda_t^{(1)} = \lambda_t^{I(\Delta A_t > 0)}, \quad \lambda_t^{(2)} = \lambda_t^{I(\Delta A_t = 0, |\lambda_t - 1| > 1/2)}$$

$$\lambda_t^{(3)} = \lambda_t^{I(\Delta A_t = 0, |\lambda_t - 1| \leq 1/2)}.$$

It is obvious that $z_t(\lambda) = z_t(\lambda^{(1)}) \cdot z_t(\lambda^{(2)}) \cdot z_t(\lambda^{(3)})$, and that the following inequalities are satisfied:

$$\sum_{t < \tau_\infty} |\lambda_t^{(1)} - 1| (1 - \Delta A_t)^+ \Delta A_t \leq C;$$

$$\int_0^{\tau_\infty} (\lambda_t^{(2)} - 1)^2 dA_t \leq C;$$

$$\int_0^{\tau_\infty} |\lambda_t^{(3)} - 1| dA_t \leq C.$$

By virtue of (1), $M z_{\tau_\infty}(\lambda^{(1)}) = 1$ and hence measure $P^{(1)}$ with $dP^{(1)} = z_{\tau_\infty}(\lambda^{(1)}) dP$ is a probability measure. By virtue of Theorem 19.4, the point process $(N_t, \mathcal{F}_t, P^{(1)})$ has the compensator $(A_t^{(1)}, \mathcal{F}_t, P^{(1)})$, where

$$A_t^{(1)} = \int_0^t \lambda_s^{(1)} dA_s = A_t^c + \sum_{s \leq t} \lambda_s \Delta A_s.$$

Hence,

$$\int_0^{\tau_\infty} (\lambda_t^{(2)} - 1)^2 dA_t^{(1)} = \int_0^{\tau_\infty} (\lambda_t^{(2)} - 1)^2 dA_t^c = \int_0^{\tau_\infty} (\lambda_t^{(2)} - 1)^2 dA_t \leq C.$$

Therefore, according to $\int_\Omega z_{\tau_\infty}(\lambda^{(2)}) dP^{(1)} = 1$ and, by virtue of Theorem 19.4, the point process $(N_t, \mathcal{F}_t, P^{(2)})$ with $dP^{(2)} = z_{\tau_\infty}(\lambda^{(2)}) dP^{(1)}$ has a compensator $(A_t^{(2)}, \mathcal{F}_t, P^{(2)})$ such that

$$A_t^{(2)} = \int_0^t \lambda_s^{(2)} dA_s^{(1)}.$$

It follows from the above that

$$\begin{aligned}
\int_0^{\tau_\infty} |\lambda_t^{(3)} - 1| dA_t^{(2)} &= \int_0^{\tau_\infty} |\lambda_t^{(3)} - 1| \lambda_t^{(2)} dA_t^{(1)} \\
&= \int_0^{\tau_\infty} |\lambda_t^{(3)} - 1| \lambda_t^{(2)} dA_t^c \\
&= \int_0^{\tau_\infty} |\lambda_t^{(3)} - 1| \lambda_t^{(2)} dA_t \\
&= \int_0^{\tau_\infty} |\lambda_t^{(3)} - 1| dA_t \leq C.
\end{aligned}$$

Hence, according to (1),

$$\int_{\Omega} z_{\tau_\infty}(\lambda^{(3)})dP^{(2)} = 1$$

and, therefore,

$$\begin{aligned} 1 &= \int_{\Omega} \xi_{\tau_\infty}(\lambda^{(3)})dP^{(2)} = \int_{\Omega} z_{\tau_\infty}(\lambda^{(3)})z_{\tau_\infty}(\lambda^{(2)})dP^{(1)} \\ &= \int_{\Omega} z_{\tau_\infty}(\lambda^{(3)})z_{\tau_\infty}(\lambda^{(2)})z_{\tau_\infty}(\lambda^{(1)})dP = Mz_{\tau_\infty}(\lambda), \end{aligned}$$

thus proving (3). □

19.3 Optimal Filtering from the Observations of Point Processes

19.3.1. Let $(\Omega, \mathcal{F}, P)$ be a complete probability space and let $(\mathcal{F}_t)$, $t \geq 0$, be a nondecreasing family of right continuous sub-σ-algebras of $\mathcal{F}$ augmented by sets from $\mathcal{F}$ of zero probability.

We shall assume that in this space a two-dimensional partially observable process (θ, N) is given where $N = (N_t, \mathcal{F}_t)$ is an observable process and $\theta = (\theta_t, \mathcal{F}_t)$, $t \geq 0$ is an unobservable component permitting the representation

$$\theta_t = \theta_0 + a_t + x_t. \tag{19.47}$$

In this case

(a) $X = (x_t, \mathcal{F}_t)$ is a uniformly integrable martingale with right continuous trajectories;

(b) $a_t = a_t^{(1)} - a_t^{(2)}$ where $a^{(i)} = (a_t^{(i)}, \mathcal{F}_t)$, $i = 1, 2, \ldots$, are nondecreasing predictable right continuous processes with

$$M(a_\infty^{(1)} + a_\infty^{(2)}) < \infty \quad a_\infty^{(i)} = \lim_{t\to\infty} a_t^{(i)}, \quad i = 1, 2, \tag{19.48}$$

(c) $M|\theta_0| < \infty$.

In the present subsection we shall obtain a representation for the conditional mathematical expectations $M(\theta_t|\mathcal{F}_t^N)$, $t \geq 0$, which is (under the assumption that $M\theta_t^2 < \infty$, $t \geq 0$) the optimal estimate of θ_t from the observations $N_0^t = \{N_s, s \leq t\}$.

Let $\alpha_t^{(i)} = M(a_t^{(i)}|\mathcal{F}_t^N)$, $i = 1, 2, \ldots$. It is easy to verify that each of these processes is a submartingale of class D and, therefore, there exist integrable increasing $\hat{\mathcal{P}}$-predictable (see Section 18.1) processes $\overline{a}^{(i)} = (\overline{a}_t^{(i)}, \mathcal{F}_t^N)$, $i = 1, 2$, such that $m^{(i)} = \alpha_t^{(i)} - \overline{a}_t^{(i)}$ are uniformly integrable martingales. It follows from the above that

$$M(a_t^{(i)} - a_s^{(i)}|\mathcal{F}_s^N) = M(\overline{a}_t^{(i)} - \overline{a}_s^{(i)}|\mathcal{F}_s^N),$$

and (Theorem 3.6)

$$M(a_\tau^{(i)} - a_\sigma^{(i)}|\mathcal{F}_\sigma^N) = M(\overline{a}_\tau^{(i)} - \overline{a}_\sigma^{(i)}|\mathcal{F}_\sigma^N),$$

where σ and τ are Markov times (with respect to $F^N = (\mathcal{F}_t^N)$, $t \geq 0$), with $P(\sigma \leq \tau) = 1$.

Setting $\overline{a}_t = \overline{a}_t^{(1)} - \overline{a}_t^{(2)}$, we find that

$$M(a_\tau - a_\sigma|\mathcal{F}_\sigma^N) = M(\overline{a}_\tau - \overline{a}_\sigma|\mathcal{F}^N) \quad (P\text{-a.s.}). \tag{19.49}$$

19.3.2. To formulate and prove the main theorem we shall introduce the notation:

$$\pi_t(\theta) = M(\theta_t|\mathcal{F}_t^N), \quad \hat{\pi}_t(\theta) = M_{\mathcal{N}}(\pi_t(\theta)|\hat{\mathcal{P}}); \tag{19.50}$$

$$\hat{\theta}_t = M_{\mathcal{N}}(\theta_t|\hat{\mathcal{P}}). \tag{19.51}$$

Theorem 19.5. *Let the process $\theta = (\theta_t, \mathcal{F}_t)$, $t \geq 0$, permit the representation given by (19.47), and let (a)–(c) be satisfied.*

Then the process $(\pi_t(\theta), \mathcal{F}_t^N)$, $t \geq 0$, has a right continuous modification and permits the representation

$$\pi_t(\theta) = \pi_0(\theta) + \overline{a}_t + \int_0^t [\hat{\theta}_s - \pi_{s-}(\theta) - \Delta\overline{a}_s](1 - \Delta\overline{A}_s)^+ d[N_s - \overline{A}_s], \tag{19.52}$$

where $\overline{A} = (\overline{A}_t, \mathcal{F}_t^N)$, $t \geq 0$, is the compensator of the point process $\overline{N} = (N_t, \mathcal{F}_t^N)$, $t \geq 0$.

We shall prove the theorem using a few lemmas.

19.3.3.

Lemma 19.7. *The random process $\overline{X} = (\overline{x}_t, \mathcal{F}_t^N)$, $t \geq 0$, with*

$$\overline{x}_t = \pi_t(\theta) - \pi_0(\theta) - \overline{a}_t \tag{19.53}$$

is a uniformly integrable martingale and has a right continuous modification.

PROOF. Since $\pi_0(\theta) = M\theta_0$, then

$$\begin{aligned} |x_t| &\leq M|\theta_0| + |\overline{a}_t| + M(|\theta_0| + |a_t| + |x_t||\mathcal{F}_t^N) \\ &\leq M|\theta_0| + (\overline{a}_\infty^{(1)} + \overline{a}_\infty^{(2)}) + M(|\theta_0| + a_\infty^{(1)} + a_\infty^{(2)} + |x_t||\mathcal{F}_t^N) \end{aligned} \tag{19.54}$$

and by (a)–(c) and Theorem 2.7 the family $(\overline{x}_t)$, $t \geq 0$ is uniformly integrable.

Further, by virtue of (a) and (19.49) we have

$$\begin{aligned} M(\overline{x}_t - \overline{x}_s|\mathcal{F}_s^N) &= M[(\theta_t - \theta_s) - (a_t - a_s)|\mathcal{F}_s^N] \\ &= M(x_t - x_s|\mathcal{F}_s^N) \\ &= M[M(x_t - x_s|\mathcal{F}_s)|\mathcal{F}_s^N] = 0. \end{aligned} \tag{19.55}$$

Therefore, $\overline{X}$ is a uniformly integrable martingale which is assumed to have right continuous trajectories because of the right continuity of the family $(\mathcal{F}_t^N)$, $t \geq 0$ (Lemma 18.4), and the corollary to Theorem 3.1. □

Corollary. *The properties of the trajectories $\overline{a}_t$, (19.53), and Lemma 19.7 imply the existence of the right continuous modification $\pi_t(\theta) = (M(\theta_t|\mathcal{F}_t^N))$ having left limits.*

Note. According to Theorem 19.1, the martingale $\overline{X}$ permits the representation

$$\overline{x}_t = \int_0^t \widehat{\Delta\overline{x}_s}(1 - \Delta\overline{A}_s)^+ d[N_s - \overline{A}_s], \tag{19.56}$$

where

$$\widehat{\Delta\overline{x}_t} = M_{\mathcal{N}}(\Delta\overline{x}_t|\hat{\mathcal{P}}). \tag{19.57}$$

19.3.4.

Lemma 19.8. *Under the assumptions (a)–(c) we have*

$$\widehat{\Delta\overline{x}_t} = \hat{\pi}_t(\theta) - \pi_{t-}(\theta) - \Delta\overline{a}_t. \tag{19.58}$$

PROOF. (19.58) for $\widehat{\Delta\overline{x}_t}$ follows from (19.53) and the corollary to Lemma 19.7 since

$$\Delta\overline{x}_t = \pi_t(\theta) - \pi_{t-}(\theta) - \Delta\overline{a}_t, \tag{19.59}$$

and the processes $\pi_{t-}(\theta)$ and $\Delta\overline{a}_t$ are $\hat{\mathcal{P}}$-predictable.

19.3.5.

Lemma 19.9. *Under the assumptions (a)-(c), for each n, $n = 1, 2, \ldots$, we have*

$$\pi_{\tau_n}(\theta) = M(\theta_{\tau_n}|\mathcal{F}_{\tau_n}^N) \quad (\{\tau_n < \infty\};\ (P\text{-a.s.})). \tag{19.60}$$

PROOF. The random variables $\pi_{\tau_n}(\theta)$ are defined (at each n, $n = 1, 2, \ldots$) by the relations

$$I_{(\tau_n = t)}\pi_{\tau_n}(\theta) = I_{(\tau_n = t)}\pi_t(\theta), \quad t < \theta. \tag{19.61}$$

Hence, to prove (19.60) it suffices to establish that P-a.s.

$$I_{(\tau_n = t)}\pi_{\tau_n}(\theta) = I_{(\tau_n = t)}M(\theta_{\tau_n}|\mathcal{F}_{\tau_n}^N), \quad t < \infty. \tag{19.62}$$

Due to the $\mathcal{F}_t^N$-measurability of the random variable $I_{(\tau_n = t)}$ we have

$$\begin{aligned} I_{(\tau_n = t)}\pi_t(\theta) &= N\{I_{(\tau_n = t)}\theta_t|\mathcal{F}_t^N\} \\ &= M\{I_{(\tau_n = t)}\theta_{\tau_n}|\mathcal{F}_t^N\} \\ &= I_{(\tau_n = t)}M(\theta_{\tau_n}|\mathcal{F}_t^N). \end{aligned} \tag{19.63}$$

Further, according to Lemma 1.9

$$I_{(\tau_n = t)}M(\theta_{\tau_n}|\mathcal{F}_t^N) = I_{(\tau_n = t)}M(\theta_{\tau_n}|\mathcal{F}_{\tau_n}^N). \tag{19.64}$$

The required equality given by (19.62) follows from (19.63) and (19.64).

19.3.6.

Lemma 19.10. *Under the assumptions (a)–(c)*

$$\hat{\pi}_t(\theta) = \hat{\theta}_t \quad (N\text{-a.s.}). \tag{19.65}$$

PROOF. Let $\varphi(t,\omega)$ be a $\hat{\mathcal{P}}$-predictable process such that

$$M\int_0^{\tau_\infty} |\varphi(t,\omega)\theta_t| dN_t < \infty. \tag{19.66}$$

Then, by the definition of $\hat{\theta}_t$ we find that

$$\begin{aligned} M\int_0^{\tau_\infty} \varphi(t,\omega)\hat{\theta}_t dN_t &= M\int_0^{\tau_\infty} \varphi(t,\omega)\theta_t dN_t \\ &= M\sum_{n\geq 1} \varphi(\tau_n,\omega)\theta_{\tau_n} \\ &= M\sum_{n\geq 1} \varphi(\tau_n,\omega)M(\theta_{\tau_n}|\mathcal{F}^N_{\tau_n}). \end{aligned} \tag{19.67}$$

In accord with (19.76), $M(\theta_{\tau_n}|\mathcal{F}^N_{\tau_n}) = \pi_{\tau_n}(\theta)$. Hence

$$\begin{aligned} M\sum_{n\geq 1} \varphi(\tau_n,\omega)\theta_{\tau_n} &= M\sum_{n\geq 1} \varphi(\tau_n,\omega)\pi_{\tau_n}(\theta) \\ &= M\int_0^{\tau_\infty} \varphi(t,\omega)\pi_t(\theta)dN_t. \end{aligned} \tag{19.68}$$

From (19.68), (19.67) and the definition of $\hat{\pi}_t(\theta)$ we obtain the equality

$$M\int^{\tau_\infty} \varphi(t,\omega)\hat{\theta}dN_t = M\int^{\tau_\infty} \varphi(t,\omega)\hat{\pi}_t(\theta)dN_t, \tag{19.69}$$

which is equivalent to the assertion of the lemma. □

19.3.7. The assertion of the theorem follows from (19.65), (19.58), (19.56) and the corollary to Lemma 19.7.

19.3.8. We consider now the cases where (19.52) has a more obvious structure.

Let us assume that instead of (a)–(c) the conditions which follow are satisfied:

(a′) $X = (x_t, \mathcal{F}_t)$ is a square integrable martingale;

(b′)

$$a_t = \int_0^t H_s da_s^0,$$

where $H = (H_t, \mathcal{F}_t)$, $t \geq 0$, is a predictable process $a^0 = (a_t^0, \mathcal{F}_t^N)$ is a nondecreasing right continuous predictable process with $a_0^0 = 0$, and

$$M\left(\int_0^\infty |H_s| da_s^0\right)^2 < \infty;$$

(c′) $M\theta_0^2 < \infty$;

(d′) the compensators A_t and $\overline{A}_t$ of the processes $N = (N_t, \mathcal{F}_t)$ and $\overline{N} = (N_t, \mathcal{F}_t^N)$ are related by

$$A_t = \int_0^t \Lambda_s d\overline{A}_s, \tag{19.70}$$

where $\Lambda = (\Lambda_t, \mathcal{F}_t)$, $t \geq 0$, is a nonnegative predictable process[3],

$$M(|\theta_{t-} + \Delta a_t| \Lambda_t) < \infty, \quad t \geq 0.$$

We shall show that

$$\overline{a}_t = \int_0^t M(H_s|\mathcal{F}_{s-}^N) da_s^0 \quad (P\text{-a.s.}), \tag{19.71}$$

$$\hat{\theta}_t - \pi_{t-}(\theta) - \Delta\overline{a}_t = M\left\{ \frac{d\langle x, m\rangle_t}{d\overline{A}_t} + \theta_t^p(\Lambda_t - 1) \middle| \mathcal{F}_{t-}^N \right\} \quad (N\text{-a.s.}), \tag{19.72}$$

where

$$\theta_t^p = \theta_{t-} + \Delta a_t. \tag{19.73}$$

It follows from (19.72) that $\theta_t = \theta_t^p + \Delta x_t$ and therefore

$$\hat{\theta}_t = \hat{\theta}_t^p + \widehat{\Delta x_t}. \tag{19.74}$$

In defining $\overline{a}_t$, $\hat{\theta}_t^p$ and $\widehat{\Delta x_t}$ we use the following auxiliary lemma.

Lemma 19.11. *Let $f = (f_t, \mathcal{F}_t)$ be a nonnegative predictable process with $Mf_t < \infty$, $t \geq 0$, and let $\varphi = (\varphi_t, \mathcal{F}_t^N)$ be a nonnegative predictable process. Then the process $\overline{f} = (\overline{f}_t, \mathcal{F}_{t-}^N)$ with $\overline{f}_t = M(f_t|\mathcal{F}_{t-}^N)$ is predictable and*

$$M\int_0^{\tau_\infty} f_t \varphi_t d\overline{A}_t = M\int_0^{\tau_\infty} \overline{f}_t \varphi_t d\overline{A}_t. \tag{19.75}$$

PROOF. It suffices to examine the case in which $f_t = I_{(a<t\leq b)}\xi$ and $\varphi_t = I(a' < t \leq b')\eta$ where : a, b and a', b' are numbers; ξ and η are bounded variables, $\xi \geq 0$, $\eta \geq 0$; ξ is $\mathcal{F}_a$-measurable; η is $\mathcal{F}_{a'}^N$-measurable.

In this case $\overline{f}_t = I_{(a<t\leq b)} M(\xi|\mathcal{F}_{t-}^N)$ and the predictability of the process $\overline{f}_t$ follows from Lévy's theorem (Theorem 1.5).

We can consider without loss of generality that $b' \wedge b < \tau_\infty$ and $M\overline{A}_{b'\wedge b} < \infty$. Then according to Lemma 3.2

[3] This condition will be satisfied if the compensator is given by (18.33) (Theorem 18.3).

$$
\begin{aligned}
M\int_{a'\wedge a}^{b'\wedge b} \eta M(\xi|\mathcal{F}_{t-}^N)d\overline{A}_t &= M\int_{a'\wedge a}^{b'\wedge b} M(\eta\xi|\mathcal{F}_{t-}^N)d\overline{A}_t \\
&= M\{M(\eta\xi|\mathcal{F}_{b'\wedge b}^N)[\overline{A}_{b'\wedge b} - \overline{A}_{a'\wedge a}]\} \\
&= M\xi\eta[\overline{A}_{b'\wedge b} - \overline{A}_{a'\wedge a}] = M\int_{a'\wedge a}^{b'\wedge b} \xi\eta d\overline{A}_t,
\end{aligned}
$$

which proves (19.75). □

Note. The lemma still holds true if the process $f = (f_t, \mathcal{F}_t)$, $t \geq 0$, is such that $M|f_t| < \infty$, $t \geq 0$, and

$$
M\int_0^{\tau_\infty} |f_t|\varphi_t d\overline{A}_t < \infty. \tag{19.76}
$$

19.3.9. By virtue of (b′) and the note to Lemma 19.11 it can be easily verified that the process

$$
\left(M\left\{ \int_0^t H_s da_s^0 \middle| \mathcal{F}_t^N \right\} - \int_0^t M(H_s|\mathcal{F}_{s-}^N)da_s^0, \mathcal{F}_t^N \right), \quad t \geq 0,
$$

is a martingale. From this in particular, (19.71) follows for $\overline{a}_t$ (compare with Theorem 7.12).

The representations for $\hat{\theta}_t$ and $\widehat{\Delta x_t}$ are proved in the next lemma.

Lemma 19.12. *Under the assumptions (a′)–(d′)*

$$
\hat{\theta}_t = M(\theta_t \Lambda_t|\mathcal{F}_{t-}^N) \quad (N\text{-a.s.}), \tag{19.77}
$$

$$
\widehat{\Delta x_t} = M\left(\frac{d\langle x, m\rangle_t}{d\overline{A}_t} \middle| \mathcal{F}_{t-}^N \right) \quad (N\text{-a.s.}). \tag{19.78}
$$

PROOF. We take a $\hat{\mathcal{P}}$-predictable process $\varphi(t,\omega)$ such that $M\int_0^{\tau_\infty} |\varphi(t,\omega)\theta_t|dN_t < \infty$. Then by Lemma 19.11 and the note to this lemma we find that:

$$
\begin{aligned}
M\int_0^{\tau_\infty} \varphi(t,\omega)\hat{\theta}_t^p dN_t &= M\int_0^{\tau_\infty} \varphi(t,\omega)\theta_t^p dN_t \\
&= M\int_0^{\tau_\infty} \varphi(t,\omega)\theta_t^p dA_t \\
&= M\int_0^{\tau_\infty} \varphi(t,\omega)\theta_t^p \Lambda d\overline{A}_t \\
&= M\int_0^{\tau_\infty} \varphi(t,\omega)M(\theta_t^p\Lambda|\mathcal{F}_{t-}^N)d\overline{A}_t \\
&= M\int_0^{\tau_\infty} \varphi(t,\omega)M(\theta_t^p\Lambda|\mathcal{F}_{t-}^N)dN_t.
\end{aligned}
$$

From this the required equality, (19.77) follows.

Let us establish the equality given by (19.78). Let $\varphi(t,\omega)$ be a $\hat{\mathcal{P}}$-predictable process such that

$$M\int_0^{\tau_\infty} |\varphi(t,\omega)|(1+|x_t|+|\Delta x_t|)dN_t < \infty.$$

Then

$$\begin{aligned} & M\int_0^{\tau_\infty} \varphi(t,\omega)\Delta x_t dN_t \\ &= M\int_0^{\tau_\infty} \varphi(t,\omega)x_t dN_t - M\int_0^{\tau_\infty} \varphi(t,\omega)x_{t-}dN_t \\ &= M\sum_{n\geq 1}\varphi(\tau_n,\omega)x_{\tau_n} - M\int_0^{\tau_\infty} \varphi(t,\omega)x_{t-}dA_t \\ &= Mx_{\tau_\infty}\sum_{n\geq 1}\varphi(\tau_n,\omega) - Mx_{\tau_\infty}\int_0^{\tau_\infty} \varphi(t,\omega)dA_t \\ &= Mx_{\tau_\infty}\int_0^{\tau_\infty} \varphi(t,\omega)dm_t, \end{aligned} \quad (19.79)$$

where we have made use of the predictability of the process $(x_{t-},\mathcal{F}_t)$, Lemma 3.2, and the equality $m_t = N_t - A_t$.

The process $m = (m_t,\mathcal{F}_t)$ is a τ_∞-locally square integrable martingale with

$$\langle m\rangle_t = \int_0^t (1-\Delta A_s)dA_s$$

(Lemma 18.12). Hence by virtue of (19.70)

$$\langle m\rangle_t = \int_0^t (1-\Delta A_s)\Lambda_s d\overline{A}_s. \quad (19.80)$$

As in the proof of Theorem 5.3, it can be established that

$$\langle x,m\rangle_t = \int_0^t g_s d\langle m\rangle_s, \quad (19.81)$$

where $g = (g_t,\mathcal{F}_t)$ is a predictable process with

$$\int_0^t g_s^2 d\langle m\rangle_s < \infty \quad (\{t<\tau_\infty\};\ (P\text{-a.s.})).$$

From (19.80) and (19.81) we find that $\langle x,m\rangle_t = \int_0^t g_s(1-\Delta A_s)\Lambda_s d\overline{A}_s$. Let $d\langle x,m\rangle_t/d\overline{A}_t = g_t(1-\Delta A_t)\Lambda_t$. Then from (5.70) and (19.75) we obtain ($\varphi(t,\omega)$ is $\widehat{\mathcal{P}}$- predictable bounded and compactly supported in t)

$$
\begin{aligned}
Mx_{\tau_\infty}\int_0^{\tau_\infty}\varphi(t,\omega)dm_t &= M\int_0^{\tau_\infty}\varphi(t,\omega)d\langle x,m\rangle_t \\
&= M\int_0^{\tau_\infty}\varphi(t,\omega)\frac{d\langle x,m\rangle_t}{d\overline{A}_t}d\overline{A}_t \\
&= M\int_0^{\tau_\infty}\varphi(t,\omega)M\left[\frac{d\langle x,m\rangle_t}{d\overline{A}_t}\middle|\mathcal{F}_{t-}^N\right]d\overline{A}_t \\
&= M\int_0^{\tau_\infty}\varphi(t,\omega)M\left[\frac{d\langle x,m\rangle_t}{d\overline{A}_t}\middle|\mathcal{F}_{t-}^N,\right]dN_t,
\end{aligned}
$$

from which the required formula, (19.78), follows.

To establish (19.72) it remains to show that

$$\pi_{t-}(\theta)=M(\theta_{t-}|\mathcal{F}_{t-}^N), \tag{19.82}$$

$$\Delta\bar{a}_t=M(\Delta a_t|\mathcal{F}_{t-}^N). \tag{19.83}$$

(19.82) is a corollary to Theorem 1.6 since

$$\pi_{t-}(\theta)=\lim_{s\uparrow t}\pi_s(\theta)=\lim_{s\uparrow t}M(\theta_s|\mathcal{F}_s^N)=M(\theta_{t-}|\mathcal{F}_{t-}^N),$$

whereas (19.83) follows from (19.49) and the fact that the variable $\Delta\bar{a}_t$ is $\mathcal{F}_{t-}^n$-measurable.

Thus we have proved the following theorem.

Theorem 19.6. *Let (a′)–(d′) be satisfied. Then the conditional expectation $\pi_t(\theta)$ $(=M(\theta_t|\mathcal{F}_t^N))$ permits the representation*

$$
\begin{aligned}
\pi_t(\theta) &= \pi_0(\theta)+\int_0^t M(H_s|\mathcal{F}_{s-}^N)da_s \\
&\quad+\int_0^t\left\{M\left[\frac{d\langle x,m\rangle_s}{d\overline{A}_s}+\theta_s\left(\frac{dA_s}{d\overline{A}_s}-1\right)\middle|\mathcal{F}_{s-}^N\right]\right\} \\
&\quad\times(1-\Delta\overline{A}_s)^+d[N-s-\overline{A}_s],
\end{aligned}
\tag{19.84}
$$

where

$$\theta_t^p=\theta_{t-}+H_t\Delta a_t^0,\quad \frac{da_T}{a\overline{A}_t}=\lambda_t.$$

Note. By virtue of (a′)–(c′), $\sup_t M\theta_t^2<\infty$ and therefore $\pi_t(\theta)$ is an optimal (in the mean square sense) estimate of θ_t from the observations N_s, $s\le t$. In this connection (19.84) can be naturally called an equation of (optimal nonlinear) filtering (compare with Theorem 8.1).

19.3.10. We give here a few examples[4] illustrating the equation given in (19.84).

EXAMPLE 1. Let θ be a random variable taking on values b and a with probabilities π and $1-\pi$, respectively ($b>0$, $a>0$). Let us assume that a point process N_t is to be observed with a compensator $A_t=\theta t$. Let $\pi_t = P(\theta=b|\mathcal{F}_t^N)$. Then by virtue of (19.84) and $\theta_t^p \equiv \theta$

$$d\pi_t = M\left[\delta(\theta,b)\left(\frac{dA_t}{d\overline{A}_t}-1\right)\middle|\mathcal{F}_{t-}^N\right]d[N_t-\overline{A}_t], \tag{19.85}$$

where $\delta(\cdot,\cdot)$ is the Kronecker delta. By virtue of (18.34)

$$\overline{A}_t = \int_0^t M(\theta|\mathcal{F}_{s-}^N)ds = \int_0^t [b\pi_{s-}+a(1-\pi_{s-})]ds.$$

Hence

$$\begin{aligned}
&M\left[\delta(\theta,b)\left(\frac{dA_t}{d\overline{A}_t}-1\right)\middle|\mathcal{F}_{t-}^N\right]\\
&= M\left[\delta(\theta,b)\left(\frac{\theta}{b\pi_{t-}+a(1-\pi_{t-})}-1\right)\middle|\mathcal{F}_{t-}^N\right]\\
&= \frac{b\pi_{t-}}{b\pi_{t-}+a(1-\pi_{t-})}-\pi_{t-} = \frac{(b-a)\pi_{t-}(1-\pi_{t-})}{a(1-\pi_{t-})+b\pi_{t-}}.
\end{aligned}$$

Therefore,

$$d\pi_t = \frac{(b-a)\pi_{t-}(1-\pi_{t-})}{b\pi_{t-}+a(1-\pi_{t-})}[dN_t-(b\pi_{t-}+a(1-\pi_{t-}))dt]. \tag{19.86}$$

Note. Let $N^a=(N_t^a,\mathcal{F}_t)$ and $N^b=(N_t^b,\mathcal{F}_t)$, $t\ge 0$, be two Poisson processes with parameters a and b, respectively. Let θ be a random variable (measurable with respect to $\mathcal{F}_0$) taking on values a and b. Then it is easy to verify that the process

$$N_t^\theta = I(\theta=b)N_t^b + I(\theta=a)N_t^a$$

is a point process with compensator $A=\theta t$.

EXAMPLE 2. Let $\alpha=(\alpha_t,\mathcal{F}_t)$ be a Markov process with states a and b ($a>0$, $b>0$) and (stationary) densities of transient probabilities

$$\tau_{a,a}=-\lambda,\quad \lambda_{a,b}=\lambda,\quad \lambda_{b,a}=\lambda,\quad \lambda_{b,b}=-\lambda,$$

and let $P(\alpha_0=b)=\pi$, $P(\alpha_0=a)=1-\pi$. Further, let $N^a=(N_t^a,\mathcal{F}_t)$ and $N^b=(N_t^b,\mathcal{F}_t)$ be Poisson processes with parameters a and b, respectively.

[4] Compare with the examples of Section 9.4.

We assume that the processes α, N^a and N^b are independent and that the process

$$N_t = \int_0^t I(\alpha_{s-} = a)dN_s^a + \int_0^t I(\alpha_{s-} = b)dN_s^b$$

is to be observed.

It is easy to see that the compensator of this process is the process

$$A_t = a \int_0^t I(\alpha_{s-} = a)ds + b \int_0^t I(\alpha_{s-} = b)ds.$$

Let us derive an equation for $\pi_t = P(\alpha_t = b|\mathcal{F}_t^N)$. To this end, we set $\theta_t = \delta(\alpha_t, b)$, where $\delta(\cdot,\cdot)$ is the Kronecker delta. It was proved in Lemma 9.2 that the process $x = (x_t, \mathcal{F}_t)$, $t \geq 0$, with

$$x_t = \theta_t - \theta_0 - \int_0^t \lambda_{\alpha_s, b} ds$$

is a square integrable martingale. We note that

$$\int_0^t \lambda_{\alpha_s, b} ds - \int_0^t \lambda_{\alpha_{s-}, b} ds \quad (P\text{-a.s.})$$

and

$$\int_0^t M(\lambda_{\alpha_{s-}, b}|\mathcal{F}_{s-}^N)ds = \lambda \int_0^t (1 - 2\pi_{s-})ds. \tag{19.87}$$

Because N^a, N^b and α are independent we have $\langle x, m\rangle_t = 0$ where $m_t = N_t - A_t$. By (18.34)

$$\overline{A}_t = a \int_0^t (1 - \pi_{s-})ds + b \int_0^t \pi_{s-} ds.$$

Therefore,

$$\frac{dA_t}{d\overline{A}_t} = \frac{aI(\alpha_{t-} = a) + bI(\alpha_{t-} = b)}{a(1 - \pi_{t-}) + b\pi_{t-}}.$$

Hence as in Example 1 we infer that

$$\begin{aligned} & M\left\{\theta_{t-}\left(\frac{dA_t}{d\overline{A}_t} - 1\right)\middle|\mathcal{F}_{t-}^N\right\} \\ &= M\left\{\delta(\alpha_{t-}, b)\left[\frac{dA_t}{d\overline{A}_t} - 1\right]\middle|\mathcal{F}_{t-}^N\right\} \\ &= \frac{(b-a)\pi_{t-}\cdot(1 - \pi_{t-})}{a(1 - \pi_{t-}) + b\pi_{t-}}. \end{aligned} \tag{19.88}$$

From (19.84), (19.87) and (19.88) we obtain

$$d\pi_t = \lambda(1-2\pi_t)dt + \frac{(b-a)\pi_{t-}(1-\pi_{t-})}{a(1-\pi_{t-})+b\pi_{t-}} \cdot [dN_t - (a(1-\pi_{t-})+b\pi_{t-})dt]. \tag{19.89}$$

EXAMPLE 3. Let $\alpha = (\alpha_t, \mathcal{F}_t)$ be a Markov process taking on values $a > 0$ and $b > 0$ with $P(\alpha_0 = b) = \pi$, $P(\alpha_0 = a) = 1 - \pi$, and with the single transition $a \to b$:

$$\lambda_{a,a} = -\lambda; \quad \lambda_{a,b} = \lambda; \quad \lambda_{b,a} = 0; \quad \lambda_{b,b} = 1.$$

We assume that the observable process

$$N_t = \int_0^t I(\alpha_{s-} = a)dN_s^a + \int_0^t I(\alpha_{s-} = b)dN_s^b,$$

where N^a, N^b are the same processes as those in Example 2, and the processes N^a, N^b and α are independent.

Using the same technique as in Example 2 we find that the *a posteriori* probability $\pi_t = P(\alpha_t = b|\mathcal{F}_t^N)$ satisfies the equation

$$d\pi_t = \lambda(1-\pi_t)dt + \frac{(b-a)\pi_{t-}(1-\pi_{t-})}{a(1-\pi_{t-})+b\pi_{t-}}[dN_t - (a(1-\pi_{t-})+b\pi_{t-})dt]. \tag{19.90}$$

19.3.11. The next example involves the computation of $\hat{\theta}_t$.

EXAMPLE 4. Let $W = (W_t, \mathcal{F}_t)$, $t \geq 0$, be a Wiener process and let $\tau = \inf(t : W_t = 1)$. Let $\theta_t = W_{t\wedge\tau}$ and $N_t = I(t \geq \tau)$. According to (1.42) the distribution function $F(t)$ of the stopping time τ is determined by the formula

$$F(t) = \sqrt{\frac{2}{\pi}} \int_{1/\sqrt{t}}^{\infty} e^{-y^2/2} dy.$$

Hence, by Theorem 18.2 the compensator $\overline{A}_t$ is determined by the relation

$$\overline{A}_t = \int_0^{t\wedge\tau} \frac{dF(u)}{1-F(u-)} = \ln \frac{1}{1-F(t\wedge\tau)}$$

and is an absolutely continuous (with respect to Lebesgue measure) function of time. At the same time the compensator A_t coincides with N_t since τ is a $\mathcal{P}$-predictable (as opposed to $\hat{\mathcal{P}}$-predictable–see Corollary 1 to Lemma 18.3) stopping time. (See also the example for Theorem 18.2).

Therefore (d′) for A_t is not satisfied.

We shall define $\pi_t(\theta)$ using the representation given by (19.52) which has the form:

$$\pi_t(\theta) = \pi_{t\wedge\tau}(\theta) = \int_0^{t\wedge\tau} (\hat{\theta}_s - \pi_{s-}(\theta))d[N_s - \overline{A}_s]$$

in the given case.

Let $(\varphi_t, \mathcal{F}_t^N)$ be a $\hat{\mathcal{P}}$-predictable process such that $M\int_0^{\tau_\infty} |\varphi_t \theta_t| dN_t < \infty$. Then in accord with the definition of $\hat{\theta}_t$ we find that

$$M\int_0^{\tau_\infty} \varphi_t \hat{\theta}_t dN_t = M\int_0^{\tau_\infty} \varphi_t \theta_t dN_t$$
$$= M\varphi_\tau \theta_\tau = M\varphi_\tau W_\tau = M\varphi_\tau = M\int_0^{\tau_\infty} \varphi_\tau dN_t$$
$$= M\int_0^\infty \varphi_t I(\tau \geq t) dN_t.$$

From this it follows that N-a.s., $\hat{\theta}_t = I(\tau \geq t)$.
Therefore

$$\pi_t(\theta) = \int_0^{t\wedge\tau} (1 - \pi_{s-}(\theta)) d[N_s - \overline{A}_s]$$

and then

$$\pi_t(\theta) = \begin{cases} 1 - e^{[1-F(t\wedge\tau)]^{-1}}, & t < \tau \\ 1, & t \geq \tau. \end{cases}$$

19.4 The Necessary and Sufficient Conditions for Absolute Continuity of the Measures Corresponding to Point Processes

19.4.1. Let $(\mathsf{X}, \mathcal{B})$ be the measurable space introduced at the beginning of Section 18.3 and let μ, $\tilde{\mu}$ be probability measures given on it. We shall consider the point processes $X = (x_t, \mathcal{B}_t, \mu)$ and $\tilde{X} = (x_t, B_t, \tilde{\mu})$ with the compensators $A = (A_t(x), \mathcal{B}_t, \mu)$ and $\tilde{A} = (\tilde{A}_t(x), \mathcal{B}_t, \tilde{\mu})$, respectively. The present section deals with the question as to the conditions under which the measure $\tilde{\mu}$ is absolutely continuous with respect to the measure μ. Since the compensators A and $\tilde{A}$ define uniquely the measures μ and $\tilde{\mu}$ (Theorems 18.4 and 18.5), it is natural to expect that the answer to the above question can be formulated in terms of the properties of the compensators A and $\tilde{A}$.

19.4.2. Before formulating the main results we shall make some remarks. We shall assume throughout, from now on, that the σ-algebras $\mathcal{B}_t$, $t \geq 0$, are augmented by sets from $\mathcal{B}$ of zero μ- and $\tilde{\mu}$-probability.

The compensators A and $\tilde{a}$ are functionals having the following properties ((μ-a.s.) and $\tilde{\mu}$-a.s.), respectively):

(1) $A_0(x) = 0$, $A_s(x) \leq A_t(x)$, $\tilde{A}_0(x) = 0$, $\tilde{A}_s(x) \leq \tilde{A}_t(x)$, $s \leq t$;
(2) $A_t(x) = A_{t\wedge\tau_\infty(x)}(x)$, $\tilde{A}_t(x) = \tilde{A}_{\wedge\tau_\infty(x)}(x)$, $A_{\tau_\infty(x)}(x) = A_{(\tau_\infty(x))-}(x)$, $\tilde{A}_{\tau_\infty(x)}(x) = \tilde{A}_{(\tau_\infty(x))-}(x)$;
(3) almost all trajectories of A and $\tilde{A}$ are right continuous;

(4) $\Delta A_t(x) \le 1$, $\Delta \tilde{A}_t(x) \le 1$.

It follows from (18.37) and properties (A)–(C) (Subsection 18.3.1) that the functionals $A_t(x)$ and $\tilde{A}_t(x)$ can be considered from now on to be defined so that properties (1)–(4) listed are fulfilled for each $x \in \mathsf{X}$ and $t \ge 0$.

We shall deal with the situation in which the compensators A and $\tilde{A}$ are related by

$$\tilde{A}_t(x) = \int_0^t \lambda_s(x) dA_s(x) \quad (\{t < \tau_\infty\};\ (\tilde{\mu}\text{-a.s.})) \tag{19.91}$$

and

$$\Delta A_t(x) = 1 \text{ implies } \lambda_t(x) = I(t < \tau_\infty(x)) \quad (\tilde{\mu}\text{-a.s.}) \tag{19.92}$$

where $\lambda = (\lambda_t(x), \mathcal{B}_t)$ is some nonnegative predictable process.

The process $\lambda_t(x)$ can be defined uniquely from (19.91) and (19.92). However, $\lambda_t(x)$ can always be defined so that, for all $x \in \mathsf{X}$,

$$\lambda_t(x) \le (\Delta A_t(x))^{-1}, \tag{19.93}$$

$$\int_0^{\theta(x)} \lambda_t(x) dA_t(x) = \lim_n \int_0^{\theta_n(x)} \lambda_t(x) dA_t(x), \tag{19.94}$$

where

$$\theta(x) = \inf \left\{ t \le \tau_\infty : \int_0^t \lambda_s(x) dA_s(x) = \infty \right\}$$

and

$$\theta(x) = \infty \text{ if } \int_0^\infty \lambda_s(x) dA_s(x) < \infty$$

and $\theta_n(x)$, $n = 1, 2, \ldots$, is the sequence of stopping times such that

$$\theta_n(x) < \theta(x), \quad \theta_n(x) \uparrow \theta(x),$$
$$\Delta A_t(x) = 1 \text{ implies } \lambda_t(x) = I\{t \le \theta(x)\}. \tag{19.95}$$

An example of such a process $\lambda_t(x)$ is given by:

$$\lambda_t(x) = I\{t \le \theta(x)\} \cdot \left[\left(\frac{d\tilde{A}_t}{dA_t}(x) \wedge \frac{1}{\Delta A_t(x)} \right) \times I\{\Delta A_t(x) < 1\} + I\{\Delta A_t(x) = 1\} \right].$$

19.4.3.

Theorem 19.7. *A necessary and sufficient condition for the measure $\tilde{\mu}$ to be absolutely continuous with respect to the measure μ ($\tilde{\mu} \ll \mu$) is that ($\tilde{\mu}$-a.s.)*

(I)

$$\tilde{A}_t(x) = \int_0^t \lambda_s(x) dA_s(x), \quad t < \tau_\infty; \tag{19.96}$$

(II)

$$\Delta A_t(x) = 1 \textit{ implies } \Delta \tilde{A}_t(x) = 1, \quad t < \tau_\infty; \tag{19.97}$$

(III)

$$\int_0^{\tau_\infty} (1 - \sqrt{\lambda_t(x)})^2 dA_t(x) + \sum_{\substack{t<\tau_\infty \\ 0<\Delta A_t(x)<1}} \left(1 - \sqrt{\frac{1 - \Delta \tilde{A}_t(x)}{1 - \Delta A_t(x)}}\right)^2$$
$$\times (1 - \Delta A_t(x)) < \infty. \tag{19.98}$$

PROOF. The necessity of conditions (I) and (II) was proved in Theorem 19.3. Hence Theorem 19.7 can be reformulated as follows: a necessary and sufficient condition, under assumptions (I) and (II), for the measure $\tilde{\mu}$ to be absolutely continuous with respect to the measure μ is that condition (III) be satisfied. If conditions (I) and (II) are satisfied, then by virtue of the corollary to Lemma 19.5, we can define the random process $z = (z_t(\lambda), \mathcal{B}_t, \mu)$ (see (19.24)) by

$$z_t(\lambda) = \prod_{\{n:\tau_n \le t\}} \lambda_{\tau_n} \cdot \prod_{\substack{s \le t \\ s \ne \tau_n}} [1 + (1 - \Delta A_s)^+ (1 - \lambda_s) \Delta A_s]$$
$$\cdot \exp\left[\int_0^t (1 - \lambda_s) dA_s^c\right], \tag{19.99}$$

being a nonnegative supermartingale as well as a τ_∞-local martingale.

Let σ_n, $n = 1, 2, \ldots$ be a sequence of stopping times such that $\sigma_n \uparrow \tau_\infty$ and let the processes $z^{(n)}(\lambda) = (z_{t \wedge \sigma_n}(\lambda), \mathcal{B}_t, \mu)$ for each $n = 1, 2, \ldots$, be uniformly integrable martingales with[5] $M_\mu z_{\sigma_n}(\lambda) = 1$.

We shall define on $(\mathsf{X}, \mathcal{B})$, the probability measures $\tilde{\mu}^{(n)}$, $n = 1, 2, \ldots$, with $d\tilde{\mu}^{(n)} = z_{\sigma_n}(\lambda) d\mu$.

By virtue of Theorem 19.4, the random processes $\tilde{X}^{(n)} = (x_t, \mathcal{B}_t, \tilde{\mu}^{(n)})$, $t \ge 0$, are point processes with the compensators

$$\tilde{A}_t^{(n)} = \int_0^{t \wedge \sigma_n} \lambda_s dA_s. \tag{19.100}$$

Consequently, ($\tilde{\mu}$-a.s.) $\tilde{A}_{t \wedge \sigma_n}^{(n)} = \tilde{A}_{t \wedge \sigma_n}$, $t \ge 0$, and, therefore, by the uniqueness theorem (Theorem 18.5) the restriction of the measure $\tilde{\mu}^{(n)}$ and that of the measure μ to the σ-algebra $\mathcal{B}_{\sigma_n}$ coincide.

---

[5] M_μ is the expectation under measure μ.

Denote by

$$\mu_n = \mu|\mathcal{B}_{\sigma_n}, \quad \tilde{\mu}_n = \tilde{\mu}|\mathcal{B}_{\sigma_n}, \quad \tilde{\mu}_n^{(n)} = \tilde{\mu}^{(n)}|\mathcal{B}_{\sigma_n}$$

the restrictions of the measures μ, $\tilde{\mu}$ and $\tilde{\mu}^{(n)}$ (respectively) to the σ-algebra $\mathcal{B}_{\sigma_n}$, Since $\tilde{\mu}_n = \tilde{\mu}_n^{(n)}$ and $\tilde{\mu}_n^{(n)} \ll \mu_n$, then $\tilde{\mu}_n \ll \mu_n$ and

$$\frac{d\tilde{\mu}_n}{d\mu_n} = M_\mu(z_{\sigma_n}(\lambda)|\mathcal{B}_{\sigma_n}) = z_{\sigma_n}(\lambda). \tag{19.101}$$

The sequence $(z_{\sigma_n}(\lambda), \mathcal{B}_{\sigma_n}, \mu)$, $n = 1, 2, \ldots$, is a nonnegative martingale with $M_\mu z_{\sigma_n}(\lambda) = 1$. Hence (Theorem 2.6) there exists $\lim_n z_{\sigma_n}(\lambda)$ $(= z(\lambda))$.

It is a well-known fact (see, for example, [74]) that the measure $\tilde{\mu}$ is absolutely continuous with respect to μ if and only if $\tilde{\mu}_n \ll \mu_n$, $n \geq 1$, and $M_\mu z(\lambda) = 1$.

In the case of interest to us, it is rather difficult to verify that condition (III) is equivalent to the condition $M_\mu z(\lambda) = 1$. Hence we shall use another criterion for absolute continuity of the two measures given in the lemma which follows.

Lemma 19.3. *Let $(\Omega, \mathcal{F})$ be a measurable space, let $(\mathcal{F}_n)$, $n = 1, 2, \ldots$, be a nondecreasing system of sub-σ-algebras of $\mathcal{F}$, and let $\sigma(\cup_n \mathcal{F}_n) = \mathcal{F}$. Let $\tilde{\mu}$ and μ be probability measures on $(\Omega, \mathcal{F})$ and let $\tilde{\mu}_n, \mu_n$ be the restrictions of these measures to $(\Omega, \mathcal{F}_n)$.*

For $\tilde{\mu} \ll \mu$ it is necessary and sufficient that the following conditions be satisfied:

(I) $\tilde{\mu}_n \ll \mu_n$, $n = 1, 2, \ldots$;
(II) $\lim_n d\tilde{\mu}_n/d\mu_n$ exists ($\tilde{\mu}$-a.s.) and is finite.

PROOF. Necessity. If $\tilde{\mu} \ll \mu$, then, obviously, $\tilde{\mu}_n \ll \mu_n$, $n = 1, 2, \ldots$. Let $\rho = d\tilde{\mu}_n/d\mu_n$. As noted above, the sequence $(\rho_n, \mathcal{F}_n, \mu)$, $n = 1, 2, \ldots$, forms a nonnegative martingale with $M_\mu \rho_n = 1$. Hence (μ-a.s.) $\lim_n \rho_n$ exists and is finite. But $\tilde{\mu} \ll \mu$, hence this limit exists and is finite under the measure $\tilde{\mu}$.

As to sufficiency, let $\tilde{\mu}_n \ll \mu_n$ and $\tilde{\rho} = \lim_n \rho_n$ ($\tilde{\mu}$-a.s.). Denote $\nu = \frac{1}{2}(\mu + \tilde{\mu})$, $\tilde{a} = d\tilde{\mu}/d\nu$, $a = d\mu/d\nu$, $\tilde{a}_n = d\tilde{\mu}_n/d\nu$, $a_n = d\mu_n/d\nu$. It is clear that $\mu\{a = 0\} = 0$, $\tilde{\mu}\{\tilde{a} = 0\} = 0$ and $a_n \to a$, $\tilde{a}_n \to \tilde{a}$, $n \to \infty$ (ν-, μ- and $\tilde{\mu}$-a.s.).

Further, for $\Gamma \in \mathcal{F}$,

$$\begin{aligned} \tilde{\mu}(\Gamma) &= \int_\Gamma \tilde{a} d\nu = \int_\Gamma \tilde{a} a^+ a d\nu + \int_\Gamma \tilde{a}(1 - a^+ a) d\nu \\ &= \int_\Gamma \tilde{a} a^+ d\mu + \int_\Gamma (1 - a^+ a) d\tilde{\mu} = \int_\Gamma \tilde{a} a^+ d\mu + \tilde{\mu}\{\Gamma \cap (a = 0)\}. \end{aligned}$$

Hence if $\tilde{\mu}\{a = 0\} = 0$, then $\tilde{\mu} \ll \mu$.

To verify that $\tilde{\mu}\{a = 0\} = 0$ under the conditions considered we note that since $\tilde{\mu}_n \ll \mu_n \ll \nu$,

$$\frac{d\tilde\mu_n}{d\nu}=\frac{d\tilde\mu_n}{d\mu_n}\frac{d\mu_n}{d\nu}, \quad (\nu\text{-a.s.}) \text{ and } (\tilde\mu\text{-a.s.}), \tag{19.102}$$

i.e., $\tilde a_n = (d\tilde\mu_n/d\mu_n)a_n$. According to the assumption made, $\lim_n d\tilde\mu_n/d\mu_n$ exists ($\tilde\mu$-a.s.). Hence $\lim_n \tilde a_n = \lim_n d\tilde\mu_n/d\mu_n \cdot \lim_n a_n$ or, which is the same,

$$\tilde a = \tilde\rho\cdot a \quad (\tilde\mu\text{-a.s.}). \tag{19.103}$$

Since by assumption $\tilde\mu\{\tilde\rho<\infty\}=1$ and $\tilde\mu\{\tilde a>0\}=1$, then $\tilde\mu\{a=0\}=0$, thus proving the lemma. □

Since instead of the sequence $\{\sigma_n\}$ we can consider without loss of generality the stopping times $\sigma_{n\wedge t}$, $n=1,2,\dots$, $t\to\tau_\infty$, it can be deduced from Lemma 19.13 that

$$\tilde\mu\left(\lim_{t\to\tau_\infty} z_t(\lambda)<\infty\right)=1 \tag{19.104}$$

is a necessary and sufficient condition for absolute continuity of $\tilde\mu\ll\mu$ if conditions (I) and (II) are satisfied.

To prove that (19.104) is equivalent to (19.98) (condition (III)) we note that due to (19.96) and (19.99) $z_t(\lambda)$ can be expressed as

$$\begin{aligned} z_t(\lambda) &= \exp\Bigg(\int_0^t I(\Delta A_s=0)\ln\lambda_s dN_s+\int_0^t(1-\lambda_s)dA_s^c \\ &\quad +\sum_{s\le t}\Delta N_s\ln\frac{\Delta\tilde A_s}{\Delta A_s}+(1-\Delta N_s)\ln\frac{1-\Delta\tilde A_s}{1-\Delta A_s}\Bigg). \end{aligned} \tag{19.105}$$

Setting

$$u(y)=\begin{cases} y & |y|\le 1,\\ \operatorname{sign} y & |y|>1,\end{cases} \tag{19.106}$$

we consider the values $z_t^u(\lambda)$ given by the formula (compare with (19.105)):

$$\begin{aligned} z_t^u(\lambda) &= \exp\Bigg(\int_0^t I(\Delta A_s=0)u(\ln\lambda_s)dN_s+\int_0^t(1-\lambda_s)dA_s^c \\ &\quad +\sum_{\substack{s\le t\\ 0<\Delta A_s<1}} u\left(\ln\frac{\Delta\tilde A_s}{\Delta A_s}\right)\Delta N_s+u\left(\ln\frac{1-\Delta\tilde A_s}{1-\Delta A_s}\right)(1-\Delta N_s)\Bigg). \end{aligned} \tag{19.107}$$

It is seen that the sets $\{0<\lim_{t\to\tau_\infty} z_t(\lambda)<\infty\}$ and $\{0<\lim_{t\to\tau_\infty} z_t^u(\lambda)<\infty\}$ coincide. Therefore, to prove the theorem it suffices to show that (19.98) is equivalent to the condition

$$\tilde\mu\left\{0<\lim_{t\to\tau_\infty} z_t^u(\lambda)<\infty\right\}=1. \tag{19.108}$$

Further we note that $\ln z_t^u(\lambda)$ can be represented as

$$\ln z_t^u(\lambda) = m_t + B_t, \tag{19.109}$$

where

$$\begin{aligned} m_t &= \int_0^t I(\Delta A_s = 0)u(\ln\lambda_s)d[N_s - \tilde{A}_s] \\ &+ \sum_{\substack{s\le t \\ 0<\Delta\tilde{A}_s<1}} u\left(\ln\frac{\Delta\tilde{A}_s}{\Delta A_s}\right)[\Delta N_s - \Delta\tilde{A}_s] + u\left(\ln\frac{1-\Delta\tilde{A}_s}{1-\Delta A_s}\right) \\ &\times[(1-\Delta N_s) - (1-\Delta\tilde{A}_s)], \end{aligned} \tag{19.110}$$

and

$$\begin{aligned} B_t &= \int_0^t (\lambda_s u(\ln\lambda_s) + 1 - \lambda_s)dA_s^c \\ &+ \sum_{\substack{s\le t \\ 0<\Delta\tilde{A}_s<1}} u\left(\ln\frac{\Delta\tilde{A}_s}{\Delta A_s}\right)\Delta\tilde{A}_s + u\left(\ln\frac{1-\Delta\tilde{A}_s}{1-\Delta A_s}\right)(1-\Delta\tilde{A}_s). \end{aligned} \tag{19.111}$$

The process B_t is a nondecreasing (predictable) function of t, since $yu(\ln y) + 1 - y \ge 0$ for $y \ge 0$, therefore by virtue of the inequality $yu(\ln y) \ge y - 1$,

$$\Delta B_s \ge \Delta A_s\left(\frac{\Delta\tilde{A}_s}{\Delta A_s} - 1\right) + (1-\Delta A_s)\left(\frac{1-\Delta\tilde{A}_s}{1-\Delta A_s} - 1\right) = 0. \tag{19.112}$$

The process $(m_t, B_t, \tilde{\mu})$, $t \ge 0$, is a τ_∞-locally square integrable martingale (compare with Lemma 18.12 and Theorem 18.8),

$$\begin{aligned} \langle m\rangle_t &= \int_0^t u^2(\ln\lambda_s)\lambda_s dA_s^c \\ &+ \sum_{\substack{s\le t \\ 0<\Delta\tilde{A}_s<1}} \left\{u^2\left(\ln\frac{\Delta\tilde{A}_s}{\Delta A_s}\right)\frac{\Delta\tilde{A}_s}{\Delta A_s}\Delta A_s + u^2\left(\ln\frac{1-\Delta\tilde{A}_s}{1-\Delta A_s}\right)\right. \\ &\left.\times\frac{1-\Delta\tilde{A}_s}{1-\Delta A_s}(1-\Delta A_s) - (\Delta B_s)^2\right\}, \end{aligned} \tag{19.113}$$

where

$$\Delta B_s = u\left(\ln\frac{\Delta\tilde{A}_s}{\Delta A_s}\right)\Delta\tilde{A}_s + u\left(\ln\frac{1-\Delta\tilde{A}_s}{1-\Delta A_s}\right)(1-\Delta\tilde{A}_s).$$

By (19.109) and Lemma 18.14 the relation given by (19.108) is satisfied if

$$\tilde{\mu}(\langle m\rangle_{\tau_\infty} + B_{\tau_\infty} < \infty) = 1. \tag{19.114}$$

From (19.111), (19.113) and the fact that ΔB_s is nonnegative it follows that (19.114) is equivalent to the following relation

$$\tilde{\mu}\Bigg(\int_0^{\tau_\infty} \varphi(\lambda_s)dA_s^c + \sum_{\substack{s<\tau \\ 0<\Delta A_s<1}} \Bigg[\varphi\left(\frac{\Delta\tilde{A}_s}{\Delta A_s}\right)\Delta A_s + \varphi\left(\frac{1-\Delta\tilde{A}_s}{1-\Delta A_s}\right)(1-\Delta A_s)\Bigg] < \infty\Bigg) = 1. \tag{19.115}$$

where

$$\varphi(y) = yu(\ln y) + 1 - y. \tag{19.116}$$

It is easy to see that there exist constants c and C such that $(y \geq 0)$

$$c(1-y^{1/2})^2 \leq \varphi(y) \leq C(1-y^{1/2})^2. \tag{19.117}$$

Hence (19.115) is equivalent to the required relation, (19.98).

Therefore, if (19.98) is satisfied, (19.104) will be also satisfied, i.e., $\tilde{\mu} \ll \mu$.

Conversely, if $\tilde{\mu} \ll \mu$ (19.104) will hold. Since $z_{\tau_\infty}(\lambda) = d\tilde{\mu}/d\mu$ then $\tilde{\mu}(z_{\tau_\infty}(\lambda) = 0) = 0$. Therefore (19.104) can be replaced with the equivalent conditions as follows:

$$\tilde{\mu}\left(-\infty < \lim_{t\to\tau_\infty} \ln z_t(\lambda) < \infty\right) = 1, \tag{19.118}$$

$$\tilde{\mu}\left(-\infty < \lim_{t\to\tau_\infty} \ln z_t^u(\lambda) < \infty\right) = 1. \tag{19.119}$$

Next we take advantage of the fact that $\ln z_t^u(\lambda)$ permits the representation given by (19.109), also using the fact that $|\Delta m_t| \leq 2$ as well as Lemma 18.13 by which it follows from (19.119) that

$$\tilde{\mu}(B_{\tau_\infty} < \infty) = 1. \tag{19.120}$$

From (19.109), (19.119) and (19.120) it follows that

$$\tilde{\mu}\left(-\infty < \lim_{t\to\tau_\infty} m_t < \infty\right) = 1. \tag{19.121}$$

Let us show that in this case

$$\tilde{\mu}(\langle m\rangle_{\tau_\infty} < \infty) = 1. \tag{19.122}$$

To this end we define the stopping times $\theta_n = \inf(t \leq \tau_\infty : m_t^2 > n)$ assuming $\theta_n = \tau_\infty$ on the set $(\sup_{t\leq\tau_\infty} m_t^2 \leq n)$. Since $m_{\theta_n}^2 \leq n+4$ then $M_{\tilde{\mu}}\langle m\rangle_{\theta_n} = M_\mu m_{\theta_n}^2 \leq n+4$. Therefore $\langle m\rangle_{\theta_n} < \infty$ ($\tilde{\mu}$-a.s.).

19.4.4. We shall formulate some assertions which follow immediately from Theorem 19.7 and which are of interest in themselves.

Theorem 19.7. *Let the point process $X = (x_t, \mathcal{B}_t, \mu)$ have the continuous compensator $A_t(x)$ (or, equivalently, the process X is left quasi-continuous). A necessary and sufficient condition for the measure $\tilde{\mu}$ to be absolutely continuous with respect to the measure μ is that ($\tilde{\mu}$-a.s.) the following conditions be satisfied:*

(A)

$$\overline{A}_t(x) = \int_0^t \lambda_s(x) dA_s(x), \quad t > \tau_\infty, \tag{19.123}$$

(B)

$$\int_0^{\tau_\infty} (1 - \sqrt{\lambda_s(x)})^2 dA_s(x) < \infty, \tag{19.124}$$

where $(\lambda_t(x), \mathcal{B}_t)$ is some nonnegative predictable process. In this case

$$\frac{d\tilde{\mu}}{d\mu}(t, x) = \exp\left\{\int_0^t \ln \frac{d\tilde{A}_s(x)}{dA_s(x)} dx_s - [\tilde{A}_t(x) - A_t(x)]\right\}. \tag{19.125}$$

Corollary. *Let the compensators $A_t(x)$ and $\tilde{A}_t(x)$ have the densities*

$$A_t(x) = \int_0^t a_s(x) ds, \quad \tilde{A}_t(x) = \int_0^t \tilde{a}_s(x) ds,$$

and

$$\int_0^t \tilde{a}_s(x)[1 - a_s^+(x) a_s(x)] ds = 0, \quad t \geq 0. \tag{19.126}$$

Then $\tilde{\mu} \ll \mu$ if and only if ($\tilde{\mu}$-a.s.)

$$\int_0^t (1 - \sqrt{\tilde{a}_s(x) a_s^+(x)})^2 a_s(x) ds < \infty. \tag{19.127}$$

EXAMPLE. Let X be the renewal process with $\tau_n = \sigma_1 + \ldots + \sigma_n$ where (σ_i) is a sequence of independent uniformly distributed random variables with the continuous distribution function $F(t) = P(\sigma_i \leq t)$. Then by virtue of Theorem 18.2 the compensator $A_t(x)$ of this process can be defined by the formula

$$A_t(x) = -\ln \prod_{k \geq 1} \{1 - F[(\tau_k \wedge t)] - F[(\tau_{k-1} \wedge t)]\} \tag{19.128}$$

and, if we assume in addition that the function $F(t)$ has the density $f(t)$, then

$$A_t(x) = \int_0^t a_s(x)ds,$$

where

$$a_s(x) = \frac{f(s - \tau_n(x))}{1 - F(s - \tau_n(x))}, \quad \tau_n(x) \le s < \tau_{n+1}(x).$$

Assume that $\tilde{X}$ is another renewal process with $\tau_n = \tilde{\sigma}_1 + \ldots + \tilde{\sigma}_n$, $\tilde{F}(t) = P(\tilde{\sigma}_i \le t)$, $\tilde{F}(t) = \int_0^t \tilde{f}(s)ds$:

$$\tilde{a}_s(x) = \frac{\tilde{f}(s - \tau_n(x))}{1 - \tilde{F}(s - \tau_n(x))}, \quad \tau_n(x) \le s < \tau_{n+1}(x).$$

Then the condition (19.127) can be rewritten as

$$\sum_{n=1}^{\infty} \int_0^{\sigma_n} (1 - \sqrt{\tilde{\nu}(s)\nu^+(s)})^2 \nu(s)ds < \infty \quad (\tilde{\mu}\text{-a.s.}), \tag{19.129}$$

where

$$\nu(s) = \frac{f(s)}{1 - F(s)}, \quad \tilde{\nu}(s) = \frac{\tilde{f}(s)}{1 - \tilde{F}(s)}.$$

From (19.129) it follows that the measure $\tilde{\mu} \ll \mu$ if and only if $F(t) = \tilde{F}(t)$, $t \ge 0$.

Theorem 19.9. *Let the compensators $A_t(x)$ and $\tilde{A}_t(x)$ of the point process $X = (x_t, \mathcal{B}_t, \mu)$ and $\tilde{X} = (x_t, \mathcal{B}_t, \tilde{\mu})$ be such that*

$$\mu(A_{\tau_\infty(x)}(x) < \infty) = 1, \quad \tilde{\mu}(\tilde{A}_{\tau_\infty(x)}(x) < \infty) = 1.$$

Then a necessary and sufficient condition for $\tilde{\mu} \ll \mu$ to be absolutely continuous is that ($\tilde{\mu}$-a.s.)

(A)

$$\tilde{A}_t(x) = \int_0^t \lambda_s(x)dA_s(x), \quad t < \tau_\infty(x);$$

(B)

$$\Delta A_t(x) = 1 \Rightarrow \Delta \tilde{A}_t(x), \quad t < \tau_\infty(x);$$

(C)

$$A_{\tau_\infty(x)}(x) < \infty.$$

PROOF. We need only show that in this case condition (C) is equivalent to assumption (III) of Theorem 19.7. But condition (C) together with the assumption $\tilde{\mu}\{\tilde{A}_{\tau_\infty}(x) < \infty\} = 1$ implies that ($\tilde{\mu}$-a.s.) the number of jumps in the compensators $A_t(x)$ and $\tilde{A}_t(x)$ is finite. From this, condition (III) of Theorem 19.7 follows obviously and, therefore, $\tilde{\mu} \ll \mu$.

Conversely, let $\tilde{\mu} \ll \mu$. Then by virtue of the assumption $\mu(A_{\tau_\infty}(x) < \infty) = 1$ condition (C) is obviously satisfied. Conditions (A) and (B) follow from Theorem 19.7 by virtue of the assumption $\tilde{\mu} \ll \mu$. □

19.4.5. Theorem 19.7 enables us to describe in terms of compensators all the point processes whose measure is absolutely continuous with respect to the Poisson measure μ_π (compare with Theorem 7.11).

Theorem 19.10. *Let $X = (x_t, \mathcal{B}_t, \mu_\pi)$ be a Poisson process with unit parameter and let $\tilde{X} = (x_t, \mathcal{B}_t, \tilde{\mu})$ be a point process with the measure $\tilde{\mu} \ll \mu_\pi$. Then the process $\tilde{X}$ has the compensator*

$$\tilde{A}_t(x) = \int_0^t \lambda_s(x)ds, \quad t < \infty,$$

where $(\lambda_t(x), \mathcal{B}_t)$ is a nonnegative predictable process such that ($\tilde{\mu}$-a.s.)

$$\int_0^\infty (1 - \sqrt{\lambda_s(x)})^2 ds < \infty.$$

In this case

$$\frac{d\tilde{\mu}}{d\mu_\pi}(t, x) = \exp\left\{\int_0^t \ln \lambda_s(x)dx_s + \int_0^t (1 - \lambda_s(x))ds\right\}. \tag{19.130}$$

The proof follows immediately from Theorem 19.7; we have only to note that $A_t(x) \equiv t$ and $\tau_\infty(x) \equiv \infty$ for the Poisson process with unit parameter. □

19.5 Calculation of the Mutual Information and the Cramer–Rao–Wolfowitz Inequality (the Point Observations)

19.5.1. Let $(\Omega, \mathcal{F}, P)$ be a probability space, let $(\mathcal{F}_t)$, $t \geq 0$, be a nondecreasing family of sub-σ-algebras of $\mathcal{F}$. We shall assume that $\alpha = (\alpha_t, \mathcal{F}_t)$ is the 'transmitted information' (with values in the measurable space $(\mathsf{A}, \mathcal{A})$ to be transmitted with the help of the point process $N = (N_t, \mathcal{F}_t)$ with range in the measurable space $(\mathsf{X}, \mathcal{B})$ (see Section 18.3).

Let us consider on $(\mathsf{X}, \mathcal{B})$, $(\mathsf{A}, \mathcal{A})$ and $(\mathsf{A} \times \mathsf{X}, \mathcal{A} \times \mathcal{B})$ the probability measures $\mu_N(b) = P(N \in B)$, $\mu_\alpha(A) = P(\alpha \in A)$, $\mu_{\alpha,N}(A \times B) = P(\alpha \in A, N \in B)$ where $A \in \mathcal{A}$, $B \in \mathcal{B}$ and assume that the conditions which follow are satisfied.

(I) There exists (μ_α-a.s.) a regular version (to be denoted $\mu_N^a(B)$) of the conditional probability $P(N \in B|\alpha)_{\alpha=a}$.

(II) There exist (measurable) predictable functionals $\lambda_t(a,x)$ and $A_t(x)$ such that:

(1) $A_t(a,x) = \int_0^t \lambda_s(a,x)dA_s(x)$ (for μ_α-almost every $a \in \mathsf{A}$) is the compensator of the point process $(x_t, \mathcal{B}_t, \mu_n^a)$;

(2) $\Delta A_t(x) = 0$, $t \geq 0$, $x \in \mathsf{X}$, and $A_t(x)$ is the compensator of the process $(x_t, \mathcal{B}_t, \mu_N)$;

(3) $A_t(\alpha, N)$ is the compensator of the point process $(N_t, \mathcal{F}_t, P)$.

(III)

$$M \int_0^{\tau_\infty(N)} |\lambda_t(\alpha,N)\ln\lambda_t(\alpha,N)|dA_t(N) < \infty.$$

Let

$$I(\alpha, N) = M \ln \frac{d\mu_{\alpha,N}}{d[\mu_\alpha \times \mu_N]}(\alpha, N) \tag{19.131}$$

be the Shannon information about the *transmitted message* α contained in the *received signal* N where, as usual, $I(\alpha, N)$ is assumed to be equal to ∞ if the measure $\mu_{\alpha,N}$ is not absolutely continuous with respect to the measure $[\mu_\alpha \times \mu_N]$ (compare with (16.64)).

Theorem 19.11. *In assumptions (I)–(III) the mutual information*

$$I(\alpha, N) = M \int_0^{\tau_\infty(N)} \lambda_t(\alpha,N)\ln\lambda_t(\alpha,N)dA_t(N). \tag{19.132}$$

PROOF. We shall show first that for μ_α-almost every $a \in \mathsf{A}$ the measure $\mu_N^a \ll \mu_N$. By virtue of Theorem 19.7 We need only show that

$$M \int_0^{\tau_\infty(N)} (1 - \sqrt{\lambda_t(\alpha,N)})^2 dA_t(N) < \infty. \tag{19.133}$$

Since for $y \geq 0$, $y \ln y + 1 - y \geq 0$ and for $n = 1, 2, \dots (\tau_n < \tau_\infty)$

$$\begin{aligned} M \int_0^{\tau_n} (1 - \lambda_t(\alpha,N))dA_t(N) &= M[A_{\tau_n}(N) - A_{\tau_n}(\alpha, N) \\ &= M[N_{\tau_n} - N_{\tau_n}] = 0, \end{aligned} \tag{19.134}$$

we then obtain by condition (III) that

$$\begin{aligned} \infty > M &\int_0^{\tau_\infty(N)} \lambda_t(\alpha,N)\ln\lambda_t(\alpha,N)dA_t(N) \\ = \lim_n M &\int_0^{\tau_n} \lambda_t(\alpha,N)\ln\lambda_t(\alpha,N)dA_t(N) \end{aligned}$$

$$= \lim_n M \int_0^{\tau_n} [\lambda_t(\alpha, N) \ln \lambda_t(\alpha, N) + 1 - \lambda_t(\alpha, N)] dA_t(N)$$
$$= M \int_0^{\tau_\infty} [\lambda_t(\alpha, N) \ln \lambda_t(\alpha, N) + 1 - \lambda_t(\alpha, N)] dA_t(N) \geq 0. \tag{19.135}$$

It is clear that there exists a constant C such that for $y \geq 0$

$$C(y \ln y + 1 - y) \geq (1 - \sqrt{y})^2. \tag{19.136}$$

Consequently, (19.133) follows from (19.135), (19.136), i.e., $\mu_N^a \ll \mu_N$ (μ_α-a.s.). Set $z(a, x) = (d\mu_N^a / d\mu_N)(x)$ and show that ($\mu_\alpha \times \mu_N$-a.s.)

$$z(a, x) = \frac{d\mu_{\alpha,N}}{d[\mu_\alpha \times \mu_N]}(a, x). \tag{19.137}$$

In fact, let $\varphi(a, x)$ be a bounded $\mathcal{A} \times \mathcal{B}$-measurable functional. Then due to assumption (I) and the Fubini theorem

$$\begin{aligned}
\int_{\mathsf{A}\times\mathsf{X}} \varphi(a, x) d\mu_{\alpha,N}(a, x) &= M\varphi(\alpha, N) = MM[\varphi(\alpha, N)|\mathcal{A}] \\
&= \int_{\mathsf{A}} M[\varphi(\alpha, N)|\alpha]_{\alpha=a} d\mu_\alpha(a) \\
&= \int_{\mathsf{A}\times\mathsf{X}} \left[\int \varphi(a, x) d\mu_N^a(x)\right] d\mu \qquad (19.138) \\
&= \int_{\mathsf{A}\times\mathsf{X}} \varphi(a, x) z(a, x) d[\mu_\alpha \times \mu_N](a, x),
\end{aligned}$$

thus proving (19.137).

By virtue of (19.125) and (II),

$$\begin{aligned}
\ln z(a, x) &= \int_0^{\tau_\infty(x)} \ln \lambda_t(a, x)[dx_t - dA_t(a, x)] \qquad (19.139) \\
&\quad + \int_0^{\tau_\infty(x)} [\lambda_t(a, x) \ln \lambda_t(a, x) + 1 - \lambda_t(a, x)] dA_t(x).
\end{aligned}$$

Hence by Theorem 18.7, (19.139), (19.135) and (III),

$$I(\alpha, N) = M \ln z(\alpha, N) = M \int_0^{\tau_\infty(N)} \lambda_t(\alpha, N) \ln \lambda_t(\alpha, N) dA_t(N). \qquad \square$$

Corollary. *Let $A_t(a, x) = \int_0^t \nu_s(a, x) db_s$ where $\nu_s = \nu_s(a, x)$ is a nonnegative predictable process and b_s is a deterministic nonnegative right continuous and nondecreasing function (compare with (18.33)). Then by virtue of Theorem 18.3, $A_t(x) = \int_0^t \overline{\nu}_s(x) db_s$ where $\overline{\nu}_s = \overline{\nu}_s(x)$ is such that (P-a.s.) $\overline{\nu}_s(N) = M[\nu_s(\alpha, N)|\mathcal{F}_{t-}^N]$. Then*

$$\begin{aligned} I(\alpha, N) &= M \int_0^{\tau_\infty(N)} (\nu_s \ln \nu_s - \nu_s \ln \overline{\nu}_s) db_s \\ &= M \int_0^{\tau_\infty(N)} (\nu_s \ln \nu_s - \overline{\nu}_s \ln \overline{\nu}_s) db_s. \end{aligned} \tag{19.140}$$

Let $b_s = s$ and $P(\tau_\infty = \infty) = 1$, and let

$$I_T(\alpha, N) = M \ln \frac{d\mu^T_{\alpha,N}}{d[\mu^T_\alpha \times \mu^T_N]}$$

where $\mu^T_{\alpha,N}$, μ^T_α and μ^T_α are the restrictions of the measures $\mu_{\alpha,N}$, μ_α and μ_N (respectively) to the σ-algebras

$$\mathcal{A}_t \times \mathcal{B}_t, \quad \mathcal{A}_t = \sigma\{a_s, s \le t\}, \quad \mathcal{B}_t = \sigma\{x_s, s \le t\}.$$

Then

$$I_T(\alpha, N) = M \int_0^T (\nu_s \ln \nu_s - \overline{\nu}_s \ln \overline{\nu}_s) ds.$$

Assume now that $\nu_s(a, x) = 1 + \Lambda_s(a, x)$ where $\Lambda_s(a, x)$ is subject to the power constraints $0 \le \Lambda_s(a, x) \le P$, P being a given constant. Let us consider the channel capacity (compare with (16.72))

$$\mathsf{C} = \sup \frac{1}{T} I_T(\alpha, N), \quad T > 0,$$

where 'sup' is taken over all α with values in $(\mathsf{A}, \mathcal{A})$ and the codings $\Lambda = (\Lambda_s(a, x), s \le T)$ satisfying the restrictions $0 \le \Lambda_s(a, x) \le p$. Then (for the proof, see [110])

$$\mathsf{C} = \frac{Q}{e} - \ln Q, \tag{19.141}$$

where $Q = (p+1)^{(p+1)/p}$. In this case the presence of the feedback does not imply an increase in the channel capacity (compare with Theorem 16.4).

EXAMPLE. Let $\alpha = (\alpha_t, \mathcal{F}_t)$ be the Markov process with two states a and b ($a > 0$, $b > 0$) considered in Example 2 of Subsection 19.3.12. Let the observation be:

$$N_t = \int_0^t I(\alpha_{s-} = a) dN^a_s + \int_0^t I(\alpha_{s-} = b) dN^b_s,$$

where N^a and N^b are two Poisson processes with parameters a and b, respectively. In this case it is assumed that the processes α, N^a and N^b are mutually independent. Then

$$I_T(\alpha, N) = M \int_0^T \lambda_s(\alpha, N) \ln \lambda_s(\alpha, N)(a(1 - \pi_{s-}) + b\pi_{s-}) ds$$

where

$$\lambda_s(\alpha, N) = \frac{aI(\alpha_{s-} = a) + bI(\alpha_{s-} = b)}{a(1 - \pi_{s-}) + b\pi_{s-}}$$

and $\pi_s = P(\alpha_s = b|\mathcal{F}_s^N)$ satisfies Equation (19.89).

19.5.2. Let there be observed the point process $X = (x_t, \mathcal{B}_t, \mu_\theta)$, $t \geq 0$, with the compensator $A_t^\theta(x)$, where θ is an unknown parameter: $\theta \in (a, b)$, $-\infty \leq a < b \leq \infty$.

By relying on the previous results for the measure densities of the point processes we can show how to find lower bounds for mean square errors in the problems of estimating the functions $f(\theta)$ from the observations of the process X.

Let $\Delta = (\tau(x), \delta(\tau(x), x))$ be some sequential estimation scheme (for details, see Section 7.8). Assume that[6]

(1)

$$M_\theta \delta^2(\tau(x), x) < \infty;$$

(2)

$$A_t^\theta(x) = \int_0^t \lambda_s(\theta, x) ds, \quad x \in \mathsf{X},$$

where $(\lambda_t(\theta, x), \mathcal{B}_t)$ is a predictable process, $0 < \lambda_t(\theta, x) < \infty$;

(3)

$$\frac{\partial}{\partial \theta} \int_0^{\tau(x)} \ln \lambda_t(\theta, x) dx_t = \int_0^{\tau(x)} \frac{\partial}{\partial \theta} \ln \lambda_t(\theta, x) dx_t,$$

$$\frac{\partial}{\partial \theta} \int_0^{\tau(x)} \lambda_t(\theta, x) dt = \int_0^{\tau(x)} \frac{\partial}{\partial \theta} \lambda_t(\theta, x) dt, \quad x \in \mathsf{X};$$

(4)

$$M \int_0^{\tau(x)} \left(\frac{\partial}{\partial \theta} \lambda_t(\theta, x) \right)^2 \lambda_t^{-1}(\theta, x) dt < \infty;$$

(5)

$$\int_0^{\tau(x)} |\lambda_t(\theta, x) - 1| dt < \infty \quad (\mu_\theta\text{-a.s}).$$

[6] M_θ is the expectation under the measure μ_θ.

Next, let μ be the measure corresponding to the Poisson process with unit parameter and let μ^τ, μ_θ^τ be the restrictions of the measures μ and μ_θ to the σ-algebra $\mathcal{B}_t$. By (5), Theorem 19.10, and the inequality $(1-\sqrt{y})^2 \leq |1-y|$ for $y \geq 0$,

$$\frac{d\mu_\theta^\tau}{d\mu^\tau}(x) = \exp\left\{\int_0^{\tau(x)} \ln \lambda_t(\theta, x)dx_t + \int_0^{\tau(x)} [1-\lambda_t(\theta,x)]dt\right\}. \qquad (19.142)$$

Assume also that:

(6) the function $b(\theta) = M_\theta\delta(\tau(x),x) - f(\theta)$ is differentiable in θ and

$$\frac{d}{d\theta}M\frac{d\mu_\theta^\tau}{d\mu^\tau}(x)\delta(\tau(x),x) = M\frac{\partial}{\partial\theta}\left(\frac{d\mu_\theta^\tau}{d\mu^\tau}(x)\right)\delta(\tau(x),x),$$

where M denotes the expectation under the measure μ.

Theorem 19.12. *Under assumptions (1)–(6) for all $\theta \in (a,b)$ we have the Cramer–Rao–Wolfowitz inequality*

$$M_\theta[\delta(\tau(x),x) - f(\theta)]^2 \geq \frac{[\frac{d}{d\theta}(f(\theta)+b(\theta))]^2}{M_\theta\int_0^{\tau(x)}\left(\frac{\partial}{\partial\theta}\lambda_t(\theta,x)\right)^2\lambda_t^{-1}(\theta,x)dt} + b^2(\theta). \qquad (19.143)$$

PROOF. By (19.142), (3) and (6),

$$\begin{aligned}\frac{d}{d\theta}[b(\theta)+f(\theta)] &= \frac{d}{d\theta}M_\theta\delta(\tau(x),x)\\ &= M\frac{d\mu_\theta^\tau}{d\mu^\tau}(x)\delta(\tau(x),x)\left[\int_0^{\tau(x)}\lambda_t^{-1}(\theta,x)\frac{\partial}{\partial\theta}\lambda_t(\theta,x)dx_t\right.\\ &\quad\left.-\int_0^{\tau(x)}\frac{\partial}{\partial\theta}\lambda_t(\theta,x)dt\right]\\ &= M_\theta\delta(\tau(x),x)\int_0^{\tau(x)}\lambda_t^{-1}(\theta,x)\frac{\partial}{\partial\theta}\lambda_t(\theta,x)[dx_t-\lambda_t(\theta,x)dt].\end{aligned} \qquad (19.144)$$

By (1), (4) and Theorem 18.8,

$$M_\theta\delta(\tau(x),x)M_\theta\int_0^{\tau(x)}\lambda_t^{-1}(\theta,x)\frac{\partial}{\partial\theta}(\lambda_t(\theta,x))[dx_t-\lambda_t(\theta,x)dt] = 0, \qquad (19.145)$$

which together with (19.144) yields the relation

$$\begin{aligned}\frac{d}{d\theta}[b(\theta)+f(\theta)] &= M_\theta[\delta(\tau(x),x) - M_\theta\delta(\tau(x),x)]\int_0^{\tau(x)}\lambda_t^{-1}(\theta,x)\frac{\partial}{\partial\theta}\lambda_t(\theta,x)\\ &\quad\times[dx_t-\lambda_t(\theta,x)dt].\end{aligned}$$

From this, by (4), (5.82), Theorem 18.8 and the Cauchy–Schwarz inequality, we have

$$\begin{aligned}\left(\frac{d}{d\theta}[b(\theta)+f(\theta)]\right)^2 &\le M_\theta[\delta(\tau(x),x)-M_\theta\delta(\tau(x),x)]^2\\ &\quad\times M_\theta\left[\int_0^{\tau(x)}\lambda_t^{-1}(\theta,x)\frac{\partial}{\partial\theta}\lambda_t(\theta,x)[dx_t-\lambda_t(\theta,x)dt]\right]^2\\ &= M_\theta[\delta(\tau(x),x)-M_\theta\delta(\tau(x),x)]^2\\ &\quad\times M_\theta\int_0^{\tau(x)}\lambda_t^{-1}(\theta,x)\left(\frac{\partial}{\partial\theta}\lambda_t(\theta,x)\right)^2dt\\ &= M_\theta[\delta(\tau(x),x)-f(\theta)-b(\theta)]^2\\ &\quad\times M_\theta\int_0^{\tau(x)}\lambda_t^{-1}(\theta,x)\left(\frac{\partial}{\partial\theta}\lambda_t(\theta,x)\right)^2dt.\end{aligned}$$

The required inequality, (19.143), follows from the above by simply noting that

$$M_\theta[\delta(\tau(x),x)-f(\theta)-b(\theta)]^2=M_\theta[\delta(\tau(x),x)-f(\theta)]^2-b^2(\theta).$$

Corollary. *If $\delta(\tau(x),x)$ is an unbiased estimate for $f(\theta)$, then*

$$M_\theta[\delta(\tau(x),x)-f(\theta)]^2\ge\frac{1}{M_\theta\int_0^{\tau(x)}\lambda_t^{-1}(\theta,x)\left(\frac{\partial}{\partial\theta}\lambda_t(\theta,x)\right)^2dt}. \tag{19.146}$$

EXAMPLE 1. Let $0<\theta<\infty$, $\lambda_t(\theta,x)=\theta$, $\tau(x)=T<\infty$.

In this case the point process $X=(x_t,\mathcal{B}_t,\mu_\theta)$ is a Poisson process with the parameter θ. If $\delta(x)$ is an unbiased estimate of the parameter θ, then, by virtue of (19.146),

$$M_\theta[\delta(x)-\theta]^2\ge\frac{\theta}{T}. \tag{19.147}$$

In this case:

$$\frac{d\mu_\theta}{d\mu}(x)=\exp(x_T\ln\theta+(1-\theta)T)$$

and, therefore, the maximum likelihood estimate is $\hat\theta_T=x_T/T$. It is clear that this estimate is unbiased, $M_\theta\hat\theta_T=\theta$, and it is easily seen that $M_\theta(\hat\theta_T-\theta)^2=\theta/T$. It follows from this and (19.147) that the maximum likelihood estimate $\hat\theta_T$ is the optimal (in the mean square sense) estimate of θ in the class of all unbiased estimates.

EXAMPLE 2. Let $\lambda_t(\theta,x)=\theta\cdot a_t(x)$, $0<\theta<\infty$, where $a_t(x)>0$, $\int_0^t a_s(x)ds<\infty$, $t<\infty$, and $\int_0^\infty a_s(x)ds=\infty$ (μ_θ-a.s.). We shall estimate θ

from the observations x_t, $0 \le t \le \tau(x)$, where $\tau(x)$ is a Markov time with respect to the family $(\mathcal{B}_t)$, $t \ge 0$, such that

$$\int_0^{\tau(x)} (1 + a_s(x))ds < \infty \quad (\mu_\theta\text{-a.s.}, \theta > 0).$$

It is easy to verify that $\mu_\theta^\tau \ll \mu^\tau$ where μ is the Poisson measure with a unit parameter, and μ^τ is the restriction of μ to the σ-algebra $\mathcal{B}_t$. Then

$$\frac{d\mu_\theta^\tau}{d\mu^\tau}(x) = \exp\left(\int_0^{\tau(x)} \ln(\theta a_t(x))dx_t + \int_0^{\tau(x)} [1 - \theta a_t(x)]dt\right)$$

and, therefore, the maximum likelihood estimate is

$$\hat{\theta}_\tau = \frac{x_{\tau(x)}}{\int_0^{\tau(x)} a_t(x)dt}.$$

Assume that the time $\tau(x) = \tau_H(x)$ (compare with Section 17.5) where

$$\tau_H(x) = \inf\left\{t : \int_0^t a_s(x)ds > H\right\}.$$

In this case the estimate $\hat{\theta}_{\tau_H} = x_{\tau_H(x)}/H$ is unbiased and

$$\begin{aligned} M_\theta(\hat{\theta}_{\tau_H} - \theta)^2 &= M\left(\frac{\theta \int_0^{\tau_H(x)} a_t(x)dt + m_\tau^\theta(x)}{H} - \theta\right)^2 \\ &= \frac{1}{H^2} M_\theta (m_{\tau_H(x)}^\theta)^2 \\ &= \frac{1}{H^2}\theta M_\theta \int_0^{\tau_H(x)} a_t(x)dt = \frac{\theta}{H}, \end{aligned} \tag{19.148}$$

where $m_t^\theta(x) = x_t - \int_0^t \theta a_s(x)ds$ is a locally square integrable martingale with $\langle m^\theta\rangle_t = \int_0^t \theta a_s(x)ds$ (Lemma 18.12). By virtue of (19.146),

$$M_\theta(\delta(\tau(x), x) - \theta)^2 \ge \frac{1}{M_\theta \int_0^{\tau(x)} a_t^2(x)(\theta a_t(x))^{-1}dt} = \frac{\theta}{M_\theta \int_0^{\tau(x)} a_t(x)dt},$$

which together with (9.148) shows that, in the class of all unbiased estimates $\delta(\tau(x), x)$ satisfying the condition

$$M_\theta \int_0^{\tau(x)} a_t(x)dt \le H,$$

the sequential maximum likelihood estimate is optimal (compare with Theorem 17.6).

Notes and References. 1

19.1. The structure of local martingales for point processes was investigated by Davis [48], Boel, Varaiya and Wong [22], Chou and Meyer [37], Dellacherie [50], Kabanov [108, 109] and Grigelionis [88].

19.2. The problems of the transformation of compensators by absolutely continuous substitution of the measure were discussed in Brémaud [26], Jacod [101], Boel, Varaiya and Wong [22], Kabanov, Liptser and Shiryaev [113, 119], Skorokhod [291], Grigelionis [86, 88] and Gikhman and Skorokhod [72].

19.3. The equations of optimal nonlinear filtering from the observation of the point processes were also deduced (under various assumptions of generality) in Segal and Kailath [273], Brémaud [28,29], Davis [48], Segal, Davis and Kailath [272], Boel, Varaiya and Wong [22], Khadgijev [145, 146] Grigelionis [86], Van Schuppen [302], Galtchouk [68], Yashin [319, 320], Snyder [292] and Snyder and Fishman [293].

19.4. The necessary and sufficient conditions for the absolute continuity of two measures corresponding to point processes (Theorem 19.7) were obtained by Kabanov, Liptser and Shiryaev. Theorem 19.8 is due to Jacod and Memin [105], and Kabanov, Liptser and Shiryaev [119]. Theorem 19.9 was proved in Kabanov, Liptser and Shiryaev [113, 119].

19.5. The computations of the mutual information for jumplike processes were made in Grigelionis [87], Boel, Varaiya and Wong [22], and Kabanov [110], (19.141) is due to Kabanov [110]. Formula (19.143) was obtained by Kutoyants.

Notes and References. 2

19.1. A modern description of the structure for local martingales can be found in Liptser and Shiryaev [214], Jacod and Shiryaev [106], Elliott [59].

19.2. A Girsanov type theorem and the problem of change of probability measures, which are distributions of semimartingales are given in [204, 214].

19.3. A derivation of the filtering equation under observation not only of the point process but also a general semimartingale can be found in [214].

19.4. The necessary and sufficient conditions for absolute continuity of measures corresponding to multivariate point processes and semimartingales are presented in Kabanov, Liptser and Shiryaev [114–117] and Jacod and Shiryaev [106].

20. Asymptotically Optimal Filtering

In previous chapters, a number of filtering models, for which the 'filtering equation' admits a closed form, like the Kalman–Bucy filter (Chapter 10), the conditionally Gaussian filter (Chapters 11 and 13), the Wonham type filter and the Kushner–Zakai filter (Chapter 8), were presented. However in applications, realistic filtering models have a more complicated structure than those to which the filters mentioned above are immediately applicable. In this chapter, we consider examples for which the following approximation technique might be successful. To construct nearly optimal filters, instead of the original model a new model, where the underlying processes are replaced by simple ones, is applied. To explain such an approach in more detail, let us consider the filtering problem for a pair of random processes $(X_t^\varepsilon, Y_t^\varepsilon)_{t\ge 0}$, where X_t^ε represents an unobservable signal and Y_t^ε is a corresponding observation, and where ε is a small parameter. Suppose the probabilistic structure of $(X_t^\varepsilon, Y_t^\varepsilon)_{t\ge 0}$ is too complicated for us to find the optimal (in the mean square sense) filtering estimate, but as $\varepsilon \to 0$, the pair $(X_t^\varepsilon, Y_t^\varepsilon)_{t\ge 0}$ converges (in some sense) to the limit pair $(\overline{X}_t, \overline{Y}_t)_{t\ge 0}$ which has a simpler description than the prelimit one, for example, it is a Markov diffusion process or, more specifically, a Gaussian diffusion. A natural procedure for creating a successful filter for a prelimit model involves finding the optimal filter for the limit model and then using it for prelimit observations.

The main problem in such an approach is the verification of the asymptotic optimality for the filters obtained. We give in this chapter two examples for which the asymptotic optimality can be checked effectively.

20.1 Total Variation Norm Convergence and Filtering

20.1.1. Consider the filtering problem for a pair of random processes

$$(X^\varepsilon, Y^\varepsilon) = (X_t^\varepsilon, Y_t^\varepsilon)_{0\le t\le T}$$

with right continuous trajectories having limits to the left. For fixed ε denote by Q^ε the distribution of $(X^\varepsilon, Y^\varepsilon)$, i.e., Q^ε is a probabilistic measure on the Skorokhod space $D(\mathbb{R}^2; [0,T])$. Assume Q^ε converges, as $\varepsilon \to 0$, in the total variation norm for a limit $\overline{Q}$:

$$\lim_{\varepsilon\to 0} \|Q^\varepsilon - \overline{Q}\| = 0. \tag{20.1}$$

In other words, (20.1) means that $(X^\varepsilon, Y^\varepsilon)$ converges, as $\varepsilon \to 0$, in the aforementioned sense for distributions, to a limit process $(\overline{X}, \overline{X}) = (\overline{X}_t, \overline{X}_t)_{0\le t\le T}$ having trajectories in $D(\mathbb{R}^2; [0,T])$ and the distribution $\overline{Q}$.

Consider a filtering problem for a signal $u(X_t^\varepsilon)$ and observation Y_t^ε provided that $u(x)$ is a measurable function such that $Mu^2(X_t^\varepsilon) < \infty$ and $Mu^2(\overline{X}_t) < \infty$, $t \le T$. Assume that the probabilistic structure of $(\overline{X}, \overline{Y})$ is simple, in the sense that the conditional expectation $M(u(\overline{X}_t)|\overline{Y}_{[0,t]})$ can be computed with the help of one of the aforementioned classical filters. For convenience of notation, introduce a family of measurable functionals $(\overline{\pi}_t(Y), Y \in D(\mathbb{R}; [0,T]))$, $0 \le t \le T$ such that

$$\overline{\pi}_t(\overline{Y}) = M\left(u(\overline{X}_t)\middle|\,\overline{Y}_{[0,t]}\right), \quad (\overline{Q}\text{-a.s.}).$$

Analogously define $(\pi_t^\varepsilon(Y)$: $\pi_t^\varepsilon(Y^\varepsilon) = M(u(X_t^\varepsilon)|Y_{[0,t]}^\varepsilon)$, $(Q^\varepsilon\text{-a.s.})$. It is clear that the use of $\overline{\pi}_t(Y^\varepsilon)$ as a filtering estimate makes sense if

$$\lim_{\varepsilon\to 0} M\left(\pi_t^\varepsilon(Y^\varepsilon) - \overline{\pi}_t(Y^\varepsilon)\right)^2 = 0 \tag{20.2}$$

since (20.2) is equivalent to the asymptotic optimality: for every t

$$\begin{aligned}\lim_{\varepsilon\to 0} M\left(u(X_t^\varepsilon) - \pi_t^\varepsilon(Y^\varepsilon)\right)^2 &= \lim_{\varepsilon\to 0} M\left(u(X_t^\varepsilon) - \overline{\pi}_t(Y^\varepsilon)\right)^2 \\ &= M\left(u(\overline{X}_t) - \overline{\pi}_t(\overline{Y})\right)^2. \end{aligned}\tag{20.3}$$

We show that (20.3), for a bounded function $u(x)$, is implied by (20.1). For unbounded $u(x)$, verification of (20.3) requires the uniform integrability condition. For the sake of simplicity, we restrict ourselves to consideration of δ-asymptotically optimal filters. To this end, the following definition is useful.

Definition. The filtering estimate $\widetilde{\pi}_t^\delta(Y^\varepsilon)$, $t \le T$, is *δ-asymptotically optimal*, if for every $\delta > 0$

$$\lim_{\varepsilon\to 0} M\left(u(X_t^\varepsilon) - \widetilde{\pi}_t^\delta(Y^\varepsilon)\right)^2 \le \liminf_{\varepsilon\to 0} M\left(u(X_t^\varepsilon) - \pi_t^\varepsilon(Y^\varepsilon)\right)^2 + \delta. \tag{20.4}$$

Theorem 20.1. *Assume (20.1) and for some constant ℓ, $|u(x)| \le \ell$. Then $\overline{\pi}_t(Y^\varepsilon)$ is asymptotically optimal in the sense of (20.3).*

Corollary. *For every fixed $t > 0$, $\pi_t^\varepsilon(Y^\varepsilon) \xrightarrow{\text{law}} \overline{\pi}_t(\overline{Y})$, $\varepsilon \to 0$.*

PROOF. We choose versions of functionals $\overline{\pi}_t(Y)$ and $\pi_t^\varepsilon(Y)$ bounded by the same constant ℓ. In what follows, the notation '$\int$' is used instead of '$\int_{D^2}$'. Note that since $\pi_t^\varepsilon(Y^\varepsilon)$ is the optimal (in the mean square sense) filtering estimate, we have

$$\int (u(X_t) - \pi_t^\varepsilon(Y))^2 \, dQ^\varepsilon \leq \int (u(X_t) - \overline{\pi}_t(Y))^2 \, dQ^\varepsilon$$

and therefore, by (20.1),

$$\limsup_{\varepsilon \to 0} \int (u(X_t) - \pi_t^\varepsilon(Y))^2 \, dQ_T^\varepsilon \leq \int (u(X_t) - \overline{\pi}_t(Y))^2 \, d\overline{Q}. \tag{20.5}$$

Show now that

$$\begin{aligned} \lim_{\varepsilon \to 0} \int (u(X_t) - \overline{\pi}_t(Y))^2 \, dQ^\varepsilon &= \int (u(X_t) - \overline{\pi}_t(Y))^2 \, d\overline{Q} \\ \liminf_{\varepsilon \to 0} \int (u(X_t) - \pi_t^\varepsilon(Y))^2 \, dQ^\varepsilon &\geq \int (u(X_t) - \overline{\pi}_t(Y))^2 \, d\overline{Q}. \end{aligned} \tag{20.6}$$

Since $u(x)$ and $\pi_t^\varepsilon(\cdot)$, $\overline{\pi}_t(\cdot)$ are bounded by ℓ, we find

$$\begin{aligned} &\left| \int (u(X_t) - \overline{\pi}_t(Y))^2 \, dQ^\varepsilon - \int (u(X_t) - \overline{\pi}_t(Y))^2 \, d\overline{Q} \right| \\ &\leq 4\ell^2 \|Q^\varepsilon - \overline{Q}\| \to 0, \quad \varepsilon \to 0. \end{aligned}$$

The filtering estimate $\overline{\pi}_t(\overline{Y})$ is optimal in the mean square sense for the signal $u(\overline{X}_t)$ given the observation $\overline{Y}_s$, $s \leq t$. Therefore

$$\begin{aligned} \int (u(X_t) - \pi_t^\varepsilon(Y))^2 \, dQ_T^\varepsilon &= \int (u(X_t) - \pi_t^\varepsilon(Y))^2 \, d\overline{Q}_T \\ &\quad + \int (u(X_t) - \pi_t^\varepsilon(Y))^2 \, d(Q^\varepsilon - \overline{Q}) \\ &\geq \int (u(X_t) - \overline{\pi}_t(Y))^2 \, d\overline{Q} - 4\ell^2 \|Q^\varepsilon - \overline{Q}\| \\ &\to \int (u(X_t) - \overline{\pi}_t(Y))^2 \, d\overline{Q}, \quad \varepsilon \to 0. \end{aligned}$$

Thus, the statement of the theorem is implied by (20.5) and (20.6). □

PROOF OF THE COROLLARY. It suffices to show the convergence of the characteristic functions ($i = \sqrt{-1}$): for every $v \in \mathbb{R}$

$$\lim_{\varepsilon \to 0} M e^{iv\pi_t^\varepsilon(Y^\varepsilon)} = M e^{iv\overline{\pi}_t(\overline{Y})}.$$

Write

$$\begin{aligned} &\left| M e^{iv\pi_t^\varepsilon(Y^\varepsilon)} - M e^{iv\overline{\pi}_t(\overline{Y})} \right| \leq M \left| e^{iv\pi_t^\varepsilon(Y^\varepsilon)} - e^{iv\overline{\pi}_t(Y^\varepsilon)} \right| \\ &+ \left| M \left(e^{iv\overline{\pi}_t(Y^\varepsilon)} - e^{iv\overline{\pi}_t(\overline{Y})} \right) \right| \end{aligned}$$

and note that the first term on the right-hand side of this inequality is bounded above by $\sqrt{M(\pi_t^\varepsilon(Y^\varepsilon) - \overline{\pi}_t(Y^\varepsilon))^2}$ while the second is bounded above by $2\|Q^\varepsilon - \overline{Q}\|$. □

We show now the existence of the δ-asymptotically optimal filter. For

$$g_n(x) = \begin{cases} x, & |x| \le n \\ n\operatorname{sign} x, & |x| > n \end{cases}$$

put $u_n(x) = g_n(u(x))$. For fixed n, define the functional $\overline{\pi}_t^n(Y)$ such that $\overline{Q}$-a.s.

$$\overline{\pi}_t^n(\overline{Y}) = M\left(u_n(\overline{X}_t)\big|\,\overline{Y}_{[0,t]}\right), \quad (\overline{Q}\text{-a.s.}).$$

Proposition 20.1. *Assume that for some $\gamma > 0$ and any ε*

$$\sup_{t\le T} M|u(X_t^\varepsilon)|^{2+\gamma} < \infty. \tag{20.7}$$

Then for every $\delta > 0$ there exists n_δ such that $\overline{\pi}_t^{n_\delta}(Y^\varepsilon)$ is a δ-asymptotically optimal filtering estimate.

PROOF. Since (20.7) implies the uniform integrability of $(u(X_t^\varepsilon) - \overline{\pi}^{n_\delta}(\overline{Y}))^2$, by virtue of Theorem 20.1 we obtain

$$\lim_{\varepsilon\to 0} M\left(u(X_t^\varepsilon) - \overline{\pi}_t^{n_\delta}(Y^\varepsilon)\right)^2 = M\left(u(\overline{X}_t) - \overline{\pi}_t^{n_\delta}(\overline{Y})\right)^2. \tag{20.8}$$

Therefore, it remains to show that one can choose appropriate n_δ to arrive at (20.4). Write

$$\begin{aligned} M\left(u(X_t^\varepsilon) - \overline{\pi}_t^{n_\delta}(Y^\varepsilon)\right)^2 &= M\left(u(X_t^\varepsilon) - \pi_t^\varepsilon(Y^\varepsilon) + \pi_t^\varepsilon(Y^\varepsilon) - \overline{\pi}_t^{n_\delta}(Y^\varepsilon)\right)^2 \\ &= M\left(u(X_t^\varepsilon) - \pi_t^\varepsilon(Y^\varepsilon)\right)^2 \\ &\quad + M\left(\pi_t^\varepsilon(Y^\varepsilon) - \overline{\pi}_t^{n_\delta}(Y^\varepsilon)\right)^2. \end{aligned}$$

Consequently a relevant choice of n_δ has to guarantee

$$\limsup_{\varepsilon\to 0} M\left(\pi_t^\varepsilon(Y^\varepsilon) - \overline{\pi}_t^{n_\delta}(Y^\varepsilon)\right)^2 \le \delta. \tag{20.9}$$

Put $\pi_t^{\delta,\varepsilon}(Y^\varepsilon) = M(u_{n_\delta}(X_t^\varepsilon)|Y_{[0,t]}^\varepsilon)$ and note that by Theorem 20.1 we have

$$\lim_{\varepsilon\to 0} M\left(\pi_t^{\delta,\varepsilon}(Y^\varepsilon) - \overline{\pi}_t^{n_\delta}(Y^\varepsilon)\right)^2 = 0.$$

We establish now the validity of (20.9), with

$$n_\delta \ge \left(\frac{\sup_{t\le T}\limsup_{\varepsilon\to 0} M|u(X_t^\varepsilon)|^{2+\gamma}}{\delta}\right)^\gamma,$$

using a chain of upper bounds and (20.7)

$$\begin{aligned} M\left(\pi_t^\varepsilon(Y^\varepsilon) - \overline{\pi}_t^{n_\delta}(Y^\varepsilon)\right)^2 &\le M\left(\pi_t^\varepsilon(Y^\varepsilon) - \pi_t^{\delta,\varepsilon}(Y^\varepsilon)\right)^2 \\ &= M\left(M\left[u(X_t^\varepsilon) - u_{n_\delta}(X_t^\varepsilon)\right] \mid Y_{[0,t]}^\varepsilon\right)^2 \\ &\le M\left(u(X_t^\varepsilon) - u_{n_\delta}(X_t^\varepsilon)\right)^2 \\ &\le Mu^2(X_t^\varepsilon) I(|u(X_t^\varepsilon)| > n_\delta) \\ &\le \frac{M|u(X_t^\varepsilon)|^{2+\gamma}}{(n_\delta)^\gamma}. \end{aligned} \tag{20.10}$$

□

20.1.2.

MODEL 1. Let $(X_t^\varepsilon, Y_t^\varepsilon)_{t\ge 0}$ be defined by the Itô equations with respect to independent Wiener processes $(W_t^x)_{t\ge0}$ and $(W_t^y)_{t\ge0}$:

$$\begin{aligned} dX_t^\varepsilon &= a(X_t^\varepsilon, Y_t^\varepsilon, \eta_{t/\varepsilon})dt + b(X_t^\varepsilon)dW_t^x \\ dY_t^\varepsilon &= A(X_t^\varepsilon, Y_t^\varepsilon, \eta_{t/\varepsilon})dt + B(Y_t^\varepsilon)dW_t^y \end{aligned} \tag{20.11}$$

subject to the initial condition (X_0, Y_0) which is a random vector independent of ε, where $\eta_{t/\varepsilon}$ is a contamination affecting drifts. Functions $a = a(x,y,z)$, $A = A(x,y,z)$, $b(x)$, $B(y)$ are continuous and Lipschitz continuous in (x,y) uniformly in z and $a(0,0,z), A(0,0,z)$ are bounded, and for some $c > 0$, $b^2(x) \ge c$, $B^2(y) \ge c$. The random process $(\eta_t)_{t\ge0}$ is assumed to be a homogeneous ergodic Markov process with trajectories in D, independent of $\{(W_t^x)_{t\ge0}, (W_t^y)_{t\ge0}, (X_0,Y_0)\}$, having the unique invariant measure μ. The main assumption on $(\eta_t)_{t\ge0}$ is that its transition probability $\lambda_{y,t} = \lambda(y,t,dz)$ converges in the total variation norm to its invariant measure μ:

$$\lim_{t\to\infty} \|\lambda_{y,t} - \mu\| = 0, \quad \forall y \in \mathbb{R}. \tag{20.12}$$

Put

$$\overline{a}(x,y) - \int u(x,y,z)\mu(dz) \quad \overline{A}(x,y) = \int A(x,y,z)\mu(dz). \tag{20.13}$$

A candidate for a limit $(\overline{X}_t, \overline{Y}_t)_{t\ge0}$ is defined by the Itô equations

$$\begin{aligned} d\overline{X}_t &= \overline{a}(\overline{X}_t, \overline{Y}_t)dt + b(\overline{X}_t)dW_t^x \\ d\overline{Y}_t &= \overline{A}(\overline{X}_t, \overline{Y}_t)dt + B(\overline{Y}_t)dW_t^x \end{aligned} \tag{20.14}$$

subject to $\overline{X}_0 = X_0$, $\overline{Y}_0 = Y_0$. For fixed $T > 0$, denote by Q^ε and $\overline{Q}$ the distributions of $(X_t^\varepsilon, Y_t^\varepsilon)_{0\le t\le T}$ and $(\overline{X}_t, \overline{Y}_t)_{0\le t\le T}$, respectively.

Theorem 20.2. *For every fixed $T > 0$*

$$\lim_{\varepsilon\to0} \|Q^\varepsilon - \overline{Q}\| = 0.$$

The proof of this theorem is based on a number of auxiliary results formulated below as lemmas.

Lemma 20.1. *For every bounded and continuous function* $f = f(x)$

$$P\text{-}\lim_{\varepsilon \to 0} \int_0^1 f(\eta_{t/\varepsilon})dt = \int f(z)\mu(dz).$$

PROOF. Without loss of generality one can assume that the initial point η_0 is fixed, the function $f(z)$ is bounded, $|f(z)| \le 1$, and $\int f(z)\mu(dz) = 0$. Then the statement of the lemma holds if

$$\lim_{\varepsilon \to 0} M \left(\int_0^t f(\eta_{s/\varepsilon})ds \right)^2 = 0.$$

By direct calculation we find

$$\begin{aligned}
& M \left(\int_0^t f(\eta_{s/\varepsilon})ds \right)^2 \\
&= 2 \int_0^t \int_0^s M \left[M(f(\eta_{s/\varepsilon})|\eta_{s'/\varepsilon}) f(\eta_{s'/\varepsilon}) \right] ds'ds \\
&\le 2 \int_0^t \int_0^s M \left| M(f(\eta_{s/\varepsilon})|\eta_{s'/\varepsilon}) f(\eta_{s'/\varepsilon}) \right| ds'ds \\
&= 2 \int_0^t \int_0^s \int \left| \int f(z)[\lambda(x, (s-s')/\varepsilon, dz) f(x) \right| \lambda(y, s'/\varepsilon, dx)ds'ds \\
&= 2 \int_0^t \int_0^s \int \left| \int f(z)[\lambda(x, (s-s')/\varepsilon, dz) - \mu(dz)] f(x) \right| \lambda(y, s'/\varepsilon, dx)ds'ds \\
&\le 2 \int_0^t \int_0^s \int \|\lambda_{x,(s-s')/\varepsilon} - \mu\| \lambda(y, s'/\varepsilon, dx)ds'ds \\
&\le 2 \int_0^t \int_0^s \int \|\lambda_{x,(s-s')/\varepsilon} - \mu\| \mu(dx)ds'ds + 4 \int_0^t \int_0^s \|\lambda_{y,s'/\varepsilon} - \mu\| ds'ds. \\
&\to 0, \quad \varepsilon \to 0.
\end{aligned}$$

□

Let the function f be the same as in the proof of Lemma 20.1. Consider now an auxiliary filtering problem with the unobservable signal $f(\eta_{t/\varepsilon})$ and the observation $(X_t^\varepsilon, Y_t^\varepsilon)$.

Lemma 20.2. *For every* $t > 0$

$$P\text{-}\lim_{\varepsilon \to 0} M \left(f(\eta_{t/\varepsilon}) \middle| X_{[0,t]}^\varepsilon, Y_{[t,0]}^\varepsilon \right) = \int f(z)\mu(dz).$$

PROOF. Introduce the filtration $(\mathcal{F}_t^\eta)_{t\ge 0}$, generated by trajectories of the process (η_t)[1], and the square-integrable martingale $(N_s^\varepsilon, \mathcal{F}_{s/\varepsilon}^\eta)_{0\le s\le t}$:

$$N_s^\varepsilon = M\left(f(\eta_{t/\varepsilon}) \middle| \mathcal{F}_{s/\varepsilon}^\eta \right). \tag{20.15}$$

Consider the filtering problem for N_s^ε given the observation $(X_s^\varepsilon, Y_s^\varepsilon)$. By Theorem 8.5 we find ($\mathcal{F}_s^\varepsilon = \sigma\{X_{s'}^\varepsilon, Y_{s'}^\varepsilon, s' \le s\}$)

$$\begin{aligned} & M(N_t^\varepsilon|\mathcal{F}_s^\varepsilon) \\ &= M(N_t^\varepsilon|\mathcal{F}_0^\varepsilon) \\ &\quad + \int_0^s \frac{1}{b(X_{s'}^\varepsilon)} M\left(M\left(N_t^\varepsilon|\mathcal{F}_{s'/\varepsilon}^\eta\right) \left[a(X_{s'}^\varepsilon, Y_{s'}^\varepsilon, \eta_{s'/\varepsilon}) - a_{s'}^\varepsilon\right] \middle| \mathcal{F}_{s'}^\varepsilon \right) d\widetilde{W}_{s'}^{x,\varepsilon} \\ &\quad + \int_0^s \frac{1}{B(Y_{s'}^\varepsilon)} M\left(M\left(N_t^\varepsilon|\mathcal{F}_{s'/\varepsilon}^\eta\right) \left[A(X_{s'}^\varepsilon, Y_{s'}^\varepsilon, \eta_{s'/\varepsilon}) - A_{s'}^\varepsilon\right] \middle| \mathcal{F}_{s'}^\varepsilon \right) d\widetilde{W}_{s'}^{y,\varepsilon}, \end{aligned} \tag{20.16}$$

where

$$a_s^\varepsilon = M\left(a(X_s^\varepsilon, Y_s^\varepsilon, \eta_{s/\varepsilon}) \middle| \mathcal{F}_s^\varepsilon\right) \quad \text{and} \quad A_s^\varepsilon = M\left(A(X_s^\varepsilon, Y_s^\varepsilon, \eta_{s/\varepsilon}) \middle| \mathcal{F}_s^\varepsilon\right)$$

and

$$\widetilde{W}_t^{x,\varepsilon} = \int_0^t \frac{dX_s^\varepsilon - a_s^\varepsilon ds}{b(X_s^\varepsilon)} \quad \text{and} \quad \widetilde{W}_t^{y,\varepsilon} = \int_0^t \frac{dY_s^\varepsilon - A_s^\varepsilon ds}{B(Y_s^\varepsilon)}$$

are independent innovation Wiener processes. Let us show that for $s = t$ all terms on the right-hand side of (20.16) converge, as $\varepsilon \to 0$, to zero in probability. Since $|f| \le 1$ and $\int f(z)\mu(dz) = 0$

$$\begin{aligned} \left|M\left(N_t^\varepsilon|\mathcal{F}_0^\varepsilon\right)\right| &= \left| \int f(z)\lambda(y, t/\varepsilon, dz) \right| \\ &= \left| \int f(z)[\lambda(y, t/\varepsilon, dz) \quad \mu(dz)] \right| \\ &\le \|\lambda_{y,t/\varepsilon} - \mu\| \to 0, \quad \varepsilon \to 0. \end{aligned}$$

For brevity of notation let us use $\alpha_{s'}^\varepsilon$ and $\widetilde{W}_{s'}$ to denote either of $\{a(X_{s'}^\varepsilon, Y_{s'}^\varepsilon, \eta_{s'/\varepsilon}) - a_{s'}^\varepsilon\}/b(X_s^\varepsilon)$ or $\{A(X_{s'}^\varepsilon, Y_s^\varepsilon, \eta_{s'/\varepsilon}) - A_{s'}^\varepsilon\}/B(Y_s^\varepsilon)$ and either of the Wiener processes $\widetilde{W}_{s'}^{x,\varepsilon}$ or $\widetilde{W}_{s'}^{y,\varepsilon}$, respectively. Using this notation, we show also that $\int_0^t M(N_t^\varepsilon|\mathcal{F}_{s'/\varepsilon}^\eta)\alpha_{s'}^\varepsilon d\widetilde{W}_{s'}$ converges to zero in probability as $\varepsilon \to 0$. To this end, we apply the Lenglart–Rebolledo inequality (see, for example, Chapter 1, Section 9 in [214]) which, being adapted to the case considered, states that for any $\gamma, \delta > 0$

[1] All processes are defined on some probability space $(\Omega, \mathcal{F}, P)$; $\mathcal{F}_t = \mathcal{F}_{t+}$, $t \ge 0$ and $\mathcal{F}_0$ is completed by P-zero sets from $\mathcal{F}$.

$$P\left(\left|\int_0^t M\left\{M\left(N_t^\varepsilon|\mathcal{F}_{s'/\varepsilon}^\eta\right)\alpha_{s'}^\varepsilon\middle|\mathcal{F}_{s'}^\varepsilon\right\}d\widetilde{W}_{s'}\right|\geq\gamma\right)$$
$$\leq\frac{\delta}{\gamma^2}+P\left(\int_0^t\left(M\left\{M\left(N_t^\varepsilon|\mathcal{F}_{s'/\varepsilon}^\eta\right)\alpha_{s'}^\varepsilon\middle|\mathcal{F}_{s'}^\varepsilon\right\}\right)^2ds'\geq\delta\right).$$

Therefore, by virtue of the arbitrariness of γ,δ, the required convergence holds provided that

$$P\text{-}\lim_{\varepsilon\to0}\int_0^t\left(M\left\{M\left(N_t^\varepsilon|\mathcal{F}_{s'/\varepsilon}^\eta\right)\alpha_{s'}^\varepsilon\middle|\mathcal{F}_{s'}^\varepsilon\right\}\right)^2ds'=0. \qquad (20.17)$$

Under the assumptions made on the functions $a(x,y,z)$, $A(x,y,z)$, $b(x)$, $B(y)$ the random variable $|\alpha_{s'}^\varepsilon|$ is bounded above by the $\mathcal{F}_{s'}^\varepsilon$-measurable random variable

$$\beta_{s'}^\varepsilon=\text{constant}\times\left(1+\sup_{s''\leq s'}|X_{s''}^\varepsilon|+\sup_{s''\leq s'}|Y_{s''}^\varepsilon|\right).$$

Therefore, it suffices to prove that

$$\begin{aligned}P\text{-}\lim_{\varepsilon\to0}\int_0^t\left\{M\left(M\left(N_t^\varepsilon|\mathcal{F}_{s'/\varepsilon}^\eta\right)\middle|\mathcal{F}_{s'}^\varepsilon\right)\right\}^2ds' &= 0\\ \lim_{c\to\infty}\limsup_{\varepsilon\to0}P(\beta_t^\varepsilon>c) &= 0.\end{aligned} \qquad (20.18)$$

To check the validity of the first part of (20.18), we use the Markov property of the process η_t: $M(N_t^\varepsilon|\mathcal{F}_{s'/\varepsilon}^\eta)=M(f(\eta_{t/\varepsilon})|\eta_{s'/\varepsilon})$ and verify

$$\lim_{\varepsilon\to0}M\int_0^t\left\{M\left(f(\eta_{t/\varepsilon})\middle|\eta_{s'/\varepsilon}\right)\right\}^2ds'=0. \qquad (20.19)$$

In fact,

$$\begin{aligned}\left|M\left(f(\eta_{t/\varepsilon})\middle|\eta_{s'/\varepsilon}\right)\right| &= \left|\int f(z)\lambda_{\eta_{(t-s')/\varepsilon},s'/\varepsilon}(dz)\right|\\ &= \left|\int f(z)[\lambda_{\eta_{(t-s')/\varepsilon},s'/\varepsilon}(dz)-\mu(dz)]\right|\\ &\leq \|\lambda_{\eta_{(t-s')/\varepsilon},s'/\varepsilon}-\mu\|,\end{aligned}$$

so that

$$\begin{aligned}M\int_0^t\left\{M\left(f(\eta_{t/\varepsilon})|\eta_{s'/\varepsilon}\right)\right\}^2ds' &\leq \int_0^t\int\|\lambda_{z,(t-s')/\varepsilon}-\mu\|^2\lambda_{y,s'/\varepsilon}(dz)ds'\\ &\leq 2\int_0^t\int\|\lambda_{z,(t-s')/\varepsilon}-\mu\|\mu(dz)ds'\\ &\quad+4\int_0^t\|\lambda_{y,s'/\varepsilon}-\mu\|ds'\\ &\to 0 \text{ as } \varepsilon\to0.\end{aligned}$$

We check now the validity of the second part of (20.18). Applying the Itô formula to $V_s^\varepsilon = (X_s^\varepsilon)^2+(Y_s^\varepsilon)^2$ and taking into account the assumptions made on all the functions involved in (20.11) we arrive at (ℓ is some appropriate constant)

$$\begin{aligned} V_s^\varepsilon &= X_0^2 + Y_0^2 \\ &\quad +2\int_0^s [X_{s'}^\varepsilon a(X_{s'}^\varepsilon, Y_{s'}^\varepsilon, \eta_{s'/\varepsilon}) + Y_{s'}^\varepsilon A(X_{s'}^\varepsilon, Y_{s'}^\varepsilon, \eta_{s'/\varepsilon})]ds' \\ &\quad + \int_0^s [b^2(X_{s'}^\varepsilon) + B^2(X_{s'}^\varepsilon)]ds' \\ &\quad +2\int_0^s [X_{s'}^\varepsilon b(X_{s'}^\varepsilon)dW_{s'}^x + Y_{s'}^\varepsilon B(Y_{s'}^\varepsilon)dW_{s'}^y] \\ &\le X_0^2 + Y_0^2 + \ell t + \ell\int_0^s V_{s'}^\varepsilon ds' + M_s^\varepsilon, \end{aligned}$$

where $M_s^\varepsilon = 2\int_0^s [X_{s'}^\varepsilon b(X_{s'}^\varepsilon)dW_{s'}^x + Y_{s'}^\varepsilon B(Y_{s'}^\varepsilon)dW_{s'}^y]$. If X_0, Y_0 are bounded random variables, then, applying the method of the proof for the last statement of Theorem 4.6, one can conclude that $\sup_{s\le t} MV_s^\varepsilon \le C$, where C is independent of ε. On the other hand, by the Gronwall–Bellman inequality $\sup_{s\le t} V_s^\varepsilon \le [X_0^2+Y_0^2+\ell t+\sup_{s\le t}|M_s^\varepsilon|]e^{\ell t}$ and since by the Cauchy–Schwarz and Doob inequalities

$$\begin{aligned} M\sup_{s\le t}|M_s^\varepsilon| \le \sqrt{M\sup_{s\le t}|M_s^\varepsilon|^2} &\le 2\sqrt{M|M_t^\varepsilon|^2} \\ &= 2\sqrt{\int_0^t M[b^2(X_{s'}^\varepsilon) + B^2(Y_{s'}^\varepsilon)]ds'}, \end{aligned}$$

it can be shown that $M\sup_{s\le t} V_s^\varepsilon \le$ constant and so $M\beta_t^\varepsilon \le$ constant as well. For unbounded X_0, Y_0 the required property for β_t^ε is implied by the Chebyshev inequality.

Thus,

$$\limsup_{c\to\infty}\limsup_{\varepsilon\to 0} P(\beta_t^\varepsilon > c) \le P(|X_0| + |Y_0| > L) \to 0, \quad L \to \infty. \qquad \square$$

PROOF OF THEOREM 20.2. Denote by $a_s^\varepsilon(x,y;\{X^\varepsilon,Y^\varepsilon\}) = M(a(x,y,\eta_{s/\varepsilon})|\mathcal{F}_s^\varepsilon)$ and note that (P-a.s.) $a_s^\varepsilon(X_s^\varepsilon, Y_s^\varepsilon;\{X^\varepsilon,Y^\varepsilon\}) = M(a(X_s^\varepsilon, Y_s^\varepsilon, \eta_{s/\varepsilon})|\mathcal{F}_s^\varepsilon)$. Lemma 20.2 implies: for fixed s, x, y,

$$P\text{-}\lim_{\varepsilon\to 0} a_s^\varepsilon(x,y;\{X^\varepsilon,Y^\varepsilon\}) = \overline{a}(x,y). \qquad (20.20)$$

For fixed x, y, $a_s^\varepsilon(x,y;\{X^\varepsilon,Y^\varepsilon\})$ is bounded as well. Therefore, due to (20.20),

$$P\text{-}\lim_{\varepsilon\to 0}\int_0^T [a_s^\varepsilon(x,y;\{X^\varepsilon,Y^\varepsilon\}) - \overline{a}(x,y)]^2\,ds = 0. \tag{20.21}$$

Next, we show that (20.21) remains true upon replacing x, y by $X_s^\varepsilon, Y_s^\varepsilon$:

$$P\text{-}\lim_{\varepsilon\to 0}\int_0^T [a_s^\varepsilon(X_s^\varepsilon,Y_s^\varepsilon;\{X^\varepsilon,Y^\varepsilon\}) - \overline{a}(X_s^\varepsilon,Y_s^\varepsilon)]^2\,ds = 0. \tag{20.22}$$

Denote by $[x]$ the greatest integer function and define the following random processes: $X_t^{\varepsilon,n} = X_{[nt]/n}^\varepsilon$ and $X_t^{\varepsilon,n,m} = [mX_t^{\varepsilon,n}]/m$. Put $X_T^{\varepsilon *} = \sup_{t\le T}|X_t^\varepsilon|$. In the same way define $Y_t^{\varepsilon,n}$, $Y_t^{\varepsilon,n,m}$, and $Y_T^{\varepsilon *}$. We show that for every $C>0$, and $m,n\ge 1$ on the set $\{X_T^{\varepsilon,*}+Y_T^{\varepsilon,*}\le C\}$

$$\int_0^T [a_s^\varepsilon(X_s^{\varepsilon,n,m},Y_s^{\varepsilon,n,m};\{X^\varepsilon,Y^\varepsilon\}) - \overline{a}(X_s^{\varepsilon,n,m},Y_s^{\varepsilon,n,m})]^2\,ds \to 0, \tag{20.23}$$

in probability as $\varepsilon\to 0$. The proof of (20.23) is based on the fact that on the set $\{X_T^{\varepsilon,*}+Y_T^{\varepsilon,*}\le C\}$ the processes $X_t^{\varepsilon,n,m}$ and $Y_t^{\varepsilon,n,m}$ have a finite number (independent of ε) of trajectories. Thus, (20.23) holds if for any continuous functions $x_s, y_s, 0\le s\le T$ and $x_s^{n,m}$, $y_s^{n,m}$, $x_{j/n}^m$, and $y_{j/n}^m$, defined similarly to $X_t^{\varepsilon,n,m}$, $Y_t^{\varepsilon,n,m}$, $X_{j/n}^{\varepsilon,m}$, and $Y_{j/n}^{\varepsilon,m}$,

$$P\text{-}\lim_{\varepsilon\to 0}\int_0^T [a_s^\varepsilon(x_s^{n,m},y_s^{n,m};\{X^\varepsilon,Y^\varepsilon\}) - \overline{a}(x_s^{n,m},y_s^{n,m})]^2\,ds = 0. \tag{20.24}$$

In turn, the validity of (20.24) follows from the chain of upper bounds:

$$\begin{aligned}
&\int_0^T [a_s^\varepsilon(x_s^{n,m},y_s^{n,m};\{X^\varepsilon,Y^\varepsilon\}) - \overline{a}(x_s^{n,m},y_s^{n,m})]^2\,ds\\
&\le \sum_{j=1}^{[nT]}\int_{(j-1)/n}^{j/n} [a_s^\varepsilon(x_s^{n,m},y_s^{n,m};\{X^\varepsilon,Y^\varepsilon\}) - \overline{a}(x_s^{n,m},y_s^{n,m})]^2\,ds\\
&\le 2\sum_{j=1}^{[nT]}\int_0^T \Big[a_s^\varepsilon(x_{j/n}^m,y_{j/n}^m;\{X^\varepsilon,Y^\varepsilon\}) - \overline{a}(x_{j/n}^m,y_{j/n}^m)\Big]^2\,ds.
\end{aligned}$$

For fixed n, m, each summand in the last sum converges to zero in probability as $\varepsilon\to 0$. Thus, (20.23) holds. Consequently, (20.22) holds if

$$\lim_{C\to\infty}\limsup_{\varepsilon\to 0} P(X_T^{\varepsilon,*}+Y_T^{\varepsilon,*}\ge C) = 0 \tag{20.25}$$

and for every $C>0$ on the set $\{X_T^{\varepsilon,*}+Y_T^{\varepsilon,*}\le C\}$

$$\begin{aligned}
&\int_0^T [a_s^\varepsilon(X_s^\varepsilon,Y_s^\varepsilon;\{X^\varepsilon,Y^\varepsilon\}) - a_s^\varepsilon(X_s^{\varepsilon,n,m},Y_s^{\varepsilon,n,m};\{X^\varepsilon,Y^\varepsilon\})]^2\,ds \to 0\\
&\int_0^T [\overline{a}(X_s^\varepsilon,Y_s^\varepsilon) - \overline{a}(X_s^{\varepsilon,n,m},Y_s^{\varepsilon,n,m})]^2\,ds \to 0
\end{aligned} \tag{20.26}$$

in probability as the limit $\lim_{n,m}\limsup_{\varepsilon\to 0}$ is taken. It is clear that (20.24) is implied by Lemma 20.1. On the other hand, each of functions $a_s^\varepsilon(x,y;\{X^\varepsilon,Y^\varepsilon\})$ and $\overline{a}(x,y)$ inherits the Lipschitz property in x,y (with an absolute constant) which provides (20.26) under

$$P\text{-}\lim_{n,m}\limsup_{\varepsilon\to 0}\sup_{s\le T}\left(|X_s^\varepsilon - X_s^{\varepsilon,n,m}| + |Y_s^\varepsilon - Y_s^{\varepsilon,n,m}|\right) = 0.$$

The above holds by virtue of Lemma 20.1 as well.

Thus, the validity of (20.22) is proved.

In the same way, for $A_s^\varepsilon(X_s^\varepsilon,Y_s^\varepsilon;\{X^\varepsilon,Y^\varepsilon\}) = M(A(X_s^\varepsilon,Y_s^\varepsilon,\eta_{s/\varepsilon})|\mathcal{F}_s^\varepsilon)$ and $\overline{A}(x,y)$, defined in (20.13), we obtain

$$P\text{-}\lim_{\varepsilon\to 0}\int_0^T \left[A_s^\varepsilon(X_s^\varepsilon,Y_s^\varepsilon;\{X^\varepsilon,Y^\varepsilon\}) - \overline{A}(X_s^\varepsilon,Y_s^\varepsilon)\right]^2 ds = 0. \tag{20.27}$$

For brevity, in what follows, a_s^ε and A_s^ε are used to designate $a_s^\varepsilon(X_s^\varepsilon, Y_s^\varepsilon;\{X^\varepsilon,Y^\varepsilon\})$ and $A_s^\varepsilon(X_s^\varepsilon,Y_s^\varepsilon;\{X^\varepsilon,Y^\varepsilon\})$, respectively. It is known from Theorem 7.12 that

$$\widetilde{W}_t^{x,\varepsilon} = \int_0^t \frac{dX_s^\varepsilon - a_s^\varepsilon ds}{b(X_s^\varepsilon)}, \quad \widetilde{W}_t^{y,\varepsilon} = \int_0^t \frac{dY_s^\varepsilon - A_s^\varepsilon ds}{B(Y_s^\varepsilon)} \tag{20.28}$$

are independent Wiener processes with respect to the filtration $(\mathcal{F}_s^\varepsilon)_{s\ge 0}$, that is the process $(X_t^\varepsilon, Y_t^\varepsilon)_{t\ge 0}$ is defined by the past-dependent Itô equations

$$\begin{aligned} X_t^\varepsilon &= X_0 + \int_0^t a_s^\varepsilon ds + \int_0^t b(X_s^\varepsilon)d\widetilde{W}_s^{x,\varepsilon} \\ Y_t^\varepsilon &= X_0 + \int_0^t A_s^\varepsilon ds + \int_0^t B(Y_s^\varepsilon)d\widetilde{W}_s^{y,\varepsilon}. \end{aligned} \tag{20.29}$$

Hence, by Theorems 7.19 and 7.20, for every $T>0$ the measures Q^ε and $\overline{Q}$ are equivalent to the distribution of a pair of diffusion processes with respect to independent Wiener processes W_t', W_t'' :

$$X_t = X_0 + \int_0^t b(X_s)dW_s', \quad Y_t = Y_0 + \int_0^t b(Y_s)dW_s''.$$

Thus, these measures are equivalent ($Q^\varepsilon \sim \overline{Q}$) and the density $d\overline{Q}/dQ^\varepsilon$ at the point '$X^\varepsilon, Y^\varepsilon$' is given by the formula (see Theorems 7.19, 7.20)

$$\begin{aligned} \frac{d\overline{Q}}{dQ^\varepsilon}(X^\varepsilon,Y^\varepsilon) &= \exp\Bigg(\int_0^T \frac{\overline{a}(X_s^\varepsilon,Y_s^\varepsilon) - a_s^\varepsilon}{b(X_s^\varepsilon)}dX_s^\varepsilon + \int_0^T \frac{\overline{A}(X_s^\varepsilon,Y_s^\varepsilon) - A_s^\varepsilon}{B(Y_s^\varepsilon)}dY_s^\varepsilon \\ &\quad -\frac{1}{2}\int_0^T \frac{[(\overline{a}(X_s^\varepsilon,Y_s^\varepsilon))^2 - (a_s^\varepsilon)^2]}{b^2(X_s^\varepsilon)}ds \\ &\quad -\frac{1}{2}\int_0^T \frac{[(\overline{A}(X_s^\varepsilon,Y_s^\varepsilon))^2 - (A_s^\varepsilon)^2]}{B^2(Y_s^\varepsilon)}ds\Bigg). \end{aligned} \tag{20.30}$$

Taking into account that $X_t^\varepsilon, Y_t^\varepsilon$ satisfy (20.29), formula (20.30) can be rewritten in the form

$$\frac{d\overline{Q}}{dQ^\varepsilon}(X^\varepsilon, Y^\varepsilon) = \exp\left(M_T^\varepsilon - \frac{1}{2}\langle M^\varepsilon\rangle_T\right) \tag{20.31}$$

with the continuous martingale $(M_t^\varepsilon)_{t\le T}$ and its predictable quadratic variation $(\langle M^\varepsilon\rangle_t)_{t\le T}$:

$$\begin{aligned} M_T^\varepsilon &= \int_0^T \frac{\overline{a}(X_s^\varepsilon, Y_s^\varepsilon) - a_s^\varepsilon}{b(X_s^\varepsilon)} d\widetilde{W}_s^{x,\varepsilon} + \int_0^T \frac{\overline{A}(X_s^\varepsilon, Y_s^\varepsilon) - A_s^\varepsilon}{B(Y_s^\varepsilon)} d\widetilde{W}_s^{y,\varepsilon} \\ \langle M^\varepsilon\rangle_T &= \int_0^T \left\{ \frac{[\overline{a}(X_s^\varepsilon, Y_s^\varepsilon) - a_s^\varepsilon]^2}{b^2(X_s^\varepsilon)} + \frac{[\overline{A}(X_s^\varepsilon, Y_s^\varepsilon) - A_s^\varepsilon]^2}{B^2(Y_s^\varepsilon)} \right\} ds. \end{aligned}$$

By (20.22) and (20.27) $\langle M^\varepsilon\rangle_T$ converges to zero in probability as $\varepsilon \to 0$. Therefore, the same convergence holds for M_T^ε since by the Lenglart–Rebolledo inequality (see, for example, Chapter 1, Section 9 in [214]) for any $\gamma, \delta > 0$ we have

$$P(|M_T^\varepsilon| \ge \gamma) \le \frac{\delta}{\gamma^2} + P(\langle M^\varepsilon\rangle_T \ge \delta). \tag{20.32}$$

Consequently, the density $\frac{d\overline{Q}}{dQ^\varepsilon}(X^\varepsilon, Y^\varepsilon)$ converges to 1 in probability as $\varepsilon \to 0$.

The statement of the theorem is now provided by the Scheffe theorem [19] since

$$\|Q^\varepsilon - \overline{Q}\| = M\left|1 - \frac{d\overline{Q}_T}{dQ_T^\varepsilon}(X^\varepsilon, Y^\varepsilon)\right| \to 0. \qquad \square$$

We give here two examples of homogeneous ergodic Markov processes for which (20.12) holds.

EXAMPLE 1. Let η_t be a diffusion Markov process defined by the Itô equation with respect to the Wiener process W_t^η:

$$d\eta_t = a(\eta_t)dt + b(\eta_t)dW_t^\eta,$$

where the drift and diffusion $a(z)$ and $b(z)$ are assumed to be continuously differentiable, having bounded derivatives and, for some constants ℓ and L,

$$0 < \ell \le b^2(z) \le L, \quad \liminf_{|z|\to\infty} a(z)\operatorname{sign} z < 0. \tag{20.33}$$

It is well known that the transition probability $\lambda(y, t, dz)$ of (η_t) admits the density $p(y,t,z)$ (with respect to dz) being a solution of the forward Fokker–Planck–Kolmogorov equation ($\mathcal{L}^\star\{\cdot\} = -\frac{\partial}{\partial z}(a(z)\{\cdot\}) + \frac{1}{2}\frac{\partial^2}{\partial z^2}(b^2(z)\{\cdot\})$)

$$\frac{\partial p(y,t,z)}{\partial t} = \mathcal{L}^\star p(y,t,z). \tag{20.34}$$

Under assumptions (20.33), the invariant measure $\mu(dz)$ exists and admits a density $m(z)$ (with respect to dz) which is the solution of the ordinary differential equation $\mathcal{L}^\star m(z) = 0$. It is well known (see, Chapter 4, Section 9, Lemma 9.5 and Chapter 3, Section 8, Example 2 of [148]) that the solution of the Cauchy problem for the partial differential equation (20.34) is stabilized in the sense that for every fixed y, z

$$\lim_{t\to\infty} p(y,t,z) = m(z).$$

Then, taking into account that $\int p(y,t,z)dz = 1$, $\int m(z)dz = 1$, by the Scheffe theorem (see [19]) we obtain

$$\lim_{t\to\infty} \int |p(y,t,z) - m(z)|dz = 0,$$

that is nothing but (20.12).

EXAMPLE 2. Let a homogeneous Markov process η_t take values in a finite state space $\{\alpha_1, \ldots, \alpha_N\}$ and η_0 be fixed. Denote by $\mathcal{P}$ the matrix of transition intensities of η_t and by $p(t) = (p_1(t), \ldots, p_N(t))$ the vector of transition probabilities ($p_{i,j}(t)$ is the transition probability from α_j to α_i over the time interval $[0,t]$). The vector $p_j(t)$ is defined by the Fokker–Planck–Kolmogorov equation

$$\frac{dp(t)}{dt} = p(t)\mathcal{P}. \tag{20.35}$$

Assume that '0' is the simple eigenvalue of the matrix $\mathcal{P}$. Then there exists a unique invariant distribution $p = (p_1, \ldots, p_N)$ such that $\lim_{t\to\infty} p_j(t) = p$, $j = 1, \ldots, N$. Hence, (20.12) holds since

$$\|\lambda_{y,t} - \mu\| = \sum_{j=1}^{N} |p_j(t) - p|$$

20.1.3. The asymptotic (δ-asymptotic) optimality does not hold for many models which formally have a structure similar to that of Model 1. We give an example below.

EXAMPLE 3. Consider a filtering model with the deterministic contamination $\eta_{t/\varepsilon} \equiv \sin(t/\varepsilon)$:

$$X_t^\varepsilon = W_t^x, \quad Y_t^\varepsilon = \int_0^t \sin(s/\varepsilon)X_s^\varepsilon ds + W_t^y.$$

Since $\lim_{t\to\infty}(1/t)\int_0^t \sin(s)ds = 0$, the limit model is defined as:

$$\overline{X}_t = W_t^x \quad \text{and} \quad \overline{Y}_t = W_t^y$$

and provides $\overline{\pi}_t(Y^\varepsilon) \equiv 0$ and $M(X_t^\varepsilon - \overline{\pi}_t(Y^\varepsilon))^2 \equiv t$. At the same time, $\pi_t^\varepsilon(Y^\varepsilon)$ is defined by the Kalman filter (see Chapter 10) with $P^\varepsilon(t) = M(X_t^\varepsilon - \pi_t^\varepsilon(Y^\varepsilon))^2$ defined by the Ricatti equation:

$$\dot{P}^\varepsilon(t) = 1 - \sin^2(t/\varepsilon)(P^\varepsilon(t))^2(t), \quad P^\varepsilon(0) = 0.$$

By the Bogolubov averaging principle [23], there exists $\lim_{\varepsilon\to 0} P^\varepsilon(t) \equiv P^\circ(t)$ defined by the Ricatti equation with the averaged coefficient $\lim_{\varepsilon\to 0}\int_0^1 \sin^2(s/\varepsilon)ds = \frac{1}{2}$:

$$\dot{P}^\circ(t) = 1 - \frac{1}{2}(P^\circ(t))^2(t).$$

Evidently, for any $t > 0$, $P^\circ(t) < t$ and is bounded above by $\sqrt{2}$, i.e., the asymptotic optimality for the filtering estimate $\overline{\pi}_t(Y^\varepsilon)$ is lost.

The above example shows that under the deterministic contamination the loss of the asymptotic optimality can be expected. We introduce below another class of filtering models for which the asymptotic (δ-asymptotic) optimality holds under deterministic contamination as well.

MODEL 2. Let the observation $Y^\varepsilon = (Y_t^\varepsilon)_{t\geq 0}$ be a diffusion-type process

$$Y_t^\varepsilon = \int_0^t A(X_s^\varepsilon)ds + \int_0^t B(Y_s^\varepsilon)dW_s, \tag{20.36}$$

where the unobservable process X_t^ε (with trajectories in D) is independent of the Wiener process $W = (W_t)_{t\geq 0}$. The functions $A(x)$ and $B(y)$ are assumed to be Lipschitz continuous and $B^2(y) \geq c > 0$.

We assume that there exists a random process $\overline{X} = (\overline{X}_t)_{t\geq 0}$, independent of W, with trajectories in D such that for every fixed $T > 0$

$$P\text{-}\lim_{\varepsilon\to 0}\int_0^T (X_t^\varepsilon - \overline{X}_t)^2 dt = 0. \tag{20.37}$$

In parallel to Y^ε, let us introduce the new diffusion-type process

$$\overline{Y}_t = \int_0^t A(\overline{X}_s)ds + \int_0^t B(\overline{Y}_s)dW_s. \tag{20.38}$$

As before, consider the filtering problem for the signal $u(X_t^\varepsilon)$ ($Mu^2(X_t^\varepsilon) < \infty$, $Mu^2(\overline{X}_t) < \infty$) and the observation Y^ε. Let the filtering estimates $\pi_t^\varepsilon(Y^\varepsilon)$, $\overline{\pi}_t(\overline{Y})$, and $\overline{\pi}_t(Y^\varepsilon)$ be defined similarly to those for the previous model (Model 1).

Theorem 20.3. *Assume $u(x)$ is a continuous function and (20.37) holds. Then for each $t \leq T$:*

1. *$\overline{\pi}_t(Y^\varepsilon)$ is asymptotically optimal and $\pi_t^\varepsilon(Y^\varepsilon)\xrightarrow{\text{law}}\overline{\pi}_t(\overline{Y})$, $\varepsilon\to 0$ if u is bounded;*
2. *$\overline{\pi}_t^{n_\delta}(Y^\varepsilon)$, defined in Proposition 20.1, is asymptotically δ-optimal if there exists $\ell>0$ such that $|u(x)|\le \ell(1+|x|)$ and for some $\gamma>0$ and any ε*

$$\sup_{t\le T} M|X_t^\varepsilon|^{2+\gamma}<\infty.$$

The proof of this theorem uses also the convergence in the total variation norm.

Lemma 20.3. *Let R^ε and $\overline{R}$ be distributions of random processes $(\overline{X}_t, Y_t^\varepsilon)_{t\le T}$ and $(\overline{X}_t,\overline{Y}_t)_{t\le T}$, respectively. Under (20.37)*

$$\lim_{\varepsilon\to 0}\|R^\varepsilon-\overline{R}\|=0. \tag{20.39}$$

PROOF. Since (W_t) and $(X_t^\varepsilon,\overline{X}_t)$ are independent, we assume, without loss of generality, that the pair $(X_t^\varepsilon,\overline{X}_t)$ is defined on the probability space $(\Omega',\mathcal{F}',P')$, while (W_t) is defined on its copy $(\Omega'',\mathcal{F}'',P'')$ (the notation M' and M'' is used for expectations with respect to P' and P'', respectively). Thus, both processes (Y_t^ε) and $(\overline{Y}_t)$ are determined on $(\Omega'\times\Omega'',\mathcal{F}'\otimes\mathcal{F}'',P'\times P'')$:

$$\begin{aligned} Y_t^\varepsilon(\omega',\omega'') &= \int_0^t A(X_s^\varepsilon(\omega'))ds+\int_0^t B(Y_s^\varepsilon(\omega',\omega''))dW_s(\omega'') \\ \overline{Y}_t(\omega',\omega'') &= \int_0^t A(\overline{X}_s(\omega'))ds+\int_0^t B(\overline{Y}_s(\omega',\omega''))dW_s(\omega''). \end{aligned} \tag{20.40}$$

Let us introduce

$$\begin{aligned} Z_t^\varepsilon(\omega,\omega') &= \exp\Bigg(\int_0^t \frac{A(\overline{X}_s(\omega'))-A(X_s^\varepsilon(\omega'))}{B(Y_s^\varepsilon(\omega',\omega''))}dW_s(\omega'') \\ &\quad -\frac{1}{2}\int_0^t\frac{\left[A(\overline{X}_s(\omega'))-A(X_s^\varepsilon(\omega'))\right]^2}{B^2(Y_s^\varepsilon(\omega',\omega''))}ds\Bigg), \end{aligned} \tag{20.41}$$

and show that for every $T>0$

$$(M'\times M'')Z_T^\varepsilon(\omega',\omega'')=1. \tag{20.42}$$

For fixed ω' and T,

$$\int_0^t\frac{\left[A(\overline{X}_s(\omega'))-A(X_s^\varepsilon(\omega'))\right]^2}{B^2(Y_s^\varepsilon(\omega',\omega''))}\le C(\omega')<\infty,\quad (P'\text{-a.s.}).$$

Then by Theorem 6.1 $M''Z_T^\varepsilon(\omega',\omega'')=1$ (P'-a.s.) and, in turn, (20.42) holds. Denote by $\mathcal{G}_T^\varepsilon$ the σ-algebra generated by $(X_t^\varepsilon,\overline{X}_t,Y_t^\varepsilon)_{t\le T}$, and by P^ε the restriction of $P'\times P''$ to $\mathcal{G}_T^\varepsilon$. Define the new probability measure $\overline{P}^\varepsilon$ with

$$d\overline{P}^{\varepsilon} = Z_T^{\varepsilon}(\omega', \omega'')dP^{\varepsilon}, \tag{20.43}$$

i.e., $Z_T^{\varepsilon}(\omega', \omega'')$ is the density of $\overline{P}^{\varepsilon}$ with respect to P^{ε}. The structure of this density provides the following property for $(\overline{X}_t, Y_t^{\varepsilon})_{t \le T}$. The process $(\overline{X}_t(\omega'))_{t \le T}$ has the same distribution with respect to both measures P^{ε} and $\overline{P}^{\varepsilon}$ while the distribution of $(Y_t^{\varepsilon}(\omega', \omega''))_{t \le T}$ is changed. In fact, by Theorem 7.12

$$\overline{W}_t^{\varepsilon}(\omega', \omega'') = W_t(\omega'') - \int_0^t \frac{A(\overline{X}_s(\omega')) - A(X_s^{\varepsilon}(\omega'))}{B(Y_s^{\varepsilon}(\omega', \omega''))} ds$$

is a Wiener process with respect to $\overline{P}^{\varepsilon}$, so that the process $(Y_t^{\varepsilon}(\omega', \omega''))_{t \le T}$ possesses the new representation with respect to $\overline{P}^{\varepsilon}$:

$$Y_t^{\varepsilon}(\omega', \omega'') = \int_0^t A(\overline{X}_s(\omega'))ds + \int_0^t B(Y_s^{\varepsilon}(\omega', \omega''))d\overline{W}_s^{\varepsilon}(\omega', \omega'').$$

Comparing this Itô equation with the second one from (20.40) we conclude that the distribution of $(Y_t^{\varepsilon}(\omega', \omega''))_{0 \le t \le T}$ with respect to $\overline{P}^{\varepsilon}$ coincides with the distribution of $(\overline{Y}_t(\omega', \omega''))_{0 \le t \le T}$ with respect to P^{ε}. Hence, repeating arguments from the proof of Theorem 7.1, we find that $R^{\varepsilon} \sim \overline{R}$ and

$$\frac{d\overline{R}}{dR^{\varepsilon}}(Y^{\varepsilon}) = (M' \times M'')\left(Z_T^{\varepsilon}(\omega', \omega'') | \overline{X}_{[0,T]}, Y_{[0,T]}^{\varepsilon}\right) \quad (P' \times P'')\text{-a.s.} \tag{20.44}$$

We show now that (20.39) holds. Equation (20.44) and the Jensen inequality imply $\|R^{\varepsilon} - \overline{R}\| \le (M' \times M'')|1 - Z_T^{\varepsilon}(\omega', \omega'')|$. Further, $Z_T^{\varepsilon}(\omega', \omega'') = \exp(M_T^{\varepsilon} - \frac{1}{2}\langle M^{\varepsilon}\rangle_T)$ with

$$\begin{aligned} M_T^{\varepsilon} &= \int_0^T \frac{A(\overline{X}_s(\omega')) - A(X_s^{\varepsilon}(\omega'))}{B(Y_s^{\varepsilon}(\omega', \omega''))} dW_s(\omega'') \\ \langle M^{\varepsilon}\rangle_T &= \int_0^T \frac{\left[A(\overline{X}_s(\omega')) - A(X_s^{\varepsilon}(\omega'))\right]^2}{B^2(Y_s^{\varepsilon}(\omega', \omega''))} ds. \end{aligned}$$

By virtue of (20.37), $\langle M^{\varepsilon}\rangle_T \to 0$ in probability as $\varepsilon \to 0$ and the same convergence for M_T^{ε} holds as well (see (20.32) or Problem 1.9.2 in [214]). Hence, $Z_T^{\varepsilon}(\omega', \omega'') \to 1$ in probability as $\varepsilon \to 0$ and by the Scheffe theorem (see [19]) the desired conclusion holds . □

PROOF OF THEOREM 20.3. Part 1. Since u is a bounded and continuous function

$$\lim_{\varepsilon \to 0} \left| M(u(X_t^{\varepsilon}) - \pi_t^{\varepsilon}(X^{\varepsilon}))^2 - M(u(\overline{X}_t) - \pi_t^{\varepsilon}(X^{\varepsilon}))^2 \right| = 0. \tag{20.45}$$

Using (20.45) and repeating the proof of (20.6), we obtain

$$\lim_{\varepsilon\to 0}\int (u(X_t)-\overline{\pi}_t(Y))^2\,dR^\varepsilon = \int (u(X_t)-\overline{\pi}_t(Y))^2\,d\overline{R}$$

$$\liminf_{\varepsilon\to 0}\int (u(X_t)-\pi_t^\varepsilon(Y))^2\,dQ^\varepsilon \geq \int (u(X_t)-\overline{\pi}_t(Y))^2\,d\overline{R}. \quad (20.46)$$

The asymptotic optimality of $\overline{\pi}_t(Y^\varepsilon)$ follows now from (20.46) and (20.5).

The proof of the second statement of part 1 is the same as for the corollary to Theorem 20.1.

Part 2. The proof is the same as for Proposition 20.1. □

20.2 Robust Diffusion Approximation for Filtering

In this section, we consider the filtering problem for a nonlinear model in which the unobservable signal X_t is a diffusion process defined by the Itô equation with respect to a Wiener process V_t:

$$dX_t = a(X_t)dt + b(X_t)dV_t \quad (20.47)$$

subject to the random initial condition X_0, $MX_0^2 < \infty$. We obtain measurements at fixed time values t_k, $k = 0, 1 \ldots$, $(t_{k+1} - t_k \equiv \varepsilon)$, so that the observation process Y_{t_k} is defined as:

$$Y_{t_0} = 0, \quad Y_{t_k} - Y_{t_{k-1}} = h(X_{t_{k-1}})\varepsilon + \sqrt{\varepsilon}\xi_k, \quad (20.48)$$

where $h(x)$ is some continuous function and $(\xi_k)_{k\geq 1}$ is an independent identically distributed sequence of zero-mean random variables independent of $(V_t), X_0$. Under $M\xi_1^2 < \infty$, the attractiveness of this model is based on the following fact. For any distribution for ξ_1, the sequence of random processes $(Y_t^\varepsilon)_{t\geq 0}, \varepsilon > 0$, with $Y_t^\varepsilon = Y_{t_k}, t_k \leq t < t_{k+1}$, converges in the distribution, as $\varepsilon \to 0$, to a diffusion-type process with respect to the Wiener process W_t independent of (X_t)

$$Y_t = \int_0^t h(X_s)ds + BW_t,$$

where $B = \sqrt{M\xi_1^2}$. Such an approximation result allows one to apply the Kushner–Zakai filter (Chapter 8), corresponding to the limit model, for the prelimit observation. Namely, let us fix some continuous bounded function $f(x)$ and, applying the Bayes formula, define the functional $\pi_t(y)$, $y \in D$ (D is the Skorokhod space) such that (P-a.s.) $\pi_t(Y) = M(f(X_t)|Y_{[0,t]})$. If $\pi_t(y)$ is a continuous (in some sense) function of the arguments, it makes sense to take $\pi_t(Y^\varepsilon)$ as the filtering estimate for the signal $f(X_t)$ under given Y_s^ε, $s \leq t$. Due to the weak convergence

$$(X_t, Y_t^\varepsilon) \xrightarrow{\text{law}} (X_t, Y_t), \quad \varepsilon \to 0$$

and the continuity of the functional $\pi_t(\cdot)$, the distributions of $(f(X_t), \pi_t(Y^\varepsilon))$ converge to the distribution of a limit $(f(X_t), \pi_t(Y))$. Therefore, we arrive at the asymptotic equivalence for prelimit and limit models:

$$\lim_{\varepsilon \to 0} M\left(f(X_t) - \pi_t(Y^\varepsilon)\right)^2 = M\left(f(X_t) - \pi_t(Y)\right)^2.$$

However, if the distribution of ξ_1 is not Gaussian, the resulting filtering estimate $\pi_t(Y^\varepsilon)$ might be far from the optimal one even if ε is too small. On the other hand, using the Bayes formula, one can find the optimal (in the mean square sense) filtering estimate $\pi_t^\varepsilon(Y^\varepsilon) = M(f(X_t)|Y^\varepsilon_{[0,t]})$ for the prelimit model, which may be asymptotically better than $\pi_t(Y^\varepsilon)$, i.e., it may happen that

$$\limsup_{\varepsilon \to 0} M\left(f(X_t) - \pi_t^\varepsilon(Y^\varepsilon)\right)^2 < M\left(f(X_t) - \pi_t(Y)\right)^2.$$

If ξ_1 is not Gaussian and, say, $M\xi_1^2 = \infty$, to remedy this situation, we make a preliminary nonlinear transformation of the observation with some smooth function $G(x)$, hereinafter called the 'limiter', and show that filtering via the diffusion approximation implemented for the transformed signal might be asymptotically better and even optimal. The main assumption here is

$$MG(\xi_1) = 0, \quad MG^2(\xi_1) < \infty.$$

Letting $Y^G_{t_0} = 0$ and $Y^G_{t_k} - Y^G_{t_{k-1}} = \sqrt{\varepsilon}G(\frac{1}{\sqrt{\varepsilon}}[Y_{t_k} - Y_{t_{k-1}}])$ and taking into account that $Y^G_{t_k} - Y^G_{t_{k-1}} \approx \sqrt{\varepsilon}G(\xi_k) + G'(\xi_k)h(X_{t_{k-1}})\varepsilon$, we arrive at another diffusion limit for the sequence of random processes $(Y^{\varepsilon,G}_t)_{t \geq 0}$, with $Y^{\varepsilon,G}_t = Y^G_{t_k}$, $t_k < t \leq t_{k+1}$:

$$Y^G_t = \int_0^t A_G h(X_s)ds + B_G W_t,$$

where $B_G = \sqrt{MG^2(\xi_1)}$ and $A_G = MG'(\xi_1)$. This type of diffusion approximation allows one to compare limiters using the parameter

$$SN_G = \frac{A_G^2}{B_G^2}$$

which naturally can be called the 'signal-to-noise' ratio.

In what follows, we fix the following assumptions.

1. (A-1) G is a differentiable function and G' is Lipschitz continuous.
2. (A-2) $a(x)$ and $b(x)$ are Lipschitz continuous; $b(x)$ is bounded.
3. (A-3) $h(x)$ is twice continuously differentiable, having bounded derivatives $h'(x)$, $h''(x)$.

4. (A-4) $(\xi_k)_{k\geq 1}$ is a sequence of independent identically distributed random variables, independent of X_0 ,(V_t), such that $M(G'(\xi_1))^2 < \infty$, $MG^2(\xi_1) < \infty$, and $MG(\xi_1) = 0$.
5. (A-5) For every $\lambda > 0$, there exists a constant $C(\lambda, B_G^2))$, depending on λ and $B_G^2 = MG^2(\xi_1)$, such that $(\mathcal{L} = a(x)\frac{d}{dx} + \frac{1}{2}b^2(x)\frac{d^2}{dx^2})$

$$\mathcal{D}^\lambda h(x) := \lambda|\mathcal{L}h(x)| - \frac{B_G^2}{2}h^2(x) \leq C(\lambda, B_G^2).$$

6. (A-6) For every $\lambda > 0$, $Me^{\lambda|h(X_0)|} < \infty$.

20.2.1 Diffusion Approximation with Limiter. For brevity, $\mathcal{W}\text{-}\lim_{\varepsilon\to\infty}$ denotes weak convergence in the Skorokhod–Lindvall and the local supremum topologies (see, for example, Chapter 6 in [214]).

Theorem 20.4. *Assume (A-1)–(A-4). Then*

$$\mathcal{W}\text{-}\lim_{\varepsilon\to 0}(X_t, Y_t^{\varepsilon,G})_{t\geq 0} = (X_t, Y_t^G)_{t\geq 0},$$

with $Y_0^G = 0$, $dY_t^G = A_G h(X_t)dt + B_G dW_t$, *where* (W_t) *is a Wiener process independent of* X_0, (V_t), *and* $A_G = MG'(\xi_1)$, $B_G = \sqrt{MG^2(\xi_1)}$.

PROOF. Let us define the increasing function $L_t^\varepsilon = \varepsilon[t/\varepsilon]$, where $[t]$ is the greatest integer function, and random processes

$$\begin{aligned}
M_t^{\varepsilon,G} &= \sqrt{\varepsilon}\sum_{k=1}^{[t/\varepsilon]} G(\xi_k) \\
u^{\varepsilon,G}(t) &= G'(\xi_k), \quad t_{k-1} < t \leq t_k \\
U_t^{\varepsilon,G} &= \varepsilon\sum_{k=1}^{[t/\varepsilon]} h(X_{t_{k-1}})\left[G'(\theta_k h(X_{t_{k-1}})\sqrt{\varepsilon}+\xi_k) - G'(\xi_k)\right].
\end{aligned} \tag{20.49}$$

Taking into account the mean value theorem and choosing appropriate random values $0 \leq \theta_k \leq 1$, we arrive at $Y_t^{\varepsilon,G} = \int_0^t u^{\varepsilon,G}(s)h(X_{s-\varepsilon})dL_s^\varepsilon + M_t^{\varepsilon,G} + U_t^{\varepsilon,G}$. We show now that for every $T > 0$

$$\begin{aligned}
&P\text{-}\lim_{\varepsilon\to 0}\sup_{t\leq T}\left|U_t^{\varepsilon,G}\right| = 0 \\
&P\text{-}\lim_{\varepsilon\to 0}\sup_{t\leq T}\left|\int_0^t [u^{\varepsilon,G}(s) - A_G]h(X_{s-\varepsilon})dL_s^\varepsilon\right| = 0.
\end{aligned} \tag{20.50}$$

In fact, by virtue of assumption (A-1) the function G' is Lipschitz continuous and therefore $\sup_{t\leq T}|U_t^{\varepsilon,G}| \leq T\sup_{t\leq T} h^2(X_t)\sqrt{\varepsilon} \to 0$, $\varepsilon \to 0$. To verify the validity of the second part of (20.50) note that

$$N_t^\varepsilon = \int_0^t [u^{\varepsilon,G}(s) - A_G] h(X_{s-\varepsilon}) dL_s^\varepsilon = \varepsilon \sum_{k=1}^{[t/\varepsilon]} h(X_{t_{k-1}}) \left(G'(\xi_k) - MG'(\xi_k)\right)$$

is a square integrable martingale with respect to the filtration $(\mathcal{F}_t^\varepsilon)_{t\geq 0}$ generated by $\{X_{t_{k-1}}, \xi_k; t_k \leq t, k \leq [t/\varepsilon]\}$. It is clear that the predictable quadratic variation of this martingale is defined as

$$\langle N^\varepsilon \rangle_t = \varepsilon^2 \sum_{k=1}^{[T/\varepsilon]} h^2(X_{t_{k-1}}) M \left(G'(\xi_1) - MG'(\xi_1)\right)^2.$$

Then, by Problem 1.9.2 in [214], $\sup_{t\leq T} |N_t^\varepsilon|$ converges to zero in probability since

$$\langle N^\varepsilon \rangle_t \leq \varepsilon T \sup_{t\leq T} h^2(X_t) MG'^2(\xi_1) \to 0, \quad \varepsilon \to 0.$$

For $s < \varepsilon$, letting $X_{s-\varepsilon} = X_0$, introduce now the process

$$\widetilde{Y}_t^{\varepsilon,G} = \int_0^t A_G h(X_{s-\varepsilon}) dL_s^\varepsilon + M_t^{\varepsilon,G}$$

and note that (20.50) implies: for every $T > 0$, $P\text{-}\lim_{\varepsilon\to 0} \sup_{t\leq T} |Y_t^{\varepsilon,G} - \widetilde{Y}_t^{\varepsilon,G}| = 0$. Obviously, the above also holds if we replace $\widetilde{Y}_t^{\varepsilon,G}$ by $\widehat{Y}_t^{\varepsilon,G} = \int_0^t A_G h(X_s) ds + M_t^{\varepsilon,G}$. Thus, by Theorem 4.1 of [19] (Chapter 1, Section 4), the statement of the theorem is fulfilled provided that

$$\mathcal{W}\text{-}\lim_{\varepsilon\to 0} (X_t, \widehat{Y}_t^{\varepsilon,G})_{t\geq 0} = (X_t, Y_t^G)_{t\geq 0}. \tag{20.51}$$

Under the assumptions made, $(\widehat{Y}_t^{\varepsilon,G})_{t\geq 0}$ is defined by a continuous mapping, in the local supremum topology, of $(X_t, M_t^{\varepsilon,G})_{t\geq 0}$, so that, taking into account the independence of (X_t) and $(M_t^{\varepsilon,G})$, only convergence

$$\mathcal{W}\text{-}\lim_{\varepsilon\to 0} (M_t^{\varepsilon,G})_{t\geq 0} = (B_G W_t)_{t\geq 0} \tag{20.52}$$

has to be established. To this end, we note that

$$M \left(M_t^{\varepsilon,G}\right)^2 \equiv L_t^\varepsilon B_G^2 \to t B_G^2 \tag{20.53}$$

and apply the Donsker theorem (see, for example, Theorems 9.1.1 and 9.1.2 in [214]) which states that (20.53) guarantees the weak convergence (in the Lindvall–Skorokhod topology)

$$(M_t^{\varepsilon,G})_{t\geq 0} \xrightarrow{\text{law}} (M_t)_{t\geq 0},$$

where $(M_t)_{t\geq 0}$ is a zero-mean continuous Gaussian martingale with variance $B_G^2 t$. Therefore, by the Doob–Lévy theorem (Theorem 4.1) $W_t = \frac{1}{B_G} M_t$ is the required Wiener process. □

20.2.2 Functional $\pi_t^G(y)$. Let the limiter G be chosen and the assumptions (A-1)–(A-4) fulfilled. Then, by Theorem 20.4, the pair $(X_t, Y_t^{\varepsilon,G})$ converges in distribution (in the Skorokhod–Lindvall topology) to (X_t, Y_t^G). We create now a functional $\pi_t^G(y)$, $t \in \mathbb{R}_+, y \in D$, as a $\mathcal{B} \otimes \mathcal{D}$-measurable function, being $\mathcal{D}_t$-measurable for fixed t, where $\mathcal{B}$ and $\mathcal{D}$ are the Borel σ-algebras on $\mathbb{R}_+$ and D respectively and $\mathcal{D}_t = \sigma\{y \in D : y_s, s \leq t\}$, such that for every $t \geq 0$, $\pi_t^G(Y^G) = M(f(X_t)|Y^G_{[0,t]})$ (P-a.s). Let us assume the pair $(X_t, Y_t^G)_{t\geq 0}$ is defined on the probability space $(\Omega, \mathcal{F}, P)$ where $(\widetilde{\Omega}, \widetilde{\mathcal{F}}, \widetilde{P})$ is its copy ($\widetilde{M}$ denotes the expectation with respect to $\widetilde{P}$, $\widetilde{X}_t$ is the copy of X_t). On $(\widetilde{\Omega} \times \Omega, \widetilde{\mathcal{F}} \otimes \mathcal{F}, \widetilde{P} \times P)$ define the new pair $(\widetilde{X}_t, Y_t^G)_{t\geq 0}$. Then, by the Kallianpur–Striebel formula [136],

$$M(f(X_t)|Y^G_{[0,t]}) = \frac{\widetilde{M} f(\widetilde{X}_t) \exp\left\{\left(\int_0^t h(\widetilde{X}_s)dY_s^G - \frac{B_G^2}{2}\int_0^t h^2(\widetilde{X}_s)ds\right)\right\}}{\widetilde{M} \exp\left\{\left(\int_0^t h(\widetilde{X}_s)dY_s^G - \frac{B_G^2}{2}\int_0^t h^2(\widetilde{X}_s)ds\right)\right\}}. \tag{20.54}$$

The right-hand side of (20.54) is presented via the Itô integral with respect to 'dY_s^G' and so it does not determine explicitly the required functional $\pi_t^G(y)$. To construct it, in what follows, we assume (A-2), (A-3), (A-5), and (A-6). By the Itô formula we find

$$\begin{aligned} h(\widetilde{X}_t)Y_t^G &= \int_0^t h(\widetilde{X}_s)dY_s^G + \int_0^t Y_s^G \mathcal{L}h(\widetilde{X}_s)ds \\ &\quad + \int_0^t Y_s^G h'(\widetilde{X}_s)b(\widetilde{X}_s)d\widetilde{V}_s, \end{aligned}$$

where $\widetilde{V}_t$, defined on $(\widetilde{\Omega}, \widetilde{\mathcal{F}}, \widetilde{P})$, is a copy of the Wiener process V_t and the operator $\mathcal{L}$ is defined in (A-5). Again applying the Itô formula we find

$$h(\widetilde{X}_t) = h(\widetilde{X}_0) + \int_0^t \mathcal{L}h(\widetilde{X}_s)ds + \int_0^t h'(\widetilde{X}_s)b(\widetilde{X}_s)d\widetilde{V}_s.$$

Due to the independence of the processes $\widetilde{X}_t, \widetilde{V}_t$ and Y_t^G, for each fixed trajectory of Y_t^G the Itô integral $\int_0^t [Y_t^G - Y_s^G]h'(\widetilde{X}_s)b(\widetilde{X}_s)d\widetilde{V}_s$ is well defined, so that we arrive at

$$\begin{aligned} \int_0^t h(\widetilde{X}_s)dY_s^G &= h(\widetilde{X}_0)Y_t^G + \int_0^t [Y_t^G - Y_s^G]\mathcal{L}h(s, \widetilde{X}_s)ds \\ &\quad + \int_0^t [Y_t^G - Y_s^G]h'(\widetilde{X}_s)b(\widetilde{X}_s)d\widetilde{V}_s. \end{aligned}$$

For $y \in D$, put

$$\begin{aligned} \Phi_t(\widetilde{X}, y) &= \exp\left(y_t h(\widetilde{X}_0) + \int_0^t \left[(y_t - y_s)\mathcal{L}h(\widetilde{X}_s) - \frac{B_G^2}{2}h^2(\widetilde{X}_s)\right] ds\right) \\ &\quad \times \exp\left(\int_0^t (y_t - y_s)h'(\widetilde{X}_s)b(\widetilde{X}_s)d\widetilde{V}_s\right) \end{aligned} \tag{20.55}$$

and then put

$$\pi_t^G(y) := \frac{\widetilde{M} f(\widetilde{X}_t)\Phi_t(\widetilde{X}, y)}{\widetilde{M}\Phi_t(\widetilde{X}, y)}. \tag{20.56}$$

It is clear that this functional satisfies the required properties.

Lemma 20.4. *Assume (A-1)–(A-6) and suppose $f(x)$ is a bounded continuous function. Then for fixed t, $\pi_t^G(\cdot)$ is continuous on D in the local supremum topology.*

PROOF. Let us show that both $\widetilde{M} f(\widetilde{X}_t)\Phi_t(\widetilde{X}, y)$ and $\widetilde{M}\Phi_t(\widetilde{X}, y)$ are continuous functions in y. Since $\Phi_t(\widetilde{X}, y)$ is uniformly continuous in probability at any point $y \in D$ and f is bounded, it suffices to check the uniform integrability of $\Phi_t(\widetilde{X}, y)$. To this end, we show that

$$\widetilde{M}\Phi_t^2(\widetilde{X}, y) < \infty, \quad y \in D. \tag{20.57}$$

Denoting $\beta(s) = 2(y_t - y_s)h'(\widetilde{X}_s)b(\widetilde{X}_s)$ and noticing that there exists a constant ℓ such that $|\beta(s)| \le \sup_{s \le t}|y_t - y_s|\ell$, we obtain

$$\widetilde{M} \exp\left\{\int_0^t \beta(s)d\widetilde{V}_s - \frac{1}{2}\int_0^t \beta^2(s)ds\right\} = 1.$$

Coupled with $\mathcal{D}^\lambda h \le C(\lambda, B_G^2)$, $\lambda = \sup_{s \le t}|y_t - y_s|$ (see (A-5)), we arrive at the following upper bound: there exists a constant $q(\lambda)$ such that

$$\widetilde{M}\Phi_t^2(\widetilde{X}, y) \le e^{q(\lambda)t}\widetilde{M} \exp\left\{2|y_t||h(\widetilde{X}_0|\right\},$$

that is (20.57) is implied by (A-6). □

20.2.3 Analysis of the Limit Model. For a bounded continuous function f and limiter G, Theorem 20.4 and Lemma 20.4 guarantee the asymptotic filtering equivalence of the prelimit model to the limit one

$$\lim_{\varepsilon \to 0} M\left(f(X_t) - \pi_t^G(Y^{\varepsilon,G}\right)^2 = M\left(f(X_t) - \pi_t^G(Y^G)\right)^2, \tag{20.58}$$

that is the asymptotic filtering accuracy depends on the chosen limiter G. For a fixed limiter G, the limit filtering model is characterized by two parameters A_G, B_G which define the 'signal-to-noise' ratio

$$SN_G = \frac{A_G^2}{B_G^2}. \tag{20.59}$$

The next lemma states that to a larger value of the 'signal-to-noise' ratio there corresponds a smaller value of the filtering error $\mathcal{E}_G(t) = M(f(X_t) - \pi_t^G(Y^G))^2$.

Lemma 20.5. *The following implication holds: for every $t > 0$,*

$$SN_{G^1} \leq SN_{G^2} \Longrightarrow \mathcal{E}_{G^1}(t) \geq \mathcal{E}_{G^2}(t).$$

PROOF. Let the limiter G be fixed. We define a new observable process

$$\widetilde{Y}_t^G = \frac{Y_t^G}{A_G}$$

and note that, since the σ-algebras $\sigma\{Y_s^G, s \leq t\}$ and $\sigma\{\widetilde{Y}_s^G, s \leq t\}$ coincide, the mean square filtering errors, corresponding to $\{Y_s^G, s \leq t\}$ and $\{\widetilde{Y}_s^G, s \leq t\}$, coincide as well. Therefore, to compare $\mathcal{E}_{G^1}(t)$ and $\mathcal{E}_{G^2}(t)$ one can use $\widetilde{Y}_t^{G^1}$ and $\widetilde{Y}_t^{G^2}$ as observation processes. From the description of the process Y_t^G (see Theorem 20.4), we find that

$$d\widetilde{Y}_t^G = h(X_t)dt + \frac{B_G}{A_G}dW_t, \quad \widetilde{Y}_0^G = 0. \tag{20.60}$$

To simplify the notation, put $\gamma' = B_{G^1}/A_{G^1}$ and $\gamma'' = B_{G^2}/A_{G^2}$. Since

$$(\gamma')^2 = \frac{1}{SN_{G'}} \quad \text{and} \quad (\gamma'')^2 = \frac{1}{SN_{G''}},$$

we have that $(\gamma')^2 \geq (\gamma'')^2$. Take $(\gamma')^2 > (\gamma'')^2$ and consider two observable processes:

$$\begin{aligned} dY_t'' &= h(X_t)dt + \gamma'' dW_t \\ dY_t' &= h(X_t)dt + \gamma'' dW_t + \sqrt{(\gamma')^2 - (\gamma'')^2}\,d\widetilde{W}_t, \end{aligned}$$

where $\widetilde{W}_t$ is a Wiener process independent of (X_t, W_t). It is clear that the diffusion parameter for the first model is $(\gamma'')^2$ while for the second it is $(\gamma')^2$. Denote

$$\begin{aligned} \mathcal{E}_{\gamma'}(t) &= M\left(f(X_t) \quad M\left(f(X_t)|\, Y'_{[0,t]}\right)\right)^2 \\ \mathcal{E}_{\gamma''}(t) &= M\left(f(X_t) - M\left(f(X_t)|\, Y''_{[0,t]}\right)\right)^2 \end{aligned}$$

and note that $\mathcal{E}_{\gamma'}(t) \equiv \mathcal{E}_{G^1}(t)$, $\mathcal{E}_{\gamma''}(t) \equiv \mathcal{E}_{G^2}(t)$. Therefore, it remains to check only the validity of the following implication:

$$\gamma' > \gamma'' \Longrightarrow \mathcal{E}_{\gamma'}(t) \geq \mathcal{E}_{\gamma''}(t). \tag{20.61}$$

Taking into account that $\sigma\{Y_s', \widetilde{W}_s, s \leq t\} \supseteq \sigma\{Y_s', s \leq t\}$, we obtain $\mathcal{E}_{\gamma'}(t) \geq \mathcal{E}(t)$, where $\mathcal{E}(t) = M(f(X_t) - M(f(X_t)|Y'_{[0,t]}, \widetilde{W}_{[0,t]}))^2$. Next, we prove

$$\mathcal{E}(t) \equiv \mathcal{E}_{\gamma''}(t). \tag{20.62}$$

In fact, noticing that $\sigma\{Y_s', \widetilde{W}_s, s \leq t\} = \sigma\{Y_s'', \widetilde{W}_s, s \leq t\}$, we conclude that

$$M\left(f(X_t)|\,Y'_{[0,t]},\widetilde{W}_{[0,t]}\right)=M\left(f(X_t)|\,Y''_{[0,t]},\widetilde{W}_{[0,t]}\right).$$

Moreover, the independence of the processes (X_t, Y''_t) and $(\widetilde{W}_t)$ implies the following chain of equalities: for all bounded random variables α and β, which are measurable with respect to σ-algebras $\sigma\{Y''_s, s \le t\}$ and $\sigma\{\widetilde{W}_s, s \le t\}$ respectively, and every bounded and measurable function $g(x)$,

$$\begin{aligned}
M\left(\alpha\beta M\left(g(X_t)|\,Y''_s,\widetilde{W}_{[0,t]}\right)\right) &= M\left(\alpha\beta g(X_t)\right)\\
&= M\left(\alpha g(X_t)\right)M(\beta)\\
&= M\left(\alpha M\left(g(X_t)|\,Y''_{[0,t]}\right)\right)M(\beta)\\
&= M\left(\alpha\beta M\left(g(X_t)|\,Y''_{[0,t]}\right)\right).
\end{aligned}$$

Hence, by virtue of the arbitrariness of α, β we obtain (P-a.s.)

$$M\left(f(X_t)|\,Y''_{[0,t]},\widetilde{W}_{[0,t]}\right)=M\left(f(X_t)|\,Y''_{[0,t]}\right).$$

Hence, (20.62) holds. □

Lemma 20.6. *Assume that the distribution of the random variable ξ_1 admits the twice continuously differentiable density $p(x)$ for which the Fisher information is finite:*

$$I_p=\int\frac{\left(p'(x)\right)^2}{p(x)}dx<\infty.$$

Then the limiter $G^\circ(x) = -p'(x)/p(x)$ has maximal 'signal-to-noise' ratio among all limiters G which are smooth functions with $\int G(x)p(x)dx = 0$ and $\int G^2(x)p(x)dx < \infty$:

$$\frac{A^2_{G^\circ}}{B^2_{G^\circ}}\ge\frac{A^2_G}{B^2_G}.$$

PROOF. Let G be an admissible limiter. Under the assumptions of the lemma

$$SN_G=\frac{\left(\int G'(x)p(x)dx\right)^2}{\int G^2(x)p(x)dx}.$$

Integrating by parts and applying the Cauchy–Schwarz inequality we obtain

$$\left(\int G'(x)p(x)dx\right)^2=\left(\int G(x)p'(x)dx\right)^2\le I_p\int G^2(x)p(x)dx,$$

that is $SN_G \le I_p$. On the other hand, $\int (G^\circ(x))^2 p(x)dx = I_p$ and, moreover, since

$$(G^\circ)'(x) = \frac{p''(x)p(x) - (p'(x))^2}{p^2(x)},$$

it holds that

$$SN_{G^\circ} = \frac{\left[\int \left(p''(x) - \frac{(p'(x))^2}{p(x)}\right) dx\right]^2}{\int (G^\circ)(x))^2 p(x) dx} = \frac{\left[\int \frac{(p'(x))^2}{p(x)} dx\right]^2}{\int (G^\circ(x))^2 p(x) dx} = I_p. \qquad \square$$

20.2.4 Asymptotically Optimal Filter. Assume now that the distribution of the random variable ξ_1 has a smooth positive density $p(x)$ such that the limiter $G^\circ(x) = -p'(x)/p(x)$ satisfies conditions (A-1) and (A-4). Then in the class of limiters G, satisfying these conditions, $G^\circ(x)$ is the 'best' in the sense that it guarantees the following lower bound for the mean square filtering error: for every continuous and bounded function f

$$\lim_{\varepsilon \to 0} M\left(f(X_t) - \pi_t^{G^\circ}(Y^{\varepsilon, G^\circ})\right)^2 \le \lim_{\varepsilon \to 0} M\left(f(X_t) - \pi_t^{G}(Y^{\varepsilon, G})\right)^2.$$

Let us compare now this lower bound with the asymptotically optimal one corresponding to the conditional expectation $\pi_t^\varepsilon(Y^\varepsilon) = M(f(X_t)\big|Y^\varepsilon_{[0,t]})$.

Let all random objects be defined on the probability space $(\Omega, \mathcal{F}, P)$ where $(\widetilde{\Omega}, \widetilde{\mathcal{F}}, \widetilde{P})$ is its copy ($\widetilde{M}$ is the expectation with respect to the measure $\widetilde{P}$; all random objects defined on $(\widetilde{\Omega}, \widetilde{\mathcal{F}}, \widetilde{P})$ are denoted by $\widetilde{\ }$). On $(\Omega \times \widetilde{\Omega}, \mathcal{F} \otimes \widetilde{\mathcal{F}}, P \times \widetilde{P})$, let us define random variables

$$\phi_k(\widetilde{X}, Y^\varepsilon) = \frac{p\left((\Delta Y^\varepsilon_{t_k}/\sqrt{\varepsilon}) - h(\widetilde{X}_{t_{k-1}})\sqrt{\varepsilon}\right)}{p\left(\Delta Y^\varepsilon_{t_k}/\sqrt{\varepsilon}\right)}, \quad k \ge 1,$$

where $\Delta Y^\varepsilon_{t_k} = Y^\varepsilon_{t_k} - Y^\varepsilon_{t_{k-1}}$, and

$$\Phi_t^\varepsilon(\widetilde{X}, Y^\varepsilon) = \prod_{k: t_k \le t} \phi_k(\widetilde{X}, Y^\varepsilon). \tag{20.63}$$

By the Bayes formula

$$\pi_t^\varepsilon(Y^\varepsilon) = \frac{\widetilde{M} f(\widetilde{X}_t)\Phi_t^\varepsilon(\widetilde{X}, Y^\varepsilon)}{\widetilde{M}\Phi_t^\varepsilon(\widetilde{X}, Y^\varepsilon)}. \tag{20.64}$$

The same type of formula is applied to generate the filtering estimate $\pi_t^{G^\circ}(Y^{\varepsilon, G^\circ})$. In fact, the explicit formula for the conditional expectation $\pi_t^{G^\circ}(Y^{G^\circ})$ is defined as (see (20.56)):

$$\begin{aligned}
\pi_t^{G^\circ}(Y^{G^\circ}) &= \frac{\widetilde{M} f(\widetilde{X}_t)\Phi_t(\widetilde{X}, Y^{G^\circ})}{\widetilde{M}\Phi_t(\widetilde{X}, Y^{G^\circ})}, \\
\Phi_t(\widetilde{X}, Y^{G^\circ}) &= \exp\left(\int_0^t h(\widetilde{X}_s) dY_s^{G^\circ} - \frac{I_p}{2}\int_0^t h^2(\widetilde{X}_s) ds\right).
\end{aligned}$$

Since the random process $(Y_t^{\varepsilon,G^\circ})$ has piecewise-constant right continuous trajectories of bounded variation for every finite interval, the Itô integral with respect to '$dY_t^{\varepsilon,G^\circ}$' coincides with the Stieltjes one with respect to '$dY_t^{\varepsilon,G^\circ}$'. This property allows us to conclude that

$$\begin{aligned}\pi_t^{G^\circ}(Y^{\varepsilon,G^\circ}) &= \frac{\widetilde{M}f(\widetilde{X}_t)\Phi_t(\widetilde{X},Y^{\varepsilon,G^\circ})}{\widetilde{M}\Phi_t(\widetilde{X},Y^{\varepsilon,G^\circ})},\\ \Phi_t(\widetilde{X},Y^{\varepsilon,G^\circ}) &= \exp\left(\int_0^t h(\widetilde{X}_s)dY_s^{\varepsilon,G^\circ} - \frac{I_p}{2}\int_0^t h^2(\widetilde{X}_s)ds\right). \qquad (20.65)\end{aligned}$$

Using (20.64) and (20.65), we show in the next theorem that the filtering estimate $\pi_t^{G^\circ}(Y^{\varepsilon,G^\circ})$ is asymptotically optimal. For sake of simplicity of the proof, we use in this theorem slightly restricted conditions.

Theorem 20.5. *Assume $I_p < \infty$ and (A-1)–(A-5) with the limiter G°. Assume also*

1. *$p(x)$ is three times continuously differentiable, and $p'(x)/p(x)$ and $p''(x)/p(x)$ are continuous and bounded;*
2. *$p'''(x+y)/p(x)$ is continuous and for small y it is bounded in x.*

If h is a bounded function, then for any bounded continuous function f and any fixed $t > 0$

$$\lim_{\varepsilon\to 0} M\left(f(X_t) - \pi_t^\varepsilon(Y^\varepsilon)\right)^2 = \lim_{\varepsilon\to 0} M\left(f(X_t) - \pi_t^{G^\circ}(Y^{\varepsilon,G^\circ})\right)^2.$$

PROOF. It is clear that the statement of the theorem is equivalent to

$$\lim_{\varepsilon\to 0} M\left(\pi_t^\varepsilon(Y^\varepsilon) - \pi_t^{G^\circ}(Y^{\varepsilon,G^\circ})\right)^2 = 0.$$

Since the function f is bounded we choose both $\pi_t^\varepsilon(Y^\varepsilon)$ and $(\pi_t^{G^\circ}(Y^{\varepsilon,G^\circ}))^2$ bounded as well. Therefore the required convergence is implied by

$$P\text{-}\lim_{\varepsilon\to 0}\left[\pi_t^\varepsilon(Y^\varepsilon) - \pi_t^{G^\circ}(Y^{\varepsilon,G^\circ})\right] = 0.$$

It is clear that this convergence is provided by

$$P\text{-}\lim_{\varepsilon\to 0}\widetilde{M}\left|\Phi_t(\widetilde{X},Y^{\varepsilon,G^\circ}) - \Phi_t^\varepsilon(\widetilde{X},Y^\varepsilon)\right|. \qquad (20.66)$$

We prove the validity of (20.66) in two steps.

Step 1. Since $\Delta Y_{t_k}^{\varepsilon,G^\circ} := \sqrt{\varepsilon}G^\circ(\Delta Y_{t_k}^\varepsilon/\sqrt{\varepsilon})$, by the mean value theorem, with appropriate random variables $\theta_k^1, \theta_k^2 \in [0,1]$, we obtain

$$
\begin{aligned}
\psi_k(\widetilde{X}, Y^\varepsilon) &= 1 + h(\widetilde{X}_{t_{k-1}})\Delta Y_{t_k}^{\varepsilon,G^\circ} + \frac{p''}{2p}(\xi_k)\, h^2(\widetilde{X}_{t_{k-1}})\varepsilon \\
&\quad + \left(\frac{p''}{2p}\right)' (\xi_k + \theta_k^1 h(X_{t_{k-1}}\sqrt{\varepsilon})\, h^3(\widetilde{X}_{t_{k-1}})\varepsilon^{3/2} \\
&\quad - \frac{p'''\left(\xi_k + h(X_{t_{k-1}})\sqrt{\varepsilon} - \theta_k^2 h(\widetilde{X}_{t_{k-1}})\sqrt{\varepsilon}\right)}{6p\left(\xi_k + h(X_{t_{k-1}})\sqrt{\varepsilon}\right)} h^3(\widetilde{X}_{t_{k-1}})\varepsilon^{3/2}. \\
&= 1 + h(\widetilde{X}_{t_{k-1}})\Delta Y_{t_k}^{\varepsilon,G^\circ} + o_k.
\end{aligned}
$$

Using the above and (20.63), we arrive at the multiplicative decomposition:

$$
\Phi_t^\varepsilon(\widetilde{X}, Y^\varepsilon) = U_t^\varepsilon \Phi_t(\widetilde{X}, Y^{\varepsilon,G^\circ}),
$$

where

$$
\begin{aligned}
U_t^\varepsilon &= \exp\Bigg(\sum_{k:t_k \le t} \ln[1 + h(\widetilde{X}_{t_{k-1}})\Delta Y_{t_k}^{\varepsilon,G^\circ} + o_k] \\
&\quad - \int_0^t h(\widetilde{X}_s) dY_s^{\varepsilon,G^\circ} + \frac{I_p}{2}\int_0^t h^2(\widetilde{X}_s)\Bigg).
\end{aligned}
$$

Hence $|\Phi_t^\varepsilon(\widetilde{X}, Y^\varepsilon) - \Phi_t(\widetilde{X}, Y^{\varepsilon,G^\circ})| \le \Phi_t(\widetilde{X}, Y^{\varepsilon,G^\circ})|1 - V_t^\varepsilon|$.

Step 2. In the proof of Lemma 20.4 it has been shown that on the set $\{\sup_{s\le t}|Y_s^{\varepsilon,G^\circ}| \le C\}$ the value $\Phi_t(\widetilde{X}, Y^{\varepsilon,G^\circ})$ is bounded by a constant depending on C. We use this to prove (20.66). In fact, (20.66) is implied by

$$
\lim_{C\to\infty}\lim_{\varepsilon\to 0} P\left(\sup_{s\le t}\left|Y_s^{\varepsilon,G^\circ}\right| > C\right) = 0
$$

$$
\lim_{\varepsilon\to 0} P\left(|1 - V_t^\varepsilon| \ge \zeta, \sup_{s\le t}\left|Y_s^{\varepsilon,G^\circ}\right| \le C\right) = 0, \quad \forall \zeta > 0, C > 0. \quad (20.67)
$$

The first part of (20.67) is nothing but one of the necessary conditions (see [19]) for the weak convergence

$$
\mathcal{W}\text{-}\lim_{\varepsilon\to 0}(Y_t^{\varepsilon,G^\circ})_{t\ge 0} = (Y_t^{G^\circ})_{t\ge 0}
$$

which has been proved in Theorem 20.4.

To check the validity of the second part of (20.67), let us denote

$$
\begin{aligned}
\Lambda_t^\varepsilon(\omega, \widetilde{\omega}) &= \exp\Bigg(\sum_{k:t_k \le t} \ln[1 + h(\widetilde{X}_{t_{k-1}})\Delta Y_{t_k}^{\varepsilon,G^\circ} + o_k] \\
&\quad - \int_0^t h(\widetilde{X}_s) dY_s^{\varepsilon,G^\circ} + \frac{I_p}{2}\int_0^t h^2(\widetilde{X}_s) ds\Bigg)
\end{aligned}
$$

and note that $V_t^\varepsilon(\omega) = \widetilde{M}\Lambda_t^\varepsilon(\omega,\widetilde{\omega})$. Applying now arguments of the same type as those used in proving Theorem 20.4, it can be shown that $P\text{-}\lim_{\varepsilon\to 0}\Lambda_t^\varepsilon(\omega, \widetilde{\omega}) = 1$ and, therefore, for large n, $P\text{-}\lim_{\varepsilon\to 0}\widetilde{M}\Lambda_t^\varepsilon(\omega,\widetilde{\omega})I(\Lambda_t^\varepsilon(\omega,\widetilde{\omega}) \le n) = 1$.

Thus, the required conclusion holds by virtue of the uniform integrability which implies, under the assumptions of the theorem, that for every $C > 0$

$$P\text{-}\lim_{n\to\infty}\sup_{\varepsilon\le 1} I(\sup_{s\le t}|Y_s^{\varepsilon,G^\circ}(\omega)| \le C)\widetilde{M}\left(I(\Lambda_t^\varepsilon(\omega,\widetilde{\omega}) > n)\Lambda_t^\varepsilon(\omega,\widetilde{\omega})\right) = 0. \quad \square$$

Notes and References

20.1. The asymptotic filtering equivalence, under weak convergence of the distributions of signals and observations, was established in Kushner [175], Kushner and Runggaldier [176], Liptser and Runggaldier [204]. The fact that the asymptotic filtering equivalence does not imply the asymptotic filtering optimality was emphasized by Goggin [80]. The proof of the asymptotic optimality is given in Kleptsina, Liptser and Serebrovski [156]. The general approach to the convergence in the total variation norm for distributions of random processes can be found in Kabanov, Liptser, and Shiryaev [118, 120]; see also Jacod and Shiryaev [106].

20.2. Theorem 20.4 and Lemma 20.4 were proved in Liptser and Zeitouni [217]. Results similar to Lemma 20.4, especially for a continuous function $y \in D$, are well known from Rozovskii [265], Chaleyat-Maurel and Michel [36], and Picard [254]. Lemma 20.5 was proved in Zeitouni and Zakai [332] and in [217]. A statement similar to Lemma 20.6 is well known from Huber [92]. In the filtering setting under the diffusion approximation with collared noise a result of this type can be found in Liptser and Lototsky [202].

20.2.4. Theorem 20.5 is related to results of Goggin [80]. The general approach to the approximation of the conditional expectation (asymptotic optimality) can also be found in Goggin [81, 82].

Bibliography

Note. A consistent transcription of author names has been selected for this book. Other variants are also possible in some cases, for example, Hasminskii (for Khasminskii), Ventzel (for Wentzell).

1. Albert, A. (1972): Regression and the Moore–Penrose Pseudoinverse. Academic, New York London
2. Albert, A. and Sittler, R.W. (1965): A method of computing least squares estimators that keep up with the data. SIAM J. Control Optimization, **3**, 384–417
3. Anderson, T. (1958): Introduction to Multivariate Analysis. Wiley, New York
4. Anulova, A., Veretennikov, A., Krylov, N., Liptser, R.S. and Shiryaev, A.N. (1998): Stochastic Calculus. Encyclopedia of Mathematical Sciences. vol 45, Springer-Verlag, Berlin Heidelberg New York
5. Aoki, M. (1958): Optimization of Stochastic Systems. Academic, New York
6. Arato, M. (1968): Calculation of confidence bounds for the damping parameter of the complex stationary Gaussian Markov process. Teor. Veroyatn. Primen, **13**, 1, 326–33
7. Arato, M., Kolmogorov, A.N. and Sinai, Ya.G. (1962): On parameter estimation of a complex stationary Gaussian Markov process. Dokl. Akad. Nauk SSSR, **146**, 4, 747–50
8. Astrom, K.I. (1965): Optimal control of Markov processes with incomplete state information. J. Math. Anal. Appl. **10**, 174–205
9. Atar, R. and Zeitouni, O. (1997): Lyapunov exponents for finite-state nonlinear filtering. SIAM J. Control Optimization, **35**, 1, 36–55
10. Atar, R. and Zeitouni, O. (1997): Exponential stability for nonlinear filtering. Ann. Inst. Henri Poincaré, Probab. Stat., **33** (1997), 697–725
11. Balakrishnan, A.V. (1971): Stochastic Differential Systems. Lecture Notes, Department of System Science, UCLA
12. Balakrishnan, A.V. (1972): A martingale approach to linear recursive state estimation. SIAM J. Control Optimization, **10**, 754–66
13. Baras, J.S. (1991): Real-time architecture for the Zakai equations and applications. In: Stochastic Analysis. E. Mayer-Wolf *et al.* (eds), Academic, Boston, MA
14. Beckenbach, A. and Bellman R. (1961): Inequalities. Springer-Verlag, Berlin Heidelberg New York
15. Bellman, R. and Cook, K.L. (1963): Differential-Difference Equations. Academic, New York
16. Benesh, V.E. (1981): Exact finite-dimensional filters for certain diffusions with nonlinear drift. Stochastics, 5, 65–92
17. Bensoussan, A. (1971): Filtrage Optimal des Systèmes Linéaires. Dunod, Paris
18. Bensoussan, A., Glowinski, R. and Rascanu, A. (1990): Approximation of the Zakai equation by the splitting-up method. SIAM J. Control Optimization, **28**, 1420–31
19. Billingsley, P. (1979): Convergence of Probability Measures. Wiley, New York

20. Blackwell, D. and Dubins L. (1962): Merging of options with increasing information. Ann. Math. Stat., **33**, 882–6
21. Blumental, R.M. and Getoor, R.K. (1968): Markov Processes and Potential Theory. Academic, New York
22. Boel, R., Varaiya, P. and Wong E. (1975): Martingales on jump processes, III. Representation results. SIAM J. Control Optimization, **13**, 5, 999–1060
23. Bogolubov, N. and Mitropolskii, A. (1961): Asymptotics Methods in Nonlinear Mechanics. Gordon and Breach, New York
24. Bolshakov, I.A. and Repin, V.G. (1961): Problems of nonlinear filtering. Avtom. Telemekh., **22**, 4, 466–78.
25. Breiman, L. (1968): Probability. Addison-Wesley, Reading, MA
26. Brémaud, P. (1972): Martingale Approach to Point Processes. ERL Memo **M 345**, University of California, Berkeley
27. Brémaud, P. (1975): An extension of Watanabe's theorem of characterization of a Poisson process over the positive real half time. J. Appl. Probab., **12**, 2, 369–99
28. Brémaud, P. (1975): Estimation de l'état d'une file d'attente et du temps de panne d'une machine par la méthode des semi-martingales. Adv. Appl. Probab., 7, 845–63
29. Brémaud, P. La méthode des semi-martingales et filtrage quand l'observation est un processus ponctuel marqué. Preprint.
30. Bucy, R.S. (1965): Nonlinear filtering theory. IEEE Trans. Autom. Control, **AC-10**, 198–204
31. Bucy, R.S. and Joseph, P.D. (1968): Filtering with Application to Guidance. Interscience, New York
32. Budhiraja, A. and Kallianpur, G. (1996): Approximations to the solution of the Zakai equation using multiple Wiener and Stratonovich integral expansions. Stochastics Stochastics Rep., **56**, 3–4, 271–315
33. Budhiraja, A. and Ocone D. (1997): Exponential stability of discrete time filters without signal ergodicity. Syst. Control Lett., **30**, 4, 185–93
34. Cameron, R.H. and Martin, W.T. (1945): Transformation of Wiener integrals under a general class of linear transformations. Trans. Am. Math. Soc., **58**, 184–219
35. Cameron, R.H. and Martin, W.T. (1949): Transformation of Wiener integrals by nonlinear transformation. Trans. Am. Math. Soc., **66**, 253–83
36. Chaleyat-Maurel, M. and Michel, D. (1986): Une propriété de continuité en filtrage non linéaire..Stochastics, **19**, 11–40
37. Chou, C.S. and Meyer, P.A. (1974): La représentation des martingales relatives à un processus ponctuel discret. C. R. Acad. Sci., Paris, Sér. A-B, **278**, 1561–3
38. Chow, P.L., Khasminskii, R.Z. and Liptser, R.S. (1997): Tracking of a signal and its derivatives in Gaussian white noise. Stochastic Processes Appl., **69**, 2, 259–73
39. Chow, Y.S., Robbins, H. and Siegmund, D. (1971): Great Expectation: The Theory of Optimal Stopping. Houghton Mifflin, Boston, MA
40. Clark, I.M. (1970): The representation of functionals of Brownian motion by stochastic integrals. Ann. Math. Stat., **41**, 4, 1282–95
41. Courrège, Ph. (1962–3): Intégrales stochastiques et martingales de carré intégrable. Sémin. Brelot–Choquet–Deny (7th year)
42. Courrège, Ph. (1963): Intégrales stochastiques associées à une martingale de carré intégrable. C. R. Acad. Sci., Paris, **256**, 867–70
43. Cramer, H. (1940): On the theory of stationary random process. Ann. Math., **41**, 2, 215–30.

44. Cramer, H. and Leadbetter M. (1967): Stationary and Related Stochastic Processes. Wiley, New York
45. Dashevsky, M.L. (1968): The method of semi-invariants in the problems of nonlinear filtering of Markov processes. Avtom. Telemekh., 7, 24–32
46. Dashevsky, M.L. and Liptser, R.S. (1967): Application of conditional semi-invariants in the problems of nonlinear filtering of Markov processes. Avtom. Telemekh., 6, 63–74.
47. Daum, F.E. (1988): New exact nonlinear filters. In: Bayesian Analysis of Time Series and Dynamic Models, J.C. Spall (ed), Marcel Dekker, New York, 199–226
48. Davis, M.H.A. (1974): The Representation of Martingales of Jump Processes. Research Report **74/78**, Imperial College, London
49. Dellacherie, C. (1972): Capacités et Processus Stochastiques. Springer-Verlag, Berlin Heidelberg New York
50. Dellacherie, C. (1973): Intégrales stochastiques par rapport aux processus de Wiener et de Poisson. Séminaire de Probabilités **8**. In: Lecture Notes in Mathematics 381, Springer-Verlag, Berlin Heidelberg New York
51. Dellacherie, C. and Meyer, P.A. (1980): Probabilités et Potentiel. Théorie des Martingales. Hermann, Paris
52. Delyon, B. and Zeitouni, O. (1991): Lyapunov exponents for filtering problem. In: Applied Stochastic Analysis, M.H.A Davis and R.J. Elliot (eds), Gordon and Breach, London, 411–521
53. Dermoune, A. and Kutoyants, Yu.A. (1995): Expansion of distribution of maximum likelihood estimate for unspecified diffusion type observation. Stochastics Stochastics Rep., **52**, 121–45
54. Djashkov, A.G. and Pinsker, M.S. (1971): On optimal linear transmission through a memoryless Gaussian channel with complete feedback. Probl. Peredachi Inf., **7**, 2, 38–46
55. Djuge, D. (1972): Theoretical and Applied Statistics. Nauka, Moscow
56. Doléans-Dade, C. (1970): Quelques applications de la formule de changement de variables pour les semimartingales locales. Z. Wahrsch. Verw. Gebiete, **16**, 181–94
57. Doob, J.L. (1954): Stochastic Processes. Wiley, New York
58. Dynkin, E.B. (1965): Markov Processes. Springer-Verlag, Berlin Göttingen Heidelberg
59. Elliott, R. (1982): Stochastic Calculus and Applications. Springer-Verlag, New York Berlin Heidelberg
60. Elliott, R.J., Aggoun, L. and Moore, J.B. (1995): Hidden Markov Models. Springer-Verlag, New York Berlin Heidelberg
61. Feldbaum, A.A. (1965): Optimal Control Systems. Academic, New York
62. Ferguson, T.S. (1967): Mathematical Statistics. Academic, New York
63. Florchinger, P. and Le Gland, F. (1990): Time-quantization of the Zakai equation for diffusion processes observed in correlated noise. In: 9th Conference on Analysis and Optimization of Systems, A. Bensoussan and J.L. Lions (eds), Lecture Notes in Control and Information Sciences 144, Springer-Verlag, Berlin Heidelberg New York
64. Friedman, A. (1969): Partial Differential Equations. Holt, Rinehart and Winston, New York
65. Frost P. and Kailath, T. (1971): An innovation approach to least-squares estimation, Part III. IEEE Trans. Autom. Control, **AC-16**, 217–26
66. Fujisaki, M., Kallianpur, G. and Kunita, H. (1972): Stochastic differential equations for the nonlinear filtering problem. Osaka J. Math., **9**, 1, 19–40

67. Galtchouk, L.I. (1969): On representation of jump processes. Soviet–Japanese Symposium on Probability (Khabarovsk), **2**, 52–61
68. Galtchouk, L.I. (1970): Filtering of jump Markov processes. Usp. Mat. Nauk, **25**, 5, 237–8
69. Gantmacher, F.R. (1967): Theory of Matrices. Nauka, Moscow
70. Gikhman, I.I. (1950): On the theory of stochastic differential equations of random processes. Ukr. Math. J., **2**, 3, 45–69
71. Gikhman, I.I. and Dorogovtsev, A.Ya. (1965): On stability of solutions of stochastic differential equations. Ukr. Math. J., **17**, 6, 3–21
72. Gikhman, I.I. and Skorokhod, A.V. (1966): On densities of probability measures on function spaces. Usp. Mat. Nauk, **21**, 83–152.
73. Gikhman, I.I. and Skorokhod, A.V. (1969): Introduction to the Theory of Random Processes. Saunders, Philadelphia
74. Gikhman, I.I. and Skorokhod, A.V. (1974): Theory of Random Processes. I. Springer-Verlag, Berlin Heidelberg New York
75. Gikhman, I.I. and Skorokhod, A.V. (1980): Stochastic Differential Equations. Springer-Verlag, Berlin Heidelberg New York
76. Girsanov, I.V. (1960): On transformation of one class of random processes with the help of absolutely continuous change of the measure. Teor. Veroyatn. Primen., **5**, 1, 314–30
77. Glonti, O.A. (1969): Sequential filtering and interpolation of Markov chain components. Litov. Mat. Sb., **9**, 2, 263–79
78. Glonti, O.A. (1969): Extrapolation of Markov chain components. Litov. Mat. Sb., **9**, 4, 741–54
79. Glonti, O.A. (1970): Sequential filtering of Markov chain components with singular diffusion matrices. Teor. Veroyatn. Primen., **15**, 4, 736–40
80. Goggin, E.M. (1992): Convergence of filters with applications to the Kalman–Bucy case. IEEE Trans. Inf. Theory., **38**, 1091–100
81. Goggin, E.M. (1994): Convergence in distribution of conditional expectations. Ann. Probab., **22**, 2, 1097–114
82. Goggin, E.M. (1997): An L^1 approximation for conditional expectations. Stochastics Stochastics Rep., **60**, 85–106
83. Golubev, G. and Khasminskii, R. (1998): Asymptotically optimal filtering for hidden Markov models. Math. Methods Stat., **7**, 2, 192–209
84. Grigelionis, B. (1972): On stochastic equations of nonlinear filtering of random processes. Litov. Mat. Sb., **12**, 4, 37–51
85. Grigelionis, B. (1973): On the structure of densities of measures corresponding to random processes. Litov. Mat. Sb., **13**, 1, 48–52
86. Grigelionis, B. (1974): On nonlinear filtering theory and absolute continuity of measures corresponding to stochastic processes. In: Lecture Notes in Mathematics 330, Springer-Verlag, Berlin Heidelberg New York
87. Grigelionis, B. (1974): On mutual information for locally infinitely divisible random process. Litov. Mat. Sb., **14**, 1, 5–11
88. Grigelionis, B. (1975): Random point processes and martingales. Litov. Mat. Sb., **15**, 4
89. Gulko, F.B. and Novoseltseva, Zh.A. (1966): Solution of nonstationary problems of filtering and prediction. Avtom. Telemekh., 4, 122–41
90. Haddad, Z.S. and Simanca, S.R. (1995): Filtering image records using wavelets and the Zakai equation. IEEE Trans. Pattern Anal. Mach. Intell., **17**, 11, 1069–78
91. Hitsuda, M. (1968): Representation of Gaussian processes equivalent to Wiener processes. Osaka J. Math., 5, 299–312
92. Huber, P.J. (1981): Robust Statistics. Wiley, New York

93. Ibragimov, I.A. and Khasminskii, R.Z. (1972): Information inequalities and supereffective estimates. Dokl. Akad. Nauk SSSR, **204**, 6
94. Ibragimov, I.A. and Rozanov, Yu.A. (1978): Gaussian Processes. Springer-Verlag, New York Berlin Heidelberg
95. Ihara, S. (1974): Coding theory in white Gaussian channel with feedback. J. Multivariate Anal. **4**, 74–87
96. Itô, K. (1944): Stochastic integrals. Proc. Imp. Acad. Tokyo, **20**, 519–24
97. Itô, K. (1959): On a formula concerning stochastic differentials. Nagoya Mat. J., **3**, 5, 55–65
98. Itô, K. (1996): Approximation of Zakai equation for nonlinear filtering. SIAM J. Control Optimization, **34**, 620–34
99. Itô, K. and McKean, H. (1965): Diffusion Processes and their Sample Paths. Springer-Verlag, New York Berlin Heidelberg
100. Itô, K. and Nisio, M. (1964): On stationary solutions of stochastic differential equations. J. Math. Kyoto Univ., **4**, 1, 1–79.
101. Jacod, J. (1975): Multivariate point processes: predictable projection, Radon–Nikodym derivatives, representation martingales. Z. Wahrsch. Verw. Gebiete, **31**, 235–53
102. Jacod, J. (1976): Un théorème de représentation pour les martingales discontinues. Z. Wahrsch. Verw. Gebiete, **34**, 225–44
103. Jacod, J. (1976): A General Theorem of Representation for Martingales. Dept. Math. Inform., *Université de Rennes*
104. Jacod, J. (1979): Calcul Stochastique et Problèmes des Martingales, Lecture Notes in Mathematics 714, Springer-Verlag, Berlin Heidelberg New York
105. Jacod, J. and Memin J. (1975): Caractéristiques Locales et Conditions de Continuité Absolue Pour les Semi-Martingales. Dept. Math. Inform., *Université de Rennes*
106. Jacod, J. and Shiryaev, A.N. (1987): Limit Theorems for Stochastic Processes. Springer-Verlag, Berlin Heidelberg New York
107. Jazwinski, A.H. (1970): Stochastic Processes and Filtering Theory. Academic, New York
108. Kabanov, Yu.M. (1973): Representation of the functionals of Wiener and Poisson processes as stochastic integrals. Teor. Veroyatn. Primen., **18**, 2, 376–80
109. Kabanov, Yu.M. (1974): Integral representation of the functionals of processes with independent increments. Teor. Veroyatn. Primen., **19**, 4, 889–93
110. Kabanov, Yu.M. (1978): Channel capacity of a Poisson type channel. Teor. Veroyatn. Primen., **23**, 1
111. Kabanov, Yu.M. (1982): On the existence of an optimal solution in a control problem for a counting process. Mat. Sb., **119**, 3
112. Kabanov, Yu.M. and Di Masi, G.B. (1997): On sensitive probabilistic criteria for LQG problems with infinite horizon. In: Statistics and Control of Random Process, Proc. Steklov Mathematical Institute Seminar
113. Kabanov, Yu.M., Liptser, R.S. and Shiryaev, A.N. (1975): Martingale methods in the point process theory. Tr. Shcoly–Seminara po Teorii Sluchainykh Protsessov, II. (Druskeninkai, 1974), 296–353
114. Kabanov, Yu.M., Liptser, R.S. and Shiryaev, A.N. (1977): 'Predictable' criteria for absolute continuity and singularity of probability measures (the continuous time case). Sov. Math., Dokl., **18**, 6, 1515–8
115. Kabanov, Yu.M., Liptser R.S. and Shiryaev, A.N. (1977): On the question of absolute continuity and singularity of probability measures. Math. USSR Sb., **33**, 2, 203–21

116. Kabanov, Yu.M., Liptser, R.S. and Shiryaev, A.N. (1979): Absolute continuity and singularity of locally absolutely continuous probability distributions, I. Math. USSR Sb., **35**, 5, 631–80
117. Kabanov, Yu.M., Liptser, R.S. and Shiryaev, A.N. (1980): Absolute continuity and singularity of locally absolutely continuous probability distributions, II. Math. USSR Sb., **36**, 1, 31–58
118. Kabanov, Yu.M., Liptser, R.S. and Shiryaev, A.N. (1983): On convergence in variation of the distributions of multivariate point processes. Z. Wahrsch. Verw. Gebiete, **63**, 475–85
119. Kabanov, Yu.M., Liptser, R.S. and Shiryaev, A.N. (1983): Strong and weak convergence of distributions of counting processes. Theor. Probab. Appl., **2**, 288–319
120. Kabanov, Yu.M., Liptser, R.S. and Shiryaev, A.N. (1986): On the variation distance for probability measures defined on filtered space. Probab. Theory Relat. Fields, **71**, 19–35
121. Kabanov, Yu.M. and Pergamenshchikov, S.M. (1991): Optimal control of singularly perturbed linear systems. Stochastics Stochastics Rep., **32**, 109–35
122. Kabanov, Yu.M. and Pergamenshchikov, S.M. (1991): On optimal control of singularly perturbed stochastic differential equations. In: Modelling, Estimation and Control of Systems with Uncertainty. G.B. Di Masi *et al.* (eds), Birkhäuser
123. Kabanov, Yu.M. and Runggaldier, W. (1996): On control of two-scale stochastic systems with linear dynamics in fast variables. Math. Control Signals Syst., 9, 107–22
124. Kadota, T.T. (1970): Nonsingular detection and likelihood ratio for random signals in white Gaussian noise. IEEE Trans. Inf. Theory, **IT-16**, 291–8
125. Kadota, T.T. and Shepp L.A. (1970): Conditions for the absolute continuity between a certain pair of probability measures. Z. Wahrsch. Verw. Gebiete, **16**, 3, 250–60
126. Kadota, T.T., Zakai, M. and Ziv, I. (1971): Mutual information of the white Gaussian channel with and without feedback. IEEE Trans. Inf. Theory, **IT-17**, 4, 368–71
127. Kagan, A.M., Linnik, Yu.V. and Rao, C.R. (1973): Characterization Problems in Mathematical Statistics. Wiley, New York
128. Kailath, T. (1968): An innovation approach to least-squares estimation, Parts I, II. IEEE Trans. Autom. Control, **AC-13**, 646–60.
129. Kailath, T. (1970): An innovation approach to detection and estimation theory. Proc. IEEE, **58**, 680–95
130. Kailath, T. (1971): The structure of Radon–Nikodym derivatives with respect to Wiener and related measures. Ann. Math. Stat., **42**, 1054–67
131. Kailath, T. and Greesy, R. (1971): An innovation approach to least-squares estimation, Part IV. IEEE Trans. Autom. Control, **AC-16**, 720–27
132. Kailath, T. and Zakai M. (1971): Absolute continuity and Radon–Nikodym derivatives for certain measures relative to Wiener measure. Ann. Math. Stat., **42**, 1, 130–40
133. Kalachev, M.G. (1972): Analytical calculation of the Kalman–Bucy filter in a multivariate problem of filtering. Avtom. Telemekh., 1, 46–50
134. Kalachev, M.G. and Petrovsky, A.M. (1972): Repeated differentiation of a signal with bounded spectrum. Avtom. Telemekh., 3, 28–34
135. Kallianpur, G. (1980): Stochastic Filtering Theory. Springer-Verlag, New York Berlin Heidelberg

136. Kallianpur, G. and Striebel, C. (1968): Estimation of stochastic systems: arbitrary system process with additive noise observation errors. Ann. Math. Stat., **39**, 785–801
137. Kallianpur, G. and Striebel, C. (1969): Stochastic differential equations occurring in the estimation of continuous parameter stochastic processes. Teor. Veroyatn. Primen., **14**, 4, 597–622
138. Kalman, R.E. (1960): A new approach to linear filtering and prediction problems. J. Basic Eng., 1, 35–45
139. Kalman, R.E. (1960): Contribution in the theory of optimal control. Bol. Soc. Mat. Mex., **5**, 102–19
140. Kalman, R.E. and Bucy, R.S. (1961): New results in linear filtering and prediction theory. Trans. ASME, **83D**, 95–100
141. Kalman, R.E., Falb, P. and Arbib, M. (1969): Topics in Mathematical Systems Theory. McGraw-Hill, New York
142. Karatzas, I. and Shreve, S.E. (1991): Brownian Motion and Stochastic Calculus. Springer-Verlag, New York Berlin Heidelberg
143. Kazamaki, N. (1977): On a problem of Girsanov. Tôhoku Math. J., **29**, 597–600
144. Kerstan, J., Matthes, K. and Mecke, J. (1974): Unbegrenzte teilbare Punktprozesse. Akademie-Verlag, Berlin
145. Khadgijev, D.I. (1976): On estimation of random processes from the observation of point processes. Usp. Mat. Nauk, **31**, 2, 233–6
146. Khadgijev, D.I. Filtering of semimartingales with observation of point processes. Preprint. Steklov Institute, Moscow
147. Khasminskii, R.Z. (1969): Stability of Differential Equation Systems Given Random Disturbances of their Parameters. Nauka, Moscow
148. Khasminskii, R. (1980): Stochastic Stability of Differential Equations. Sijthoff and Noordhoff
149. Khasminskii, R. and Lazareva, B. (1991): On simplified procedure of filtering for partially observed Markov chain. Sov. Math. Survey, **46**, 3, 191–2
150. Khasminskii, R. and Lazareva, B. (1992) On some filtering procedures for jump Markov process observed in white Gaussian noise. Ann. Stat., **20**, 4, 2153–60
151. Khasminskii, R., Lazareva, B. and Stapleton, J. (1994): Some procedures for state estimation of hidden Markov chain with two states. Statistical Decision Theory and Related Topics. Springer-Verlag, New York, 477–87
152. Khasminskii, R. and Zeitouni, O. (1996): Asymptotic filtering for finite-state Markov chains. Stochastic Processes Appl., **63**, 1–10
153. Kitsul, P.I. (1970): Non-linear Filtering on the Basis of Continuous and Discrete Observations. Transactions of the II All-Union Conference on Stochastic Control Problems (Tashkent). Nauka, 52–7
154. Kitsul, P.I. (1970): On filtering of Markov diffusion processes via discrete time observations. Avtom. Telemekh., 11, 29–37
155. Kitsul, P.I. (1971): A problem of optimal linear filtering. Avtom. Telemekh., 11, 46–52
156. Kleptsina, M.L., Liptser, R.S. and Serebrovski, A.P. (1997): Nonlinear filtering problem with contamination. Ann. Appl. Probab., **7**, 4, 917–34
157. Kolmogorov, A.N. (1936): The Foundations of Probability Theory. ONTI, Moscow Leningrad
158. Kolmogorov, A.N. (1941): Interpolation and extrapolation of stationary random sequences. Izv. Akad. Nauk SSSR, Ser. Mat., **5**, 5
159. Kolmogorov, A.N. (1956): Theory of Information Transmission. Izd. Akad. Nauk SSSR

160. Kolmogorov, A.N. and Fomin, S.V. (1968): Elements of the Theory of Functions and Functional Analysis. Nauka, Moscow
161. Kramkov, D.O. and Shiryaev, A.N. (1996): Sufficient conditions for the uniform integrability of exponential martingales. Proc. European Congress of Mathematics (Budapest)
162. Krasovsky, N.N. and Lidsky, A.A. (1961): Analytical design of regulators in systems with random properties, III. Avtom. Telemekh., 11, 1425–31
163. Krylov, N.V. (1967): On Itô stochastic integral equations. Nagoya Math. J., 30, 209–45
164. Krylov, N.V. (1980): Controlled Diffusion Processes Springer-Verlag, New York Berlin Heidelberg (Russian edition 1977)
165. Krylov, N.V. and Rozovskii, B.L. (1977): On the Cauchy problem for linear stochastic partial differential equations. Mat. SSSR Izv., 11, 1267–84
166. Krylov, N.V. and Rozovskii, B.L. (1981): Stochastic evolution equations. J. Sov. Math., 16, 1233–76
167. Krylov, N.V. and Zatezalo, A. (1998): Filtering of finite-state time-non homogeneous Markov processes, a direct approach. Appl. Math. Optimization. Submitted
168. Kunita, H. (1971): Ergodic properties of nonlinear filtering processes. In: Spatial Stochastic Processes. K. Alexander and J. Watkins (eds). Birkhäuser, Boston, 233–56
169. Kunita, H. (1971): Asymptotic behavior of the nonlinear filtering errors of Markov processes. J. Multivariate Anal., **1**, 4, 365–93
170. Kunita, H. (1981): Cauchy problem for stochastic partial differential equation arising in nonlinear filtering theory. Syst. Control Lett., **1**, 1, 37–41
171. Kunita, H. and Watanabe, Sh. (1967): On square integrable martingales. Nagoya Math. J., **30**, 209–45
172. Kushner, H.J. (1964): On the dynamical equations of conditional probability density functions with applications to optimal stochastic control theory. J. Math. Anal. Appl., 8, 332–44
173. Kushner, H.J. (1964): On the differential equations satisfied by conditional probability densities of Markov processes. SIAM J. Control Optimization, **2**, 106–19
174. Kushner, H.J. (1967): Dynamical equations for nonlinear filtering. J. Differ. Equations, **3**, 179–90
175. Kushner, H.J. (1990): Weak Convergence Method and Singular Perturbed Stochastic Control and Filtering Problems. Birkhäuser, Boston Basel Berlin
176. Kushner, H.J. and Runggaldier, W.J. (1987): Filtering and control for wide-bandwidth noise-driven systems. IEEE Trans. Autom. Control, **AC-23**, 123–33
177. Kutoyants, Yu.A. (1984): Estimation of the intensity parameter of a non-homogeneous Poisson process. Probl. Peredachi Inf., **13**, 4, 253–8
178. Kutoyants, Yu.A. (1994): Non-consistent estimation by diffusion type observations. Stat. Probab. Lett., **20**, 1–7
179. Kutoyants, Yu.A. (1994): Identification of Dynamical Systems with Small Noise. Kluwer, Dordrecht
180. Kutoyants, Yu.A., Mourid, T. and Bosq, D. (1992): Estimation paramétrique d'un processus de diffusion avec retards. Ann. Inst. Henri Poincaré, **28**, 1, 95–106
181. Kutoyants, Yu.A. and Vostrikova, L.Yu. (1995): On non-consistency of estimators. Stochastics Stochastic Rep., **53** 53–80
182. Kuznetsov, N.A., Liptser, R.S. and Serebrovski, A.P. (1980): Optimal control and information processing in continuous time. Avtom. Telemekh., 10, 47–53

183. Laidain, M. (1971): Test entre deux hypothèses pour un processus défini par une équation différentielle stochastique. Rev. Cethedec, **8**, 26, 111–21
184. Laning, G.H. and Battin, R.G. (1956): Random Processes in Automatic Control Problems. McGraw-Hill, New York
185. Lelekov, R.A. (1998): Necessary and sufficient conditions for the convergence of conditional mathematical expectation with respect to σ-algebras, Usp. Mat. Nauk, **53**, 6(324), 255–256
186. Leman, E. (1959): Testing Statistical Hypotheses. Wiley, New York
187. Leonov, V.P. and Shiryaev, A.N. (1959): Methods for calculating semi-invariants. Teor. Veroyatn. Primen., **4**, 2, 342–55
188. Letov, A.M. (1960): Analytical design of regulators, I. Avtom. Telemekh., 4, 436–41
Letov, A.M. (1960): Analytical design of regulators, II. Avtom. Telemekh., 5, 561–8
Letov, A.M. (1960): Analytical design of regulators, III. Avtom. Telemekh., 6, 661–3
Letov, A.M. (1961): Analytical design of regulators, IV. Avtom. Telemekh., 4, 425–35
189. Levin, B.R. (1966–7): Theoretical Foundations of Statistical Radio Engineering, I, II. Sov. Radio, Moscow
190. Lévy, P. (1965): Processus stochastiques et mouvement brownien (Second edition). Gauthier-Villars, Paris
191. Lie, R. (1966): Optimal Estimates. Definition of Characteristic. Control. Nauka, Moscow
192. Linnik, Yu.V. (1966): Statistical Problems with Nuisance Parameters. Nauka, Moscow
193. Liptser, R.S. (1966): Comparison of nonlinear and linear filtering of Markov processes. Teor. Veroyatn. Primen., **11**, 3, 528–33
194. Liptser, R.S. (1967): On filtering and extrapolation of the components of diffusion type Markov processes. Teor. Veroyatn. Primen., **12**, 4, 754–6
195. Liptser, R.S. (1968): On extrapolation and filtering of some Markov processes, I. Kibernetika (Kiev), 3, 63–70
196. Liptser, R.S. (1968): On extrapolation and filtering of some Markov processes, II. Kibernetika (Kiev), 6, 70–6
197. Liptser, R.S. (1974): Optimal coding and decoding in the transmission of a Gaussian Markov signal with noiseless feedback. Probl. Peredachi Inf., **10**, 4, 3–15
198. Liptser, R.S. (1975): Gaussian martingales and generalization of the Kalman–Bucy filter. Teor. Veroyatn. Primen., **20**, 2, 292–308
199. Liptser, R.S. (1976): On the representation of local martingales. Teor. Veroyatn. Primen., **21**, 4, 718–26
200. Liptser, R.S., Di Masi, G.B. and Kitsul, P.I. (1991): Minimal dimension linear filters for stationary Markov processes with finite state space. Stochastics Stochastics Rep., **36**, 4–19
201. Liptser, R.S., Krichagina, E.V. and Rubinovich, E.Ya. (1984): Kalman filter for Markov processes. In: Proc. Steklov Seminar 'Statistics and Control of Stochastic Processes'. N.V. Krylov *et al.* (eds). Optimization Software, Inc., New York, 197–213
202. Liptser, R.S. and Lototsky, S.V. (1992): Diffusion approximation and robust Kalman filter. J. Math. Syst. Estim. Control, **2**, 3, 263–74
203. Liptser, R. and Muzhikanov, P. (1998): Filtering with a limiter. J. Appl. Math. Stochastic Anal., **11**, 3, 289–300

204. Liptser, R.S. and Runggaldier, W.J. (1991): Diffusion approximations for filtering. Stochastic Processes Appl., **38**, 205–38
205. Liptser, R.S. and Shiryaev, A.N. (1968): Nonlinear filtering of diffusion type Markov processes. Tr. Mat. Inst. Steklova, **104**, 135–80
206. Liptser, R.S. and Shiryaev, A.N. (1968): On filtering, interpolation, and extrapolation of diffusion type Markov processes with incomplete data. Teor. Veroyatn. Primen., **13**, 3, 569–70
207. Liptser, R.S. and Shiryaev, A.N. (1968): Extrapolation of multivariate Markov processes with incomplete data. Teor. Veroyatn. Primen., **13**, 1, 17–38
208. Liptser, R.S. and Shiryaev, A.N. (1968): On the case of effective solution of the problems of optimal nonlinear filtering, interpolation, and extrapolation. Teor. Veroyatn. Primen., **13**, 3, 570–1
209. Liptser, R.S. and Shiryaev, A.N. (1968): Nonlinear interpolation of the components of diffusion type Markov processes (forward equations, effective formulae). Teor. Veroyatn. Primen., **13**, 4, 602–20
210. Liptser, R.S. and Shiryaev, A.N. (1969): Interpolation and filtering of the jump component of a Markov process. Izv. Akad. Nauk SSSR, Ser. Mat., **33**, 4, 901–14
211. Liptser, R.S. and Shiryaev, A.N. (1969): On the densities of probability measures of diffusion type processes. Izv. Akad. Nauk SSSR, Ser. Mat., **33**, 5, 1120–31
212. Liptser, R.S. and Shiryaev, A.N. (1972): On absolute continuity of measures corresponding to diffusion type processes with respect to a Wiener measure. Izv. Akad. Nauk SSSR, Ser. Mat., **36**, 4, 874–89
213. Liptser, R.S. and Shiryaev, A.N. (1972). Statistics of conditionally Gaussian random sequences. In: Proc. Sixth Berkeley Symposium on Mathematics, Statistics and Probability (1970), vol II. University of California Press, 389–422
214. Liptser, R.S. and Shiryaev, A.N. (1989): Theory of Martingales. Kluwer, Dordrecht (Russian edition 1986)
215. Liptser, R.S., Shiryaev, A.N. and Pukelsheim, F. (1982): Necessary and sufficient conditions for contiguity and entire asymptotic separation of probability measures. Russ. Math. Surv., **37**, 6, 107–36
216. Liptser, R.S., Steinberg, Y., Bobrovski, B.Z. and Schuss, Z. (1996): Fixed lag smoothing of scalar diffusions. Part I: the filtering–smoothing equation. Stochastic Processes Appl., **64**, 237–55
217. Liptser, R.S. and Zeitouni, O. (1998): Robust diffusion approximation for nonlinear filtering. J. Math. Syst. Estim. Control, **8**, 1, 138–42
218. Loève, M. (1977): Probability Theory. Springer-Verlag, New York Berlin Heidelberg
219. Lototsky, S.V., Mikulevicius, R. and Rozovskii, B.L. (1997): Nonlinear filtering revisited. A spectral approach. SIAM J. Control Optimization, **35**, 2, 435–61
220. Lototsky, S.V., Rao, C. and Rozovskii, B.L. (1996): Fast nonlinear filter for continuous–discrete time multiple models. Proc. 35th IEEE Conf. on Decision and Control (Kobe, Japan), 4060–4
221. Lototsky, S.V. and Rozovskii, B.L. (1996): Recursive multiple Wiener integral expansion for nonlinear filtering of diffusion processes. In: A Festschrift in Honor of M.M. Rao, Marcel Dekker
222. Lototsky, S.V. and Rozovskii, B.L. (1996): Nonlinear filtering revisited: separation of parameters and observations, II. Proc. 35th IEEE Conf. on Decision and Control, (Kobe, Japan)
223. Lototsky, S.V. and Rozovskii, B.L. (1998): Recursive nonlinear filter for a continuous–discrete time model. IEEE Trans. Autom. Control, **43**, 8, 1154–8

224. Makowski, A.M. and Sowers, R.B. (1992): Discrete-time filtering for linear systems with non-Gaussian initial conditions: asymptotic behaviors of the difference between the MMSE and the LMSE estimates. IEEE Trans. Autom. Control, **37**, 114–21
225. Marsaglia, G. (1964): Conditional means and covariance of normal variables with singular covariance matrix. J. Am. Stat. Assoc., **308**, 59, 1203–4
226. McKean, H. (1969): Stochastic Integrals. Academic, New York
227. Meditch, J.S. (1969): Stochastic Optimal Linear Estimation and Control. McGraw-Hill, New York
228. Métivier, M. and Pellaumail, J. (1980): Stochastic Integration. Academic, New York
229. Meyer, P.A. (1966): Probabilités et Potentiel. Hermann, Paris
230. Meyer, P.A. (1989): A short presentation of stochastic calculus. In: Stochastic Calculus in Manifolds, M. Emery (ed). Springer-Verlag, Berlin Heidelberg New York
231. Mikulevicius, R. and Rozovskii, B.L. (1993): Separation of observations and parameters in nonlinear filtering. Proc. 32nd IEEE Conf. on Decision and Control, vol 2. IEEE Control Systems Society, 1559–64
232. Mikulevicius, R. and Rosovskii,B.L. (1999): Martingale problems for stochastic PDE's. In: Stochastic partial differential equations: six perspectives, Carmona, Rene A. *et al.* (eds), American Mathematical Society. Math. Surv. Monogr. **64**, 243–325
233. Miller, B.M. (1985): Optimal control of observations in the filtering of diffusion processes, I. Avtom. Telemekh., **46**, 2, 207–14
234. Miller, B.M. (1985): Optimal control of observations in the filtering of diffusion processes, II. Avtom. Telemekh., **46**, 6, 745–54
235. Miller, B.M. (1991): Generalized optimization in problems of observation control. Avtom. Telemekh., **52**, 10, 83–92
236. Miller, B.M. (1993): Observation control for discrete–continuous systems. Proc. 12th IFAC World Congress (Sydney), 165–8
237. Miller, B.M. and Rubinovich, E.Ya. (1992): Optimal impulse control problem with a constrained number of impulses. Math. Comput. Simul., **34**, 1, 23–49
238. Miller, B.M. and Rubinovich, E.Ya. (1995): Regularization of a generalized Kalman filter. Math. Comput. Simul., **39**, 87–108
239. Miller, B.M. and Rubinovich, E.Ya. (1997): Stochastic control for the optimization of observations. SIAM J. Control Optimization, **35**, 3, 1030–52
240. Miller, B.M. and Runggaldier, W.J. (1997) Kalman filtering for linear systems with coefficients driven by a hidden Markov jump process. Syst. Control Lett., **31**, 93–102
241. Miller, B.M. and Serebrovski, A.P. (1979): Optimal Control of Linear System with Singular Quadratic Performance Criterion. Avtom. Telemekh., **40**, 3, 339–48
242. Nahi, N.E. (1969): Estimation Theory and Applications. Wiley, New York
243. Nevelson, M.V. and Khasminskii, R.Z. (1973): Stochastic Approximation and Recursive Estimation of Stochastic Processes. Translations of Mathematical Monographs, Vol 47. American Mathematical Society, Providence, RI
244. Neveu, J. (1965): Mathematical Foundations of the Calculus of Probability. Holden-Day, San Francisco
245. Neveu, J. (1975): Discrete Parameter Martingales. North-Holland, Amsterdam
246. Novikov, A.A. (1971): Sequential estimation of the parameters of diffusion type processes. Teor. Veroyatn. Primen., **16**, 2, 394–6
247. Novikov, A.A. (1971): On stopping times of Wiener process. Teor. Veroyatn. Primen., **16**, 3, 548–50

248. Novikov, A.A. (1972): On an identity for stochastic integrals. Teor. Veroyatn. Primen., **17**, 4, 761–5
249. Ocone, D. and Pardoux, E. (1986): Asymptotic stability of the optimal filter with respect to its initial condition. SIAM J. Control Optimization, **34**, 1, 226–43
250. Øksendal, B. (1985): Stochastic Differential Equations. Springer-Verlag, Berlin Heidelberg, New York (5th edition 1998).
251. Orey, S. (1974): Radon–Nikodym Derivative of Probability Measures: Martingale Methods. Dept. Found. Sci., Tokyo University of Education
252. Pardoux, E. (1989): Filtrage non linéaire et equations aux dérivées partielles stochastiques associées. Ecole d'été de Probabilités de Saint-Flour. Springer-Verlag, Berlin Heidelberg New York
253. Picard, J. (1991): Efficiency of the extended (Kalman) filter for nonlinear systems with small noise. SIAM J. Appl. Math., **51**, 3, 843–85
254. Picard, J. (1994): Robustesse de la solution des problèmes de filtrage avec bruit blanc indépendant. Stochastics Stochastics Rep., **13**, 229–45
255. Prokhorov, Yu.V. (1956): Convergence of random processes and limit theorems of probability theory. Teor. Veroyatn. Primen., **1**, 2, 177–238
256. Prokhorov, Yu.V. and Rozanov, Yu.A. (1969): Probability Theory: Basic Concepts, Limit Theorems, Random Processes. Springer-Verlag, New York Berlin Heidelberg
257. Protter, P. (1990): Stochastic Integration and Differential Equations. A New Approach. Springer-Verlag, Berlin Heidelberg New York
258. Pugachev, V.S. (1962): Theory of Random Functions and its Application to Automatic Control Problems. Fizmatgiz, Moscow
259. Rao, C.M. (1969): On decomposition theorems of Meyer. Math. Scand. **24**, 1, 66–78
260. Rao, C.R. (1968): Linear Statistical Methods and their Applications. Nauka, Moscow
261. Revuz, D. and Yor, M. (1991, 1994, 1998): Continuous Martingales and Brownian Motion. Springer-Verlag, Berlin Heidelberg New York
262. Rogers, C. and Williams, D. 1987: Diffusions, Markov Processes and Martingales: Itô Calculus. Wiley, New York
263. Rozanov, Yu.A. (1967): Stationary Random Processes. Holden-Day, San Francisco
264. Rozovskii, B.L. (1972): Stochastic partial differential equations arising in nonlinear filtering problems. Usp. Mat. Nauk SSSR, **27**, 3, 213–4
265. Rozovskii, B.L. (1990): Stochastic Evolution Systems. Linear Theory and Application to Nonlinear Filtering. Reidel, Dordrecht
266. Rozovskii, B.L. (1991): A simple proof of uniqueness for Kushner and Zakai equations. In: Stochastic Analysis. E. Mayer-Wolf *et al.* (eds). Academic, New York, 449–58
267. Rozovskii, B.L. and Shiryaev, A.N. (1972): On infinite systems of stochastic differential equations arising in the theory of optimal nonlinear filtering. Teor. Veroyatn. Primen., **17**, 2, 228–37
268. Rubinovich, E.Ya. (1980): Trajectory control of observations in discrete stochastic optimization problems. Avtom. Telemekh., **41**, 6, 93–102
269. Rubinovich, E.Ya. (1997): Generalized linear–quadratic stochastic control problem with incomplete information. Avtom. Telemekh., **58**, 7, 1243–62
270. Ruymgaart, P.A. (1971): A note on the integral representation of the Kalman–Bucy estimate. Indag. Math., **33**, 4, 346–60.
271. Segal, A. (1973): A martingale approach to modeling, estimation and detection of jump processes. Tech. Report. No. **7050-21**, MIT

272. Segal, A., Davis, M.H.A. and Kailath, T. (1975): Nonlinear filtering with counting observation. IEEE Trans. Inf. Theory, **IT-21**, 2, 143–9
273. Segal, A. and Kailath, T. (1975): The modeling of random modulated jump processes. IEEE Trans. Inf. Theory, **IT-21**, 2, 135–42
274. Shalkwijk, J.P.M. and Kailath, T. (1966): A coding scheme for additive noise channels with feedback, Part I. IEEE Trans. Inf. Theory, **IT-12**, 172–82
275. Shatashvili, A.D. (1970): Nonlinear filtering for some stochastic differential equations. Kibernetika (Kiev), 3, 97–102
276. Shepp, L.A. (1966): Radon–Nikodym derivatives of Gaussian measures. Ann. Math. Stat., **37**, 321–54
277. Shiryaev, A.N. (1960): Problems of spectral theory of higher moments, I. Teor. Veroyatn. Primen., **5**, 3, 293–313
278. Shiryaev, A.N. (1966): On stochastic equations in the theory of conditional Markov processes. Teor. Veroyatn. Primen., **11**, 1, 200–6
279. Shiryaev, A.N. (1966): Stochastic equations of nonlinear filtering of jump Markov processes. Probl. Peredachi Inf., **2**, 3, 3–22
280. Shiryaev, A.N. (1967): New results in the theory of controlled random processes. In: Trans. 4th Prague Conference on Information Theory (1965), 131–203
281. Shiryaev, A.N. (1968): Studies in statistical sequential analysis. Mat. Zametki, **3**, 6, 739–54
282. Shiryaev, A.N. (1969): Statistical Sequential Analysis. Nauka, Moscow
283. Shiryaev, A.N. (1970): Sur les Équations Stochastiques aux Dérivées Partielles. Actes Congrès Int. Math.
284. Shiryaev, A.N. (1971): Statistics of diffusion type processes. In: Proc. Second Japan–USSR Symposium on Probability Theory (Kyoto), vol I, 69–87
285. Shiryaev, A.N. (1976): Statistical Sequential Analysis 2nd edn, Nauka, Moscow
286. Shiryaev, A.N. (1978): Optimal Stopping Rules. Springer-Verlag, New York Berlin Heidelberg
287. Shiryaev, A.N. (1996): Probability, 2nd edn. Springer-Verlag, Berlin Heidelberg New York
288. Shiryaev, A.N. (1999): Essentials of Stochastic Finance. World Scientific, Singapore
289. Siegmund, D., Robbins, H. and Wendel, J. (1968): The limiting distribution of the last time $S_n \geq n\varepsilon$. Proc. Natl. Acad. Sci. USA, **62**, 1
290. Skorokhod, A.V. (1965): Studies in the Theory of Random Processes. Addison-Wesley, Reading, MA
291. Skorokhod, A.V. (1971): Random Processes with Independent Increments. Kluwer, Dordrecht
292. Snyder, D.L. (1972): Filtering and detection for double stochastic Poisson processes. IEEE Trans. Inf. Theory, **IT-18**, 91–102
293. Snyder, D.L. and Fishman, P.M. (1975): How to track a swarm of fireflies by observing their flashes. IEEE Trans. Inf. Theory, **IT-21**, 6, 692–5
294. Stettner, L. (1989): On invariant measure of filtering processes. In: Stochastic Differential Systems, Lecture Notes in Control and Information Science 126. Springer-Verlag, New York Berlin Heidelberg
295. Stratonovich, R.L. (1960): Conditional Markov processes. Teor. Veroyatn. Primen., **5**, 2, 172–95
296. Stratonovich, R.L. (1966): Conditional Markov Processes and their Applications to Optimal Control Theory. Izd. MGU, Moscow
297. Striebel, C.T. (1965): Partial differential equations for the conditional distribution of a Markov process given noisy observations. J. Math. Anal. Appl., 11, 151–9

298. Tikhonov, A.N. (1965): On a stable algorithm for solution of singular systems of linear algebraic equations. Zh. Vychisl. Mat. Mat. Fiz., 4, 718–22
299. Tsypkin, Ya.Z. (1971): Adaptation and Learning in Automatic Systems. Academic, New York
300. Tsyrelson, B.S. (1975): An example of a stochastic equation having no strong solution. Teor. Veroyatn. Primen., **20**, 2, 427–30
301. Turin, G. (1969): Notes on Digital Communication. Van Nostrand Reinhold, New York
302. Van Schuppen, I.L. (1973): Filtering for counting processes. Martingale approach. In: Proc. 4th Symposium on Nonlinear Estimation and Applications (San Diego, CA)
303. Wentzell, A.D. (1961): Additive functionals of a multivariate Wiener process. Dokl. Akad. Nauk SSSR, **130**, 1, 13–6
304. Wentzell, A.D. (1965): On equations of the conditional Markov processes. Teor. Veroyatn. Primen., **10**, 2, 390–3
305. Whittle, P. (1963): Prediction and Regulation. Van Nostrand Reinhold, New York
306. Wiener, N. (1923): Differential space. J. Math. Phys., **58**, 131–74
307. Wiener, N. (1949): Extrapolation, Interpolation and Smoothing of Stationary Time Series. Wiley, New York
308. Williams, D. (ed) (1981): Proc. Durham Symposium on Stochastic Integrals (1980). Lecture Notes in Mathematics 851, Springer-Verlag, New York Berlin Heidelberg
309. Wolfowitz, J. (1946): On sequential binomial estimation. Ann. Math. Stat., **17**, 489–93
310. Wolfowitz, J. (1947): The efficiency of sequential games, estimates and Wald's equation for sequential processes. Ann. Math. Stat., **18**, 215–30
311. Wonham, W.M. (1963): Stochastic Problems in Optimal Control. Technical Report **63-14**, Research Institute for Advanced Study, Baltimore, MD
312. Wonham, W.M. (1965): Some applications of stochastic differential equations to optimal nonlinear filtering. SIAM J. Control Optimization, **2**, 347–69
313. Wonham, W.M. (1968): On the separation theorem of stochastic control. SIAM J. Control Optimization, **6**, 312–26
314. Wonham, W.M. (1968): On a matrix Ricatti equation of stochastic control. SIAM J. Control Optimization, **6**, 681–97
315. Yaglom, A.M. (1952): Introduction to the theory of stationary random functions. Usp. Mat. Nauk, **7**, 5, 3–168
316. Yamada, T. and Watanabe Sh. (1971): On the uniqueness of solution of stochastic differential equations. J. Math. Kyoto Univ., **11**, 1, 155–67
317. Yashin, A.I. (1968): Filtering of a jump Markov process with unknown probability characteristics given additive noise. Avtom. Telemekh, 12, 25–30
318. Yashin, A.I. (1969): On distinguishing jump parameters of multivariate processes. Avtom. Telemekh., 10, 60–7
319. Yashin, A.I. (1975): On constructive algorithms of optimal nonlinear filtering, I. Avtom. Telemekh., 11, 33–9
Yashin, A.I. (1975): On constructive algorithms of optimal nonlinear filtering, II. Avtom. Telemekh., 12, 108–13
320. Yashin, A.I. (1976): Estimation of characteristics of jump random processes. Avtom. Telemekh., 4, 55–60
321. Yashin, A.I. (1980): Conditional Gaussian estimation of dynamic systems under jumping observations Avtom. Telemekh., 5, 38–47
322. Yashin, A.I. (1983): On a problem of sequential hypothesis testing. Teor. Veroyatn. Primen., **28**, 1, 149–56

323. Yashin, A.I. (1985): Dynamics in survival analysis: conditional Gaussian property versus the Cameron–Martin formula. In: Statistics and Control of Stochastic Processes. N.V. Krylov *et al.* (eds), Translation series in Mathematics and Engineering, Optimization Software. Springer-Verlag, New York Berlin Heidelberg
324. Yashin, A.I. and Kuznetsov, N.A. (1981): On consistent parameter estimation in adaptive filtering. Probl. Control Inf. Theory, **10**, 5, 317–27
325. Yershov, M.P. (1969): Nonlinear filtering of Markov processes. Teor. Veroyatn. Primen., **14**, 4, 757–8
326. Yershov, M.P. (1970): Sequential estimation of diffusion processes. Teor. Veroyatn. Primen., **15**, 4, 705–17
327. Yershov, M.P. (1972): On representation of Itô processes. Teor. Veroyatn. Primen., **17**, 1, 167–72
328. Yershov, M.P. (1972). On absolute continuity of measures corresponding to diffusion type processes. Teor. Veroyatn. Primen., **17**, 1, 173–8.
329. Yershov, M.P. (1972): Stochastic Equations. In: Proc. Second Japan–USSR Symposium on Probability Theory (Kyoto), vol I, 101–6
330. Yor, M. (1967): Représentation intégrales des martingales de carré intégrable. C. R. Acad. Sci., Paris, Sér. A-B, **264**, 799–800
331. Zakai, M. (1969): On the optimal filtering of diffusion processes. Z. Wahrsch. Verw. Gebiete, **11**, 230–33
332. Zeitouni, O. and Zakai, M. (1992): On the optimal tracking problem. SIAM J. Control Optimization, **30**, 426-439; Erratum **32** (1994), 1194
333. Zheng, W. (1980): A note on the convergence of sequences of conditional expectations of random variables. Z. Wahrsch. Verw. Gebiete, **53**, 3, 291–2
334. Zhukovsky, E.L. and Liptser, R.S. (1972): On recursive computing schemes for normal solution of linear algebraic equations. Zh. Vychisl. Mat. Mat. Fiz., **12**, 4, 843–57
335. Zigangirov, K.Sh. (1967): Transmission of messages through a binary Gaussian channel with feedback. Probl. Peredachi Inf., **3**, 2, 98–101

Index

Printing: Mercedes-Druck, Berlin
Binding: Stürtz AG, Würzburg